Applied Biotechnology in Genetic Engineering, Pharmaceuticals and Agriculture

Applied Biotechnology in Genetic Engineering, Pharmaceuticals and Agriculture

Edited by **Joy Adam**

SYRAWOOD
PUBLISHING HOUSE
New York

Published by Syrawood Publishing House,
750 Third Avenue, 9th Floor,
New York, NY 10017, USA
www.syrawoodpublishinghouse.com

Applied Biotechnology in Genetic Engineering, Pharmaceuticals and Agriculture
Edited by Joy Adam

International Standard Book Number: 978-1-68286-276-6 (Hardback)

Printed in China.

Contents

Preface

This book has been an outcome of determined endeavour from a group of educationists in the field. The primary objective was to involve a broad spectrum of professionals from diverse cultural background involved in the field for developing new researches. The book not only targets students but also scholars pursuing higher research for further enhancement of the theoretical and practical applications of the subject.

Biotechnology has applications in multiple disciplines. This book is focused on applied biotechnology in the areas of genetic engineering, drug discovery and development, and agriculture. The chapters included in this book elucidate new techniques and their applications in a multidisciplinary approach with their current status and prospective developments. It aims to equip students and experts with the advanced topics and upcoming concepts in this area.

It was an honour to edit such a profound book and also a challenging task to compile and examine all the relevant data for accuracy and originality. I wish to acknowledge the efforts of the contributors for submitting such brilliant and diverse chapters in the field and for endlessly working for the completion of the book. Last, but not the least; I thank my family for being a constant source of support in all my research endeavours.

Editor

Characterization of a modified ROCK2 protein that allows use of N6-ATP analogs for the identification of novel substrates

Amber L Couzens[1,2], R Montgomery Gill[1] and Michael P Scheid[1*]

Abstract

Background: The Rho-associated coiled-coil kinase-2 (ROCK2) is an important signaling transducer in the transmission of extracellular signals effecting organization of the actin cytoskeleton. ROCK2 has been implicated in numerous pathologies and the current focus is on understanding the molecular events that couple ROCK2 activity to biological function. To aid in the search for new ROCK2 substrates, we have developed an analog-sensitive (AS) ROCK2 protein that allows the use of selective ATP analogs that are not efficiently utilized by other protein kinases.

Results: The analog sensitive protein, M160A ROCK2, was highly active and could phosphorylate proteins from a cellular homogenate with $\gamma^{32}P$-N_6 (benzyl)ATP. We show the utility of this approach by identifying a putative ROCK2 substrate, elongation initiation factor-1-α1. We further show that the major site of ROCK2 phosphorylation of EIF1α1 is Thr^{432}.

Conclusions: Our work demonstrates that AS-ROCK2 could be useful in a systematic proteomic approach for identifying novel ROCK2 substrates.

Keywords: ROCK2, Protein kinase, Chemical engineering, Cell signaling

Background

The Rho-associated coiled-coil kinase-2 (ROCK2) is a large serine/threonine kinase that plays diverse roles in the cell including contraction, motility and morphology (reviewed in [1]). In the absence of Rho-GTP, ROCK2, and its closely related homologue ROCK1, adopts an auto-inhibited form [2], and upon Rho-GTP binding auto-inhibition is relieved and the kinase adopts an open, active conformation [3]. Rho-GTP activates ROCK2 to alter the activity of the actin reorganization machinery. For example, ROCK2 activation indirectly increases myosin light chain (MLC) phosphorylation, through the phosphorylation and inactivation of MLC phosphatase (MLCP) [4]. ROCK2 can also directly phosphorylate MLC at Ser^{19} [5], the same site targeted by MLC kinase. ROCK2 further causes the disruption of the head-to-tail association of ERM proteins, through the phosphorylation of ezrin, radixin and moesin [6]. In addition, LIM-kinase 2 is activated by ROCK2, which then phosphorylates its downstream target, cofilin [7]. Phosphorylation of cofilin inhibits its actin depolymerizing function, thus increasing the number of actin filaments and leads to reorganization of the cytoskeleton [8].

Many human cancers demonstrate increased ROCK2 activity, which can augment tumor invasiveness [9,10]. Animal models have revealed that ectopic ROCK2 activation in established tumors is sufficient to drive metastasis of tumor cells into the surrounding stroma [9]. ROCK2 has also been implicated in the pathogenesis of hypertension, since ROCKs play a crucial role in smooth muscle contraction [11], through phosphorylation of MLC and MLCP. Furthermore, ROCK2 has been shown to influence the expression of genes that are important in vascular function, such as plasminogen activator inhibitor-1 (PAI-1) [12] and osteopontin [13]. Since ROCK2 plays a role in a number of human diseases, this kinase has received considerable interest as a potential therapeutic target.

ROCK2 is a member of the AGC kinase family and shares homology within the catalytic domain with other

* Correspondence: mscheid@yorku.ca
[1]Department of Biology, York University, 4700 Keele Street, Toronto, ON M3J1P3, Canada
Full list of author information is available at the end of the article

AGC kinase members including PKA, PKB, PKC, S6K, and SGK. This has led to the realization that ROCK2, like other AGC kinases, could target sequences that fall within a characteristic phosphorylation motif of R/KXS/T or R/KXXS/T [14]. Substrate preference for ROCK2 phosphorylation of this consensus motif is likely governed by spatial and temporal constraints; for example PKA distinguishes *bona fide* substrates through a mechanism in which bridging between kinase and substrate is provided by adapter and scaffold molecules [15].

To identify new ROCK2 substrates, a potentially useful approach is to modify the kinase so that it utilizes an ATP molecule restricted from use by other kinases. For example, an ATP with a bulky hydrophobic group attached to the N_6 position of the purine base prevents entry of the analog into the catalytic site of most protein kinases [16,17]. The kinase domain can be engineered to accept this analog by introducing a modification to sterically accommodate the ATP analog. This chemical engineering approach was first demonstrated in the prototypical protein tyrosine kinase v-Src [17] and has been adapted to various kinases leading to the discovery of many important new substrates [18,19].

Here, we have adapted this chemical engineering approach to ROCK2. We show that a modified ROCK2 harboring a single amino acid substitution in the catalytic domain resulted in a 100-fold decrease in the K_m for N_6(Benzyl)ATP utilization compared with wildtype kinase. Phosphorylation of a cellular homogenate with $\gamma^{32}P$-N_6(Benzyl)ATP led to a mixture of highly phosphorylated proteins that were separated by 2D-gel electrophoresis. Mass spectral identification and biochemical analysis of one of these phosphorylated proteins, eukaryotic elongation factor-1-α-1 (eEF1α1), demonstrated the utility of this approach and provides an important reagent for the future identification of ROCK2 signaling targets.

Results and discussion

Generation of AS-ROCK2

We were interested in identifying ATP binding-pocket mutations within ROCK2 that permitted the use of bulky ATP analogs. To screen for these mutations, we developed an *in vitro* nonradioactive assay based on phosphorylation of a biotinylated ROCK2 substrate peptide matching the consensus ROCK2 phosphorylation site in LIMK. Following phosphorylation *in vitro* by ROCK2, the biotinylated LIMK peptides were bound to a 96 well streptavidin-coated plate and phosphorylation was measured by probing with a commercially available phospho-specific antibody coupled to a secondary antibody conjugated to an 680 nm wavelength fluorochrome (IRDye680). Detection of the phospho-peptide antibody complex was performed directly on-plate using the Licor Odyssey infrared laser scanner. The assay was simple, rapid, and had a wide dynamic range comparing phospho-Thr^{505} LIMK fluorescence with the non-phosphorylated LIMK peptide (Figure 1). The unphosphorylated peptide was unreactive to the phosphospecific antibody up to the maximal concentration tested of 4 μg/ml.

For ROCK2, Met^{160} is analogous to Ile^{338} in v-Src (Figure 2A), where this single bulky residue of v-Src was shown to prevent the acceptance of N6-modified ATP analogues [20]. Mutation of the Met^{160} residue in ROCK2 to an alanine or glycine was modeled to yield the space required to accommodate N6(Benzyl)ATP. This mutation was introduced into the W1161A ROCK2 background as we have previously shown that this protein exhibits high kinase activity levels [21]. The Met^{160} to Ala substitution resulted in a 4-fold increase in substrate phosphorylation over wildtype ROCK2 at an N_6(Benzy)ATP concentration of 100 μM (Figure 2B). The Met^{160} to Gly substitution had a 50% reaction velocity compared with the wildtype sequence, while Met^{135} to Val^{135} substitution also resulted in a major decrease in reaction velocity, indicating that these substitutions are inhibitory.

Since the Met^{160} mutation resulted in a greater velocity at 100 μM N_6(Benzy)ATP than wildtype ROCK2, we analyzed this protein over a concentration range of analog to estimate K_m for N_6(Benzyl)ATP. We saw no difference in the K_m concentration of ATP between

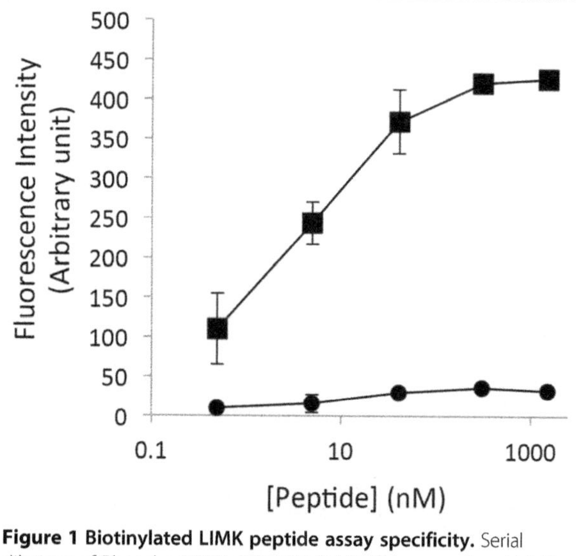

Figure 1 Biotinylated LIMK peptide assay specificity. Serial dilutions of Phospho-LIMK1 (DRKKRYpTVVGNPY; squares) and LIMK1 (DRKKRYTVVGNPY; circles) peptide with concentrations ranging from 4 μg/ml to 90 ng/ml were bound to streptavidin plates (Pierce) for 2 hours at room temperature (24°C). Wells were then washed and probed with phospho-LIMK (Cell Signaling) over night at room temperature. Secondary decoration with IRDye®680 anti-rabbit(Li-Cor) was performed at room temperature for 1.5 hours. Proteins were visualized by direct fluorescence scanning with a Li-Cor Odyssey Imager. Error bars represent the standard deviation of triplicate determinations and are representative of three independent experiments.

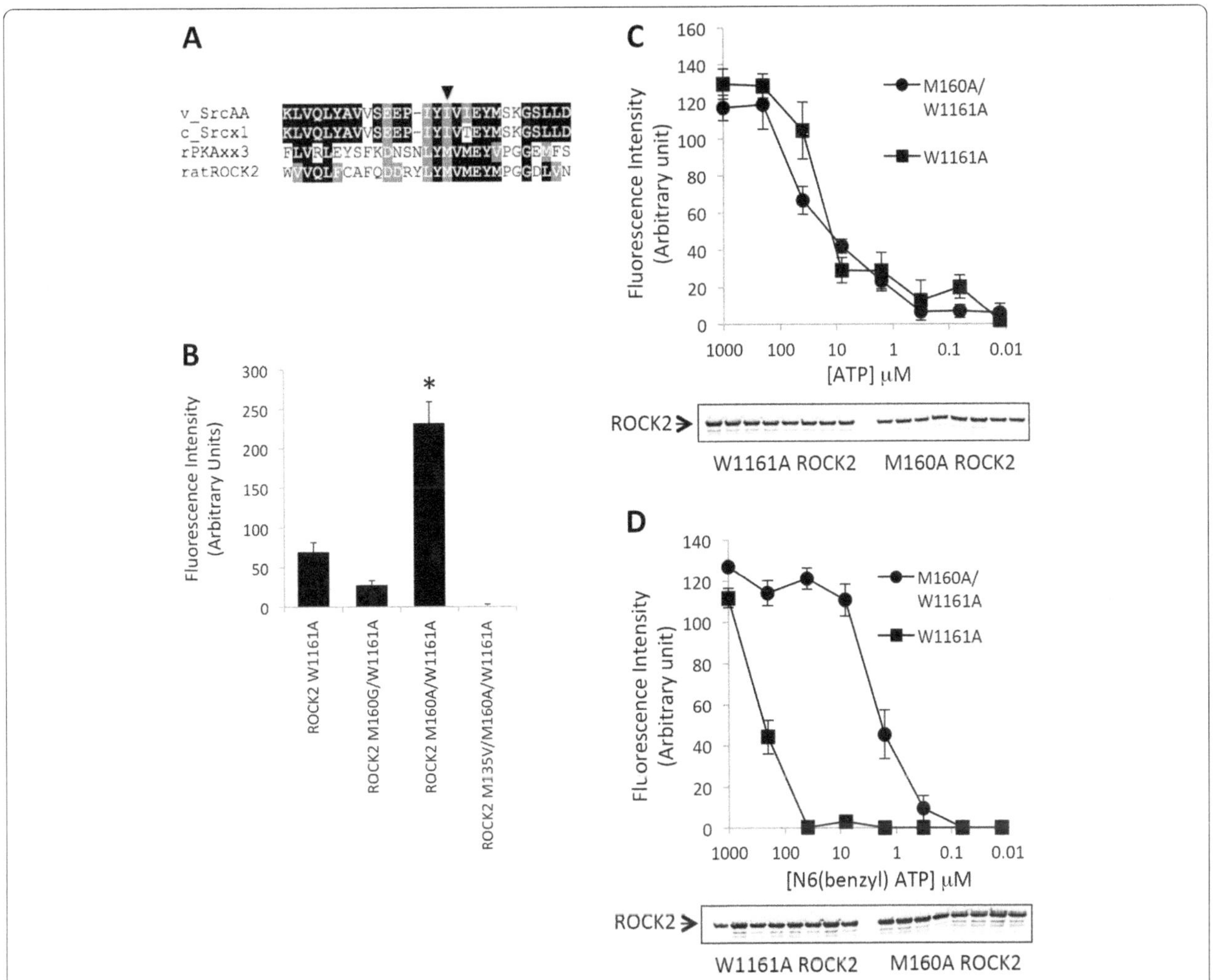

Figure 2 Creation of AS-ROCK2. A. ROCK2, V-Src, C-Src and rat PKA sequences were aligned using ClustalW. The v-Src 'gatekeeper' residue within the ATP binding pocket is Ile^{338} (indicated by arrow) and is primarily responsible for restricting the ability of protein kinases from utilizing N6-substiuted ATP analogues. Ile^{338} in v-Src corresponds to Met^{160} in ROCK2. **B**. HEK 293 cells were transfected with various FLAG-ROCK2 constructs where indicated. The following day, cells were lysed and ROCK2 was immunoprecipitated with anti-FLAG conjugated agarose. LIMK peptide was phosphorylated using N_6(Benzyl)ATP, loaded onto streptavidin plates (Pierce), and probed with P-LIMK (Cell Signaling) overnight at room temperature. Secondary decoration with IRDye® 680 anti-rabbit (Li-Cor) was performed at room temperature for 1.5 hours. LIMK peptide phosphorylation was visualized by direct fluorescence scanning with a Li-Cor Odyssey Imager. The relative LIMK phosphorylation signal is shown by the histogram for each ROCK2 mutation. Error bars represent standard deviation of triplicate determinations from three independent experiments. Asterisks indicate significant difference ($P < 0.001$) with W1161A ROCK2 as using Student's *t*-test. **C**. Kinase activity of W1161A ROCK or W1161A/M160A ROCK2 was determined as in *B*, using serial dilutions of ATP with concentrations ranging from 1 mM to 12.8 nM. **D**. As in *C*, except N_6(Benzyl)ATP was used. For *C* and *D*, error bars represent the range of duplicate determinations. Results are representative of three independent experiments.

wildtype ROCK2 and Ala^{160} ROCK2 (both ~30 μM, which is similar to the published K_m for ROCK2) (Figure 2C). In contrast, there was a 100-fold decrease in K_m for N_6(Benzyl)ATP between the two kinases (~200 μM for wildtype ROCK2 and ~2 μM for Ala^{160} ROCK2) (Figure 2D).

eEF1α1 is phosphorylated by AS-ROCK2 in vitro

Next, we utilized M160A/W1161A ROCK2 to phosphorylate HEK293 cellular homogenate. Cellular homogenate was incubated with M160A/W1161A ROCK2 in the presence $\gamma^{32}P$-N_6(Benzyl)ATP (Figure 3A). In addition to autophosphorylated ROCK2, numerous other ^{32}P-labeled proteins were clearly observed by autoradiography. The phosphorylated protein homogenate was then fractionated by 2D-electrophoresis and Figure 3B shows that at least eight proteins were phosphorylated by the exogenous M160A/W1161A ROCK2 protein. Two spots in particular were labeled strongly (*A* and *B*; Figure 3B), and one of them (*B*) was identified by mass spectroscopy to be the eukaryotic elongation initiation factor-1-α1 (eEF1α1).

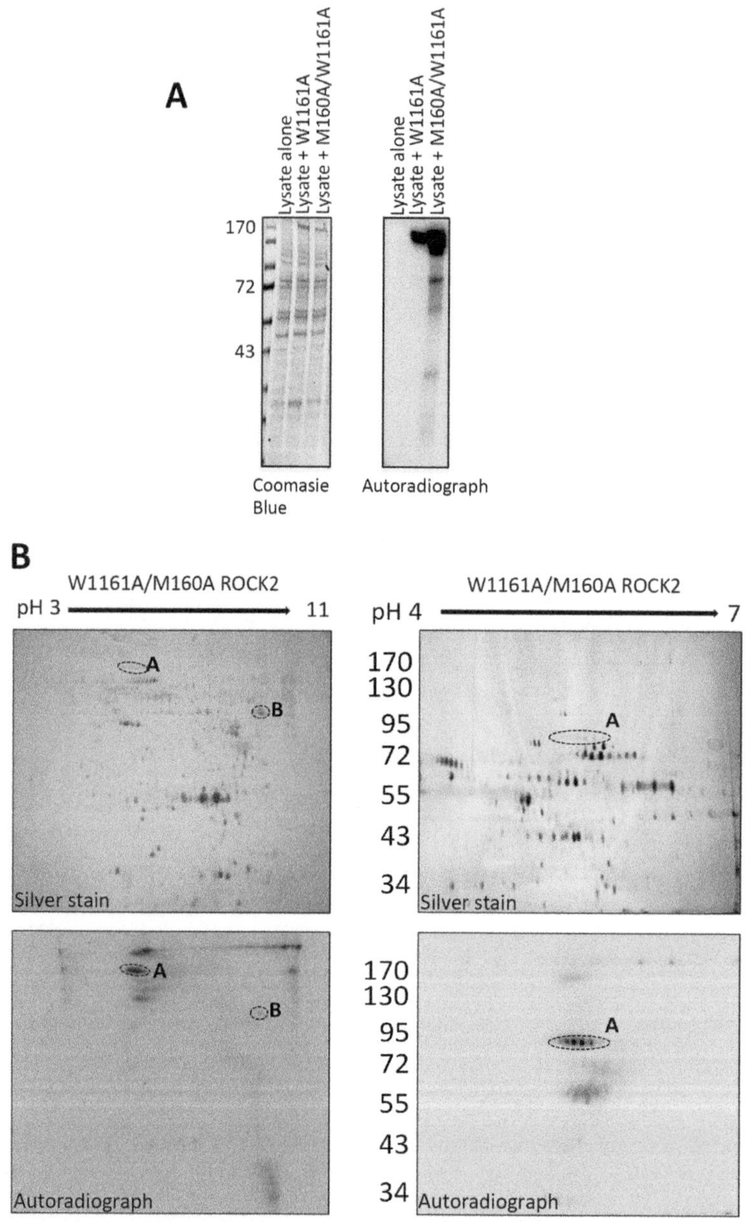

Figure 3 Two-Dimensional separation of ROCK2-phosphorylated HEK293 cellular homogenate. A. HEK 293 cells were transfected with either W1161A or M160A/W1161A ROCK2. The following day, ROCK2 was immunoprecipitated and resuspended in HEK 293 cellular homogenate (60 µg) with $\gamma^{32}P\text{-}N_6$(Benzyl)ATP in activation buffer for 30 minutes at 30°C. A small sample was taken and separated by SDS-PAGE followed by staining with Coomassie Blue and exposure to film. **B**. The remaining sample of M160A/W1161A ROCK2-phosphorylated homogenate was separated from the beads and proteins were precipitated with acetic acid/TCA, and then resuspended in 2D sample buffer. Proteins were loaded onto pH 3-11NL or pH 4-7 L IEF strips, as indicated. Following overnight IEF, the samples were separated in the 2nd dimension by 8% SDS-PAGE. Gels were then silver-stained, dried (upper panel), and exposed to Amersham Hyperfilm (lower panel). Dashed circles indicate putative ROCK2 substrates that were excised and sent for identification. Results are representative of three independent experiments.

We noted three sites of eEF1α1 (Ser^{53}, Thr^{72}, and Thr^{432}) that fell within the ROCK consensus phosphorylation motif. These residues were exchanged for alanine and subjected to an *in vitro* kinase assay with W1161A ROCK2 (Figure 4). Thr^{432} substitution of eEF1α1 resulted in a 90% loss of phosphorylation, suggesting that Thr^{432} is a major site of phosphorylation by ROCK2.

The identification of new ROCK2 substrates is important for understanding how this essential regulator of cell mobility and contraction signals to control cellular events. Our study has provided a number of useful advancements in this regard. First, the biotinylated LIMK peptide assay was developed to quickly and quantitatively assess ROCK2 activity *in vitro* with phospho-specific

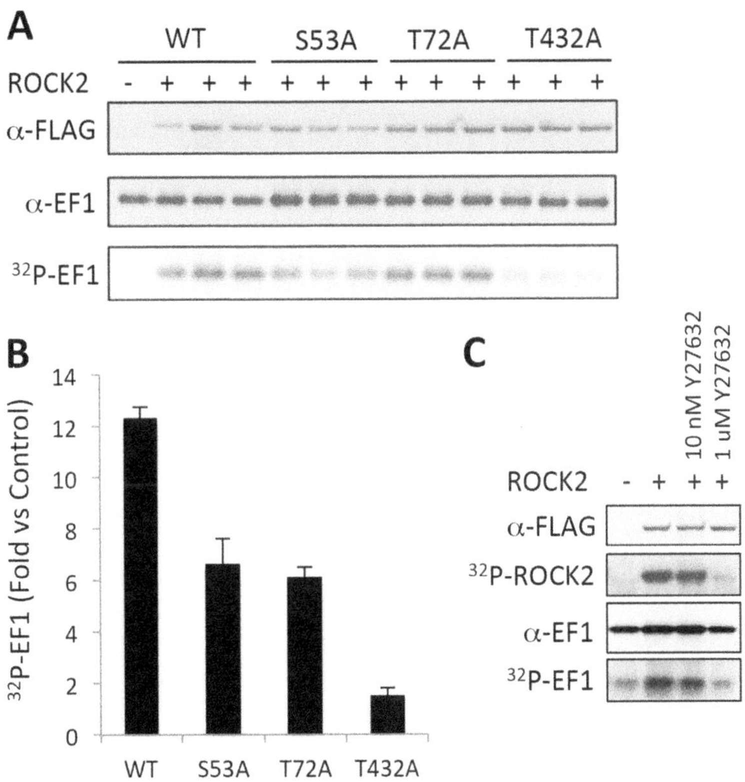

Figure 4 Phosphorylation of EF1α1 *in vitro* by ROCK2. A. EF1α1 (5 μg) was phosphorylated *in vitro* with W1161A ROCK2 as described in *Methods*. The samples were then separated by SDS-PAGE and transferred to a PVDF membrane. FLAG-ROCK2 was visualized by blotting with anti-FLAG M2 antibody, and EF1α1 was detected by anti-EF1α1 antibody. EF1α1 phosphorylation was visualized by phosphoimager. **B.** Fold increase in phosphorylation of EF1α1 by W1161A ROCK2 compared with no kinase for each of the Ser/Thr to alanine substitutions. Error bars represent the standard deviation of triplicate determinations from two independent experiments. **C.** EF1α1 was phosphorylated *in vitro* by W1161A ROCK2 in the presence of 10 nm or 1 μM Y27623. The figure is representative of two experiments.

antibodies and without radioactivity. The biotinylated LIMK peptide assay could be useful in future research applications that seek to assess ROCK2 catalytic activity *in vitro*, such as testing the catalytic effects of point mutations or the discovery of small compound inhibitors of ROCK2.

Secondly, we have utilized the biotinylated LIMK peptide assay to evaluate a ROCK2 mutation that allows utilization of N_6(Benzyl)ATP. Since this bulky ATP analog is not efficiently used by ROCK2 (K_m of 200 μM; Figure 2D) or many other protein kinases [17,18], the M160A ROCK2 protein represents a valuable tool for the future identification of novel ROCK2 substrates. This modified protein was able to phosphorylate the LIMK peptide *in vitro*, and was able to phosphorylate an array of proteins in a cellular lysate that could be separated by two-dimensional electrophoresis and identification by mass spectroscopy. This resulted in the successful identification of the putative ROCK2 substrate eukaryotic elongation initiation factor-1α-1 (eEF1α1). The eEF1α1 is a highly conserved GTP-binding protein that interacts with aminoacyl-tRNA and recruits it to the ribosome during peptide elongation [22]. In addition to this role, eEF1α1 has also been found to be a part of a diverse number of cellular activities, including interactions with actin [23] mitotic apparatus complex formation [24], association with phosphorylated PKB [25], and interactions with PDK1 [26].

Izawa and colleagues reported that eEF1α phosphorylation inhibited co-sedimentation with F-actin [27]. eEF1α binds F-actin or aminoacyl-tRNA in a competitive manner [28]. Thus, eEF1α phosphorylation by ROCK may release the elongation factor from the cytoskeleton, allowing binding of aminoacyl-tRNA, leading to localized translation [27]. Other studies have since shown that overexpression of eEF1α2 leads to filopodia formation in human breast cancer cells, which was reversed with the ROCK2 inhibitor Y-27632 [29]. These results argue that further investigation is required to elucidate whether eEF1α1 represents a physiological ROCK2 substrate.

Examination of the eEF1α1 protein sequence revealed that residues Ser^{53}, Thr^{72}, and Thr^{432} all fall within a ROCK2 consensus motif R/KXS/T or R/KXXS/T (where X is any amino acid) [14]. Individual mutation of these residues to alanine followed by *in vitro* phosphorylation by ROCK2 showed that Thr^{432} is the major site of ROCK2 phosphorylation, since minimal phosphorylation was observed with T432A eEF1α1 (Figure 4). Future work will involve the generation of a phospho-specific antibody to Thr^{432} of eEF1α1 that will aid in elucidating the role of phosphorylation at this residue in cells.

Amano and colleagues [30] have previously reported an alternative proteomic approach to identify substrates of ROCK2, by combining mass spectrometry with affinity column chromatography. This method utilized the catalytic domain of ROCK2 as bait to probe a fraction of cytosol for interacting proteins [30], where our ATP analogue approach relies on a full-length protein that specifically phosphorylates substrates. Both techniques are similar in that proteins of interest were first separated from non-substrates, followed by identification by mass spectrometry.

Further identification of ROCK2 substrates using the ATP analogue approach will focus on specifically enriching for proteins phosphorylated by the AS-ROCK2 protein. This could be achieved by utilizing an N_6-ATP-γ-S analogue and substrate enrichment with iodoacetyl-agarose resin, which binds the sulfonated group [31], prior to identification by mass spectrometry.

Conclusion

In summary, our work has shown the utility of an AS-ROCK2 mutation that allows the use of bulky ATP analogs. This model will be of significant value for future efforts aimed at identifying novel ROCK2 substrates that could play a role in human disease such as cancer and hypertension.

Methods

Cell culture

HEK (human embryonic kidney)-293 cells were obtained from the A.T.C.C. (Manassas, VA, U.S.A.) and cultured in DMEM (Dulbecco's modified Eagle's medium) supplemented with 10% fetal bovine serum and antibiotics at 37°C in a humidified incubator containing 5% CO2.

Reagents

Antibodies used were anti-FLAG M2 (Sigma-Aldrich), anti-[phosphoLIMK1(Thr^{508})/LIMK2(Thr^{505})] (Cell Signaling Technology) and anti-EF1α1 (H-300) (Santa Cruz Biotechnology).

Site-directed mutagenesis

The M160A/W1161A and M160G/W1161A ROCK2 were generated using QuikChange® mutagenesis kit (Stratagene). Mutations were sequence-verified.

cDNA transfections

HEK-293 cells were plated onto 100-mm dishes at 80% confluence and transfected with 2.5 mg of cDNA using Lipofectamine™2000 (Invitrogen) following the manufacturers protocol. 4 hours post transfection, the medium was removed and replaced with DMEM containing fetal bovine serum.

Cell lysis

Cells were lysed in radioimmunoprecipitation (RIPA) buffer, containing 10 mM sodium phosphate (pH 7.4), 150 mM NaCl, 0.1% (w/v) SDS, 1% (v/v) Triton X100, 0.25% (w/v) deoxycholate, 5 mM EDTA, 25 mM NaF, 25 mM 2-glycerolphosphate, 200 μM Na_3VO_4 and protease inhibitors.

Biotinylated LIMK1 peptide assay

FLAG-tagged ROCK2 was immunoprecipitated from transfected cell lysates with M2 monoclonal antibody conjugated to agarose (Sigma–Aldrich). Immunoprecipitates were washed three times in RIPA buffer, followed by an additional three times in activation buffer containing (50 μM Tris–HCl, 0.1 mM EGTA, 0.1% 2-mercaptoethanol), and finally resuspended in 30 μl of activation buffer containing various concentrations of ATP or N_6(Benzyl)ATP. Biotinylated (N-terminal) LIMK1 peptide (DRKKRYTVVGNPY) was used as a substrate (200 μM). Samples were incubated for 30 min at 30°C, the reaction was terminated by the addition of 7.5 mM EDTA. The samples were then loaded onto streptavidin plates (Pierce) and bound for 2 hours at room temperature (24°C). The plates were then probed with phospho-LIMK (Cell Signaling) overnight at room temperature. Secondary decoration with IRDye® 680 anti-rabbit (Li-Cor) was performed at room temperature for 1.5 hours. Proteins were visualized by direct fluorescence scanning with a Li-Cor Odyssey Imager.

Cellular homogenate labeling

3×10^6 HEK293 cells were lysed in 2 ml hypotonic lysis buffer (20 mM Hepes, pH 7.4, 2 mM MgCl2, 200 μM Na3VO4, 0.25% NP-40, and protease inhibitors). Insoluble particulates were removed by centrifugation and 10% glycerol was added to the supernatant and stored at −80°C. Prior to *in vitro* labeling, the homogenate was heat inactivated for 10 min at 60°C. Cellular homogenate was then added to immunoprecipitated ROCK2 proteins with $\gamma^{32}P$-N_6(Benzyl)ATP in activation buffer. The reaction was incubated for 30 min at 30°C and terminated by the addition of 2X LDS sample buffer.

Two-dimensional electrophoresis

Proteins were precipitated from samples with 4x the volume acetone, 10% TCA and 20 mM DTT at −20°C for

45 min. Proteins were then pelleted by centrifugation at 4°C for 15 min. The supernatant was removed and the protein pellet was subjected to an acetone wash, and then dried of residual acetone under vacuum for 2 min. The pellet was then resuspended in 260 µl 2D sample buffer (7 M Urea, 2 M Thio-Urea, 4% CHAPS, 2% DTT, 1% IPG buffer). To ensure protein solubilization, samples were incubated at room temperature for 15 min with intermittent vortexing. The sample was then transferred to a fresh Eppendorf tube and 250 µl was pipetted into a 13 cm strip holder (GE Healthcare). An immobiline dry strip (GE Healthcare) was then placed in the holder, overlayed with drystrip cover fluid. Both pH 3–11 NL and pH 4–7 L immobiline dry strips were used, see figures for details.

Proteins were then separated in the first dimension with the following isoelectric focusing (IEF) conditions: 10 hours rehydration, 5 hours 30 V, 1 hour 500 V, 1 hour 1000 V, 2.5 hour 8000 V, and 30 min 8000 V for ph3-11 NL. The conditions for pH 4–7 were as follows: 10 hours rehydration, 5 hours 30 V, 1 hour 500 V, 1 hour 1000 V, 2.5 hours 8000 V, and 0.55 hours 8000 V. Following IEF, strips were removed from the holders and pre-equilibrated in SDS-equilibration buffer (50 mM Tris–HCl, pH 8.8, 6 M Urea, 30% glycerol, 2% SDS, and 10 mg/ml DTT) for 1 hour at room temperature (24°C) with rocking. Proteins were then separated in the second dimension with 8% SDS-PAGE. After electrophoresis, gels were rinsed in ultrapure water and fixed in 50% ethanol, 10% acetic acid and stained using the Silver Quest kit (Invitrogen). ^{32}P-labeled proteins identified by autoradiography were excised from the gel with a scalpel. Gel fragments were resuspended in 1% acetic acid in microcentrifuge tubes prior to analysis by mass spectrometry.

EF1α1 *in vitro* kinase assay

FLAG-ROCK2 was immunoprecipitated from cell lysates with 20 µl of M2 FLAG monoclonal antibody conjugated to agarose (Sigma-Aldrich). Immunoprecipitates were washed three times with RIPA buffer, followed by two additional washes with kinase assay buffer containing 20 mM Tris pH 7.5, 50 mM NaCl, 5 mM MgCl2, 1 mM dithiothreitol (DTT), and finally resuspended in 30 µl of kinase assay buffer containing 10 µM ATP and 1 mCi [^{32}P]ATP. Purified EF1α1 proteins were used as substrate. Samples were incubated for 30 min at 30°C, and terminated by the addition of 2x LDS sample buffer, and resolved on 4-12% Criterion™XT precast gel (Bio-Rad Laboratories). ^{32}P-EF1α1 was visualized by phosphoimager (Bio-Rad Laboratories).

Competing interest

The authors declare that they have no competing interests.

Authors' contributions

ALC contributed to the design of the study, performed the experiments, interpreted the results, and wrote the initial draft of the manuscript. RMG contributed to the generation of reagents for the study. MPS conceived and contributed to the design of the study, interpreted the results, and edited the final draft of the manuscript. All authors read and approved the final manuscript.

Acknowledgements

We thank Professor Shuh Narumiya (Department of Cell Pharmacology, Kyoto University Graduate School of Medicine, Kyoto, Japan) for ROCK2 expression vectors. Mass Spectra were provided by the Centre for Research in Mass Spectrometry at York University. Operating funds from the Canadian Institute of Health Research supported this work. The funders had no role in study design, data collection and analysis, decision to publish or preparation of the manuscript.

Author details

[1]Department of Biology, York University, 4700 Keele Street, Toronto, ON M3J1P3, Canada. [2]Current Address: Centre for Systems Biology, Lunenfeld-Tanenbaum Research Institute, Mount Sinai Hospital, Toronto, ON M5G 1X5, Canada.

References

1. Riento K, Ridley AJ: **Rocks: multifunctional kinases in cell behaviour.** *Nat Rev Mol Cell Biol* 2003, **4**(6):446–456.
2. Amano M, Chihara K, Nakamura N, Kaneko T, Matsuura Y, Kaibuchi K: **The COOH terminus of Rho-kinase negatively regulates rho-kinase activity.** *J Biol Chem* 1999, **274**(45):32418–32424.
3. Ishizaki T, Maekawa M, Fujisawa K, Okawa K, Iwamatsu A, Fujita A, Watanabe N, Saito Y, Kakizuka A, Morii N, *et al*: **The small GTP-binding protein Rho binds to and activates a 160 kDa Ser/Thr protein kinase homologous to myotonic dystrophy kinase.** *EMBO J* 1996, **15**(8):1885–1893.
4. Kawano Y, Fukata Y, Oshiro N, Amano M, Nakamura T, Ito M, Matsumura F, Inagaki M, Kaibuchi K: **Phosphorylation of myosin-binding subunit (MBS) of myosin phosphatase by Rho-kinase in vivo.** *J Cell Biol* 1999, **147**(5):1023–1038.
5. Totsukawa G, Yamakita Y, Yamashiro S, Hartshorne DJ, Sasaki Y, Matsumura F: **Distinct roles of ROCK (Rho-kinase) and MLCK in spatial regulation of MLC phosphorylation for assembly of stress fibers and focal adhesions in 3 T3 fibroblasts.** *J Cell Biol* 2000, **150**(4):797–806.
6. Matsui T, Maeda M, Doi Y, Yonemura S, Amano M, Kaibuchi K, Tsukita S: **Rho-kinase phosphorylates COOH-terminal threonines of ezrin/radixin/moesin (ERM) proteins and regulates their head-to-tail association.** *J Cell Biol* 1998, **140**(3):647–657.
7. Sumi T, Matsumoto K, Nakamura T: **Specific activation of LIM kinase 2 via phosphorylation of threonine 505 by ROCK, a Rho-dependent protein kinase.** *J Biol Chem* 2001, **276**(1):670–676.
8. Maekawa M, Ishizaki T, Boku S, Watanabe N, Fujita A, Iwamatsu A, Obinata T, Ohashi K, Mizuno K, Narumiya S: **Signaling from Rho to the actin cytoskeleton through protein kinases ROCK and LIM-kinase.** *Science* 1999, **285**(5429):895–898.
9. Croft DR, Sahai E, Mavria G, Li S, Tsai J, Lee WM, Marshall CJ, Olson MF: **Conditional ROCK activation in vivo induces tumor cell dissemination and angiogenesis.** *Cancer Res* 2004, **64**(24):8994–9001.
10. Lane J, Martin TA, Watkins G, Mansel RE, Jiang WG: **The expression and prognostic value of ROCK I and ROCK II and their role in human breast cancer.** *Int J Oncol* 2008, **33**(3):585–593.
11. Loirand G, Guerin P, Pacaud P: **Rho kinases in cardiovascular physiology and pathophysiology.** *Circ Res* 2006, **98**(3):322–334.
12. Takeda K, Ichiki T, Tokunou T, Iino N, Fujii S, Kitabatake A, Shimokawa H, Takeshita A: **Critical role of Rho-kinase and MEK/ERK pathways for angiotensin II-induced plasminogen activator inhibitor type-1 gene expression.** *Arterioscler Thromb Vasc Biol* 2001, **21**(5):868–873.
13. Kawamura H, Yokote K, Asaumi S, Kobayashi K, Fujimoto M, Maezawa Y, Saito Y, Mori S: **High glucose-induced upregulation of osteopontin is mediated via Rho/Rho kinase pathway in cultured rat aortic smooth muscle cells.** *Arterioscler Thromb Vasc Biol* 2004, **24**(2):276–281.

14. Amano M, Ito M, Kimura K, Fukata Y, Chihara K, Nakano T, Matsuura Y, Kaibuchi K: **Phosphorylation and activation of myosin by Rho-associated kinase (Rho-kinase).** *J Biol Chem* 1996, **271**(34):20246–20249.
15. Pidoux G, Tasken K: **Specificity and spatial dynamics of protein kinase A signaling organized by A-kinase-anchoring proteins.** *J Mol Endocrinol* 2010, **5**:271–284.
16. Elphick LM, Lee SE, Gouverneur V, Mann DJ: **Using chemical genetics and ATP analogues to dissect protein kinase function.** *ACS Chem Biol* 2007, **2**(5):299–314.
17. Shah K, Liu Y, Deirmengian C, Shokat KM: **Engineering unnatural nucleotide specificity for Rous sarcoma virus tyrosine kinase to uniquely label its direct substrates.** *Proc Natl Acad Sci USA* 1997, **94**(8):3565–3570.
18. Habelhah H, Shah K, Huang L, Burlingame AL, Shokat KM, Ronai Z: **Identification of new JNK substrate using ATP pocket mutant JNK and a corresponding ATP analogue.** *J Biol Chem* 2001, **276**(21):18090–18095.
19. Eblen ST, Kumar NV, Shah K, Henderson MJ, Watts CK, Shokat KM, Weber MJ: **Identification of novel ERK2 substrates through use of an engineered kinase and ATP analogs.** *J Biol Chem* 2003, **278**(17):14926–14935.
20. Liu Y, Shah K, Yang F, Witucki L, Shokat KM: **Engineering Src family protein kinases with unnatural nucleotide specificity.** *Chem Biol* 1998, **5**(2):91–101.
21. Couzens AL, Saridakis V, Scheid MP: **The hydrophobic motif of ROCK2 requires association with the N-terminal extension for kinase activity.** *Biochem J* 2009, **419**(1):141–148.
22. Riis B, Rattan SI, Clark BF, Merrick WC: **Eukaryotic protein elongation factors.** *Trends Biochem Sci* 1990, **15**(11):420–424.
23. Yang F, Demma M, Warren V, Dharmawardhane S, Condeelis J: **Identification of an actin-binding protein from Dictyostelium as elongation factor 1a.** *Nature* 1990, **347**(6292):494–496.
24. Ohta K, Toriyama M, Miyazaki M, Murofushi H, Hosoda S, Endo S, Sakai H: **The mitotic apparatus-associated 51-kDa protein from sea urchin eggs is a GTP-binding protein and is immunologically related to yeast polypeptide elongation factor 1 alpha.** *J Biol Chem* 1990, **265**(6):3240–3247.
25. Lau J, Castelli LA, Lin EC, Macaulay SL: **Identification of elongation factor 1alpha as a potential associated binding partner for Akt2.** *Mol Cell Biochem* 2006, **286**(1–2):17–22.
26. Pecorari L, Marin O, Silvestri C, Candini O, Rossi E, Guerzoni C, Cattelani S, Mariani SA, Corradini F, Ferrari-Amorotti G, *et al*: **Elongation Factor 1 alpha interacts with phospho-Akt in breast cancer cells and regulates their proliferation, survival and motility.** *Mol Cancer* 2009, **8**:58.
27. Izawa T, Fukata Y, Kimura T, Iwamatsu A, Dohi K, Kaibuchi K: **Elongation factor-1 alpha is a novel substrate of rho-associated kinase.** *Biochem Biophys Res Commun* 2000, **278**(1):72–78.
28. Liu G, Tang J, Edmonds BT, Murray J, Levin S, Condeelis J: **F-actin sequesters elongation factor 1alpha from interaction with aminoacyl-tRNA in a pH-dependent reaction.** *J Cell Biol* 1996, **135**(4):953–963.
29. Amiri A, Noei F, Jeganathan S, Kulkarni G, Pinke DE, Lee JM: **eEF1A2 activates Akt and stimulates Akt-dependent actin remodeling, invasion and migration.** *Oncogene* 2007, **26**(21):3027–3040.
30. Amano M, Tsumura Y, Taki K, Harada H, Mori K, Nishioka T, Kato K, Suzuki T, Nishioka Y, Iwamatsu A, *et al*: **A proteomic approach for comprehensively screening substrates of protein kinases such as Rho-kinase.** *PLoS ONE* 2010, **5**(1):e8704.
31. Blethrow JD, Glavy JS, Morgan DO, Shokat KM: **Covalent capture of kinase-specific phosphopeptides reveals Cdk1-cyclin B substrates.** *Proc Natl Acad Sci USA* 2008, **105**(5):1442–1447.

Pharmacological characteristics and efficacy of a novel anti-angiogenic antibody FD006 in corneal neovascularization

Qun Wang[1], Jing Yang[2], Kun Tang[3], Longlong Luo[2], Liqiang Wang[1], Lei Tian[1], Yanming Jiang[4], Jiannan Feng[2], Yan Li[2], Beifen Shen[2], Ming Lv[2*] and Yifei Huang[1*]

Abstract

Background: Vascular endothelial growth factor (VEGF) is a key angiogenic factors. It plays an important role in both physiologic and pathologic angiogenesis and increases permeability across the vessels. Using antibody phage display technology, we obtained a novel anti-VEGFA IgG, named as FD006. In this study, the pharmacological characteristics and efficacy of FD006 in corneal neovascularization (CoNV) were evaluated.

Results: FD006 was predicted to have similar binding mode to bevacizumab. Experimental analysis showed that the binding ability of FD006 seemed a little stronger than bevacizumab, for the EC50 of FD006 to bind VEGF analyzed by ELISA was about 0.037 μg/mL while that of bevacizumab was 0.18 μg/mL. Binding kinetics assays showed similar results that FD006 possessed 2-fold higher affinity to bind VEGF than bevacizumab due to slower dissociation rate of FD006; meanwhile, FD006 inhibited the VEGF-induced proliferation of HUVEC with an IC50 value of 0.031 ± 0.0064 μg/ml, which seemed similar or a litter better than bevacizumab (0.047 ± 0.0081 μg/ml). The subconjunctival administration of FD006, bevacizumab or dexamethasone could significantly inhibit the growth of CoNV contrasting to N.S ($p < 0.01$). At the early stage, FD006 showed better inhibitory effect on the growth of CoNV compared with bevacizumab ($p < 0.05$). Western blot analysis showed that FD006 could inhibit the expression of VEGF, VEGFR-1, VEGFR-2, MMP-9 and ICAM-1, which could explain its favorable anti-angiogenic activity.

Conclusions: The pharmacological characteristics of FD006 were similar or even a little better than bevacizumab in inhibiting corneal neovascularization.

Keywords: Neovascularization, Cornea, Bevacizumab, Angiogenesis, Anti-angiogenic treatment

Background

Corneal avascularity is necessary for the preservation of corneal transparency and optimal vision [1]. There is a dynamic balance between pro- and anti-angiogenic factors in the cornea that plays a key role in maintaining avascularity. Under normal physiologic conditions, the process of angiogenesis is well controlled, reflecting a perfect balance of endogenous growth factors and suppressors. However, in the condition of persistent hypoxia, chronic inflammation, injury and hereditary stem cell deficiency, the pro-angiogenic growth factors outnumber angiogenesis inhibitors, the balance shifts in favor of angiogenesis, and then the corneal angiogenic privilege could be destroyed [2,3]. Although new blood vessel growth in response to ischemic conditions is to repair the injury, pathologic corneal neovascularization (CoNV) and hyperpermeability can cause decreased vision and failure of transplantation. CoNV is a sight-threatening condition, which is also a challenge for ophthalmologists. The inhibitory effects of steroids, cyclosporine, thalidomide, photodynamictherapy, conjunctival limbal allograft and fine needle diathermy for CoNV vary in clinical and experimental use [4-9]. They are not always effective, sometimes may induce complications.

* Correspondence: Lm62033@aliyun.com; huangyf301@gmail.com
[2]Institute of Basic Medical Sciences, P.O. Box 130 (3), Taiping Road, Beijing 100850, China
[1]Department of Ophthalmology, General Hospital of People's Liberation Army, No.28 Fuxing Road, Haidian District, Beijing 100853, China
Full list of author information is available at the end of the article

Angiogenesis, the process by which new vessels sprout from pre-existing vasculature, has become the subject of intense research in recent years and therapies that directly target key factors of neovascular formation are promising. Among the angiogenic factors, vascular endothelial growth factor (VEGF) plays an important role in both physiologic and pathologic angiogenesis and increases permeability across the vessels. Extensive research studies focus on the VEGF families and receptors [10,11]. Given that several studies have demonstrated that the secretion and receptor expression of VEGF were increased during the neovascularization of the cornea, specific molecules have been targeted for drug development and have drawn much attention recently [10-12]. The targeted inhibition of VEGF provides ophthalmologists with an angiogenesis-specific pharmacological approach to treating CoNV and deserves to be evaluated. There are many VEGF inhibitors which are in advanced clinical use for ocular neovascularization, such as Lucentis, Macugen and bevacizumab/Avastin. Administration of these drugs in CoNv has been associated with promising results [13-15].

In our previous work, we screened out a novel anti-VEGF antibody from a natural human scFv phage library which was constructed in our lab. Then, the recombinant vector to express the full antibody was constructed and transfected into mammalian cells. The full IgG, named as FD006, was prepared for further studies. The purpose of the present study was to investigate the pharmacological characteristics and efficacy of FD006 both *in vitro* and *in vivo*.

Results

The predicted binding mode of FD006 was similar to bevacizumab

Using antibody phage display library, we obtained a novel anti-VEGFA antibody, named as FD006. After sequencing, the amino acid residues of the variable region in FD006 were determined as shown in Figure 1A. Using computer-guided homology modeling method, the Stable 3-D structure of FD006 variable region was constructed and shown in Figure 1B. Based on the crystal structure of VEGFA, the complex structure of VEGFA and FD006 was constructed and optimized (Figure 1C) using computer-guided molecular docking method. Using interactive graphics and the hard-sphere approximation for atoms, we found that the FD006 variable region was fitted into the binding site of VEGFA which identified by bevacizumab. The binding mode of FD006, *e.g.* binding energy and epitope, was analyzed and shown in Table 1, which showed that the binding mode between FD006 and VEGF was similar to bevacizumab and VEGF.

FD006 possessed similar or even higher affinity than bevacizumab

ELISA assay was performed to determine the antigen-binding specificity of FD006. As shown in Figure 2A, it was shown that both bevacizumab and FD006 could bind to VEGF on a dose-dependent manner. In addition, the EC50 of FD006 was about 0.037 μg/mL while that of bevacizumab was 0.18 μg/mL, suggesting that FD006 might possess stronger binding ability than bevacizumab (5-fold). Further experiments indicated the interacting kinetics of two mAbs to bind to VEGF (Figure 2B), displaying approximately 2-fold higher affinity of FD006 than bevacizumab (Table 2) due to its slower dissociation rate; besides, the inhibitory ability of FD006 against the VEGF-induced cell proliferation in HUVEC was identified. The IC50 value of FD006 was approximately 0.031 ± 0.0064 μg/ml, indicating its somewhat better antiangiogenic activity of FD006 than bevacizumab (0.047 ± 0.0081 μg/ml) (Figure 2C).

FD006 inhibited in vivo neovascularization in CoNV

Alkali-induced CoNV model is a simple model system for the study of neovascularization. After the alkali-soaked filter paper was removed from the cornea, the central cornea appeared opaque with a clear margin. After modeling, vessels began sprouting into the cornea from the limbus. It was demonstrated that the subconjunctival administration of FD006, bevacizumab and dexamethasone could all significantly inhibit CoNV in NaOH cauterized rats compared with the control group ($p < 0.01$) (Figures 3 and 4); furthermore, compared with bevacizumab, FD006 was a little more effective on inhibiting CoNV to reduce the length and area of the corneal neovasculature at day 3 and day 7 ($0.01 < P < 0.05$). However, since day 14, there was no obvious difference between the two groups ($p > 0.05$). Meanwhile, the two antibodies were less effective than dexamethasone ($p < 0.01$). Typical HE staining images of the rats' CoNV were shown in Figure 4, which also displayed weaker vascularization in the FD006 group, bevacizumab group and dexamethasone group contrasting to that in the control group (Figure 5).

FD006 downregulated the expression of neovascularization-related molecules in the cornea

According to western blot analysis, the subconjunctival administration of FD006 or bevacizumab significantly decreased the protein expression of VEGF, VEGFR-1, VEGFR-2, MMP-9 and ICAM-1, while 0.9% NaCl treated group and solvent group displayed visibly high expression of the above factors in the cornea (Figure 6). It might be the possible mechanism of FD006's inhibitory ability against neovascularization *in vivo*, indicating that FD006 might be a favorable anti-angiogenic candidate drug.

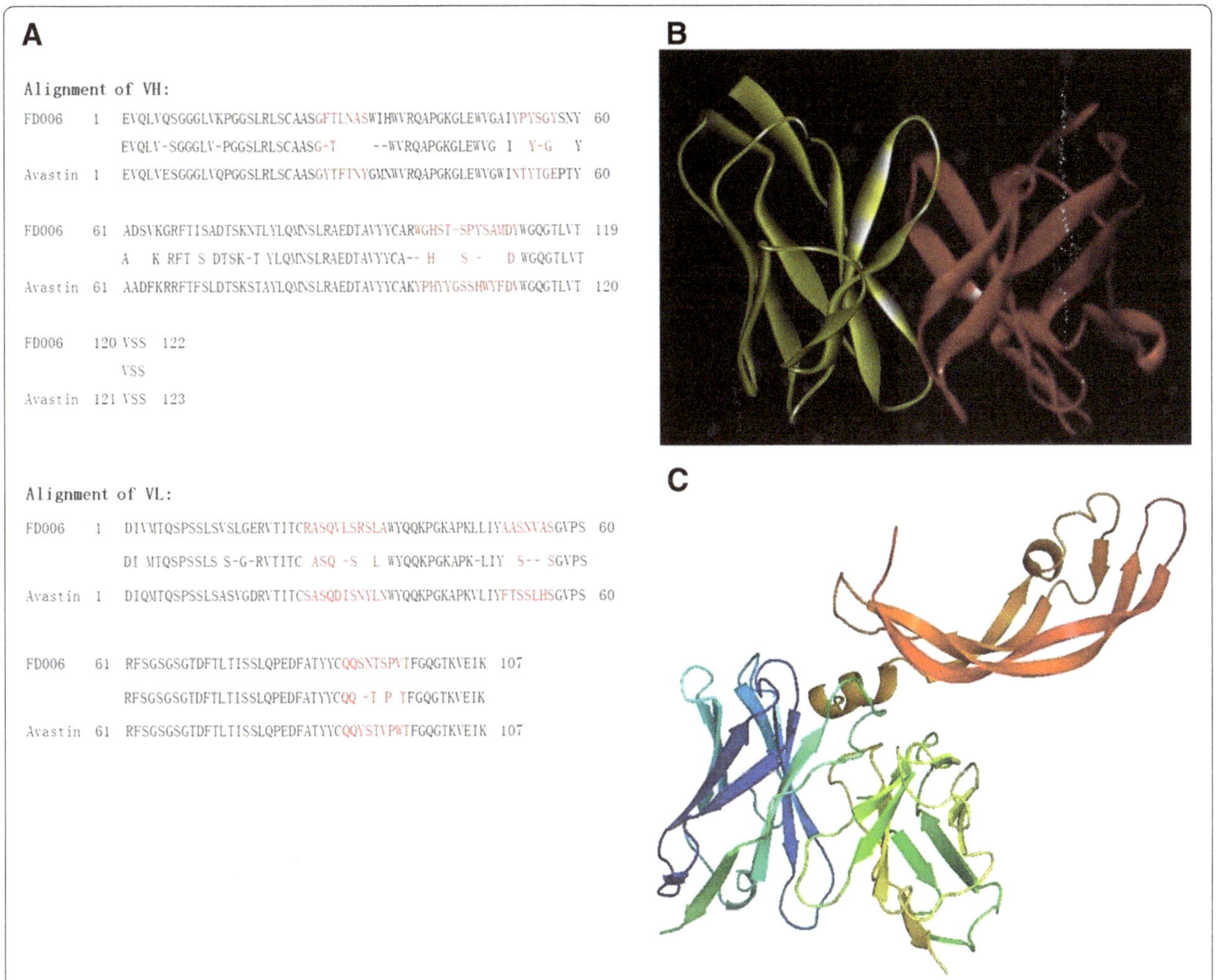

Figure 1 Theoretical analysis of FD006 to bind VEGF. A: The amino acid residues and CDR region classification of FD006 (versus Avastin/bevacizumab); **B**: The 3-D structure of variable region in FD006 using computer-guided homology modeling and molecular dynamics methods. The yellow ribbon denoted the light chain variable region and the red ribbon denoted the heavy chain variable region; **C**: The 3-D complex structure of FD006 variable region and VEGFA obtained from computer-guided molecular docking and dynamics methods. The upper red ribbon denoted VEGF, the lower left (blue) ribbon was FD006-VL and the lower right (green and yellow) denoted FD006-VH.

Discussion

The normal cornea is an avascular structure. The formation of neovascularization may be a physiological response to various stimuli; however, the chronic and persistent upregulation of pro-angiogenic factors may result in pathological CoNV, which can lead to blindness.

Table 1 The predicted binding energy (kCal/mol) and epitope between antibody (FD006 or bevacizumab) and antigen (VEGFA)

	Binding energy (kcal/mol)	Identified epitope
FD006-VEGF	−168.24	$F^{17}M^{18}D^{19}$, $Y^{21}Q^{22}R^{23}$, Y^{25}, $L^{66}E^{67}$, $N^{100}K^{101}$
bevacizumab-VEGF	−156.35	$F^{17}M^{18}D^{19}$, $Y^{21}Q^{22}R^{23}$, Y^{25}, K^{101}

For it was reported that VEGF is required for inflammatory neovascularization and a crucial endogenous corneal angiogenic factor, VEGF was confirmed as the key target for blocking the cascade of neovascular formation [16,17]. VEGF is a vascular endothelial cell mitogen and one of the key regulators of angiogenesis in various physiological and pathological conditions [14]. By binding to VEGFR, VEGF triggers its tyrosine phosphorylation and induces a signaling cascade that promotes growth, survival, migration, differentiation of endothelia cells as well as mobilizing endothelial progenitor cells from the bone marrow into the peripheral circulation [16-18]. The angiogenic activity of VEGF is believed to be regulated by two high affinity receptor tyrosine kinases, VEGFR-1 (fms-like tyrosine kinase, Flt-1) and VEGFR-2 (kinase insert domain receptor, Flk-1), which is

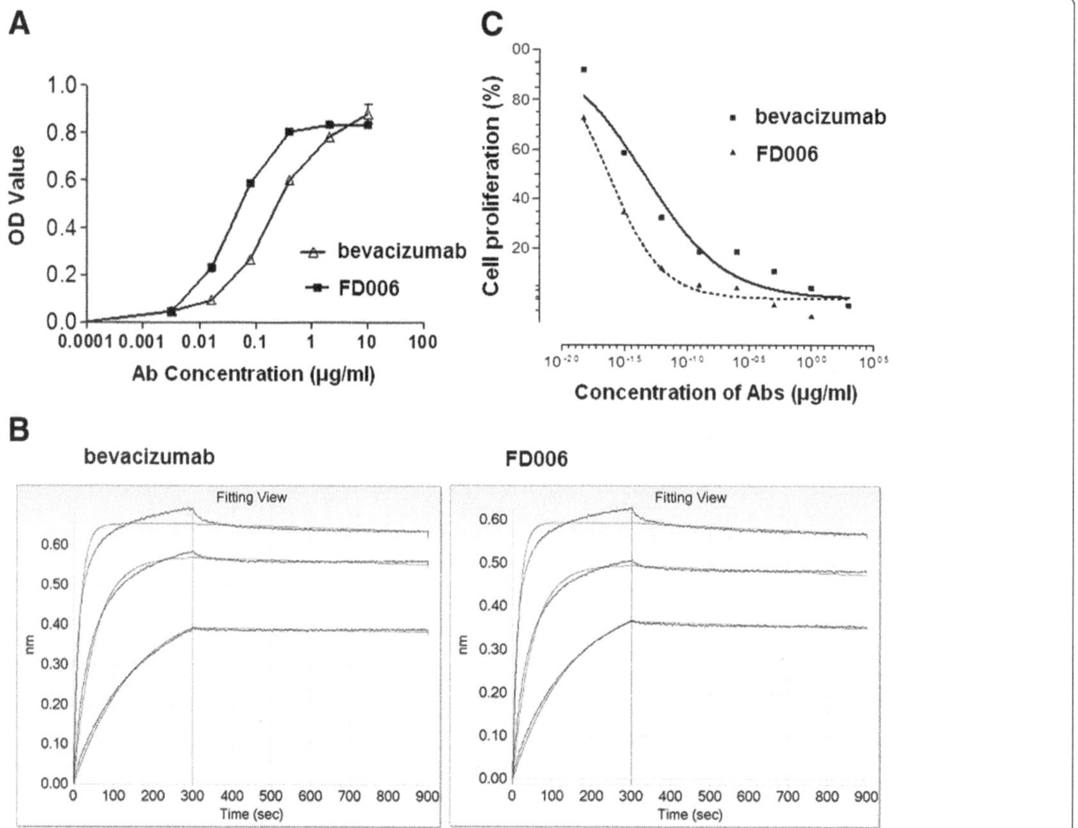

Figure 2 Antigen binding identification of FD006 to bind VEGF. (A) The binding activity of FD006 by ELISA; **(B)** Affinity detection of FD006 by binding kinetics assays; **(C)** The inhibitory effects of FD006 on VEGF-induced human umbilical vein endothelial cell (HUVEC) proliferation. The proliferation percentage was calculated using the formula $OD_{sample}/OD_{untreated} \times 100\,\%$. Data points represent the mean 6 SD of values acquired in duplicate.

reported to be the dominant angiogenic activity receptor of VEGF. Although the complications of dexamethasone such as cataract and secondary glaucoma have limited its clinical use, it is still a powerful anti-VEGF agent; bevacizumab is also considered as an effective anti-VEGF agent that can bind to soluble VEGF, preventing receptor binding [19,20].

We discovered that the protein expression of ICAM-1 and MMP-9 was significantly higher in the control groups than in the other groups. FD006 reduced the expression of ICAM-1 and MMP-9 compared with the control groups (Figure 6). ICAM-1 (Intercellular adhesion molecule 1) is one of cell surface glycoproteins that are typically expressed on endothelial cells. ICAMs may play a critical role in regulating leukocyte migration and

Table 2 Interacting kinetics of bevacizumab and FD006 binding to VEGF

	KD (M)	Kon (1/Ms)	Kdis (1/s)	Full R2
bevacizumab	6.50E-10	1.75E + 05	1.02E-04	0.992536
FD006	3.11E-10	1.48E + 05	4.59E-05	0.992964

accumulation at sites of inflammation, thus generating inflammatory response and damage to tissue [21]. Corneal fibroblasts expressed ICAM-1 and VCAM-1 when activated with IL-4 and TNF-alpha [22]. Luo *et al.* discovered that VEGF and b-FGF can facilitate the expression of ICAM-1 [23]. Therefore we inferred that the expression level of ICAM-1 was lower in FD006 group because there may be a feedback reaction where the decreased VEGF expression influences the release of ICAM-1 after FD006 blocked the VEGF signaling.

MMPs play indispensable roles in the formation of CoNV and are one of the most potent proangiogenic factors. MMP-9, also known as gelatin B, plays an important role in degrading the ECM and basement membrane during organizational restructuring and angiogenesis. MMPs can promote the migration of endothelial cells by destroying connections between the cells and the extracellular matrix [24]. Particularly, the synergistic actions of MMPs and VEGF have been discovered in angiogenesis, and studies have already demonstrated that VEGF and MMPs influence each other during angiogenesis. MMP can enhance VEGF release and modulate VEGF

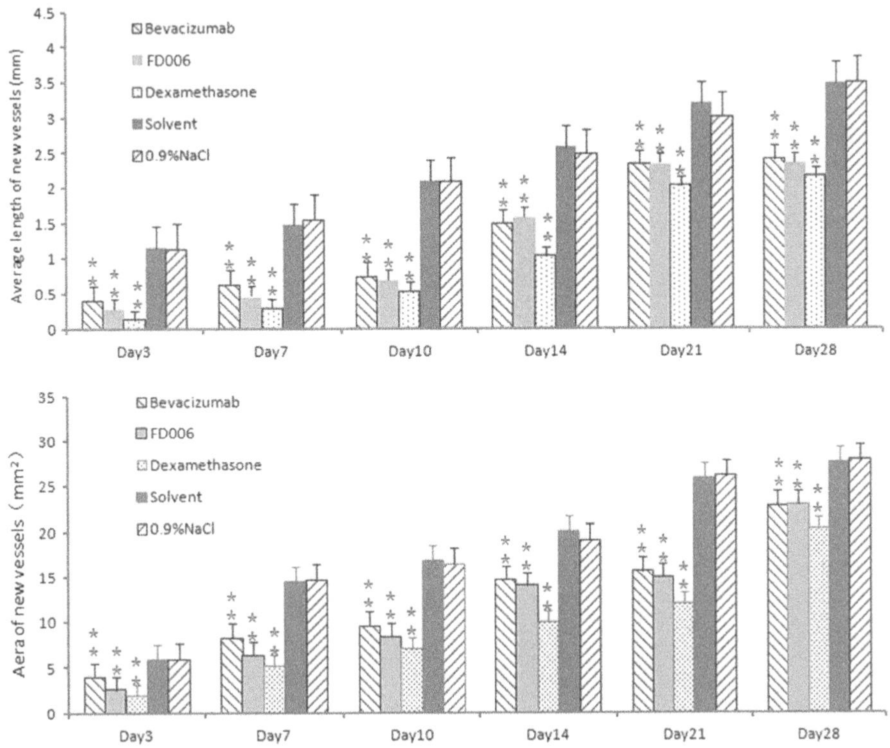

Figure 3 The mean length and area of the corneal neovascularization. 0.9% NaCl treated group and solvent group were set as controls. Results are represented as mean ± SEM. *p < 0.05, **p < 0.01.

expression [25-28]. Additionally, VEGF increases the release of the MMPs and decreases the release of the tissue inhibitor of metalloproteinase, whereas MMPs activate the angiogenic activity of VEGF. We believe that, among ICAM-1, MMP-9 and VEGF, there is a cascade of chain reactions that influence each other. However, our hypotheses require further investigation.

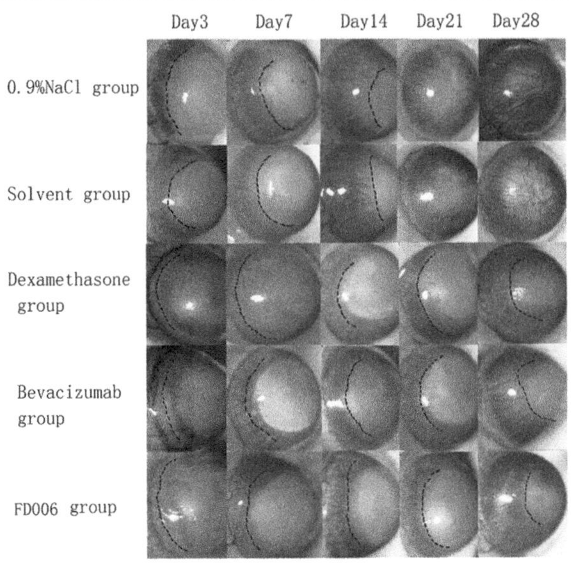

Figure 4 Typical images of rat corneal neovascularization caused by alkali burn. The dotted lines represent the margin of the corneal neovascularization.

In this study, we screened and predicted the novel anti-VEGF monoclonal antibody FD006 to have similar affinity to bevacizumab (Figure 1 and Table 1). Further experiments testified that both bevacizumab and FD006 could bind to VEGF specifically on a dose-dependent manner; meanwhile, FD006 showed somewhat stronger affinity to bind VEGF than bevacizumab by both ELISA (5-fold) and binding kinetics assays (2-fold) mainly because of its slower dissociation rate (Table 2). In principle, the higher affinity of anti-VEGF means higher efficiency to neutralize VEGF; furthermore, FD006 seemed to have a better inhibitory effect on the VEGF-induced proliferation of HUVEC than bevacizumab, which was consistent with antigen binding assays (Figure 2).

The alkali burn-induced CoNV model has been widely used to investigate the mechanism of corneal neovascular formation. Here, this method also induced CoNV successfully. Several studies reported that bevacizumab was more effective via subconjunctival administration for CoNV compared with via topical eye drops [19,20], therefore in this study FD006 was subconjunctivally injected in alkali-burn induced corneal neovascularizaiton to evaluate its biological properties *in vivo*. It was demonstrated that FD006 could inhibit corneal neovascular formation. Particularly, during the early stage after

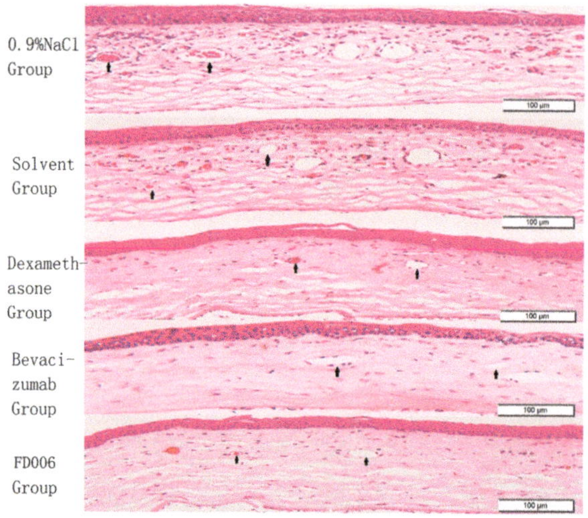

Figure 5 Hematoxylin and eosin (HE) staining observation of corneas. At day 7, corneal neovascularization of various sizes (black arrowhead) were present in the superficial central corneal stroma after alkali burn (×200) and neovascularization containing RBCs in the corneal stroma were indicated with black arrowheads. Cornea stroma exhibited edema and mononuclear inflammatory response. In this plane of section, the corneal epithelium is relatively normal. Bar = 100 μm.

alkali burn, it seemed a little better than bevacizumab (Figures 3, 4 and 5). The reason might be the higher ability of FD006 to bind VEGF. However, at the late stages, FD006 and bevacizumab had an equal therapeutic effect on inhibiting CoNV. Besides, FD006 significantly decreased the expression of VEGF, VEGFR-1, VEGFR-2, ICAM-1, and MMP-9 in cornea compared with the control group (Figure 6), explaining partially why FD006 had satisfactory anti-neovascularization effect.

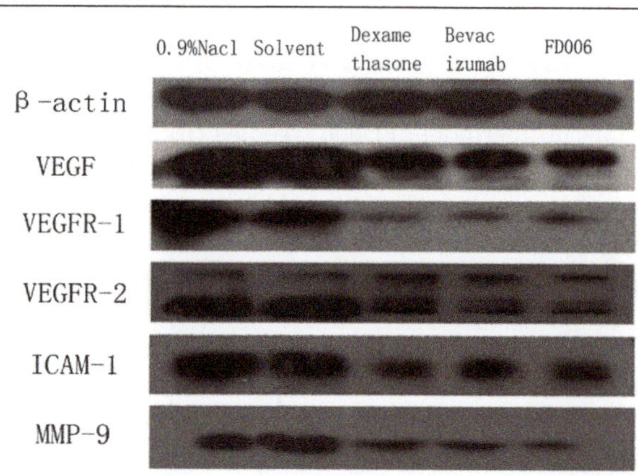

Figure 6 The protein expression of neovascularization-related signal molecules in the cornea. β-actin served as an internal standard. The protein expression of VEGF, MMP-9, VEGFR-1, VEGFR-2 and ICAM-1 in the cornea was significantly weaker in antibody- or dexamethasone- treated samples.

Conclusion

This study displayed the pharmacological characteristics of FD006, which were similar or a little superior to bevacizumab, such as higher affinity, better anti-neovascularization function *in vitro* and *in vivo*, indicating that FD006 might be a promising agent in the treatment of human CoNV in the future.

Methods

Chemicals and antibodies

FD006 is a potent recombinant humanized monoclonal antibody directed against VEGF and prepared in our lab. The commercial anti-VEGF antibody bevacizumab (Avastin, 100 mg/4 ml) was purchased from Roche Pharmaceutical (Genentech, South San Francisco, USA), analyzed in our theoretical platform and set as positive control. Bevacizumab was dialyzed with the same solvent as FD006 in this study.

Computational modeling and docking

Based on the variable heavy (VH) and light chain (VL) sequences of FD006, sequence alignmment were performed using BLASTp (http://www.ncbi.nlm.nih.gov) software with PDB database. And then, the 3-D structure of the antibody FD006 variable region was constructed using computer-guided homology modeling method based on the crystal structures of VH domain of B20-4 Fab (PDB code: 2FJH) and VL domain of B20-4 Fab (PDB code: 2FJH). After constructing the backbone of the Fv fragment, the side chains were replaced using rotamer libraries and algorithm implemented in the Homology package of InsightII 2000 (MSI, San Diego, CA). Furthermore, the structure of the FD006 variable region was optimized under CVFF force field [29] using molecular dynamics method. On the basis of the crystal structure of the extracellular domain of VEGFA (PDB code: 2FJH [30], the partial charges of the VEGFA and FD006 Fv fragment were assigned. According to the crystal complex structure of VEGFA and its functional antibody bevacizumab (PDB code: 2FJH), the FD006 Fv fragment was docked manually into the binding site of VEGFA. The antigen-antibody complex was relaxed by energy minimization and subsequent molecular dynamic simulations in vacuo using the CVFF forcefield. Molecular dynamic simulations were performed using the NVT ensemble of the Discover 3.0 package (MSI). The binding energy and identified domain between FD006 and VEGFA were analyzed with distance geometry and computer graphics technology.

ELISA analysis

ELISA plates were coated with 100 μl/well 0.5 μg/ml VEGF. Bevacizumab or FD006 were diluted (10, 2, 0.4, 0.08, 0.016 and 0.0032 μg/ml) and added for 2 hours' incubation in 37. After three times of washing, the Peroxidase-conjugated affinipure goat anti-human immunoglobulin G (IgG) was added as the secondary antibody for 45 minutes at room temperature (RT for short, the same below). Binding signals were visualized using o-phenylenediamine dihydrochloride (OPD) substrate and the light absorbance was measured with an ELISA reader at 492 nm.

Binding kinetics assays

The binding kinetics of bevacizumab or FD006 to VEGF was measured using Bio-Layer Inter-Ferometry on Octet RED (Fortebio, USA). The assay was conducted at 30°C in PBS buffer. Sensor tips were pre-wet for 15 mins in buffer immediately prior to use, and the microplates were filled with 200 µl per well of diluted samples (VEGF) or buffer and agitated at 1000 rpm. The anti-human IgG biosensor were pre-saturated with bevacizumab or FD006 (10 µg/ml) and washed in buffer for 120 seconds, and then transferred to VEGF at concentrations of 10 µg/ml, 3 µg/ml and 1 µg/ml. The VEGF association and dissociation rates were measured for 5mins and 10mins, respectively. The Kinetics parameters (Kon and Koff) and affinities (KD) were calculated from a non-linear global fit using the Octet analysis software. Multiple independent measurements were performed.

Proliferation assays

Human umbilical vein endothelial cells (HUVECs) (1×10^4 cells/100 µL/well) were seeded in 96-well plates and cultured at 37 for 14 h with Endothelial Cell Medium (ScienCell) supplemented with 5% heat-inactivated FCS, 100 U/ml penicillin, 100U/ml streptomycin, and endothelial cell growth supplement (ScienCell). After low-serum starvation overnight, cells were treated with different concentrations of FD006 or bevacizumab which were pre-incubated with 10 ng/ml VEGF for 30 minutes and incubated at 37, 5% CO2 for 72 hours. Then, 10 µl CCK8 was added to each well and incubated for another 4 hours. The absorbance was measured by spectrophotometer at 450 nm to determine the cell viability.

Animals

Ninety male Sprague–Dawley rats (6–8 weeks, 180-200 g) were obtained from the Laboratory Animal Centre in Academy of military medical sciences. The right eyes of the animals were used for experiments, and the left eyes were left intact. All rats were carefully examined before the procedure using a slit lamp microscope and the corneas with any defects, new vessels, or cataract were excluded.

The animal studies were approved by Animal Ethics Committee of Institute of Basic Medical Sciences. All procedures involving animal eye studies were conducted strictly in accordance with the ARVO Statement for the Use of Animals in Ophthalmic and Vision Research.

Alkali burn model

Prior to the procedure, the rats were anesthetized using an intraperitoneal injection of 3.5 ml/kg 10% chloral hydrate (General Hospital of PLA, Beijing), and the eyes were treated with a topical application of 0.5% proparacaine hydrochloride (Alcain; Alcon, USA). Then, a 3 mm diameter Whatman 3# filter paper was soaked in 5 µL of 1 M sodium hydroxide for 60 s and then placed in the center of the cornea for 40 s. The burned area and conjunctival sac was then irrigated with 60 ml saline for 1 minute.

Animal treatment

After modeling, ninety rats were randomly divided into five groups (eighteen rats per group) and received a subconjunctival injection with 0.05 ml per rat of (1) 0.9% NaCl, (2) solvent, (3) 0.1% dexamethasone (Dexamethasone sodium phosphate, NCPC.LTD, China), (4) 25 mg/ml bevacizumab and (5) 25 mg/ml FD006 in the superior temporal conjunctiva on the day after modeling. All chemical burns and treatments were performed by one investigator. The operator was blinded to the treatment group from which each cornea was derived. At postoperative days 3, 7, 14, 21 and 28, the eyes were harvested for further studies after the rats were sacrificed with an overdose of 10% chloral hydrate.

Quantification of CoNV

Slit lamp examinations were performed in a blinded manner on day 3, 7, 14, 21 and 28 by a same operator. The changes in length and area of the corneal neovasculature and related adverse complications were recorded using a Canon Cybershot camera (Canon Corporation, Japan). The corneas were examined and photographed routinely. The photographs were digitized, and the images were analyzed using Image J Software. The length of the CoNV was measured from the edge of the sclera using a reticule on days 3, 7, 14, 21 and 28 after modeling. The area (A) of the CoNV was calculated using the following formula: $A = C/12 \times \pi[r^2 - (r - l)^2]$, where C is the clock hours of corneal NV coverage on the cornea, l is the average length of the selected vessels, and r is the radius of the rat cornea.

Immunohistochemistry

The rat eyeballs were harvested and fixed in 4% paraformaldehyde. The corneas were excised, embedded in paraffin, cut into 5-µm sections and then prepared for hematoxylin and eosin (H&E) staining. The cornea sections were fixed in acetone. The primary antibodies were VEGF-A, fms-like tyrosine kinase (VEGFR-1, Flt-1) and kinase insert domain receptor (VEGFR-2, Flk-1), matrix metalloproteinase-9 (MMP-9) and intercellular adhesion molecule-1 (ICAM-1). Goat anti-mouse IgG was used as the secondary antibody. Then stained samples were observed and photographed using inverted microscopy.

Western blot

The rats were euthanized, and the corneas were harvested from the treated eyes. The corneas were dissected and frozen at −70°C, then homogenized in ice-cold RIPA lysis buffer solution (Santa cruz, USA). After being centrifuged for 5 minutes at 12,000 revolutions per minute (rpm), the supernatants of the samples were collected. Certain amounts of the protein were separated by electrophoresis on an 8% and 12% sodium dodecyl sulfate-polyacrylamide gel, and transferred to nitrocellulose membranes (Invitrogen, San Diego, CA). Then the membranes were blocked using 5% skim milk in TBST (20 mM Tris, 150 mM NaCl and 0.1% Tween-20) and incubated overnight with primary antibodies: anti-VEGF (Santa Cruz, USA), MMP-9 (Abcam, UK), VEGF receptor 1 (Abcam, UK), ICAM-1 (LifeSpan BioSciences, USA) and VEGF receptor 2 (LifeSpan BioSciences, USA). The secondary polyclonal antibodies were used to detect bound primary antibodies: anti-mouse (EMD Millipore, USA), anti-rabbit (EMD Millipore, USA), and anti-goat (EMD Millipore, USA). After washing, an enhanced chemiluminescence solution (ECL reagent: Santa Cruz, CA) were used to visualize specific lanes.

Statistical analysis

Statistical analysis was performed using IBM SPSS Version 20.0 software (IBM SPSS, Chicago, IL). Experimental data were analyzed using one-way ANOVA, the Dunnett multiple comparison test, and independent two-sample t-tests. All reported P values are 2-sided and $P < 0.05$ was considered statistically significant. All results are presented as the mean ± standard error of the mean. All measurements were within 95% confidence limits.

Competing interests

The authors declare that they have no competing interests.

Authors' contributions

QW did in vivo experiments and prepared the manuscript, JY did in vitro binding, cell proliferation experiments and western blot analysis, KT constructed expression vectors, LL selected FD006 from the phage library, LW established the animal model, LT helped do the animal operation, YJ helped observe animals and take photos, JF did theoretical anlaysis, YL purified FD006, BS did antibody preparation in mammalian cells, ML designed the work and did affinity identification, YH designed the work and revised the manuscript. All authors read and approved the final manuscript.

Authors' information

Qun Wang and Jing Yang are co-first authors.

Acknowledgement

This research was supported by the National Key Basic Research Program of China (973 Program, No. 2013CB967001), Natural Science Foundation of China (No.31271059, No.31370938) and Chinese People's Liberation Army general hospital support funding (No.2012-fc-1003).

Author details

[1]Department of Ophthalmology, General Hospital of People's Liberation Army, No.28 Fuxing Road, Haidian District, Beijing 100853, China. [2]Institute of Basic Medical Sciences, P.O. Box 130 (3), Taiping Road, Beijing 100850, China. [3]Department of Ophthalmology, Aerospace center Hospital, No. 15 Yuquan Road, Haidian District, Beijing 100049, China. [4]Department of Ophthalmology, The second Artillery General Hospital, No.16 Xinwai Road, Xicheng District, Beijing 100088, China.

References

1. Beebe DC: **Maintaining transparency: a review of the developmental physiology and pathophysiology of two avascular tissues.** *Semin Cell Dev Biol* 2008, **19**:125–133.
2. Chang JH, Gabison EE, Kato T, Azar DT: **Corneal neovascularization.** *Curr Opin Ophthalmol* 2001, **12**:242–249.
3. Dana MR, Schaumberg DA, Kowal VO, Goren MB, Rapuano CJ, Laibson PR, Cohen EJ: **Corneal neovascularization after penetrating keratoplasty.** *Cornea* 1995, **14**:604–609.
4. Haynes WL, Proia AD, Klintworth GK: **Effect of inhibitors of aarachidonic acid metabolism on corneal neovascularization in the rat.** *Invest Ophthalmol Vis Sci* 1989, **20**:1588–1593.
5. Lipman RM, Epstein RJ, Hendricks RL: **Suppression of corneal neovascularization with cyclosporine.** *Arch Ophthalmol* 1992, **110**:405–407.
6. Kruse FE, Joussen AM, Rohrschneider K, Becker MD, Volcker HE: **Thalidomide inhibits corneal angiogenesis induced by vascular endothelial growth factor.** *Graefes Arch Clin Exp Ophthalmol* 1998, **236**:461–466.
7. Gohto Y, Obana A, Kanai M, Nagata S, Nakajima S, Miki T: **Treatment parameters for selective occlusion of experimental corneal neovascularization by photodynamic therapy using a water soluble photosensitizer, ATX-S10(Na).** *Exp Eye Res* 2001, **72**(1):13–22.
8. Daya SM, Ilari FA: **Living related conjunctival limbal allograft treatment of stem cell the deficiency.** *Ophthalmology* 2001, **108**:126–133.
9. Pillai CT, Dua HS, Hossain P: **Fine needle diathermy occlusion of corneal vessels.** *Invest Opthalmol Vis Sci* 2000, **41**:2148–2153.
10. Philipp W, Speicher L, Humpel C: **Expression of vascular endothelial growth factor and its receptors in inflamed and vascularized human corneas.** *Invest Ophthamlol Vis Sci* 2004, **41**:2514–2522.
11. Joussen AM, Poulaki V, Mitsiades N, Stechschulte SU, Kirchhof B, Dartt DA, Fong GH, Rudge J, Wiegand SJ, Yancopoulos GD, Adamis AP: **VEGF-dependent conjunctivalization of the corneal surface.** *Invest Ophthalmol Vis Scl* 2003, **44**:117–123.
12. Cursiefen C, Cao J, Chen L, Liu Y, Maruyama K, Jackson D, Kruse FE, Wiegand SJ, Dana MR, Streilein JW: **Inhibition of hemangiogenesis and lymphangiogenesis after normal-risk corneal transplantation by neutralizing VEGF promotes graft survival.** *Invest Ophthalmol Vis Sci* 2004, **45**(8):2666–2673.
13. Adamis AP, Shima DT: **The role of vascular endothelial growth factor in ocular health and disease.** *Retina* 2005, **25**:111–118.
14. Ju M, Mailhos C, Bradley J, Dowie T, Ganley M, Cook G, Calias P, Lange N, Adamis AP, Shima DT, Robinson GS: **Simultaneous but not prior inhibition of VEGF165 enhances the efficacy of photodynamic therapy in multiple models of ocular neovascularization.** *Invest Ophthalmol Vis Sci* 2008, **49**(2):662–670.
15. Finger PT, Chin KJ: **Refractory squamous cell carcinoma of the conjunctiva treated with subconjunctival ranibizumab (Lucentis): a two-year study.** *Ophthal Plast Reconstr Surg* 2012, **28**(2):85–89.
16. Lai CM, Spilsbury K, Brankov M, Zaknich T, Rakoczy PE: **Inhibition of corneal neovascularization by recombinant adenovirus mediated antisense VEGF RNA.** *Exp Eye Res* 2002, **75**(6):625–634.
17. Ferrara N, Davis-Smyth T: **The biology of vascular endothelial growth factor.** *Endocr Rev* 1997, **18**:4–25.
18. Brown LF, Detmar M, Claffey K, Nagy JA, Feng D, Dvorak AM, Dvorak HF: **Vascular permeability factor/vascular endothelial growth factor: a multifunctional angiogenic cytokine.** *EXS* 1997, **79**:233–269.
19. Dastjerdi MH, Sadrai Z, Saban DR, Zhang Q, Dana R: **Corneal penetration of topical and subconjunctival bevacizumab.** *Invest Ophthalmol Vis Sci* 2011, **52**(12):8718–8723.

20. Dastjerdi MH, Saban DR, Okanobo A, Nallasamy N, Sadrai Z, Chauhan SK, Hajrasouliha AR, Dana R: **Effects of topical and subconjunctival bevacizumab in high-risk corneal transplant survival.** *Invest Ophthalmol Vis Sci* 2010, **51**(5):2411–2417.
21. Carmona G, Chavakis E, Koehl U, Zeiher AM, Dimmeler S: **Activation of Epac stimulates integrin-dependent homing of progenitor cells.** *Blood* 2008, **111**(5):2640–2646.
22. Okada N, Fukagawa K, Takano Y, Dogru M, Tsubota K, Fujishima H, Matsumoto K, Nakajima T, Saito H: **The implications of the upregulation of ICAM-1/VCAM-1 expression of corneal fibroblasts on the pathogenesis of allergic keratopathy.** *Invest Ophthalmol Vis Sci* 2005, **46**(12):4512–4518.
23. Luo YZ, Guo ZH, Yu GZ, Qian JX, Gu XQ, Cao CW, Wang JJ: **The Role of ICAM-1, VE-cadherin in Endothelial Progenitor Cells Taking Part in the Tumor Angiogenesis.** *J Oncol (Chinese)* 2008, **8**:651–653.
24. Muroski ME, Roycik MD, Newcomer RG, Van den Steen PE, Opdenakker G, Monroe HR, Sahab ZJ, Sang QX: **Matrix metalloproteinase-9/gelatinase B is a putative therapeutic target of chronic obstructive pulmonary disease and multiple sclerosis.** *Curr Pharm Biotechnol* 2008, **9**(1):34–46.
25. Belotti D, Paganoni P, Manenti L, Garofalo A, Marchini S, Taraboletti G, Giavazzi R: **Matrix metalloproteinases (MMP9 and MMP2) induce the release of vascular endothelial growth factor (VEGF) by ovarian carcinoma cells: implications for ascites formation.** *Cancer Res* 2003, **63**(17):5224–5229.
26. Lamoreaux WJ, Fitzgerald ME, Reiner A, Hasty KA, Charles ST: **Vascular endothelial growth factor increases release of gelatinase A and decreases release of tissue inhibitor of metalloproteinases by microvascular endothelial cells in vitro.** *Microvasc Res* 1998, **55**(1):29–42.
27. Bergers G, Brekken R, McMahon G, Vu TH, Itoh T, Tamaki K, Tanzawa K, Thorpe P, Itohara S, Werb Z, Hanahan D: **Matrix metalloproteinase-9 triggers the angiogenic switch during carcinogenesis.** *Nat Cell Biol* 2000, **2**(10):737–744.
28. Hashimoto G, Inoki I, Fujii Y, Aoki T, Ikeda E, Okada Y: **Matrix metalloproteinases cleave connective tissue growth factor and reactivate angiogenic activity of vascular endothelial growth factor 165.** *J Biol Chem* 2002, **277**(39):36288–36295.
29. Dauber-Osguthorpe P, Roberts VA, Osguthorpe DJ, Wolff J, Genest M, Hagler AT: **Structure and energetics of ligand binding to proteins: Escherichia coli dihydrofolate reductase-trimethoprim, a drug-receptor system.** *Proteins* 1988, **4**(1):31–47.
30. Fuh G, Wu P, Liang WC, Ultsch M, Lee CV, Moffat B, Wiesmann C: **Structure-function studies of two synthetic anti-vascular endothelial growth factor Fabs and comparison with the Avastin Fab.** *J Biol Chem* 2006, **281**(10):6625–6631.

Positive regulations of adipogenesis by Italian ryegrass [*Lolium multiflorum*] in 3T3-L1 cells

Soundarrajan Ilavenil[1], Mariadhas Valan Arasu[1,5], Jeong-Chae Lee[2], Da Hye Kim[3], Mayakrishnan Vijayakumar[1], Kyung Dong Lee[4] and Ki Choon Choi[1*]

Abstract

Back ground: Intramuscular fat deposition in the meat animal is relatively new strategy for developing the meat quality. Fat deposition is largely depending on the adipocyte proliferation and differentiation. Therefore, we investigated the effect of chloroform extract of *L. multiflorum* [CELM] on cell proliferation, lipid accumulation and adipocyte differentiation in 3T3-L1 cells and body weight of mouse.

Results: We identified 6,9-Octadecatrienoic acid, Hexadecanoic acid, 2-hydroxypropanoic acid, butane-2,3-diol and hexane-1,2,3,4,5,6-hexaol in CELM. *L. multiflorum* extract increased the cell viability, lipid accumulation, cell cycle progression and key transcriptional and secretory factors like PPRAγ2, C/CEBP-α, adiponectin, aP2, GLUT-4, FAS and SREBP-1 mRNA expression as compared with control cells. For *in-vivo*, mice administered with CELM significantly increased body weight throughout the experiment periods. Further, the identified fatty acids like 3, 6, 9-Octadecatrienoic acid and Hexadecanoic acid was docked with target protein [PPRAγ2] using HEX 6.12. The least binding energy considered as high affinity with target protein. The maximum affinity with the target protein was observed in the Hexadecanoic acid followed by 3, 6, 9-Octadecatrienoic acid. The binding efficacy of Hexadecanoic acid and 3, 6, 9-Octadecatrienoic acid to the active site of PPAR-γ2 may be enhanced the adipocyte differentiations.

Conclusion: These findings suggest that CELM stimulates adipogenesis via activating the PPARγ-mediated signaling pathway in adipocyte which could be useful for the development of meat quality in animals.

Background

Intramuscular fat deposition or marbling in meat animals is an important component of uniqueness which influences the eating quality like meat tenderness, juiciness and tastes. The deposition of intramuscular lipid is associated with genetic background, development and nutrition of animals. The cellular development of adipocyte is much important for deposition of intramuscular fat in meat animals, because intramuscular fat accumulation is strongly related to increases of adipocyte proliferation and differentiation which is main regulatory steps in the marbling [1]. The animals possess very good integrated systems for regulating the energy storage and their expenditure. These systems are mainly involved in utilizing the energy storage during excess food available. When the system meets an inadequate food supply, the mobilization of food from the storage place will be progressed by regulating the cellular levels and whole body organisms through its coordination action of circulating hormones and different neural signal from the central nervous system [2].

Adipogenesis is a process of adipocytes cell proliferation and differentiation. When the cells meet positive energy balance, adipogenesis stimulates the storage of energy through the generation of functional fat cells. These processes were achieved in adipocytes [3]. In adipocytes differentiation, fibroblastic like pre-adipocytes were matured into spherical shape adipocytes which contain huge lipid droplets. Specific transcriptional factors such as peroxysome proliferators activated receptor- γ [PPAR-γ], CCAAT/enhancer binding protein- β [C/CEB-β], enhancer binding protein- δ [C/CEB-δ], enhancer binding protein- α [C/CEB-α], Kriuppel like factor and sterol regulatory element binding protein-1[SREBP-1] are mainly involved in promoting the adipogenesis [4]. These adipogenic transcriptional factors have the ability to regulate gene expression. During the adipocytes differentiation,

* Correspondence: choiwh@korea.kr
[1]Grassland and forage division, National Institute of Animal Science, RDA, Seonghwan-Eup, Cheonan-Si, Chungnam 330-801, Korea
Full list of author information is available at the end of the article

the C/CEB-β and C/CEB-δ, factors are induced immediately by glucorticoids and Insulin hormones. Then it activates the PPAR-γ and C/CEB-α. The final stage of adipocytes differentiation requires the expression genes which are involved in the adipocytes phenotype and its function maintenance via lipid metabolic enzymes. Numbers of genes were involved in the lipid metabolism. They are fatty acid binding protein-4 [aP2] which is responding to PPAR-γ and C/CEB-α. aP2 is a key factor of intracellular fatty acid transport and lipid metabolism. This aP2 is regulated through the PPAR-γ and C/CEB-α. It is identified as the key marker of terminal steps in the adipocytes differentiations [5,6]. The PPAR-γ has ability in modulating the insulin-signaling pathway through up regulation of many factors involved in the signaling cascade from GLUT-4 which enhances the glucose uptake levels by the cells. The adiponectin is a hormone called as adipo-Q or adipocytes complement related protein which is specifically expressed in the adipose tissue during differentiation [7]. This hormone enhances the sensitivity in muscle and liver. It increases free fatty acid oxidation in several tissues including muscle fiber [2,8].

Lolium multiflorum (Italian Rye-grass IRG - poaceae family) is an herbaceous annual, perennial grass is being cultivated in South Korea for production of silage for animals. In South Korea IRG is a common grass for the production of silage because of it possesses the high nutritive values and fast growing nature under this climate. The nutritional quality of IRG varied with other varieties. Farmers purchases many types of forage products from companies to use as a feed for ruminants, but the cost of using a commercial feed is relatively expensive than IRG-based silage is used. Thereby, we investigate the chemical composition of *L. multiflorum* and their effects on adipogenesis using 3T3-L1 cells because adipogenesis is an important step in the intramuscular fat (marbling) accumulation.

Methods

Chemicals

Mouse 3T3-L1 pre-adipocytes were purchased from the American Type Culture Collection [ATCC, USA]. Dulbecco modified Eagle medium/high glucose [DMEM], cell viability assay kit [EZ-CYTox, Daeillab Service Co. Ltd.], mRNA extraction kit and RT-PCR kit were obtained from the Daeil Lab services Co., Ltd and Invitrogen Life technology. Other chemicals and analytical grade solvents were procured from Sigma [USA].

Plant material and extraction

The *L. multiflorum* (poaceae family) was collected during the month of September, 2011 from the farm in National Institute of Animal Science [NIAS], South Korea. The plant was authenticated by a taxonomist at NIAS. The *L. multiflorum* was harvested at flowering stage without root and ensiled. *L. multiflorum* were collected at 40 days after cultivation, dried at 60°C for 3 days and grounded by using an electric grinder to give a coarse powder. The powder obtained [5 kg] was soaked in 10 L of 70% methanol for 72 h at room temperature by intermittent mixing using orbital shaker. After that the mixture was filtered using Whatman filter paper to separate the supernatant and the sediment. The filtrate was concentrated under reduced pressure at 40°C until extraction solvent was completely removed and stored in a refrigerator at 0–4°C for further use in subsequent experiments. A green soluble crude residue was obtained [about 100 g]. Then this extract was subjected into sequential extraction with chloroform, ether and water for further purification and analysis.

LC/ESI-MS/MS analysis of plant extract

An API 4000 Q TRAP tandem mass spectrometer [Applied Biosystems, Foster City, CA], equipped with an Agilent 1200 series HPLC system [Agilent Technologies] and an electrospray ionization tandem mass spectrometry [ESI-MS/MS] source in positive ion mode [$[M + H]^+$], was used to identify the chloroform extract. The analytical conditions of mass spectrometry were described in detail as follows: range, start [100 amu], stop [1,300 amu], and scan time [4.8 s]; curtain gas, 20 psi [N_2]; heating gas temperature, 550°C; nebulizing gas, 50 psi; heating gas, 50 psi; ion spray voltage, 5500 V; declustering potential, 100 V; entrance potential, 10 V.

Cell culture

The 3T3-L1 pre-adipocytes were obtained from ATCC [Manassas, VA]. Cells inductions for differentiation from pre-adipocytes to adipocytes were carried out by Choi et al. [9] with slight modification. Briefly, 3T3-L1 pre-adipocyte were seeded in the 6 well at a density of 3×10^4 cells/well. The cells were plated and maintained for 2 days in DMEM containing 10% fetal bovine serum. The medium was replenished every 48 h. 3T3-L1 pre-adipocytes were induced the differentiation by the addition of DMEM containing 10% fetal bovine serum, 0.5 mM 3-isobutyl-1-methylxanthine, 1 μM dexamethasone, and 1 μg/ml insulin and antibiotics. After 48 h, medium was replaced with DMEM supplemented with 10% fetal bovine serum, and cells were maintained in this medium for at least 10 days after the induction of differentiation. The plates were incubated at 37°C in 5% CO_2. The effects of chloroform extract of *L. multiflorum* on adipogenesis enhancement were monitored. 10 mg of extract was dissolved in the chloroform for adipocyte experiment. Every two days, the adipocytes were treated with 10, 50 and 100 μg/ml chloroform extract of *L. multiflorum* until the end of the experiment periods [10 days], without *L. multiflorum* extract considered as control cells.

Cell viability assay

The water soluble tetrazolium [WST; 2[2-Methoxy-4-nitrophenyl] - 3[4-Nitrophenyl]-5-[2, 4- disulfophenyl]-2-H-tetrazolium monosodium salt was used for analysis of viability of 3T3-L1 pre-adipocytes cells. The cells were seeded in the 96 well at a density of 5×10^3 cells/well. The cells were exposed to the chloroform extract of *L. multiflorum*. It was incubated at the 37°C in 5% CO_2 incubator for 24 h and then the culture was treated with WST incubated for 2 h. The living cells absorbed the WST then it was converted into an orange colour product. Then intensity of colour was measured at 450 nm using spectra count ELISA reader [Packard Instrument Co., Downers Grove, IL, USA]. Based on the viability test we have selected the chloroform extract of *L. multiflorum* for adipocytes differentiation investigation.

Oil red O staining for lipid accumulation

Tested cells were fixed with 10% formalin for an hour and then removed the formalin. After that the cells were washed with 60% Isopropanol, and added 3 ml of Oil Red O working solution to the fix cells. The fixed cells were incubated at room temperature for 10 min. After that removed the Oil Red O solution and then washed 2–4 time with distilled water immediately. The cells were photographed using Olympus CKX41 microscope. For quantification, Oil Red O was extracted from wells of fixed cells with 100% isopropanol, and the extracted material was diluted in 100% isopropanol and read at 490 nm using plate reader.

Flow cytometer analysis for cell cycle

The tested cells were harvested by the trypsinization method and adjusted to a density of 5×10^5 cells. The cells were washed with ice cold PBS and re-suspended in 1 ml of ethanol. Treated cells were fixed with 25% ethanol incubated at −20°C for overnight. The fixed cells were centrifuged and re-suspended in 300 µl [50–100 µg/ml in 1.12% sodium citrate] of propidium iodine solution [containing 500 units/mL RNase] for 30 min at room temperature. The sample was stored in the dark at 4°C and analyzed the cell cycle using Flow cytometry caliber [Becton Dickinson, San Jose, CA, USA].

RT-PCR quantification of PPRAγ2, C/CEBP-α Adiponectin, aP2, GLUT-4, FAS and SREBP-1 mRNA expression

The total RNA was extracted according to the manufactures instruction [RNA lipid tissue mini kit, Qiagen USA]. The extracted RNA was measured using UVS-99 Micro volume UV/Vis Spectrophotometer-ACT Gene. A quantity of 1 µg RNA was reverse transcribed using oligo [dT] and III reverse transcriptase is a version of M-MLV RT [superscript III first stand synthesis system for RT-PCR - Invitrogen Life technology]. Real-time PCR was carried out with an ABI 7500 Real-Time PCR System. Target cDNA levels were determined by SYBR green-based real-time PCR in 20 µL reactions containing 10 µL Power SYBR Green Master Mix [Applied Biosystems, Foster City, CA], 10 pmole forward [FP] and reverse primers [RP] [PPAR-γ2 - FP: gtgctccagagaagatgacagac, RP: ggtgggactttcctg ctaa, C/EBP-α- FP: gcaggaggaagatacaggaag, RP: acagactcaaatcccaaca, Adiponectin- FP: ccgttctcttcacctacgac, RP: tccccatccccatacac, aP2- FP: tgtgtgatgcctttgtgg, RP: tgtg tgatgccttgtgg, FAS-FP: cccagcccataagagttaca, RP: atcggga agtcagcacaa, GLUT-4- FP: cccacagaaggtgattgaac, RP: gg tggagatgatgaccctttt, SREBP-1-FP: gaagtggtggagagacgcttac, RP: tatcctcaaagggctggactg]. Expression was normalized against beta-actin transcript signal.

Animals

IRC mice [Each 20 g body weight] were obtained from Orient Co., Seoul Korea. They were housed under standardized conditions [at a constant room temperature with alternating 12-h periods of light and darkness and they fed with standard laboratory diet [AIN 93G, Feed lab Co., Korea]. The use of animals in this study was approved by the Laboratory Animal Center (Permit Number: CBU 2012–0039) of Chonbuk National University (Jeonju, South Korea) and all of the experiment was carried out according to the guidelines of the Animal Care and Use Committee of the University.

Experimental design

Each group consist of 6 mouse for four groups

I. Group I Control mice that received normal food and water for seven weeks
II. Group II Mice received normal food along with chloroform extract of *L. multiflorum* 100 mg/kg of body weight for seven weeks
III. Group III Mice received normal food along with chloroform extract of *L. multiflorum* 200 mg/kg of body weight for seven weeks
IV. Group IV Mice received normal food along with chloroform extract of *L. multiflorum* 300 mg/kg of body weight for seven weeks

After experimental periods, the mice weight was measured.

Statistical analysis

Numerical data obtained from experiments were expressed as mean [mean ± SEM] standard error. Statistical difference between the Control and experimental cells were analyzed by SPSS/16 software hypothesis testing methods that including analysis of variance [ANOVA] followed by least significance difference test. P values of less than 0.05 were considered to show statistical significance.

Molecular Docking

PPAR-γ2 PDB format obtained from the RCSB Protein Data Bank [http://www.rcsb.org/pdb]. After that we have identified the possible binding site of - PPRA-γ2 using Q-Site Finder for predicting the ligand binding sites [http://www.modelling.leeds.ac.uk/qsitefinder/]. The conical smile and PDB format of the ligand had been obtained from http://pubchem.ncbi.nlm.nih.gov/, http://cactus.nci.nih.gov/services/translate/. Docking analysis was carried out using HEX 6.12version. The following parameters are considered for docking purpose Correlation type, FFD mode, Grid dimension, and receptor, ligand and twist range. The interaction of ligand with receptor [LIGPLOT] after docking has been obtained from the PDB SUM [http://www.ebi.ac.uk/pdbsum/].

Results

Effect of *L. multiflorum* extract on adipocytes viability

The chloroform extract of *L. multiflorum*[CELM] shows significantly increases the adipocyte proliferation in a concentration dependent manner. The CELM at a concentration of 100 μg/ml showed maximum adipocyte proliferation as compared with control cells. (Figure 1).

Analyses of chloroform extract of *L. multiflorum* [CELM] using LC/ESI-MS/MS

The MS spectral data led to the identification of 3,6,9-Octadecatrienoic acid, Hexadecanoic acid, 2-hydroxypropanoic acid, butane-2,3-diol and hexane-1,2,3,4,5,6-hexaol as the major compounds in the chloroform extract (Figure 2).

Effect of chloroform extract of *L. multiflorum* [CELM] on adipocyte differentiation

After two days of confluence, 3T3-L1 pre-adipocytes were treated with CELM at concentrations of 10, 50 and 100 μg/ml. Results indicated that the pre-adipocytes were differentiated into the adipocyte in a concentration dependent manner. The maximum differentiation was observed at a concentration of 100 μg/ml (Figure 3d). The morphological changes were caused by the accumulation of intracellular lipids in the cytoplasm of the adipocyte. The intracellular lipid content in adipocyte was significantly increased by the cells treated with CELM as compared with control cells. Oil red O staining as an evidence for accumulation of lipids in the cytoplasm of adipocyte (Figure 3a-b and c). Flow cytometry analysis report revealed that the CELM promotes the cell cycle progress. The histogram results revealed that the increased the cell population in the S phase at all the concentration of CELM (Figure 4 a-f).

Effect of CELM on PPRAγ2, C/CEBP-α Adiponectin, aP2, FAS, GLUT-4 and SREBP-1

The adipogenic and lipogenic genes such as PPARγ2, C/EBP –α Adiponectin, aP2, Fatty acid synthase [FAS], GLUT-4 and SREBP-1 was analyzed by qPCR. The expression of PPARγ2, C/EBP-α, Adiponectin, FAS and SREBP-1 mRNA was increased in concentration of 50 and 100 μg/ml of CELM on the 5th and 10th day of differentiation as compared with control cells. For aP2, no changes were observed on the 5th day of differentiation

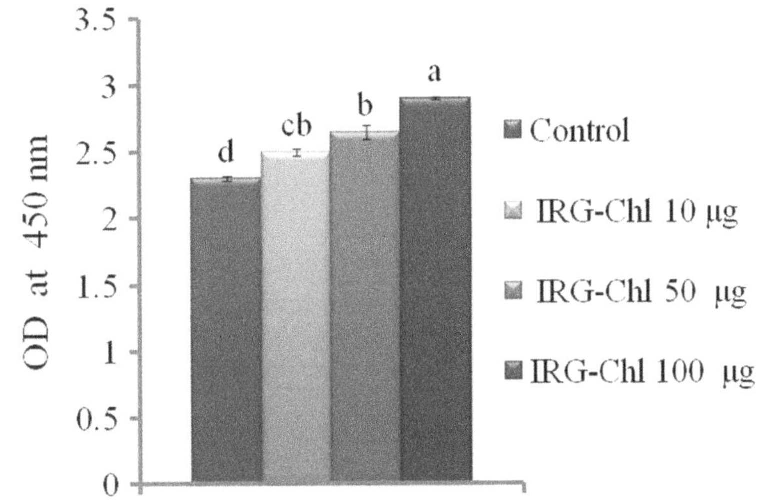

Figure 1 Effect of *L. multiflorum* extracts on adipocyte viability. The cells were treated with different concentration of CELM (10, 50,100 μg/ml) for 48 h in 5%CO2 at 37°C. The results represent the mean ± SEM of duplicate experiments. abcd different letters within treatment represent significant difference ($p < 0.05$).

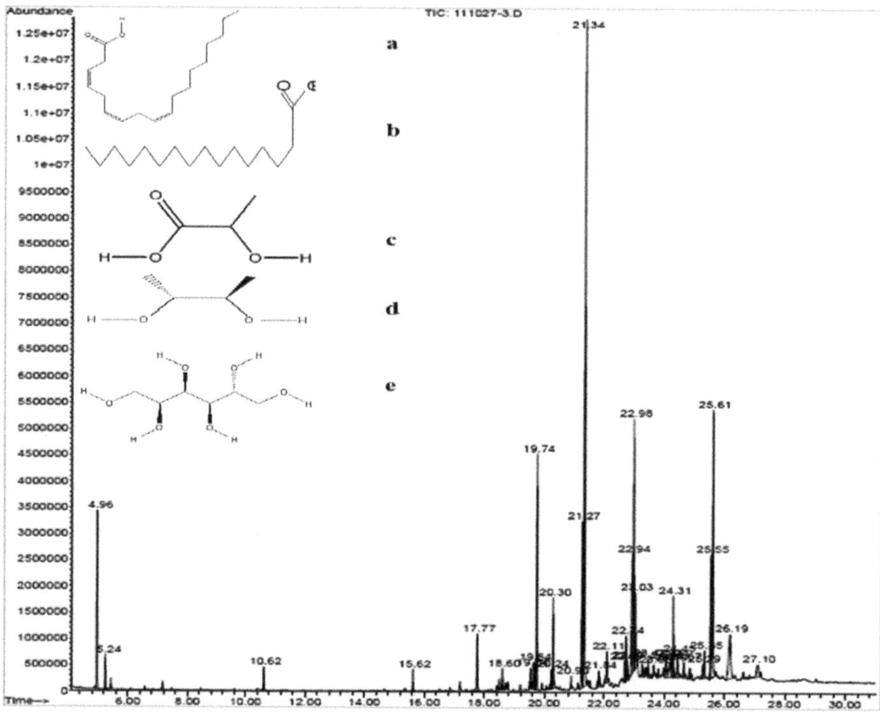

Figure 2 LC/ESI-MS/MS chromatogram of chloroform extract obtained from *L. multiflorum*. a; 3,6,9-Octadecatrienoic acid, **b**; Hexadecanoic acid, **c**; 2-hydroxypropanoic acid, **d**; butane-2,3-diol, **e**; hexane-1,2,3,4,5,6-hexaol.

at all the concentrations. But, on the 10th day shows significant increases of aP2 mRNA transcripts as compared with control cells. There is a no significant change of GLUT-4 mRNA expressions was observed in the CELM treated cells at concentrations of 50 and 100 μg/ml on the 5th day as compared with control cells. However, significant increases of GLUT-4 mRNA expressions were observed in the CELM treated cells at concentrations of 50 and100 μg/ml on the 10th day of differentiation (Figure 5).

Effect of CELM on body weight of mouse

The mice administered with three different concentrations [100, 200 and 300 mg/kg of body weight] of CELM for seven weeks after that the mice weight were measured. Results indicated that the CELM enhanced the body weight of mice from beginning to end of experiment periods (Figure 6). It shows the CELM may be stimulating the adipogenesis and increased the body weight of mice. This result concurrence with the above said molecular investigation.

Molecular docking

LC-MS analysis revealed that five major compounds such as 3,6,9-Octadecatrienoic acid, Hexadecanoic acid, 2-hydroxypropanoic acid, butane-2,3-diol and hexane-1,2,3,4,5,6-hexaol present in the CELM. Among these, two are fatty acids. Generally, many fatty acids enhance the adipogenesis and lipogenesis. In that aspect, we docked these two fatty acids with adipogenic main key transcriptional factor PPAR-γ2 using HEX 6.12 docking software. The target protein structure was docked with 3, 6, 9-Octadecatrienoic acid, Hexadecanoic acid. The binding energy for 3, 6, 9-Octadecatrienoic acid [-277.64, Emax = -203.42] and Hexadecanoic acid [-273.33, Emax = -176.01]. In docking lowest minimum energy values considered highest affinity with the target protein. We found that the lowest minimum energy in the hexadecanoic acid followed by 3, 6, 9-Octadecatrienoic acid. Hexadecanoic acid strongly binds with Aspartic acid [D-394], Arginine [R-395] and proline [P-396] and Arginine [R-441] of the PPAR-γ2 by hydrogen bond and non ligand residues involved hydrophobic contacts. 3, 6, 9-Octadecatrienoic acid binds with aspartic acid [D-473] and Leucine [L-474] of the PPAR-γ2 by non ligand residues involved hydrophobic contacts. This binding efficacy provides strong evident for CELM enhances the adipogenesis in *in-vitro* cell lines. The enhancement adipocytes differentiations may be due to fatty acids present in the CELM (Figure 7).

Discussion

Adipose tissue is a very much important for the whole body energy production. It integrates the both central and peripheral metabolic pathways and it regulates the energy balance [10]. Imbalance between energy intake and energy expanse leads to affect the development of

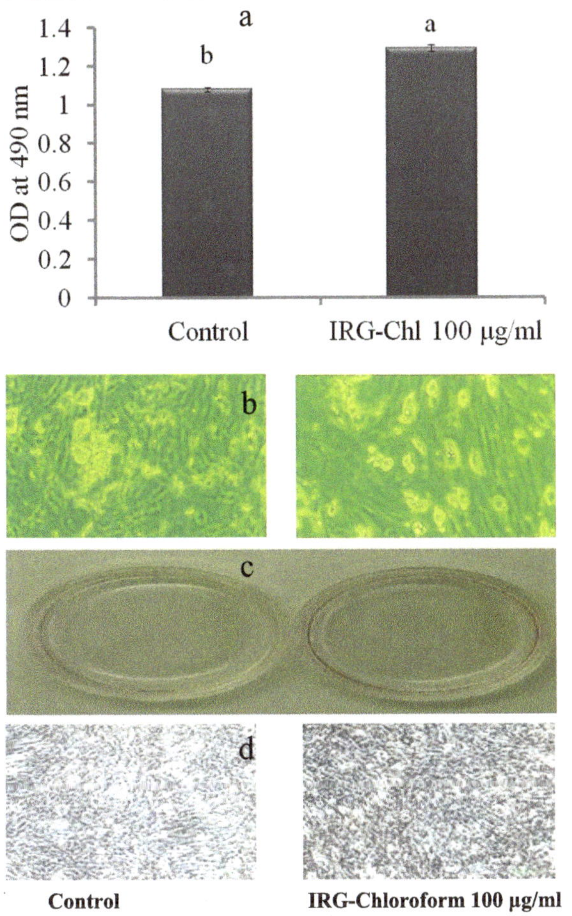

Figure 3 Effect of chloroform extract of *L. multiflorum* (CELM) on adipocytes differentiation. 3; Quantification of lipids in adipocytes, 3; Microscopic view of differentiated adipocyte, 3; Oil red O staining of lipid accumulations in cytoplasm of the adipocyte. The results represent the mean ± SEM of duplicate experiments. [abcd]different letters within treatment represent significant difference ($p < 0.05$).

the adipose tissue by combination of proliferation and differentiation. During the differentiation periods, the excess of the triglycerides was accumulated in the existing adipose tissue. Our results showed that CELM treatments was increased the proliferation of 3T3-L1 cell as compared with control cells. Han et al., 2009 [11] reported that the protocatechuic acid from aplpinia oxyphylla stimulates the proliferation of adipose tissue derived stromal cells. For differentiation of adipocyte, we observed the CELM promotes the intracellular accumulation of lipids in adipocyte as compared with control cells. It is indicated that the CELM considered as a potent stimulator for adipogenesis. The flow cytometer result revealed that the CELM increased the cell populations in the S phase as compared with control cells. The adipocyte specific transcriptional factor C/EBP-α and PPRAγ2 was increased in adipocyte by the CELM treatment as compared with control cells at concentrations of 50 and 100 µg on the 5th and 10th day. C/EBP-α and PPRAγ2 plays an important role in the adipocyte differentiation because of its key transcriptional factors for other adipogenic genes [12]. C/EBP-α is also involved for stimulating and maintaining the PPRAγ2 expression during the adipogenesis [13]. C/EBP-δ and C/EBP-β are promoters for the key transcriptional factors PPRAγ2 and C/EBP-α. Our results claimed that the increase of PPRAγ2 expression may continuously activate the adipose tissue differentiation by directly acting on the C/EBP-α which creates a positive feedback loop. In addition, C/EBP-α stimulate the genes that are involved in insulin sensitivity, lipogenesis, lipolysis and other encoding genes such as GLUT-4, ap2, lipoprotein lipase [LPL], 1 acyl glycerol-3- phosphate acyltransferase[AGPAT2], perilipin, fatty acid synthase [FAS] and some secreted factors like adiponectin and leptin [14,15]. The CELM may be joined with transcriptional factor to promote the differentiation of the adipocyte. It is consistent with previous reports; the mulberry leaf extract strongly involved the adipogenesis enhancement by stimulating the differentiation of 3T3-L1 pre-adipocyte into adipocyte by enhancing the specific transcriptional factor such as PPRAγ2 and C/EBP-α [16].

Adiponectin is a secretory factor in the adipocyte which is exclusively synthesized and regulated during the adipocyte differentiation [17]. It acts as an autocrine factor in adipose tissue by triggering the gene expression responsible for adipogenesis [18]. In the present investigation, the CELM increased the lipid accumulation in the adipocyte. It's confirmed by the Oil Red staining method. It may due to increases of adiponectin in the adipocyte by the CELM treatment during the differentiation. It is also predicted that hormones may be directly involved to enhance the intracellular lipid accumulation in the adipocyte. Similarly, the mulberry leaf extract enhances the adiponectin secretion in the differentiated adipocytes by stimulating insulin sensitivity and lipid accumulation. Therefore, our results confirmed that adiponectin could contribute directly to adipose tissue remodeling by increasing many numbers of the adipocytes [16].

Fatty acid binding protein aP2 is present in adipose tissue which is an important protein involved in the regulation of intracellular metabolism and transport of fatty acids. During the adipocyte differentiation, the expression of aP2 level was observed in adipose tissue and adipogenic cell lines [19,20]. We found that the increased levels of aP2 mRNA expression in CELM treated adipocyte as compared with control adipocyte. This expression may be due to the involvement of CELM on lipid metabolism by binding of the long chain fatty acids. During the differentiation periods the lipogenesis was increased. Oil Red O evidenced that the increased the lipid

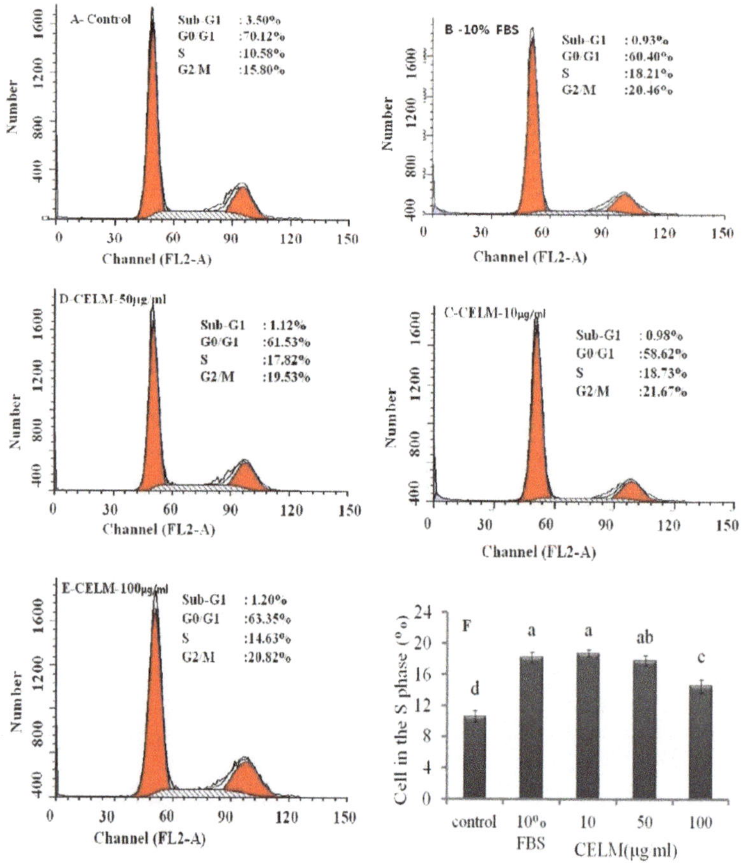

Figure 4 The adipocyte were treated with different concentration of CELM (C-10, D-50,E-100 μg/ml) for 24 h; B- the cells incubated same times in the 10% FBS as positive control; A- Control cells. The cell cycle progress was investigated using a flow cytometer; **F**- cell population in the S phase of experimental cells was calculated from triplicate experiments. [abcd]different letters within treatment represent significant difference ($p < 0.05$).

droplets in the CELM treated cells. Therefore, CELM enhances the lipogenesis via ap2 during the differentiation. The adipocytes synthesize adipocyte specific products such as aP2 an adipocyte specific fatty acid binding protein that has been very much important intermediate marker for the adipocytes differentiation [21]. Fatty acid synthase is the main key enzyme in the lipogenesis pathway [22]. It catalyses all the enzymatic steps involved in the conversion of acetyl CoA, molanyl CoA finally to palmitate. Therefore, FAS enzyme and their mRNA expressions are considered as a key marker for the lipogenesis. Increased level of FAS in the CELM treated adipocytes indicated that the extract enhanced the fatty acid metabolism via FAS. It's consistent with Oil Red O result that CELM increased the lipid accumulation in the adipocytes.

GLUT-4 plays a major role in the energetic and metabolic activities of the adipocyte by allowing the glucose transportation into the cells by the insulin signal. Activation of the insulin receptor stimulates the large increases of GLUT-4 vesicle exocytosis and endocytosis [23]. PPARγ2 also involved in stimulating the adipogenesis and may regulate the lipogenesis by imparting insulin sensitivity to fat cells in responsive adipose depots [24,25]. CELM may stimulate the insulin sensitivity via PPARγ2. This factor has ability to modulate the insulin signaling pathway through the up regulation of several factors for the signaling cascade for GLUT-4 [26].

SREBP-1 highly expressed in the liver and both white and brown adipose tissues. This ADD-1/SREBP-1 mRNA was significantly higher in the pre-adipocytes compared with other fibroblast. Further, it was increased during the differentiation periods [27]. In addition, PPARγ2 and ADD-1/SREBP-1induction stimulates the expression of several important lipogenenic genes in the adipocyte like fatty acid synthase [28] aceyl carboxylase [29,30], glycerol-3- phosphate acyltransferase [31] and the lipoprotein lipase [29]. This ADD-1/SREBP-1 target genes regulate many important steps in fatty acid metabolism by involving in production of natural fatty acids derived PPARγ2 ligands and activators. ADD-1/SREBP regulates the adipocytes differentiation by controlling the PPARγ2

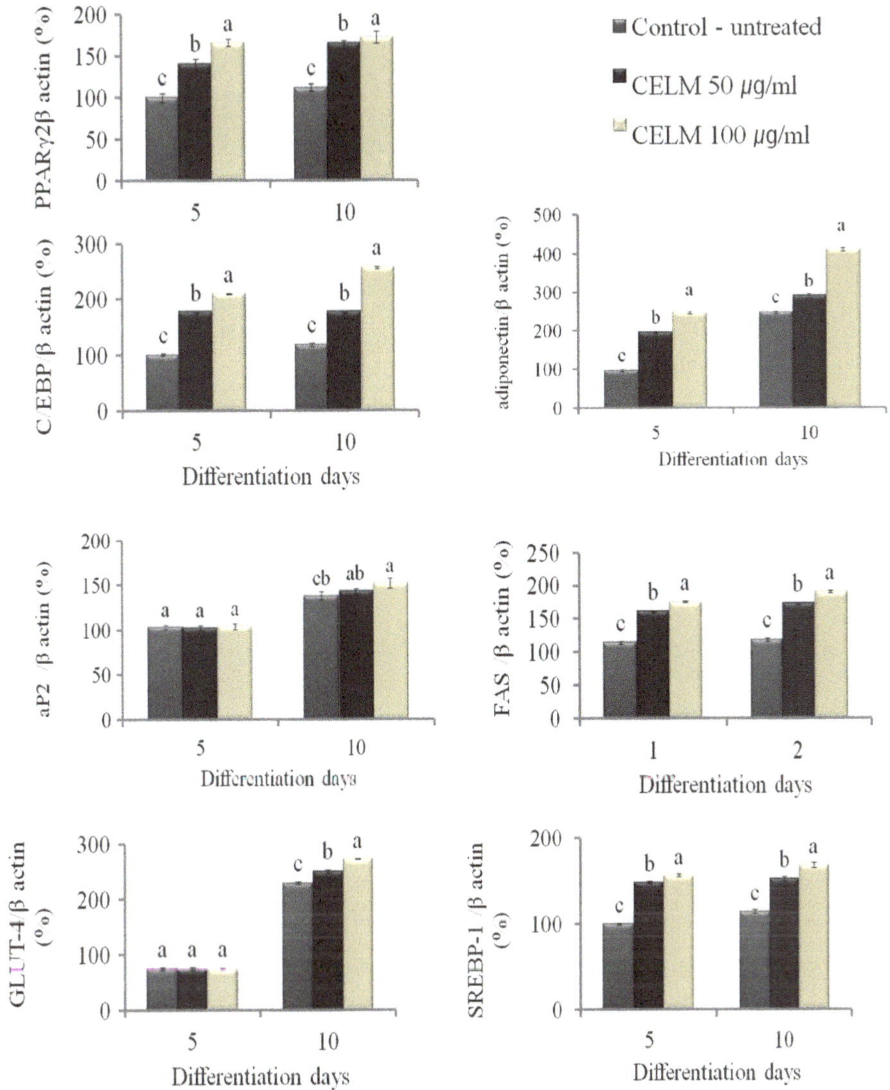

Figure 5 shows quantification target expression of PPAR- γ2, C/EBP-α, adiponectin, aP2, FAS GLUT-4and SREBP-1 transcript level in experimental adipocyte. The results represent the mean ± SEM of duplicate experiments. [abcd]different letters within treatment represent significant difference ($p < 0.05$).

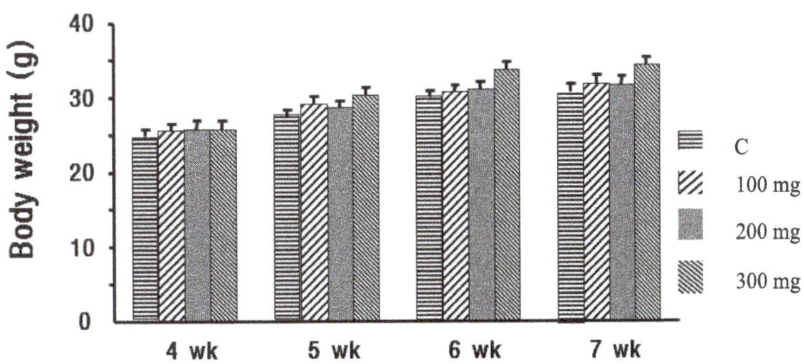

Figure 6 Effect of extract on the body weight of mice in control and experimental groups. C: control, 100 mg chloroform extract, 200 mg chloroform extract, 300 mg chloroform extract.

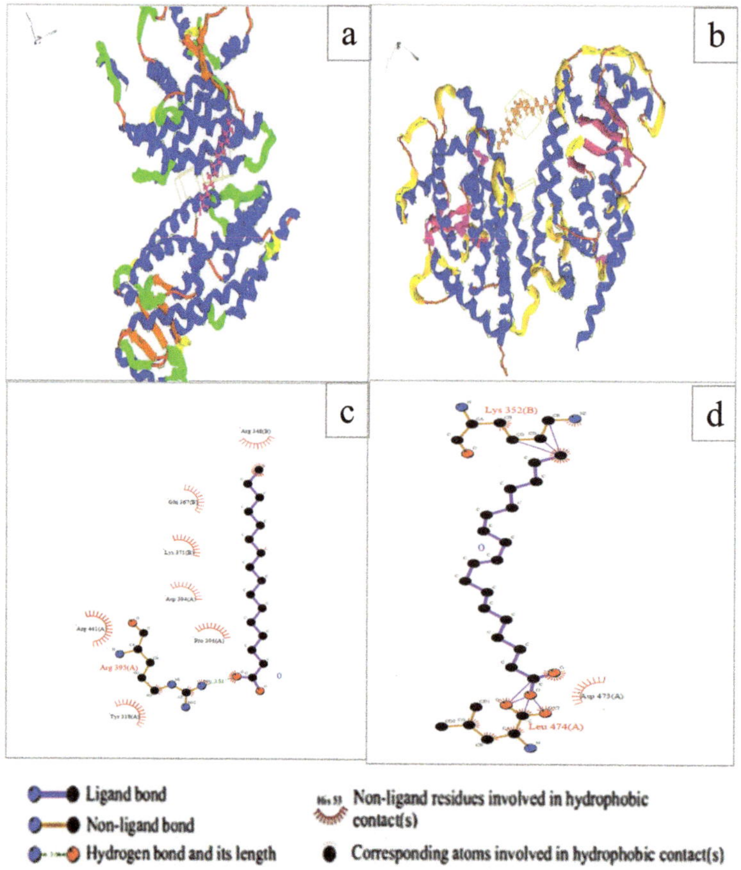

Figure 7 Docked orientation of fatty acids with target active site protein (PDB ID-3adx) and their run ligplot. a - Docked orientation of Hexadecanoic acid with active site of PPARγ2, **b** - Docked orientation of 3, 6, 9-Octadecatrienoic acid with active site of PPARγ2. **c** – run ligplot of Hexadecanoic acid with active site of PPARγ2, **d** - run ligplot of 3, 6, 9-Octadecatrienoic acid with active site of PPARγ2. It shows ligand interaction with PPARγ2 by hydrogen bond and non ligand residues involved hydrophobic contacts.

transcriptional factor [32]. In our results at 50 and100 μg/ml concentration of CELM increased the expression of SREBP-1. Therefore, these results confirmed that the CELM enhances the adipocytes differentiation by SREBP-1 gene expression with PPARγ2 induction. The main role of the SREBP-1 is activating the specific transcription of certain genes responsible for adipocytes differentiation. It also has the capacity to trans activate the promoters of the fatty acid synthase and S14 genes. The human analogue of SREBP-1 involved in the regulation of many genes which are responsible of fatty acid metabolism [33]. ADD-1/SREBP-1c may play important role in the action of insulin to regulate the adipocytes gene expression by stimulating genes that involved in the lipogenesis and reducing the genes which are involved in the fatty acid oxidation [34-36]. For confirmation of *in vitro* results, we carried out the animal experiments, mice administered with CELM through orally for seven weeks. The results clearly confirmed the increases of mice body weight as compared with control mice.

The intake of high fat content is associated with fat mass development through key transcriptional factor PPARγ2 which stimulates the adipogenesis. Addition of saturated, mono saturated, and polyunsaturated fatty acids enhance the adipocytes differentiation. Fatty acids stimulate the porcine adipocytes differentiation and their specific transcriptional and co factor genes [37]. Docosahexaenoic acid increased the PPARγ2 leading to increase expression aP2 and adiponectin. Triglycerides mixture, caprylic acid, very low, low and high density lipoprotein stimulate the proliferation and differentiations of bovine adipocytes [38]. The present investigation, we identified the five major compounds 3,6,9-Octadecatrienoic acid, Hexadecanoic acid, 2-hydroxypropanoic acid, butane-2,3-diol and hexane-1,2,3,4,5,6-hexaol, among these, two are fatty acids. These fatty acids were docked with main adipogenic transcriptional factor PPAR-γ2 using HEX docking software 12 because of many reports claimed that the fatty acids enhance the adipogenesis. This docking result also strongly supports both *In-vitro* investigations.

The highest affinity with target protein active site is hexadecanoic acid followed by 3, 6, 9-Octadecatrienoic acid.

Conclusions

Intramuscular accumulation of fat or marbling in meat animals are much important to contribute significantly for developing the quality of meat. The accumulation of intramuscular fat is largely depend on adipocyte proliferation and differentiation because of adipocyte fill the lipids in the cells through a differentiation process and it's an important regulatory process in the deposition of marbling. Present study, CELM strongly enhances the adipocyte proliferation, lipid accumulation and differentiation through activation of PPAR-γ2 and C/EBP; subsequently it activates the, Adiponectin, aP2, FAS, GLUT-4 and ADD-1/SREBP-1 and Fatty acid synthase. Therefore we suggest that the CELM induces the lipid accumulation through stimulating PPARγ2 that could be useful to make a plan to enhance the meat quality in animals.

Competing interests

There are no conflicts of interests between the authors.

Authors' contributions

SI MVA DHK MV performed the experiments. HSP KDL helped in interpretation of the results and helped draft the manuscript. JCL KCC supervised and analyzed to draft the manuscript. All authors read and approved the final manuscript.

Acknowledgments

This work was carried out with the support of "Cooperative Research Program for Agriculture Science & Technology Development (Project No. PJ007538022012)" Rural Development Administration, Republic of Korea.

Author details

[1]Grassland and forage division, National Institute of Animal Science, RDA, Seonghwan-Eup, Cheonan-Si, Chungnam 330-801, Korea. [2]Research Center of Bioactive Materials, Chonbuk National University, Jeonju 561-756, Republic of Korea. [3]The United Graduate School of Agricultural Sciences, Tottori University, Tottori-Shi 680-8553, Japan. [4]Department of Oriental Medicine Materials, Dongsin University, Naju 520-714, Korea. [5]Department of Botany and microbiology, College of Science, King Saud University, Riyadh 11451, Saudi Arabia.

References

1. Hovenier R, Brascamp EW, Kanis E, Dan Der Werf JH, Wassenberg AP: **Economic values of optimum traits: The example of meat quality in pigs.** *J Anim Sci* 1993, **71**:1429–1433.
2. Otto TC, Lane MD: **Adipose development: from stem cell to adipocyte.** *Crit Rev Biochem Mol Biol* 2005, **40**:229–242.
3. Trayhurn P, Wood IS: **Signalling role of adipose tissue: adipokines and inflammation in obesity.** *Biochem Soc Trans* 2005, **33**:1078–1081.
4. White UA, Stephens JM: **Transcriptional factors that promote formation of white adipose tissue.** *Mol Cell Endocrinol* 2009, **318**:10–14.
5. Mandrup S, Lane MD: **Regulating adipogenesis.** *J Biol Chem* 1997, **2**:5367–5370.
6. Rosen ED, Spiegelman BM: **Molecular regulation of adipogenesis.** *Annu Rev Cell Dev Biol* 2000, **16**:145–171.
7. Trujillo ME, Scherer PE: **Adiponectin–journey from an adipocytes secretory protein to biomarker of the metabolic syndrome.** *J Intern Med* 2005, **257**:167–175.
8. Yusuf S, Hawken S, Ounpuu S, Bautista L, Franzosi MG, Commerford P, Lang CC, Rumboldt Z, Onen CL, Lisheng L, Tanomsup S, Wangai PJ, Razak F, Sharma AM, Anand SS: **Study Investigators. Obesity and the risk of myocardial infarction in 27,000 participants from 52 countries: a case- control study.** *Lancet* 2005, **366**:1640–1649.
9. Choi K, Roh SG, Hong YH, Shrestha YB, Hishikawa D, Chen C, Kojima M, Kangawa K, Sasaki S: **The role of ghrelin and growth hormone secretagogues receptor on rat adipogenesis.** *Endocrinology* 2003, **144**(3):754–759.
10. Ahima RS, Lazar MA: **Adipokines and the peripheral and neural control of energy balance.** *Mol Endocrinol* 2008, **22**:1023–1031.
11. Han W, Tian-Qing L, Yan-Xia Z, Shui G, Xue-Hu M, Zhan-Feng C: **Effect of protocatechuic acid from Alpinia oxyphylla on proliferation of human adipose tissue-derived stromal cells *in vitro*.** *Mol Cell Biochem* 2009, **330**:47–53.
12. Gregoire FM, Smas CM, Sul HS: **Understanding adipocyte differentiation.** *Physiol Rev* 1998, **78**:783–809.
13. Camp HS, Ren D, Leff T: **Adipogenesis and fat-cell function in obesity and diabetes.** *Trends Mol Med* 2002, **8**(9):442–447.
14. Lefterova MI, Zhang Y, Steger DJ, Schupp M, Schug J, Cristancho A, Feng D, Zhuo D, Stoeckert CJ, Liu XS, Lazar MA: **PPAR-γ2 and C/EBP factors orchestrate adipocytes biology via adjacent binding on a genome-wide scale.** *Genes Dev* 2008, **22**:2941–2952.
15. Nielsen R, Pedersen TA, Hagenbeek D, Moulos P, Siersbaek R, Megens E, Denissov S, Borgesen M, Francoijs KJ, Mandrup S, Stunnenberg HG: **Genome-wide profiling of PPAR-γ2: RXR and RNA polymerase II occupancy reveals temporal activation of distinct metabolic pathways and changes in RXR dimer composition during adipogenesis.** *Genes Dev* 2008, **22**:2953–2967.
16. Jarinyaporn N, Choon HC, Patchareewan P, Ran C, Bo HK, Mi YL, Upa K: **Mulberry leaf extract increases adiponectin inmurine 3T3-L1 adipocytes.** *Nutr Res* 2012, **32**:39–44.
17. Korner A, Wabitsch M, Seidel B, Fischer Posovszky P, Berthold A, Stumvoll M, Blüher M, Kratzsch J, Kiess W: **Adiponectin expression in humans is dependent ondifferentiation of adipocytes and down-regulated by humoral serum components of high molecular weight.** *Biochem Biophys Res Commun* 2005, **337**:540–550.
18. Fu Y, Luo N, Klein RL, Garvey WT: **Adiponectin promotes adipocytes differentiation, insulin sensitivity, and lipid accumulation.** *J Lipid Res* 2005, **46**:1369–1379.
19. Coe NR, Bernloh DA: **Physiological properties and functions of intracellular fatty acid-binding proteins.** *Biochim Biophy Acta* 1998, **1391**(3):287 306.
20. Reese-Wagoner A, Thompson J, Banaszak L: **Structural properties of the adipocytes lipid binding protein.** *Biochim Biophys Acta* 1999, **1441**(2–3):106–116.
21. Bernlohr DA, Angus CW, Lane MD, Bolanowski MA, Kelly J: **Expression of specific mRNAs during adipose differentiation: identification of an mRNA encoding a homologue of myelin P2 protein.** *Proc Natl Acad Sci U S A* 1984, **81**(17):5468–5472.
22. Hillgartner FB, Salati LM, Goodridge AG: **Physiological and molecular mechanisms involved in nutritional regulation of fatty acid synthesis.** *Physiol Rev* 1995, **75**:47–76.
23. Pessin JE, Thurmond DC, Elmendorf JS, Coker KJ, Okada S: **Molecular basis of insulin-stimulated GLUT4 vesicle trafficking. Location! Location! Location.** *J Biol Chem* 1999, **274**:2593–2596.
24. Weng JR, Chen CY, Pinzone JJ, Ringel MD, Chen CS: **Beyond peroxisome proliferator-activated receptor gamma signaling: the multi-facets of the antitumor effect of thiazolidinediones.** *Endocr Relat Cancer* 2006, **13**:401–413.
25. Mueller E, Drori S, Aiyer A, Yie J, Sarraf P, Chen H, Hauser S, Rosen ED, Ge K, Roeder RG, Spiegelman BM: **Genetic analysis of adipogenesis through peroxisome proliferator-activated receptor gamma isoforms.** *J Biol Chem* 2002, **7**:41925–41930.
26. Gurnell M: **Peroxisome proliferator-activated receptor gamma and the regulation of adipocyte function: lessons from human genetic studies.** *Best Pract Res Clin Endocrinol Metab* 2005, **19**:501–523.
27. Tontonoz P, Kim JB, Graves RA, Spiegelman BM: **ADD I: A novel helix-loop-helix transcription factor associated with adipocyte determination and differentiation.** *Mol Cell Biol* 1993, **13**:4753–4759.
28. Kim JB, Spiegelman BM: **ADD1/SREBP1 promotes adipocytes differentiation and gene expression linked to fatty acid metabolism.** *Genes Dev* 1996, **10**:1096–1107.
29. Shimano H, Horton JD, Hammer RE, Shimomura I, Brown MS, Goldstein JL: **Overproduction of cholesterol and fatty acids causes massive liver**

enlargement in transgenic mice expressing truncated SREBP- 1a. *J Clin Invest* 1996, **8**:1575–1584.

30. Lopez JM, Bennett MK, Sanchez HB, Rosenfeld JM, Osborne TF: **Sterol regulation of acetyl coenzyme A carboxylase: a mechanism for coordinate control of cellular lipid.** *Proc Natl Acad Sci U S A* 1996, 1049–1053.
31. Ericsson J, Jackson SM, Kim JB, Spiegelman BM, Edwards PA: **Identification of glycerol-3-phosphate acyltransferase as an adipocytes determination and differentiation factor 1- and sterol regulatory element binding protein-responsive gene.** *J Biol Chem* 1997, **272**:7298–7305.
32. Kim JB, Wright HM, Wright M, Spiegelman BM: **ADD1/SREBP1 activates PPAR-γ2 through the production of endogenous ligand.** *Proc Natl Acad Sci U S A*, **95**:4333–4337.
33. Kim JB, Sarraf P, Wright M, Yao KM, Mueller E, Solanes G, Lowell BB, Spiegelman BM: **Nutritional and insulin regulation of fatty acid synthetase and leptin gene expression through ADD-1/SREBP1.** *J Clin Invest* 1995, **101**:1–9.
34. Foretz M, Guichard C, Ferré P, Foufelle F: **Sterol regulatory binding element binding protein-1c is a major mediator of insulin action on the hepatic expression of glucokinase and lipogenesis-related genes.** *Proc Natl Acad Sci U S A* 1999, **96**:12737–12742.
35. Shimomura I, Bashmakov Y, Ikemoto S, Horton JD, Brown MS, Goldstein JL: **Insulin selectively increases SREBP-1c mRNA in the livers of rats with streptozotocin-induced diabetes.** *Proc Natl Acad Sci U S A* 1999, **96**:13656–13661.
36. Florence M, Perla S, Josiane S, Takahiko M, Takuya K, Shuh N, Philippe G, Zoubir A, Raymond N, Gérard A: **Arachidonic acid and prostacyclin signaling promote adipose tissue development: a human health concern?** *J Lipid Res* 2003, **44**.
37. Shih Torng D, Harry J: **Mersmann Fatty acids modulate porcine adipocyte differentiation and transcripts for transcription factors and adipocytes characteristic proteins.** *J Nutr Biochem* 2001, **12**:101–108.
38. Peixing WU, Kota S, Funika Yoshiaki H, Katsumoto K: **Effect of lipid related factors on adipocytes differentiation of stromal vascular cells in primary culture.** *J Veter Medic Sci* 2000, **2**(9):933–939.

4

A novel self-assembled oligopeptide amphiphile for biomimetic mineralization of enamel

Quan-Li Li[1*†], Tian-Yun Ning[2†], Ying Cao[1,2], Wei-bo Zhang[1], May Lei Mei[2] and Chun Hung Chu[2*]

Abstract

Background: Researchers are looking for biomimetic mineralization of ena/mel to manage dental erosion. This study evaluated biomimetic mineralization of demineralized enamel induced by a synthetic and self-assembled oligopeptide amphiphile (OPA).

Results: The results showed that the OPA self-assembled into nano-fibres in the presence of calcium ions and in neutral acidity. The OPA was alternately immersed in calcium chloride and sodium hypophosphate solutions to evaluate its property of mineralization. Transmission electron microscopy (TEM) and scanning electron microscopy (SEM) showed nucleation and growth of amorphous calcium phosphate along the self-assembled OPA nano-fibres when it was repetitively exposed to solutions with calcium and phosphate ions. Energy dispersive spectrometry (EDS) confirmed that these nano-particles contained calcium and phosphate. Furthermore, electron diffraction pattern suggested that the nano-particles precipitated on OPA nano-fibres were comparable to amorphous calcium phosphate. Acid-etched human enamel slices were incubated at 37°C in metastable calcium phosphate solution with the OPA for biomimetic mineralization. SEM and X-ray diffraction indicated that the OPA induced the formation of hydroxyapatite crystals in organized bundles on etched enamel. TEM micrographs revealed there were 20–30 nm nano-amorphous calcium phosphate precipitates in the biomimetic mineralizing solution. The particles were found separately bound to the oligopeptide fibres. Biomimetic mineralization with or without the oligopeptide increased demineralized enamel microhardness.

Conclusions: A novel OPA was successfully fabricated, which fostered the biomimetic mineralization of demineralized enamel. It is one of the primary steps towards the design and construction of novel biomaterial for future clinical therapy of dental erosion.

Keywords: Amelogenin, Peptide, Remineralization, Enamel, Biomimetic, Mineralization, Self-assemble

Background

Dental erosion is described as tooth surface loss produced by acids of non-bacterial origin and is now recognized as a major cause of tooth wear [1]. The treatment particular in advanced cases is complicated and difficult and is a great challenge to many dentists. Regeneration by biomimetic mineralization is a non-invasive therapeutic technique that has received increasing attention in the past decades [2]. A cell-free strategy of biomimetic mineralization has been proposed [2,3]. Researchers tried to mimic the natural process for biomineralization and developed inorganic–organic hybrid materials with a hierarchical microstructure that simulates the natural mineralized tissues [4]. This biomimetic mineralization method draws attention to the design of organic molecules as templates for controlling the nucleation and growth of mineral crystals and the design of the form of mineral-ion transport. It incorporates these two factors to form hierarchical hybrid materials under physiologic conditions [5,6]. Various in situ methods of biomimetic mineralization with promising results to regenerate the hierarchical microstructure of the enamel have been reported. The methods utilized gelatine gel loaded with calcium and phosphate ions [7], hydrogen peroxide containing calcium phosphate paste [8], hydrothermal method with control release of calcium from Ca-EDTA [9], surfactant assembly hydroxyapatite [10], self-assembly of amphiphilic dendrons

* Correspondence: ql-li@126.com; chchu@hku.hk
†Equal contributors
1College & Hospital of Stomatology, Anhui Medical University, 230032 Hefei, China
2Faculty of Dentistry, the University of Hong Kong, 34 Hospital Road, Hong Kong, China

[11], electrolytical deposition taking place at 85°C [12], and amelogenin [13].

Matured enamel is composed of 95 wt% nano-rod-like hydroxyapatite. It is crucial to design the macromolecules as the template to induce the enamel microstructure regeneration under physiological conditions [14]. During enamel formation, amelogenin, the major enamel protein constituting approximately 90% of all organic matrix material [15], spontaneously self-assembles into nano-spherical supermolecular structures. Amelogenin provides an effective template to induce enamel microstructure regeneration. However, it can be easily denatured and expensive to extract. Oligopeptides are less costly to prepare and more stable than natural protein. This biomaterial has been used as a molecular tool for the synthesis, growth, and assembly of nano-structured materials [16]. Hartgerink et al. developed a peptide amphiphile that contains an alkyl tail and an ionic oligopeptide [17]. The peptide amphiphile is able to self-assemble into nano-fibre networks. It can promote the mineralization of hydroxyapatite to form a composite material in which the crystallographic c-axes of hydroxyapatite are aligned with the long axes of the fibres. Kirkham et al. developed an oligomeric β-sheet-forming peptide that spontaneously undergoes hierarchical self-assembly into fibrillar scaffolds. It can induce hydroxyapatite nucleation de novo and facilitate remineralization of caries-like lesions [18].

Amelogenin spontaneously self-assembles into nano-spherical, supermolecular structures during enamel formation. Afterwards, it self-assembles into birefringent, microribbon structures [19]. This supermolecular structure plays a vital role in controlling enamel hydroxyapatite crystal co-assembly [19]. The molecular mechanism of amelogenin self-assembly involves the primary structure of the protein maintaining a bipolar nature, which leads to unique globular monomers with flexible, hydrophilic, C-terminal (–Thr-Lys-Arg-Glu-Glu-Val-Asp) "tails" on globular monomer surfaces [19]. The hydrophilic C-terminal "tail" interacts with calcium ions and initializes hydroxyapatite nucleation [20]. This study aimed to fabricate a self-assembly anionic oligopeptide amphiphile that contains the hydrophilic functional domain of amelogenin to initialize hydroxyapatite nucleation and promote biomimetic mineralization of demineralized enamel.

Methods

Oligopeptide amphiphile fabrication

The oligopeptide amphiphile ($C_{18}H_{35}O$-Thr-Lys-Arg-Glu-Glu-Val-Asp) fabricated in this study consists of "two parts" hydrophilic C-terminal amino residue of amelogenin (–Thr-Lys-Arg-Glu-Glu-Val-Asp) and a derivative of stearic acid ($C_{18}H_{35}COOH$) (Sigma-Aldrich, St. Louis, MO, USA). The amino residue of amelogenin was used as the hydrophilic functional group, and the stearic acid derivative was used as the hydrophobic tail. The oligopeptide amphiphile was synthesized using standard, Fmoc solid-phase, peptide synthesis (SPPS, Additional file 1: Figure S1), which proceeded step by step in a C-terminal to N-terminal fashion using automated synthesizers. The synthetic oligopeptide was purified by ammonium sulphate precipitation and reverse-phase, high-performance liquid chromatography (HPLC). It was characterized by high-performance liquid chromatography (HPLC) and mass spectrometry.

Self-assembly of the oligopeptide induced by calcium ions

Oligopeptide solutions at concentrations of 1, 2.5, 5, or 10 mg/mL were prepared by dissolving the oligopeptide into 0.1 M sodium hydroxide solution (Sigma-Aldrich, St. Louis, MO, USA). The alkalinity of the oligopeptide solution was then adjusted to neutral (pH = 7.0) by adding hydrochloric acid. A 100 μL aliquot of a 1 M calcium chloride ($CaCl_2$) solution (Sigma-Aldrich, St. Louis, MO, USA) was added to 1 mL of the oligopeptide solutions at concentrations of 1, 2.5, 5.0, or 10.0 mg/mL. A suspension was formed and mixed thoroughly in a vial after each drop of the $CaCl_2$ solution to promote diffusion of the metal ions. The self-assembled oligopeptide was characterized by transmission electron microscopy (TEM) and scanning electron microscopy (SEM) as the following. A suspension of 10 mg/mL self-assembled oligopeptide was ultrasonically diluted with ethanol. The diluted suspension was smeared onto a carbon-coated TEM grid and examined under TEM (JEOL-2010, JEOL Ltd. Co., Tokyo, Japan). For SEM examination (Sirion 200, FEI Co., Hillsboro, OR, USA), the oligopeptide-coated TEM grids were fixed with 2% glutaraldehyde in phosphate-buffered saline (PBS) (pH = 7.4) at 23°C for 2 h. It was subsequently dehydrated in ethanol and critical-point dried before gold sputtering.

Biomimetic mineralization property of the oligopeptide nano-fibres

Self-assembled oligopeptide was deposited onto a porous carbon-coated TEM grid and fixed with 2% glutaraldehyde in PBS (pH = 7.4) at 23°C for 2 h. Grids coated with the oligopeptide were first immersed in a 10 mM $CaCl_2$ solution for 1 h at 23°C and then washed with deionized water. Afterwards, the samples were exposed to a 5 mM sodium hypophosphate (Na_2HPO_4) solution for 1 h at 23°C and washed with deionized water. The protocol of $CaCl_2$ followed by Na_2HPO_4 immersion was repeated five times. The specimens were examined under TEM and SEM to study the biomimetic mineralization of the oligopeptide nano-fibres. Electron diffraction was used to study the crystal structure. Energy dispersive spectrometry (EDS) was used to evaluate the calcium and phosphorus content.

Biomimetic mineralization of enamel in metastable calcium phosphate solution

Sound extracted human molars were collected with ethics approval (IRB UW 10–210). Eight enamel slices of 1 mm in thickness were prepared perpendicular to the longitudinal axis of the tooth using a low-speed IsoMet diamond saw (Buehler Ltd., Lake Bluff, IL, USA). The slices were polished with 1200-grit silicon carbide paper under running water. The slices were checked under microscope (×10) for cracks before they were cleaned in ethoxylate detergent solution, acetone, ethanol, and deionized water. The enamel slices were demineralized by acid-etching with 37% phosphoric acid (H_3PO_4) for 60 s, rinsed with deionized water, and stored in a polyethylene tube at 4°C.

The metastable calcium phosphate solution was prepared as suggested by Fan [21]. It contained 2.58 mM calcium ($CaCl_2 \cdot 2H_2O$), 1.55 mM phosphates (KH_2PO_4), 1 mg/L sodium fluoride (NaF), and 180 mM sodium chloride (NaCl). It was buffered to a pH of 7.6 by 50 mM of trihydroxymethylaminomethane (Tris)-hydrochloric acid (Sigma-Aldrich, St. Louis, MO, USA). The biomimetic mineralizing solution was developed by adding oligopeptide. It was added to the buffered metastable calcium phosphate solution to achieve a concentration of 15 μg/mL. In order to reveal the particle size distribution of Ca^{2+}-induced self-assembled peptide, the biomimetic mineralizing solution was characterized at 25°C by a nano-particle size analyser (Zetasizer Nano S90, Malvern Instruments Ltd, UK). Particle size distribution was studied using multimodal algorithms based on the size distribution of the scattering light intensity. The stability of the biomimetic mineralizing solution and the metastable calcium phosphate solution were assessed by examination for precipitation after 30 days. The precipitate formed, if any, was studied using TEM.

Two tooth slices were placed into a polyethylene tube containing 15 mL biomimetic mineralizing solution and incubated at 37°C for 1 day. Another 2 slices were incubated for 20 days. The biomimetic mineralizing solution was refreshed every 2 days. The tooth slices after incubation were washed with deionized water and air-dried before examination under SEM and X-ray diffraction (XRD) (X'Pert Pro, Philips, Almelo, The Netherlands). A microhardness tester (Type M, Shimadzu Corp., Kyoto, Japan) was used to measure the Vicker's hardness on the enamel of the tooth slices. The enamel surface was subjected to a load of 100×10^{-3}N for 10 s at each test point for microhardness assessment. The above experiments were repeated on 4 tooth slices using the biomimetic mineralizing solution without the oligopeptide as a control.

The collected data was entered into an Excel file and checked to minimize data entry errors. The data was then analysed with statistical software SPSS 17.0 for Windows (SPSS Inc. Chicago, Illinois, USA). GLM univariate analysis will be used to check normality and variance of the samples. One-way analysis of variance and Tukey's test were used to compare the microhardness of the enamel surface i) with or without H_3PO_4 etching enamel and ii) after incubation in biomimetic mineralizing solution with or without the oligopeptide. The level of significance was 5%.

Results

Oligopeptide amphiphile fabrication

The oligopeptide amphiphile consists of an anionic oligopeptide and an alkyl tail and its chemical structure was shown in Figure 1. The results of the mass spectrometry showed that molecular weight of the synthetic peptide was 1,142.4 (Figure 2), which corresponds to the designed oligopeptide amphiphile ($C_{18}H_{35}O$-Thr-Lys-Arg-Glu-Glu-Val-Asp). HPLC demonstrated that the purity of the synthetic peptide was 98.29% (Figure 3).

Self-assembly of the oligopeptide induced by calcium ions

The oligopeptide solution formed a hydrogel after adding calcium chloride. The gel strength and viscosity increased with time and oligopeptide concentration (Figure 4). A thick gel that could stick to the bottom of an inverted vial was formed at oligopeptide concentrations of 5 mg/mL and 10 mg/mL after 120 minutes.

Biomimetic mineralization property of the oligopeptide nano-fibres

SEM and TEM micrographs demonstrated that the self-assembled oligopeptide had well-defined nano-fibres with a clear helical twist (Figure 5a and b). The width of the fibres was approximately 12 nm, with a full-twist pitch of approximately 240 nm. The nano-fibres varied in length and the long fibre could be several micrometers. After the self-assembled oligopeptide was alternately immersed in a calcium ions solution and a phosphate ions solution for 5 cycles, abundant nano-particles were found precipitated along the oligopeptide fibres, forming a corn-on-the-cob appearance (Figure 5c and d). Results of EDS confirmed that these nano-particles contained calcium and phosphate (Figure 5e). There were few noticeable diffraction arcs in the resulting interference pattern, and the electron diffraction pattern suggested that the precipitates were comparable to amorphous calcium phosphate (ACP) (Figure 5f).

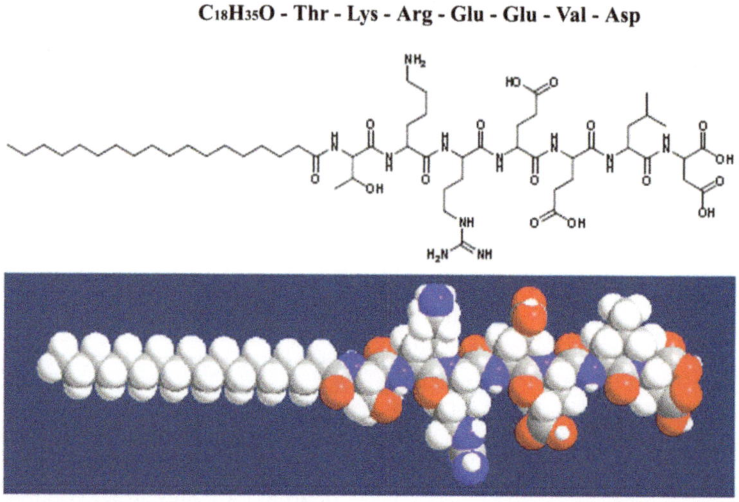

Figure 1 The chemical structure of the oligopeptide amphiphile.

Biomimetic mineralization of enamel in metastable calcium phosphate solution containing the peptide

Flocculent precipitates were observed in the metastable calcium phosphate solution after 30 days. TEM micrographs revealed that precipitates contained ACP and apatite (Figure 6a). This is probably because ACP spheres were partly transformed into apatite crystallite clusters. The calcification solution with oligopeptide was relatively clear with few precipitates over a period of 30 days. TEM micrographs revealed that the precipitates were 20–30 nm nano-ACP particles (Figure 6b). The particles were found separately bound to the oligopeptide fibres. No apatite crystal was observed. The final concentration was 2.58 mM calcium, 1.55 mM phosphate, 1mg/L NaF, 180 mM NaCl, and 15 μg/mL oligopeptide, which was buffered by 50 mM trihydroxymethylaminomethane (Tris)-hydrochloric acid to a pH of 7.6. The particle sizes of the Ca^{2+}-induced self-assembled peptide in the biomimetic mineralizing solution was characterized by a nano-particle size analyser (Figure 7). The size of the particles was mostly 338.7nm (87.4%). There are also two small peaks lying in 1.288 nm (9.2%) and 5,560 nm (3.4%), which may possibly be due to the presence of short oligopeptide nano-fibres and nano-fibres aggregation.

Figure 8 shows the precipitation of crystals on the enamel surface with or without oligopeptide after 24 h. XRD confirmed that the crystals were hydroxyapatite (Figure 9). When the partially demineralized enamel slices were incubated at 37°C in a metastable calcium phosphate solution (Figure 8a, b), hydroxyapatites were found precipitated in a homogeneous manner on the demineralized enamel surface. However, the crystal morphology on the enamel surface with metastable calcium phosphate solution contained no oligopeptide and

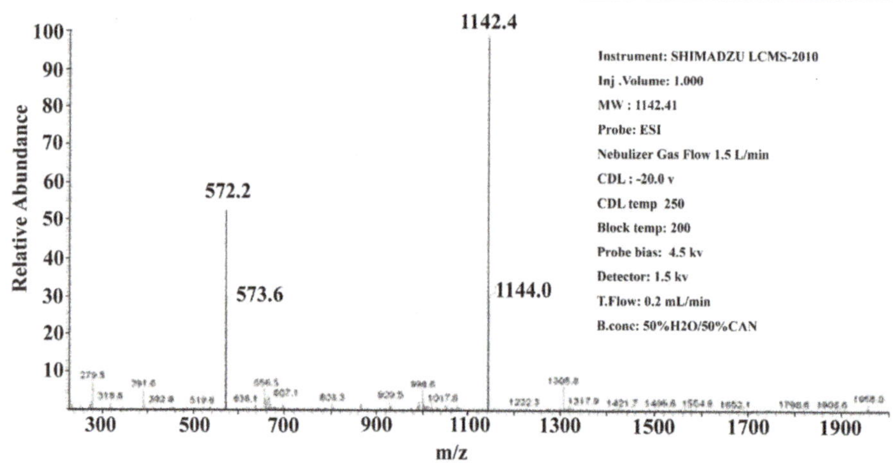

Figure 2 Mass spectrometry report of the oligopeptide.

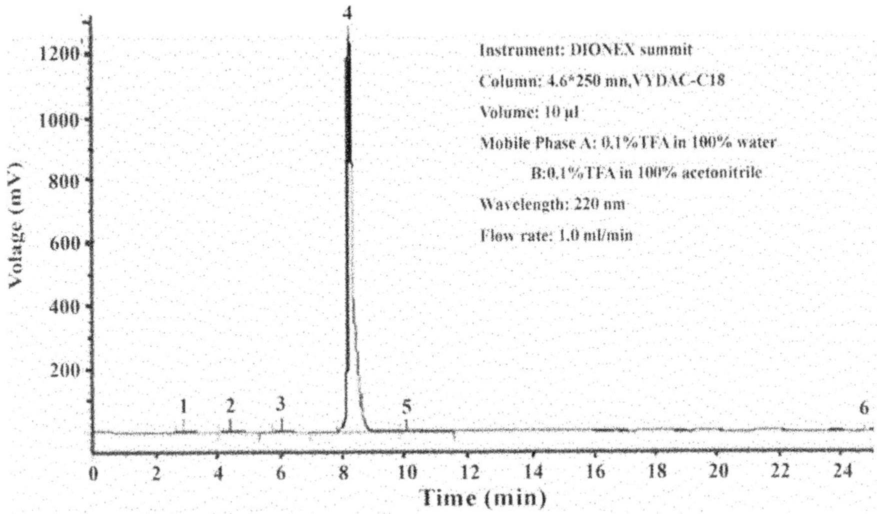

Peak No.	Ret Time (min)	Area	Conc. (%)
1	2.927	55798.633	0.3466
2	4.363	40717.374	0.2529
3	6.053	51933.032	0.3226
4	8.237	15825232.000	98.2965
5	10.043	101161.015	0.6283
6	25.153	24642.645	0.1531
Total		16099484.700	100.00

Figure 3 HPLC report of the purified oligopeptide.

was different from that with a metastable calcium phosphate solution with oligopeptide. SEM (10,000×) showed that hydroxyapatite crystals were uniformly precipitated on the pre-existing enamel crystals when there was no oligopeptide (Figure 8c). Higher magnification (100,000×) showed that they were needle-like, well defined, and precipitated in a relatively isolated and separate manner (Figure 8d). In the presence of the oligopeptide, the precipitated hydroxyapatite crystals were obtuse and rounded at both ends. These blunted, rod-like crystals were smaller in size than those crystals grown without oligopeptide (Figure 8e, f).

The morphology of the precipitated hydroxyapatite crystals after 20 days is shown in Figure 10. The precipitates in the absence of the oligopeptide were spiky, needle-like hydroxyapatite crystals (Figure 10a, b); those precipitated in the presence of the oligopeptide were rod-like hydroxyapatite crystals and tended to be more bundled (Figure 10d, e). The cross-section of the enamel slices showed that homogeneous hydroxyapatite crystals were oriented in parallel and packed together to form a dense, uniform layer. The thickness of the precipitated crystal layer in the presence of the oligopeptide was approximately half of that without oligopeptide (Figure 10b, e). The rod-like precipitated crystals with oligopeptide were more densely packed than those spiky crystals precipitated in the absence of oligopeptide.

Figure 9a showed XRD spectra of enamel samples. The diffraction peaks (002), (211), (112), and (300) planes corresponded well to the peaks for hydroxyapatite. The ratios of diffraction intensity of the c-axis (002) reflection to the diffraction intensity of the (211) or a-axis (300) reflections were considerably enhanced in all the calcium phosphate-coated samples compared to the etched tooth surface. The result suggested that the precipitated hydroxyapatite was oriented along its c-axis. The size of the precipitate crystal along the c-axis (length) and a-axis (width) was calculated using the Debye-Scherrer formula, $D = 0.89\lambda/(\beta 1/2 * \cos\theta)$ (D is the crystallite size; λ is the wavelength of the X-rays; and $\beta 1/2$ is the diffractometry (XRD) pattern broadening). The length of the crystal was shorter in the presence of the oligopeptide than that formed without the oligopeptide. The width of the crystal was affected to a lesser extent. The results match the observations from the SEM (Figure 10).

The mean values of Vicker's hardness of (A) untreated enamel, (B) demineralized enamel by acid-etching, (C) enamel coated with hydroxyapatite precipitates without

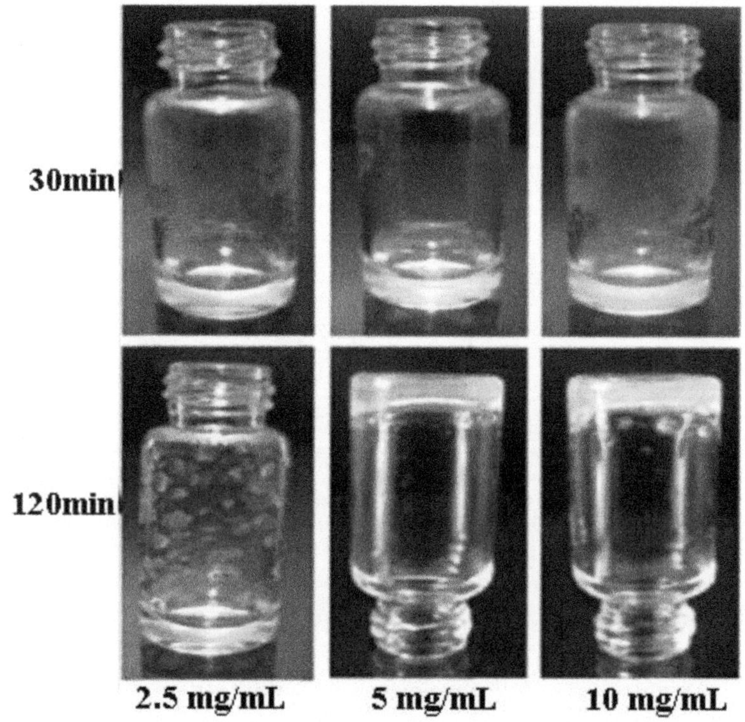

Figure 4 Effect of the oligopeptide solution after adding calcium chloride according to time and oligopeptide concentration.

the oligopeptide, and (D) enamel coated with hydroxyapatite precipitates in the presence of the oligopeptide are shown in Figure 11. The results showed a statistically significant difference among the 4 groups. A has the highest mean value of Vicker's hardness, whereas B has the lowest. The results showed that acid-etching significantly reduces enamel microhardness. Biomimetic mineralization with or without the oligopeptide increased demineralized enamel microhardness. However, the microhardness is still lower than that of the untreated enamel. There was no significant difference in hardness between the remineralization groups regardless of oligopeptide use.

Discussion

The oligopeptide used in this study is fabricated by SPPS, one the most common techniques for peptide synthesis, which proceeded step by step in a C-terminal to N-terminal fashion and ended with the alkylation of the NH_2 terminus of the peptide [22]. HPLC and mass spectrometry were used to characterize and evaluate the purity of the oligopeptide [23]. This study demonstrated that it can be synthesized and purified to a high degree of purity although it is a relatively long molecule. The synthetic oligopeptide is amphiphilic because it has a hydrophilic anionic oligopeptide with a hydrophobic tail. The stearic acid derivative has a long carbon alkyl chain which was used as the hydrophobic tail. It was reported that the alkyl tail attached to the NH_2 or COOH termini of peptide could influence their aggregation and secondary structure in water in both synthetic and natural systems [17]. The hydrophobic alkyl tails would pack together in the centre and leave the peptide segments exposed to the aqueous environment, resulting in peptide amphiphile self-assembly. The amino acid sequence of the peptide derived from the hydrophilic C-terminal amino residue of amelogenin and acts as a functional domain to induce mineralization. The oligopeptide can be dissolved in an alkaline solution but is insoluble in neutral or acidic solutions, which may contribute to its hydrophilic anionic peptide sequence. The results suggested that calcium ions induced the oligopeptide to self-assemble. It is reported that it is an essential step of supermolecular assembly for the oligopeptide to act function [18,19]. We considered that the property of calcium ions' induced self-assembly may facilitate the mineralization to capture and trap calcium ions within the mineralized areas. Although the results cannot confirm that the self-assembly oligopeptide is able to capture calcium and phosphate ions from the solutions and to initiate the nucleation, this study demonstrated the peptide can induce ACP to form along the self-assembly fibres when it is alternately immersed in a calcium ions solution and a phosphate ions solution. Quantitative assessment of the calcium and phosphate content of ACP such as atomic absorption spectroscopy is essential, but

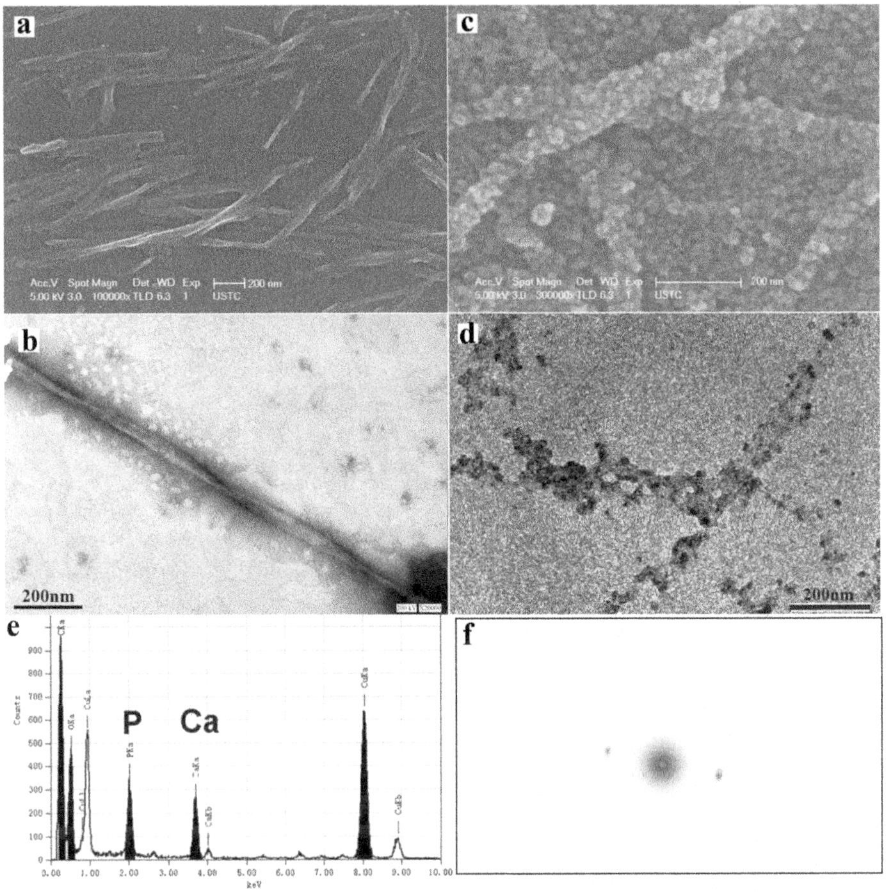

Figure 5 Biomimetic mineralization property of the self-assembled oligopeptide. (a) SEM and **(b)** TEM (negative stain with phosphotungstic acid) micrographs of the self-assembled oligopeptides. **(c)** SEM and **(d)** TEM (dense electronic images) micrograph showing precipitation of nanoparticles along the oligopeptide fibers. **(e)** EDS and **(f)** electron diffraction pattern of the precipitated nanoparticles.

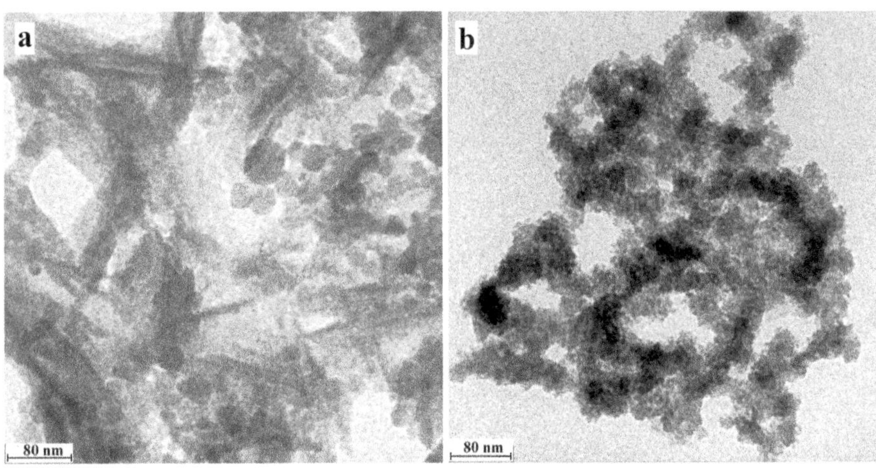

Figure 6 TEM micrographs of the precipitates from the calcification solution that settled over 30 days at 37°C. (a) without oligopeptide; **(b)** with 15 μg/mL of oligopeptide.

			Diam. (nm)	% Intensity	Width (nm)
Z-Average (d.nm):	476.9	Peak 1:	338.7	87.4	70.12
PdI:	0.642	Peak 2:	1.288	9.2	0.1388
Intercept:	0.605	Peak 3:	5560	3.4	0.000

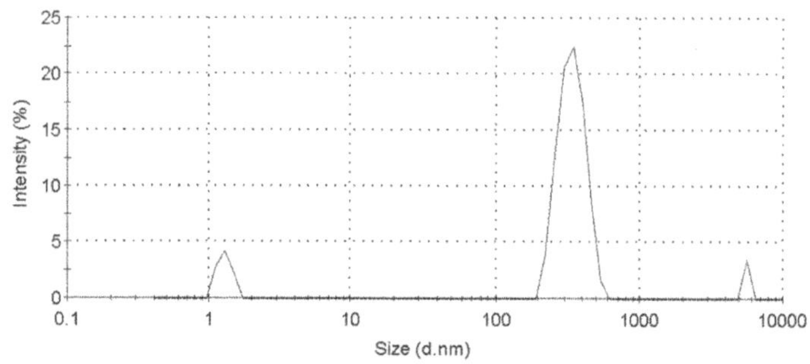

Figure 7 The particle size distribution in the metastable calcium phosphate solution with oligopeptide.

the quantity of mineralized peptide fibres was too little for assessment.

The biomineralization process is an organic matrix-mediated mineralization other than a simple thermodynamic and kinetic process of inorganic mineral ions crystallization under aqueous solution condition [24]. A key aspect of this organic matrix-mediated deposition concerns the role of molecular interactions in controlling oriented nucleation at the matrix-mineral interface. In this study, the flocculent precipitates containing ACP and apatite were observed in the metastable calcium phosphate solution without oligopeptide addition after 30 days. This is probably because ACP spheres were partly transformed into apatite crystallite clusters. However, the calcification solution with oligopeptide was relatively clear with few precipitates of 20–30 nm nano-ACP particles

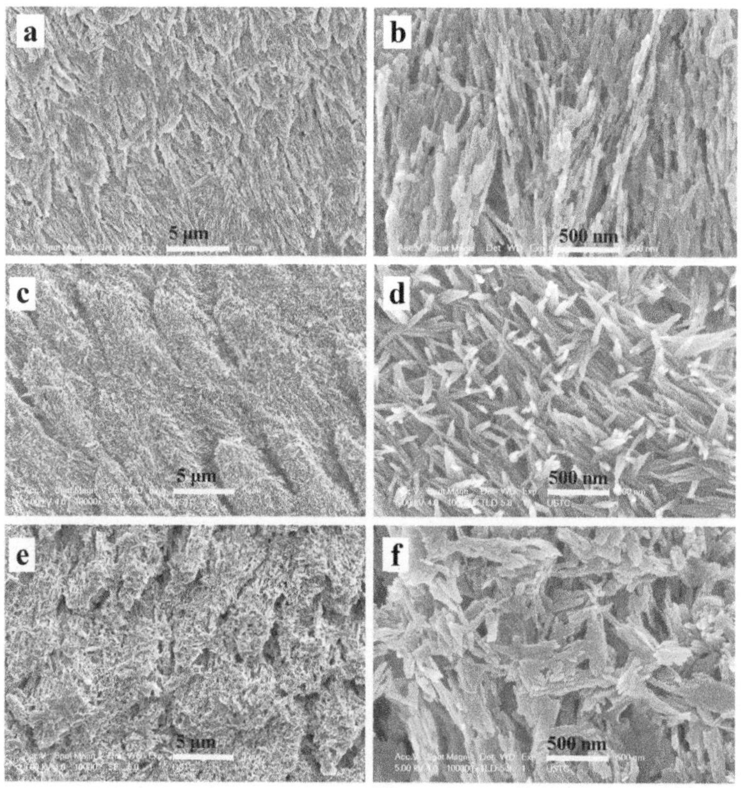

Figure 8 SEM micrographs of enamel surface at different magnifications. (a, b) Demineralized surface after acid etching. **(c, d)** Hydroxyapatites precipitated on the enamel surface after 24 hours without oligopeptide. **(e, f)** Hydroxyapatites precipitated on the enamel surface after 24 hours with oligopeptide.

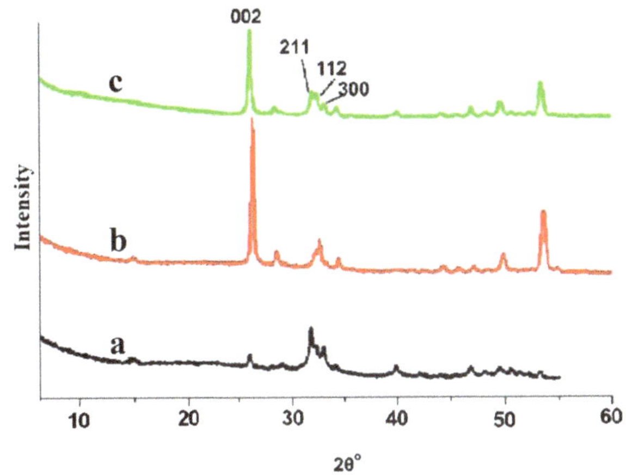

Figure 9 XRD spectra of the specimens. (a) Enamel after acid etching. **(b)** Precipitates on enamel without oligopeptide after 20 days. **(c)** Precipitates on enamel with oligopeptide after 20 days.

over 30 days. No apatite crystal was observed. This suggests that the nano-ACP precursor is obtained by adding the oligopeptide to metastable calcium phosphate solutions with fluoride. The negatively charged peptide binds to ACP nano-precursors and sequesters the ions. There is inhibition of crystal nucleation and induction of liquid-liquid-phase separation in the aqueous crystallizing solution [25,26]. The sequestered ACP presents as droplets that appear transparent to the naked eye because they are smaller than the wavelength of light. Because of the sequestering mechanism, the oligopeptide-ACP complex can prevent ACP nano-precursors from aggregation and precipitation and make them small. Therefore, the nano-ACP precursor is highly active and can be transported to the mineralizing zone to induce mineralization [27].

More and more evidence supports the notion that biomineralization is a non-classical crystallization pathway and a bottom-up approach [5]. In such cases, crystallization often proceeds by a sequential process involving structural and compositional modifications of amorphous precursors and crystalline intermediates. Crystallization often involves an initial amorphous phase (such as ACP) that may be non-stoichiometric, hydrated, and susceptible to rapid phase transformation [6]. According to this concept, a biomimetic mineralization method has been successfully utilized in the duplication of a hard tissue structure [28]. Acidic macromolecules are added to a metastable calcium phosphate solution to stabilize the amorphous mineral phases and to reduce these amorphous phases to the nanoscale [29]. Crystal growth is then controlled by the macromolecules that undergo self-assembly. In this study, we examined the initial nucleation and growth (24 h) of the precipitated crystals and the speed of HA crystal growth (20 days) induced by the oligopeptide. It is more desirable to study the samples in multiple time point. It is our primary study and communication report, and more time point should be studied to explain the HA crystal growth in the following study.

In this study, the microhardness of the samples with hydroxyapatite precipitates in the presence of oligopeptide was similar to those without oligopeptide. Homogeneous hydroxyapatite crystals were oriented in parallel and packed together to form a dense, uniform layer. As the crystals on the surface were newly mineralized, they did not reach the mature stage of the crystal that formed

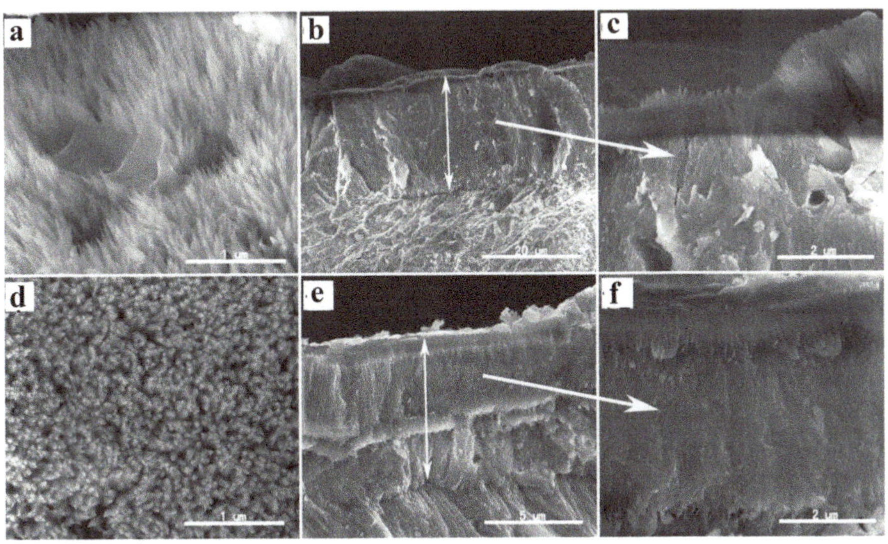

Figure 10 SEM micrographs of hydroxyapatites precipitated on enamel surface. (a) Precipitates without oligopeptide. **(b)** Transverse section of **(a)**. **(c)** High magnification of **(b)**. **(d)** Precipitates with oligopeptide. **(e)** Transverse section of **(d)**. **(f)** High magnification of **(e)**.

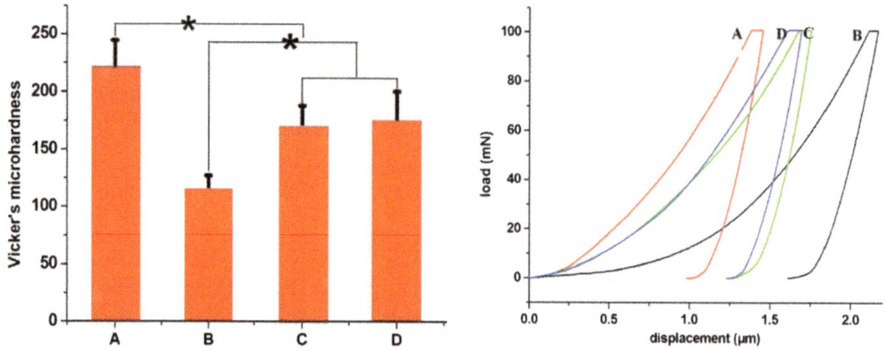

Figure 11 Vicker's hardness of the enamel specimens. A - untreated enamel, B - demineralized enamel by acid-etching, C - enamel coated with hydroxyapatite precipitates without the oligopeptide, D - enamel coated with hydroxyapatite precipitates in the presence of the oligopeptide. * $p < 0.05$. The right pane is the typical load–displacement curve of different samples under the Vicker's microhardness test.

underneath. This agrees with the findings reported by Busch [7] that there was no significant difference in hardness between the remineralization groups regardless of oligopeptide use.

Primary and secondary biomimetic growth of calcium phosphate can be distinguished on polymer surfaces. It is however not easy to distinguish the growth with the SEMs provided in this study. Mature enamel hydroxyapatite crystals bundled in parallel to form a special microstructure of repeating units of prisms. It is plausible that the structure of the precipitate was similar to the microstructure of the enamel. It is better to evaluate the remineralization with a quantitative method. However, the precipitate and the enamel had similar composite and structure; even the atomic absorption spectroscopy could not show the difference.

This study did not assess the stability or incorporation of the hydroxyapatite precipitates onto the enamel surface. In addition, the thickness of the crystal layer precipitated in the presence of the oligopeptide was less than that formed without oligopeptide. Therefore, the presence of the oligopeptide reduced the precipitation rate of hydroxyapatite crystal. It is noteworthy that the precipitated crystals with oligopeptide were more densely packed than those precipitated without oligopeptide. This is probably due to the different rate of crystal precipitation. The results showed that the morphology and the growth of the precipitated layer (remineralized layer) was different between the experimental samples (with the oligopeptide) and the control samples (without oligopeptide). Compared to the control samples, the advantage of using the oligopeptide is that the crystals are more densely packed. The disadvantage was that the deposition rate is slow in the presence of oligopeptide. However, the slow speed of crystal growth contributes to crystals packing densely. The metastable calcium phosphate solution with oligopeptide is more stable than the metastable calcium phosphate solution without oligopeptide. This advantage will facilitate its potential clinic use.

Bertassoni and his co-workers proposed that a more appropriate endpoint to evaluate the effectiveness of remineralization should be associated with the recovery of the mechanical properties of the hydrated dental tissue, which is presumed to correlate well with its overall functionality [30]. Hardness testing is an indirect method of tracking changes in the mineral content of dentin [30]. Biomechanical properties of dental tissues in general have been mostly evaluated by microhardness testing methods [31]. A disadvantage in microhardness assessment is the lack of analysis of reading error [32]. Although reading errors may have occurred in this study, the findings of microhardness measurement are consistent with the SEM and TEM observations. In addition, any reading error in this investigation should be small because the rater is experienced in microhardness testing and became fully acquainted with the procedure.

Conclusion

A novel self-assembly oligopeptide amphiphile was fabricated with high purity. The oligopeptide self-assembles into nano-fibres in the presence of calcium ions and at neutral acidity. The self-assembled oligopeptide allows precipitation of ACP along its nano-fibres when it was alternately immersed in a calcium ions solution and a phosphate ions solution. The metastable calcium phosphate solution with the oligopeptide can induce formation of a relatively stable nano-ACP complex precursor. The nano-ACP complex precursor can induce nano-rod hydroxyapatite growth which densely packed and bundled together to form enamel-like tissue on demineralized enamel and the precipitation of these crystals increases enamel hardness. This synthetic novel oligopeptide can be a potential molecular tool for biomimetic mineralization on the enamel surface. It is one of the primary steps towards the design and construction of novel biomaterial for future clinical care of dental erosion.

Additional file

Additional file 1: Figure S1. Fmoc SPPS solid-phase, peptide synthesis.

Competing interests

The authors declare that they have no competing interests.

Authors' contributions

QLL and CHC designed the oligopeptide amphiphile, supervised the project, and prepared the manuscript. WL was involved in the SPPS, HPLC, and mass spectrometry. YC prepared the tooth slides and oligopeptide solutions and performed SEM. TYN helped in all procedures, performed TEM and data analysis. All authors read and approved the final manuscript.

Acknowledgements

This study was supported by NSFC (No.30973352) and NSFC/RGC Joint Research Scheme (N_HKU 776/10 and No.81061160511), and Youth Foundation of Anhui Provincial Natural Science Foundation (No.1208085QH144).

References

1. Chu CH, Pang KK, Lo EC: **Dietary behavior and knowledge of dental erosion among Chinese adults.** *BMC Oral Health* 2010, **10**:13.
2. Featherstone JD: **Remineralization, the natural caries repair process-the need for new approaches.** *Adv Dent Res* 2009, **21**:4–7.
3. Cochrane NJ, Cai F, Huq NL, Burrow MF, Reynolds EC: **New approaches to enhanced remineralization of tooth enamel.** *J Dent Res* 2010, **89.**1187–1197.
4. Yamane S, Sugawara A, Watanabe A, Akiyoshi K: **Hybrid nanoapatite by polysaccharide nanogel-templated mineralization.** *J Bioact Compat Poly* 2009, **24**:151–168.
5. Colfen H, Mann S: **Higher-order organization by mesoscale self-assembly and transformation of hybrid nanostructures.** *Angew Chem* 2003, **42**:2350–2365.
6. Niederberger M, Colfen H: **Oriented attachment and mesocrystals: non-classical crystallization mechanisms based on nanoparticle assembly.** *Phys Chem Chem Phys* 2006, **8**:3271–3287.
7. Busch S: **Regeneration of human tooth enamel.** *Angew Chem* 2004, **43**:1428–1431.
8. Yamagishi K, Onuma K, Suzuki T, Okada F, Tagami J, Otsuki M, Senawangse P: **Materials chemistry: a synthetic enamel for rapid tooth repair.** *Nature* 2005, **433**:819.
9. Chen H, Tang Z, Liu J, Sun K, Chang SR, Peters MC, Mansfield JF, Czajka-Jakubowska A, Clarkson BH: **Acellular synthesis of a human enamel-like microstructure.** *Adv Mater* 2006, **18**:1846–1851.
10. Fowler CE, Li M, Mann S, Margolis HC: **Influence of surfactant assembly on the formation of calcium phosphate materials-A model for dental enamel formation.** *J Mater Chem* 2005, **15**:3317–3325.
11. Yang S, He H, Wang L, Jia X, Feng H: **Oriented crystallization of hydroxyapatite by the biomimetic amelogenin nanospheres from self-assemblies of amphiphilic dendrons.** *Chem Commun (Camb)* 2011, **47**:10100–10102.
12. Ye W, Wang XX: **Ribbon-like and rod-like hydroxyapatite crystals deposited on titanium surface with electrochemical method.** *Mater Lett* 2007, **61**:4062–4065.
13. Fan Y, Sun Z, Wang R, Abbott C, Moradian-Oldak J: **Enamel inspired nanocomposite fabrication through amelogenin supramolecular assembly.** *Biomaterials* 2007, **28**:3034–3042.
14. Luo J, Ning T, Cao Y, Zhu X, Xu X, Tang X, Chu CH, Li L: **Biomimic enamel remineralization by hybridization calcium- and phosphate-loaded liposomes with amelogenin-inspired peptide.** *Key Eng Mater* 2012, **512–515**:1727–1730.
15. Zhou Y-Z, Cao Y, Liu W, Chu CH, Li Q-L: **Polydopamine-induced tooth remineralization.** *ACS Appl Mater Inter* 2012, **4**:6901–6910.
16. Semino CE: **Self-assembling peptides: from bio-inspired materials to bone regeneration.** *J Dent Res* 2008, **87**:606–616.
17. Petka WA, Harden JL, McGrath KP, Wirtz D, Tirrell DA: **Reversible hydrogels from self-assembling artificial proteins.** *Science* 1998, **281**:389–392.
18. Hartgerink JD, Beniash E, Stupp SI: **Self-assembly and mineralization of peptide-amphiphile nanofibers.** *Science* 2001, **294**:1684–1688.
19. Kirkham J, Firth A, Vernals D, Boden N, Robinson C, Shore RC, Brookes SJ, Aggeli A: **Self-assembling peptide scaffolds promote enamel remineralization.** *J Dent Res* 2007, **86**:426–430.
20. Du C, Falini G, Fermani S, Abbott C, Moradian-Oldak J: **Supramolecular assembly of amelogenin nanospheres into birefringent microribbons.** *Science* 2005, **307**:1450–1454.
21. Uskokovic V, Li W, Habelitz S: **Amelogenin as a promoter of nucleation and crystal growth of apatite.** *J Cryst Growth* 2011, **316**:106–117.
22. Fan Y, Sun Z, Moradian-Oldak J: **Controlled remineralization of enamel in the presence of amelogenin and fluoride.** *Biomaterials* 2009, **30**:478–483.
23. Hosseinkhani H, Hosseinkhani M, Kobayashi H: **Design of tissue-engineered nanoscaffold through self-assembly of peptide amphiphile.** *J Bioact Compat Poly* 2006, **21**:20.
24. Fan Y, Wen TZ, Liao S, Lallier T, Hagan JL, Twomley JT, Zhang J-F, Sun Z, Xu X: **Novel amelogenin-releasing hydrogel for remineralization of enamel artificial caries.** *J Bioact Compat Poly* 2012, **27**:14.
25. Ning T-Y, Xu X-H, Zhu L-F, Zhu X-P, Chu CH, Liu L-K, Li Q-L: **Biomimetic mineralization of dentin induced by agarose gel loaded with calcium phosphate.** *J Biomed Mater Res Part B* 2012, **100**:138–144.
26. Olszta MJ, Cheng X, Jee SS, Kumar R, Kim Y, Kaufman MJ, Douglas EP, Gower LB: **Bone structure and formation: a new perspective.** *Mat Sci Eng R* 2007, **58**:77–116.
27. Cao Y, Mei ML, Xu JG, Lo EC, Li QL, Chu CH: **Biomimetic mineralisation of phosphorylated dentine by CPP-ACP.** *J Dent* 2013, **41**:818–825.
28. Gower LB: **Biomimetic model systems for investigating the amorphous precursor pathway and its role in biomineralization.** *Chem Rev* 2008, **108**:4551–4627.
29. Jee SS, Thula TT, Gower LB: **Development of bone-like composites via the polymer-induced liquid-precursor (PILP) process. Part 1: influence of polymer molecular weight.** *Acta Biomater* 2010, **6**:3676–3686.
30. Bertassoni LE, Habelitz S, Kinney JH, Marshall SJ, Jr Marshall GW: **Biomechanical perspective on the remineralization of dentin.** *Caries Res* 2009, **43**:70–77.
31. Chu CH, Lo EC: **Microhardness of dentine in primary teeth after topical fluoride applications.** *J Dent* 2008, 36:387–391.
32. Chu CH, Lam A, Lo EC: **Dentin hypersensitivity and its management.** *Gen Dent* 2011, **59**:115–122.

Biophysiochemical properties of endothelial cells cultured on bio-inspired collagen films

Eunseok Seo[1], Kyung Won Seo[2], Jung-Eun Gil[3], Young-Ran Ha[1], Eunseop Yeom[2], Seungchul Lee[2] and Sang Joon Lee[1,2*]

Abstract

Background: In this study, we investigated the effect of the extracellular matrix on endothelial dysfunction by careful observation of human umbilical vein endothelial cells (HUVECs) cultured on denatured collagen film.

Results: HUVECs on denatured collagen film showed relatively high surface roughness compared with normal HUVECs. The expression levels of MMP-1, MMP-2 and CD146 increased in the ECs on denatured collagen film. In addition, we examined the accumulation of fluorescent beads on HUVEC layers subjected to circulatory flow. The number of accumulated fluorescent beads increased on the disorganized HUVEC layers.

Conclusions: The proposed *in vitro* study using bio-inspired collagen films could potentially be used in the size- and ligand-based design of drugs to treat endothelial dysfunction caused by circulatory vascular diseases.

Background

Atherosclerosis is a major cause of morbidity and mortality [1]. Many *in vitro* experiments on the early development stage of atherogenesis have been performed using pro-inflammatory mediators, such as lipopolysaccharide in bacterial cell walls [2-9], as stimuli. Endothelial cell (EC)-leukocyte interaction of these stimuli induces changes in shape, permeability, and gene expression of ECs [10-13]. However, previous studies overlooked the effects of the extracellular matrix (ECM) on variations in morphology and growth of ECs. Thus, this study experimentally investigated the effect of ECM properties on endothelial dysfunction.

The EC layer functions as an interface between circulating blood and surrounding tissues [14,15]. ECs modulate thrombosis and inflammation, and control mural smooth muscle cells and vascular health. These cells also function in physiological processes, such as innate and adaptive immune responses. Thus, endothelial dysfunction is associated with the development of atherosclerosis and other cardiovascular disorders [16-18]. ECM degradation, which is possibly regulated by matrix-degrading metalloproteinases (MMPs) and their endogenous tissue inhibitors, is closely involved in the outbreak of atherogenesis [19-21]. MMPs belong to a family of zinc metalloendopeptidases. These enzymes, which degrade ECM proteins, exhibit important functions in embryo development, morphogenesis, and tissue remodeling, as well as in various diseases, such as arthritis, atherosclerosis, invasion, and metastasis of cancer cells [22]. A maximum of 25 MMP members are found in humans and animals. Among these MMPs, 14 are related to atherosclerosis [23]. Vascular cells, including ECs and macrophages, secrete MMPs. ECs also express MMP-1 (collagenase), MMP-2 (gelatinase), and MMP-3 (stromelysin) [24-26].

The mechanical properties of arterial walls and susceptibility to pathological vascular remodeling are mainly determined by the macromolecular structures of collagen, elastin, and proteoglycans [27-29]. Under low pressure, the high compliance of elastin dominates the wall displacement. At high pressure, the wall displacement is limited by the stiffer collagen. Degradation of these mechanical properties with aging [30], smoking [31], diabetes [32], hypertension [33], and atherosclerosis [29,34] is associated with changes in the collagen and elastin scaffold. Although these macromolecules generate the biphasic mechanical response of arterial walls

* Correspondence: sjlee@postech.ac.kr
[1]Division of Integrative Biosciences and Biotechnology, Pohang University of Science and Technology, San 31, Hyoja-dong, Nam-Gu, Pohang, Gyeongbuk 790-784, Korea
[2]Center for Biofluid and Biomimic Research, Department of Mechanical Engineering, Pohang University of Science and Technology, San 31, Hyoja-dong, Nam-Gu, Pohang, Gyeongbuk 790-784, South Korea
Full list of author information is available at the end of the article

to pressure, the specific features of their microstructure remain as a missing element in the modeling of wall mechanics [35,36]. Therefore, we hypothesized that the structural change in collagen by MMP-1 and MMP-2 expression may induce endothelial dysfunction, such as disorganization and thickening of the EC layer. To validate this hypothesis, we prepared two different collagen films as ECM models. The normal collagen film corresponded to the ECM of normal blood vessels, and the denatured collagen film was adopted to mimic the ECM of atherosclerotic blood vessels. The denatured collagen film was intentionally disrupted by collagenase treatment to simulate the denatured ECM by increasing MMP-1 and MMP-2 expression under *in vivo* conditions. Through preliminary experiments using zebrafish models for investigating the early development stage of atherogenesis, we observed an EC monolayer surrounded by a collagen basement membrane in the normal vascular endothelium. The vascular endothelium of the early development stage of atherogenesis shows the disorganization and thickening of the EC layer [37]. To simulate these morphological results, we cultured human umbilical vein endothelial cells (HUVECs) on two different types of collagen films.

In this study, morphological variations in the bio-inspired ECM were experimentally investigated. The spatial distribution of surface roughness of ECs on the ECM was observed by phase-contrast digital holographic microscopy (DHM). Biological specimens, such as living cells, are transparent. However, transparent biological samples are difficult to observe clearly by bright-field microscopy. Therefore, non-invasive high-resolution imaging of living cells under *in vivo* conditions is important to visualize biological processes. Interferometry-based DHM can be performed to determine the spatial distributions of the phase and optical path length of a test sample. This technique can also provide quantitative phase information of a sample with a spatial resolution of tens of nanometers [38-41]. Temporal variations in morphological structure and EC layer thickness can be clearly observed with a time-resolved DHM technique.

To study the mechanism of endothelial dysfunction, we investigated the expression of CD146, vascular cell adhesion molecule (VCAM), and E-selectin on the surface of abnormal ECs. These three molecules are used as biomarkers of EC injury [6,17,42-45]. We also examined the positional information of vascular endothelial cadherin (VE-cadherin) of ECs. VE-cadherin has an important function in controlling vascular organization and modulating endothelial permeability [46].

To develop an optimal delivery system for vascular targeting and maximize drug accumulation on the EC layer while avoiding entrapment in the lung and small capillaries [47], the specific ligand-receptor interaction with the vessel walls [48] and surface properties related to specific vascular adhesion and reduced macrophage uptake should be investigated. To establish an *in vitro* study for drug screening of circulatory diseases, fluorescent bead accumulation on EC layers was examined under circulating flow conditions.

Results

Collagen film and denatured collagen film

Figures 1A and 1D illustrate the collagen and denatured collagen films coated on octadecyltrichlorosilane (OTS)-self-assembled monolayer (SAM)-treated coverslips. Figures 1B and 1E show typical scanning electron microscopy (SEM) images (1000×) of HUVECs on the collagen and denatured collagen films, respectively. The surface structures of the two films were also observed using atomic force microscopy (AFM). Figures 1C and 1F are typical AFM images showing the morphological structures of the collagen and denatured collagen films, respectively. As shown in Figures 1B and 1C, the collagen film had fiber structures. However, the fibers were broken in the collagenase-treated collagen film (Figures 1E and 1F).

Roughness of ECs

The spatiotemporal thickness profiles of HUVECs cultured on collagen film (Figure 2A) or denatured collagen film (Figure 2B) were quantitatively analyzed. The biophysical properties of HUVECs grown on normal or enzyme-degraded collagen films were measured using phase-contrast DHM (Figure 2C). The typical phase-contrast DHM images in Figures 2D and 2E show the spatial distributions of the physical thickness of HUVECs cultured on collagen and denatured collagen films, respectively. We evaluated the surface roughness of HUVECs using phase-contrast DHM images. The surface roughness of ECs was represented mathematically using the root mean square (RMS) value of its fluctuations. The HUVECs initially seeded on the collagen film and denatured collagen film exhibited relatively low RMS values. Figure 2F shows that the RMS value of HUVECs cultured on collagen film slightly increased after 6 h, and then rapidly decreased in the time period (t) from 6–24 h. This result was attributed to normal cell adhesion to the ECM substrate. The average RMS value was approximately 3.6 μm (S.D. = 0.8) 24 h after initial seeding. This value corresponded to an approximate reduction of 59% from the initial RMS value (6.1 ± 2.3). The surface roughness of HUVECs cultured on collagenase-treated collagen film for 5 min gradually increased until $t = 24$ h after HUVEC seeding. The RMS value increased by approximately 178% at $t = 24$ h compared with the initial value (4.5 ± 1.9) at $t = 1$ h. The surface roughness of HUVECs cultured on collagenase-treated collagen film

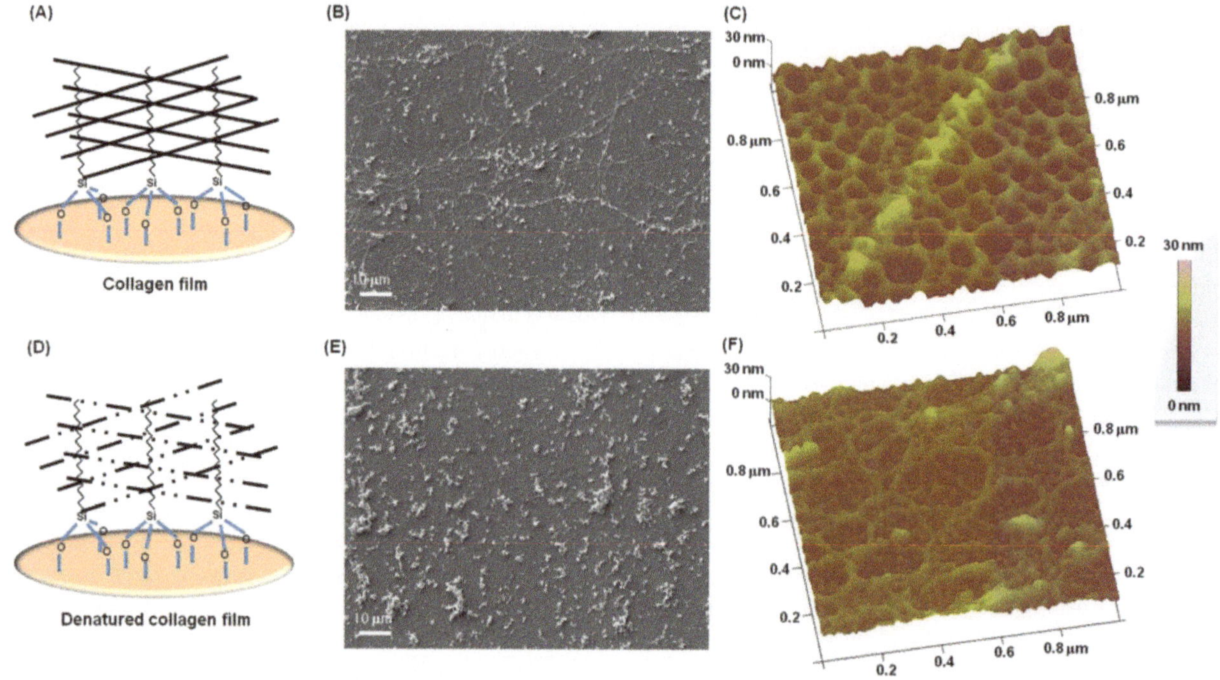

Figure 1 Two collagen films prepared for the *in vitro* study. (A) Normal collagen film, which mimics the ECM of normal blood vessels. Typical SEM image **(B)** and AFM image **(C)** of the collagen film. **(D)** Denatured collagen film, which mimics the ECM of abnormal blood vessels. The structure of the denatured collagen film was intentionally broken using collagenase to simulate the abnormal ECM. Typical SEM image **(E)** and AFM image **(F)** of the denatured collagen film.

for 15 min increased more rapidly at $t = 6$ h compared with those treated with collagenase for 5 min, and then gradually decreased as the cells detached from the ECM substrate after 6 h. The RMS value increased to 300% at $t = 3$ h compared with the initial value (4.0 ± 1.7) at $t = 1$ h, and decreased by approximately 255% at $t = 24$ h. The RMS value of HUVECs cultured on normal collagen films decreased after 24 h of seeding because the cells adhered to the ECM substrate. By contrast, the RMS values of HUVECs cultured on collagen films treated with collagenase for 5 min increased because of abnormal cell deformation. The surface roughness of HUVECs on collagen films treated with collagenase for 15 min increased more rapidly than those treated for 5 min. However, the RMS value decreased after 12 h because the abnormally clustered cells were detached from the ECM substrate.

Cell migration

We established two different collagen films as ECM models: the collagen film, which had a fiber structure similar to that of the ECM of normal blood vessels, and the denatured collagen film, which mimicked the ECM of atherosclerotic blood vessels. To observe the dynamic behaviors of cell-to-cell interactions, we cultured HUVECs on the collagen and denatured collagen films. Dynamic motions of ECs were observed using confocal laser scanning microscopy (CLSM). ECs cultured on collagen film or denatured collagen film were mounted in the flow chamber. Figures 3A and 3B show that ECs on the normal collagen film (non-treatment group) formed a monolayer, whereas those found on the denatured collagen films (treatment group) were aggregated and multi-stacked. Their movements were detected for 90 min at an interval of 10 min using CLSM. For the time-resolved tracking of ECs, cell movement was systematically analyzed by tracking the nuclei of cells. We marked the center of the nucleus of each cell to track its trajectory. The circular plots shown in Figures 3C and 3D represent the direction angle of cell migration and number of distributed cells on the collagen film and denatured collagen film. All cells were positioned at the zero point of the starting time, but these cells spread out as time progressed. As shown in the circular plots, ECs on the collagen films moved radially outward in all directions almost evenly. However, cells on the denatured collagen film moved in a preferential direction. Therefore, the non-treatment group showed a monolayer, whereas the treatment group exhibited aggregated ECs.

Growth properties of ECs

To compare the proliferation and survival of HUVECs between the two groups, DNA damage, apoptosis, and cell proliferation were analyzed using fluorescence-activated

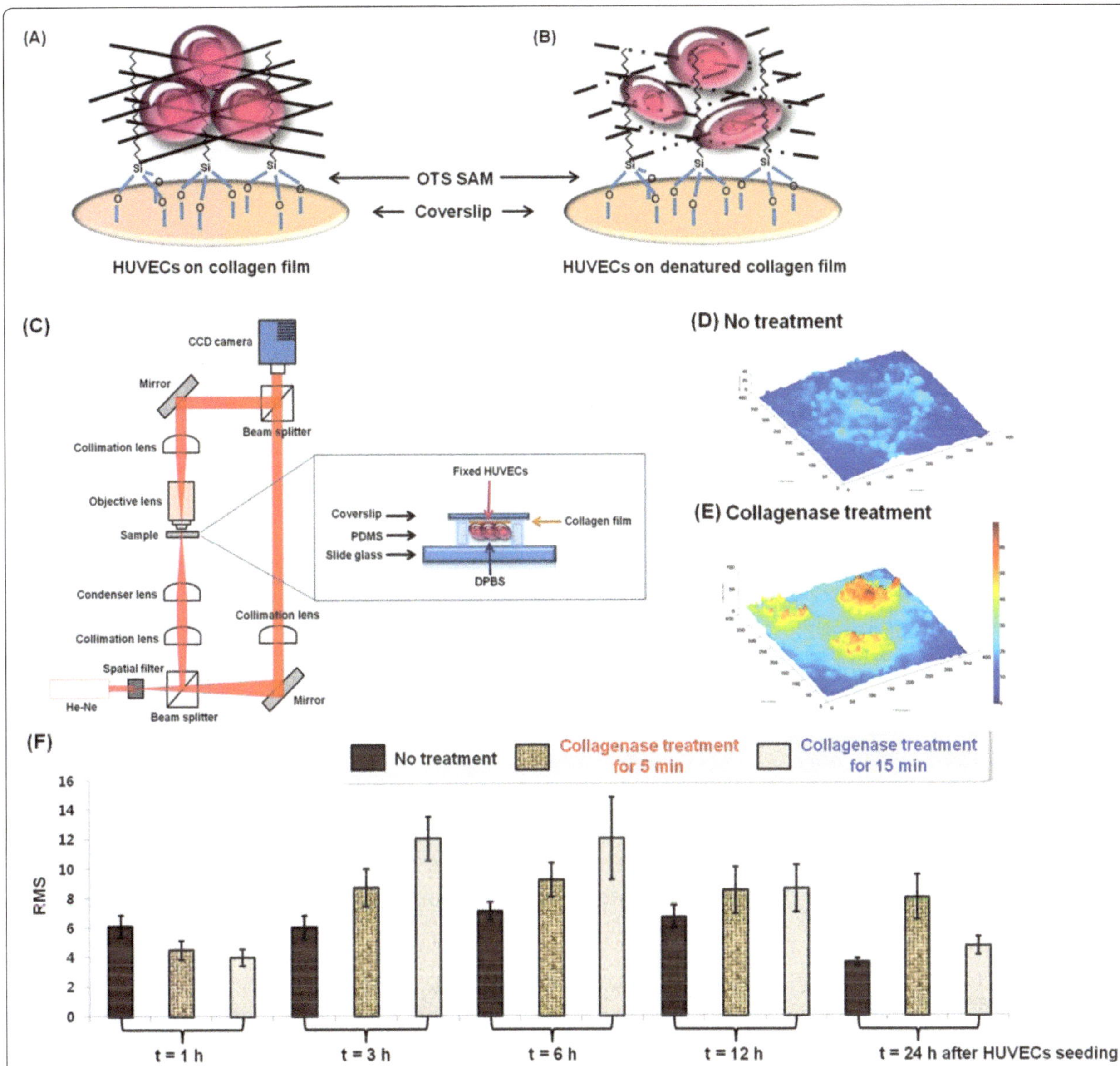

Figure 2 Surface roughness of HUVECs cultured on the two collagen films. HUVECs were incubated on top of the collagen film **(A)** and denatured collagen film **(B)**. Surface roughness of HUVECs on the two collagen films was analyzed using phase-contrast DHM. **(C)** Experimental setup of the phase-contrast DHM and cell-substrate samples used in this study. **(D and E)** Spatial distribution of physical thickness of HUVECs on the normal collagen film (non-treatment group) and denatured collagen film (treatment group). Holograms exhibit the thickness distribution of HUVECs in the non-treatment group **(D)** and treatment group **(E)**. **(F)** Temporal variations in surface roughness of HUVECs cultured on the collagen film and collagenase-treated collagen film. RMS = root mean square. *N* = 10.

cell sorting (FACS) technique. The control group (HUVECs on the collagen film) and treatment group (HUVECs on the denatured collagen film treated with collagenase for 5 min) were cultured for 6, 12, 24, and 36 h. As shown in Figures 3A–3D, cells with DNA damage (H2AX-positive population) and apoptosis (PARP-positive population) were not observed for both groups. H2AX-negative cells (Figures 4A and 4B) and PARP-negative cells (Figures 4C and 4D) did not show noticeable differences on both films. Furthermore, HUVECs on the denatured collagen film enhanced the proliferation rate compared with the control group (Figures 4E–4L). These results show that the aggregated form of HUVECs on the denatured collagen film was caused by the selective migration of HUVECs into a specific area or by enhanced proliferation of HUVECs in a specific area, such as the remaining collagen fibers in the denatured collagen film.

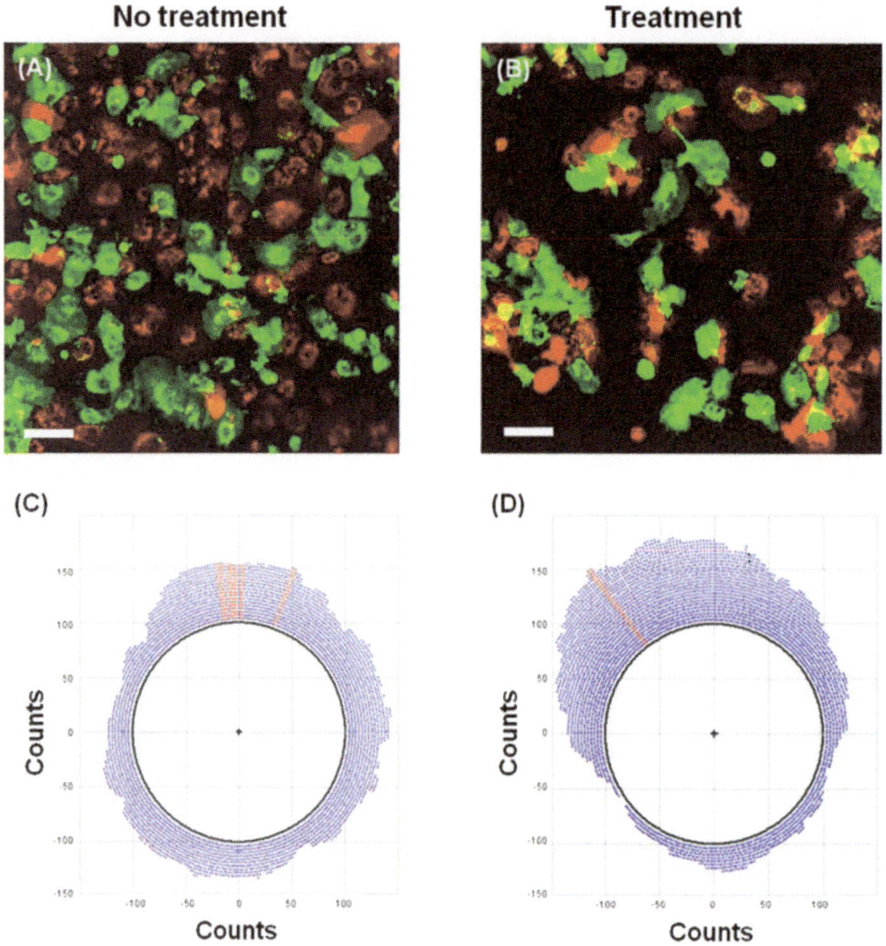

Figure 3 Confocal images of HUVECs cultured on the two collagen films. (A) Confocal images of HUVECs cultured on the collagen film (non-treatment group). **(B)** Confocal images of HUVECs cultured on the collagenase-treated collagen film (treatment group). Circular plots of the migration angle and number of HUVECs for the non-treatment group **(C)** and treatment group **(D)**. Scale bar = 100 μm.

MMP-1 and MMP-2 of ECs

HUVECs were transfected with MMP-1/enhanced green fluorescent protein (EGFP) and MMP-2/mCherry vectors to visualize the secretion of MMP-1 and MMP-2. Cells were cultured on normal (non-treatment group) and denatured collagen films (treatment group) for 24 h. These cells were then fixed with paraformaldehyde solution, and stained with 4′,6-diamidino-2-phenylindole (DAPI, blue). Figure 5A shows that MMP-1 and MMP-2 were not expressed in the non-treatment group. As shown in Figure 5B, MMP-1 and MMP-2 were observed in the treatment group. These results indicate that ECs on the denatured collagen film could secrete MMP-1 and MMP-2.

Collagen zymography

MMP is secreted as a pro-enzyme, and only activated MMP can degrade the collagen matrix. Therefore, immunofluorescent images do not show MMP activity. To overcome this limitation, we performed substrate activity using collagen zymography assay. As shown in Figure 6A, collagen zymography showed three distinct MMP bands. The lower band was MMP-1, which was about 40–60 kDa. The middle band was gelatinase-A (MMP-2), which was about 72 kDa. The upper band represented gelatinase-B (MMP-9), which was at about 95 kDa. The MMP-1 level decreased in the 5 min group compared with that in the control group (Figure 6B), and further decreased in the 15 min group. By contrast, the levels of MMP-2 and MMP-9 increased in the 5 min group compared with those in the control group. However, they significantly decreased in the 15 min group.

Expression of E-selectin, VCAM, CD146, and VE-cadherin

The expression levels of E-selectin, VCAM, CD146, and VE-cadherin on the EC surface for the two collagen films were visualized using immunostaining method and confocal imaging technique. Among the four adhesion molecules, CD146 and VE-cadherin were expressed on the EC surface cultured on both collagen films (Figure 7).

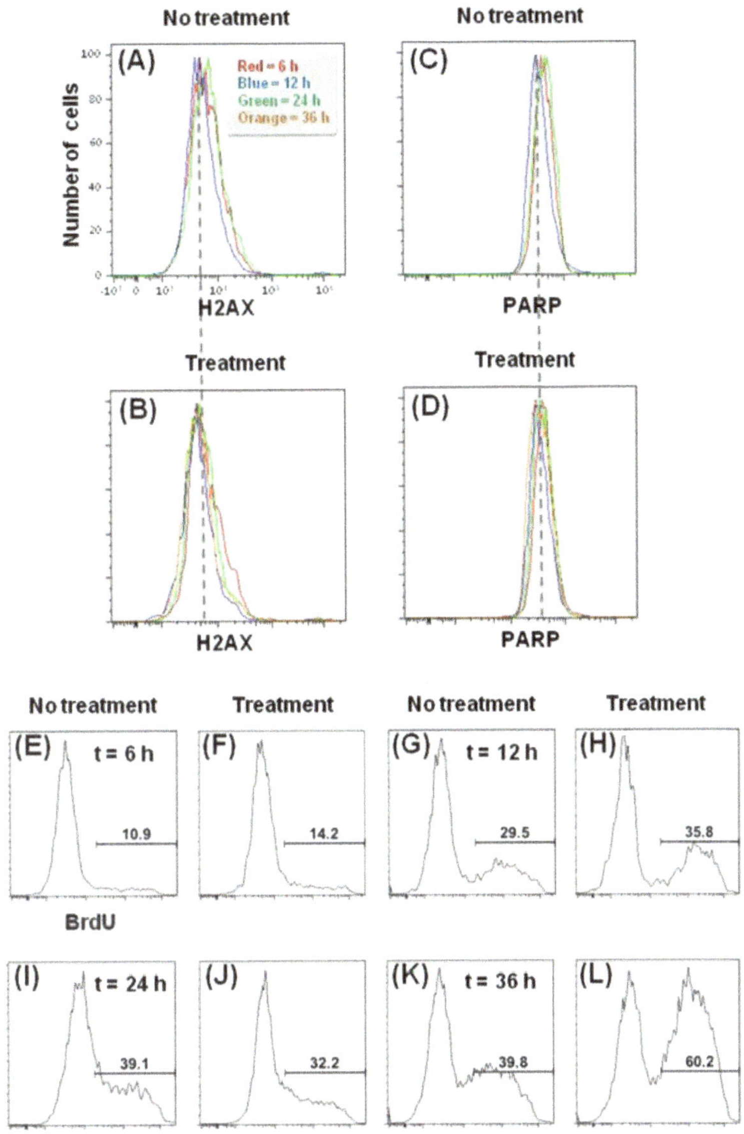

Figure 4 Survival and proliferation of HUVECs. HUVECs were cultured on both films for 6, 12, 24, and 36 h. DNA damage, apoptosis, and cell proliferation of HUVECs plated on the collagen-treated film were analyzed using flow cytometry to investigate their survival and proliferation. Comparison of histograms of H2AX-negative cells [**(A)** and **(B)**, non-DNA damage] and PARP-negative cells [**(C)** and **(D)**, non-apoptosis] on the normal collagen film (non-treatment group) and denatured collagen film (treatment group). Cell death by DNA damage (positive population of H2AX) and apoptosis (positive population of PARP) was not detected. **(E–L)** Comparison of HUVECs on the collagenase-treated collagen film showed enhanced proliferation rates.

Figure 7F shows that CD146 expression significantly increased on the EC surface of the denatured collagen film compared with that on ECs cultured on normal collagen film (Figure 7E). The results from the present *in vitro* studies on the cultured ECs suggest the important function of the biophysical properties of ECM in the formation of the EC layer. In particular, MMP-1 and MMP-2 caused weak cell interactions. Therefore, CD146 expression increased on the EC surface of the denatured collagen film.

Bead accumulation on the EC layer

ECs cultured on the collagen film or denatured collagenase-treated collagen film for 24 h were mounted on the flow chamber. The dynamic motion of beads on the cultured ECs was observed by time-lapse microscopy. Weakly adhered beads were detached under flow conditions, and some beads attached to the HUVEC layer were continuously concentrated to form EC clusters. In HUVECs cultured on the collagen film and denatured collagen film for 24 h, the size and number of

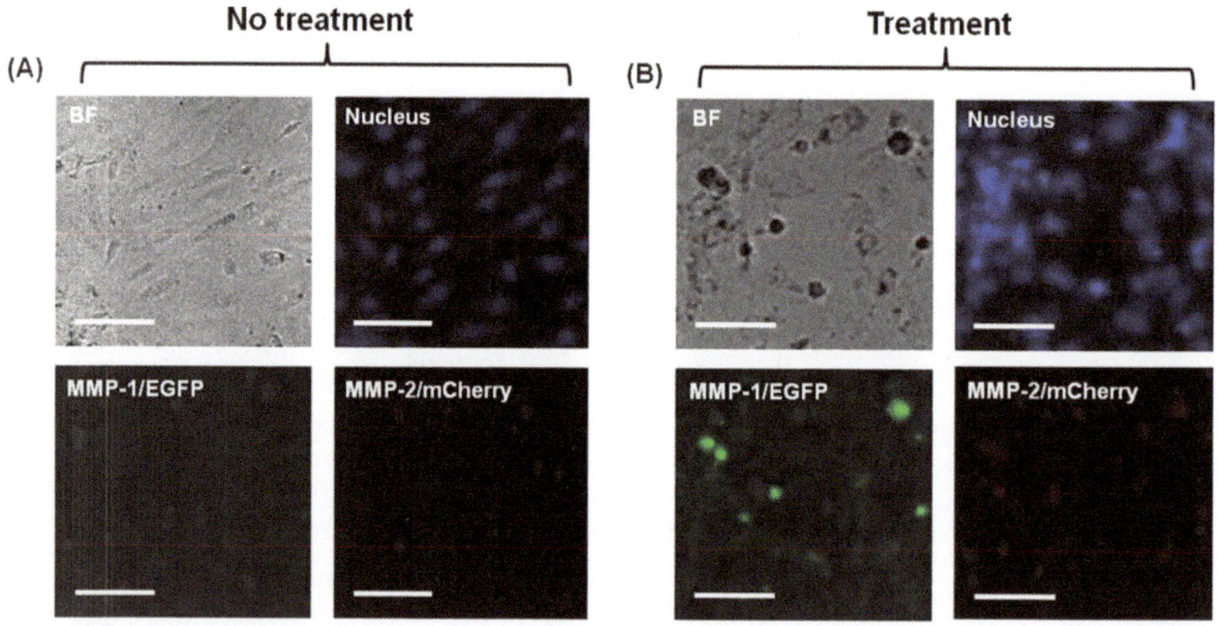

Figure 5 MMP-1 and MMP-2 of HUVECs. Bright field (gray), nucleus (blue), MMP-1/EGFP (green), and MMP-2/mCherry (red) images of normal **(A)** and abnormal HUVECs **(B)** cultured for 24 h. Scale bar = 100 μm.

clusters gradually increased (data not shown) as the flow rate increased. The beads were clustered on the HUVEC layers, and the cluster size increased. Figures 8A and 8B show the confocal images captured after 1 min at a flow rate of 1 mL/h. Figures 9A–9D show the fluorescence images captured after 1 min at the same flow rate. As shown in these figures, the beads on the normal EC layer (Figures 6A, 7B, and 9A) were less accumulated than those on the abnormal EC layers (Figures 8B, 9C, and 9D). In addition, the clusters in the treatment group were larger than those in the non-treatment group. Figures 9E and 9F show the temporal variations in the number of clusters and in the sum of cluster areas, respectively. These values were obtained from time-lapse images recorded consecutively at a flow rate of 1 mL/h. In Figure 9E, the number of clusters gradually increased, and the number of clusters on the normal EC layer was lower than that on the abnormal EC layer. Figure 9F shows that the beads on the normal EC layer were less accumulated than those on the abnormal EC layer. The slope increased

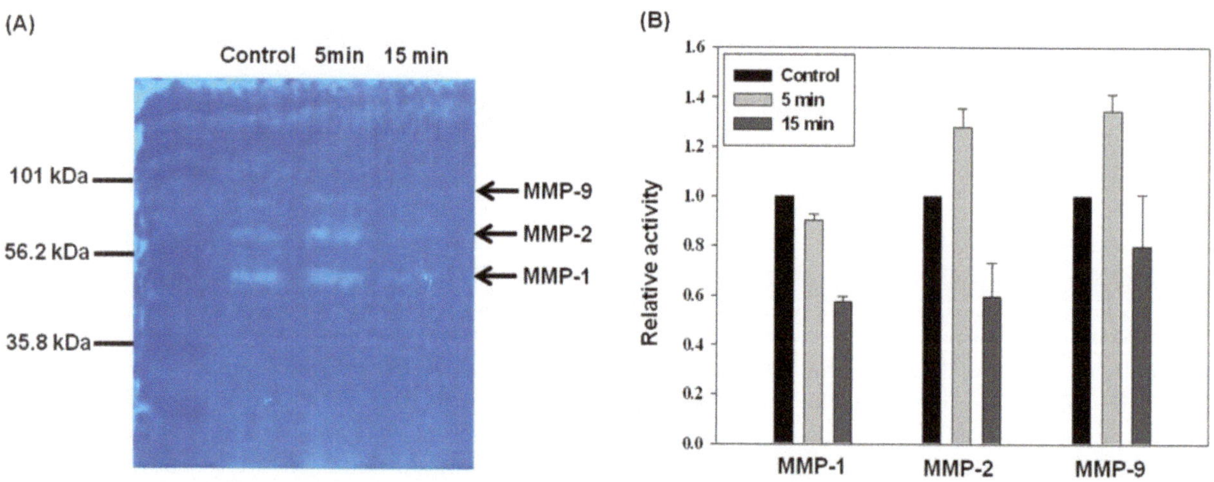

Figure 6 Comparison of the activities of MMP-1, −2, and −9 in HUVECs. MMP activity measured by collagen zymography for HUVECs cultured for 24 h. **(A)** Collagen zymography. **(B)** Densitometric analysis of MMP-1, −2, and −9 activity in HUVECs. Control, HUVECs on the collagen film; 5 min, HUVECs on the denatured collagen film treated with collagenase for 5 min; 15 min, HUVECs on the denatured collagen film treated with collagenase for 15 min. *N* = 3.

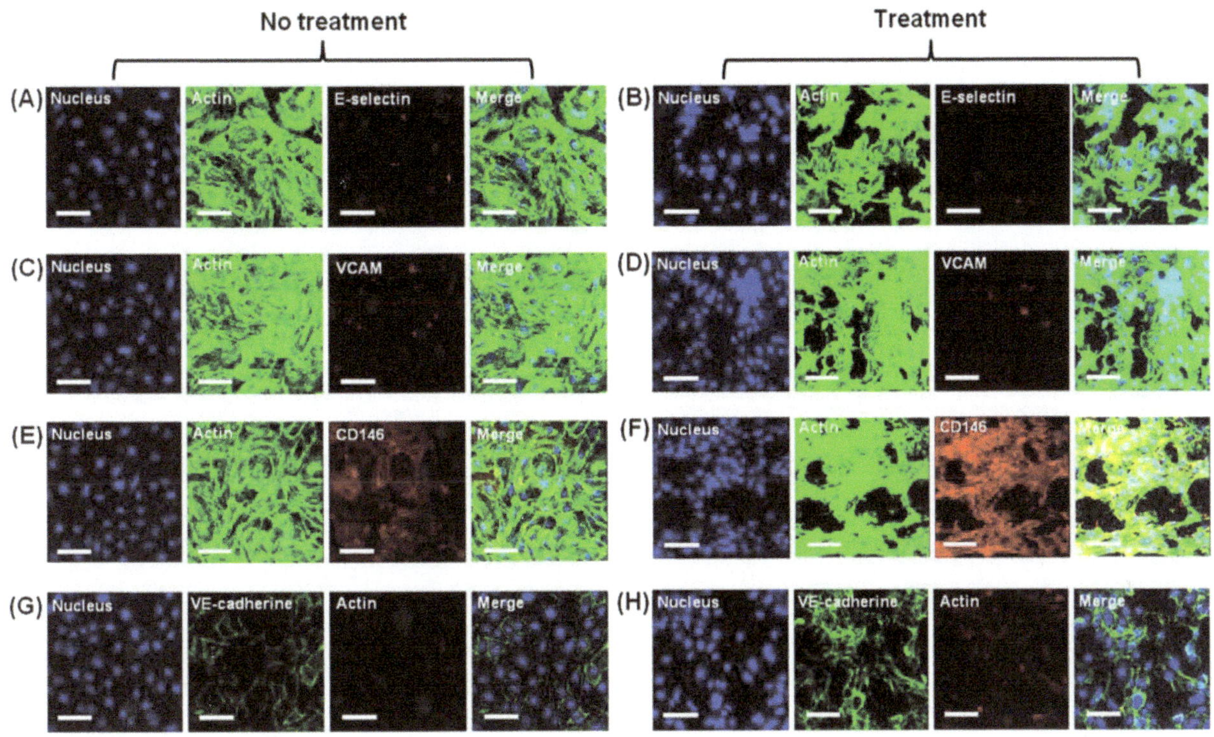

Figure 7 Comparison of the expression of E-selectin, VCAM, CD146, and VE-cadherin. HUVECs cultured for 24 h on the collagen film **(A, C, E, and G)** and denatured collagen film treated with collagenase for 5 min **(B, D, F, and H)**. Fluorescence images of actin [red in **(G)** and **(H)**, green in other figures], nucleus (blue), adhesion molecules (green in VE-cadherin and red in E-selectin, VCAM, and CD146), and their merged images. E-selectin **(A, B)**, VCAM **(C, D)**, CD146 **(E, F)**, and VE-cadherin **(G, H)**. Scale bar = 250 μm.

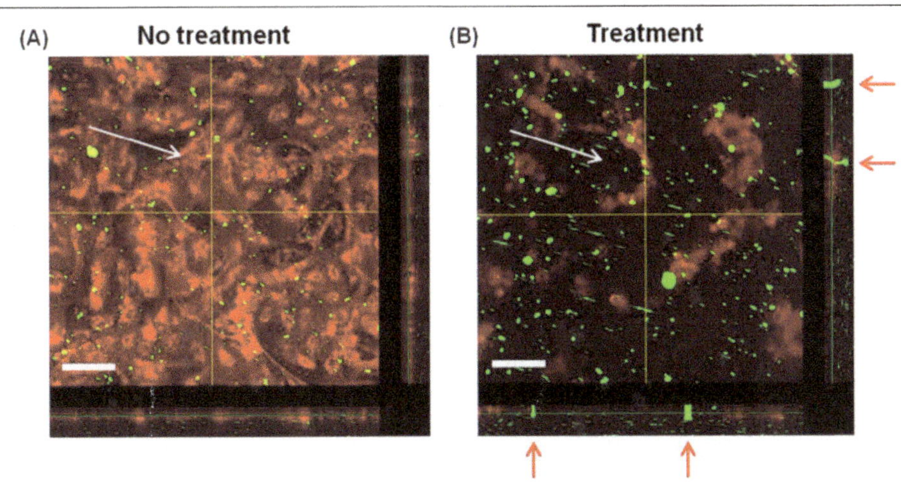

Figure 8 Confocal images of beads accumulated on the HUVEC layers. (A) Accumulated beads (green) on the normal HUVEC layer (non-treatment group). **(B)** Accumulated beads (green) on the abnormal HUVEC layer (treatment group). The images were captured after 1 min under a flow rate of 1 mL/h (shear stress: 23.96 mPa). White arrows indicate the direction of medium flow, and red arrows in **(B)** denote the bead clusters accumulated near the HUVEC clusters. The cross-sectional profiles are displayed along the yellow lines in the right ends and bottom parts. Scale bar = 100 μm.

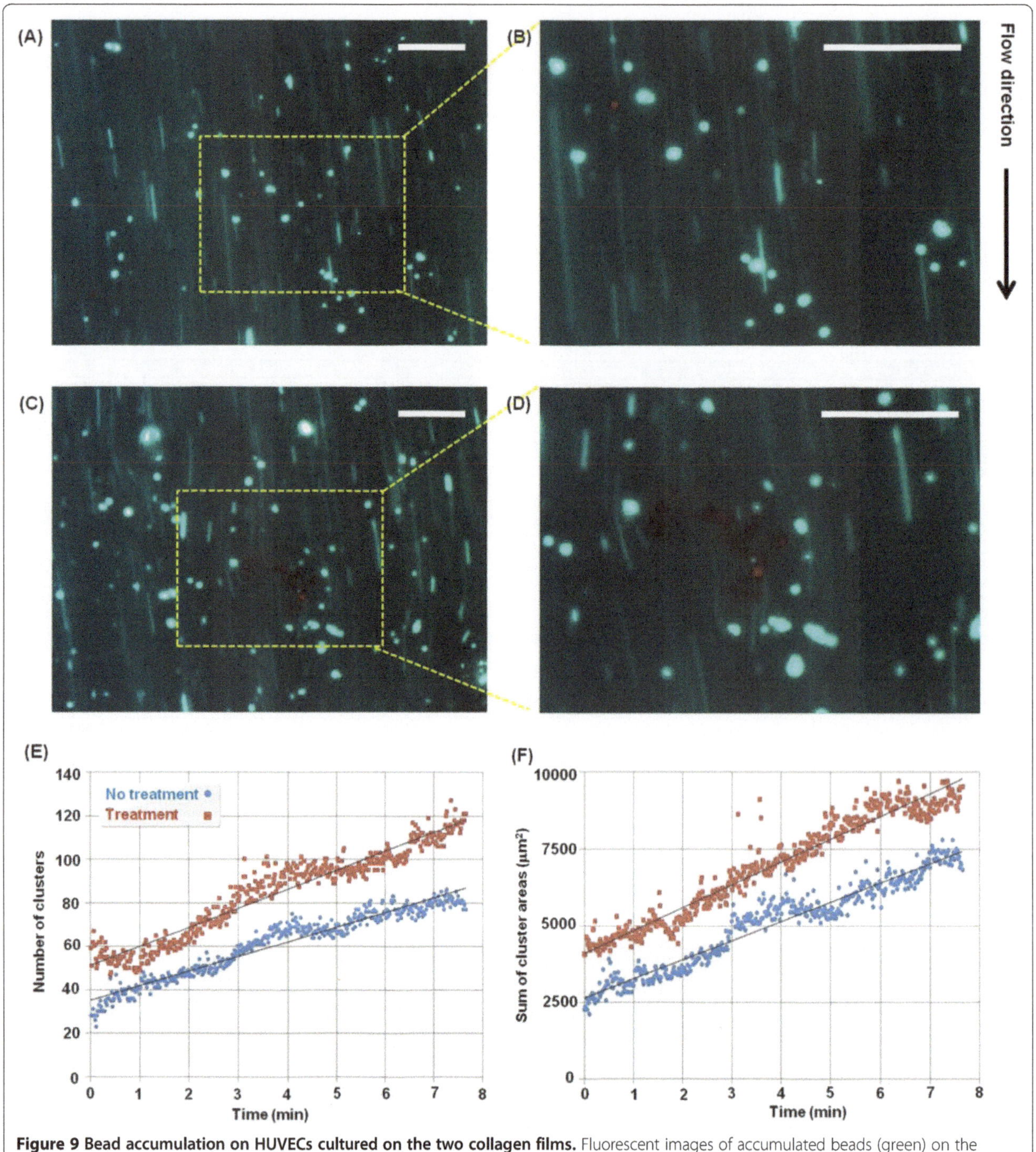

Figure 9 Bead accumulation on HUVECs cultured on the two collagen films. Fluorescent images of accumulated beads (green) on the normal **(A and B)** and abnormal HUVEC (red) layers **(C and D)**. The images were captured after 1 min under a flow rate of 1 mL/h (shear stress: 23.96 mPa). The images in the yellow dotted box were magnified two times **(B and D)**. Scale bar: 100 μm. **(E and F)** Temporal variations in the number of clusters **(E)** and sum of cluster areas **(F)**.

by approximately 30% for the abnormal EC layer group compared with that for the normal EC layer group. In addition, the sum of cluster area in the collagenase-treated group was higher than that of the non-treatment normal group. The slope was higher (18.8%) for the abnormal EC layer group than that for the normal EC layer group.

Discussion

In this study, the expression levels of MMP-1 and MMP-2 were observed in ECs on denatured collagen film. The results show that the destruction of the ECM structure was caused by MMP-1 and MMP-2, which was secreted by ECs on denatured collagen film. In addition, MMP-1 and MMP-2 could influence the formation of EC clusters and detachment of EC clusters from the EC layer. The experimental results obtained from the present *in vitro* studies on the cultured ECs suggest that the biophysical properties of ECM had an important function in the formation of the EC layer. Only HUVECs grown on the denatured collagen film expressed MMP-1 and MMP-2, whereas no MMP expression was found in HUVECs grown on the collagen film. Previous studies have shown that HUVECs grown on collagen gels express MMP-1, −2, and MT1-MMP to degrade the collagen matrix [49,50]. Consequently, MMP expression is higher on stiffer matrix (collagen film) compared with that on softer matrix (denatured collagen film) [51,52]. Given that MMP is secreted as a pro-enzyme (pro-MMP) and only activated MMP can degrade the collagen matrix, immunofluorescent images do not show much about MMP activity [53,54]. The collagen zymography data obtained in this study show that HUVECs on the three different collagen films had relatively different MMP activities. Zymography is an electrophoretic technique to detect hydrolytic enzymes based on the substrate repertoire for the enzyme. Collagen zymography allows us to examine MMP activity, whereas fluorescence images of MMP in Figure 5 distinguish whether the MMP is present or not. Thus, the fluorescence images showing the expression of MMPs in Figure 5 are different from the zymography data showing the activity of MMPs in Figure 6.

HUVECs adhered to the collagen films during collagen zymography, but the experimental results do not reflect the entire endothelium. To overcome this limitation, further collagen zymography experiments using detached ECs (circulating ECs) or MMPs secreted by ECs are required in the future.

All the experiments were performed by culturing ECs on the collagen films for up to 24 h. At 24 h after seeding of ECs, HUVECs clustered on the denatured collagen films gradually detached from the substrate. Given the difficulty in quantitatively analyzing the cells because of their complexity, we only carried out the experiments for up to 24 h.

Among several adhesion molecules, CD146 expression significantly increased on the EC surface of the denatured collagen film compared with that on ECs cultured on normal collagen film. The adhesion molecules of ECs possibly modulated gene expression by changing the morphological structures of the ECM by MMP-1 and MMP-2. These molecules may activate many signaling molecules in focal adhesion and cytoplasm by dynamically interacting with ECM proteins. The denatured ECM may cause endothelial dysfunction by mechanotransduction mechanisms. CD146 is a marker for circulating endothelial cells. The increased levels of CD146 are positively correlated with active inflammatory reactions in idiopathic myopathy [55], chronically inflamed tissues [56], inflammatory skin disease [57], rheumatoid arthritis [58], inflammatory bowel disease [59,60], chronic obstructive pulmonary disease [61], and multiple sclerosis disease [62,63]. The engagement of CD146 initiates the protein kinase phosphorylation cascade through association with Fyn, a Src family kinase. Phosphorylated Fyn subsequently transfers phosphate to the downstream kinase of PKC-γ, which triggers a Ca^{2+} burst within cells. Consequently, the induced association among proteins of P130, Pyk2, and paxillin, as well as activated p125 (FAK), promotes the polarization actions of actins. Thus, this CD146-mediated signaling pathway can be used to decipher the mechanism that CD146 promotes cell motility through extracellular signals to downstream-signaling components for cytoskeleton remodeling [64,65]. The integrins of ECs may activate many signaling molecules in focal adhesion and cytoplasm via dynamic interaction with collagen fibers. Therefore, the mechanical properties of denatured collagen fibers may contribute to the preferential EC migration on the denatured collage film. For better understanding of its function in signaling transduction, further study on the crosstalk with members of various signaling pathways is required [66].

The functional assessment of ECs can be used to identify endothelial damage and predict cardiovascular risks. However, this type of assessment does not provide sufficient information about the mechanisms underlying the development of endothelial dysfunction [67,68].

In this study, the flow rates used for bead accumulation experiment do not fully reflect the *in vivo* pathophysiological conditions. To select effective drugs for circulatory vascular diseases, the size and surface properties of the test drugs and flow condition (flow rate or low endothelial shear stress) of microfluidic devices are highly important. In particular, the physical dimensions of the microfluidic devices used in the present *in vitro* study should be optimized based on data obtained from *in vivo* disease models to maximize the throughput.

Conclusions

Using an *in vitro* disease model, we observed the endothelial dysfunction, such as disorganization and thickening, of the EC layer. HUVECs cultured on normal collagen films were thinly adhered to the substrate. By comparison, HUVECs cultured on denatured collagen films exhibited abnormal cell growth, such as aggregation and cluster formation of ECs. The expression levels of MMP-1, MMP-2, and CD146 increased in ECs on the denatured collagen film. These results indicate that the denatured ECM was closely related to endothelial dysfunction, which is one of the main pathogeneses of circulatory vascular diseases. The morphological structures of the denatured collagen film likely caused EC deformation. In addition, more fluorescent beads accumulated on the EC layer on the denatured collagen film. These *in vitro* studies would be useful for understanding the outbreak mechanism of circulatory vascular disorders. *In vitro* studies using a bio-inspired ECM could contribute in developing a unique experimental modality for optimizing the design parameters to screen effective therapeutic drugs, and determine the physical dimensions and flow rate condition of a microfluidic device used for clinical treatments of circulatory diseases by matching *in vivo* pathophysiological conditions.

Methods

Denatured collagen film

The substrates of 18 mm round cover glasses (Marlenfeld GmbH & Co. KG, Lauda-Königshofen, Germany) were coated with silicon nitride (coating thickness = 200 nm). The cover slips were then treated with reactive O_2 plasma (20 sccm, 200 mTorr, 50 W for 3 min) to simulate the morphological structures of normal collagen films. A methyl-terminated SAM was created at room temperature by incubating the substrate overnight with 3.78 μL of OTS mixed with 10 mL of anhydrous toluene solution. The samples were washed with toluene solution, and baked in an oven at 90°C for 20 min. The substrate was cleaned by sonication in toluene solution for 1 min, washed with 70% ethanol, and dried under nitrogen gas stream. To prepare the collagen films on a glass surface, we diluted 1 mg/mL collagen type I (from rat tails; Invitrogen, Carlsbad, CA, USA) to a final concentration of 0.32 mg/mL using PBS (pH 7.4) solution, which contained calcium and magnesium; the resulting mixture was incubated overnight with previously prepared methyl-terminated glass at 37°C [69]. The samples were then washed with PBS solution (without calcium and magnesium). Collagenase type I (from *Clostridium histolyticum*; Sigma-Aldrich) was diluted to a final concentration of 2 mg/mL using PBS solution containing calcium and magnesium. The prepared collagen film surface was treated for 5 or 15 min to form denatured collagen films on a glass surface. The collagen and denatured collagen-conjugated glass surfaces were immediately washed with PBS solution (without calcium and magnesium). For SEM imaging, both collagen films were rinsed with deionized water and dried with nitrogen gas for the analysis of structural properties. The samples were mounted on metal stubs and coated with platinum (SC7640 model, Quorum Technology). SEM images were captured using SEM (JEOL JSM-7401 F, Japan) at an acceleration voltage of 15 kV. Surface structures of the two films were also observed by AFM (VECO Dimension 3100). The three-dimensional (3D) reconstruction of AFM images was performed using Nanoscope V (Version 7.0) software. Based on a previous study by Elliot et al., the collagen thin film coated on the OTS-SAM-coated glass is amenable for use in engineering applications and for manipulating and studying cell behaviors. This reproducible and analytically tractable collagen matrix provides a unique tool for the systematical examination of matrix organization and composition [69].

Cell preparation

All experimental procedures complied with the act on life ethics and safety of the Ministry of Health and Welfare of South Korea. HUVECs were obtained from Invitrogen (Carlsbad, CA, USA) and used in passages 5 or 7. The cells were cultured in Medium 200 (Gibco) with low-serum growth supplement (Gibco) containing 20% fetal bovine serum (Gibco) and 1% penicillin-streptomycin (HyClone). The cells were cultured in a T75 flask (Falcon) under a humidified atmosphere of 5% CO_2 at 37°C. Trypsin/ethylenediaminetetraacetic acid solution was used to detach cells from the culture flask. The surface-treated cover glass (collagen film or denatured collagen film) was placed in a cell culture dish (diameter = 35 mm), loaded with 10^6 cells/mL cell suspension, and incubated under a humidified 5% CO_2 atmosphere at 37°C. Figures 1A and 1B illustrate the ECs cultured on the collagen and denatured collagen films, respectively. HUVECs were cultured on the collagen film or denatured collagenase-treated collagen film for 1, 3, 6, 12, and 24 h. For DHM imaging, the test samples were fixed with 4% paraformaldehyde solution for 20 min at room temperature, and washed with PBS solution. Each sample was placed on a drop of PBS solution at the center of a polydimethylsiloxane chamber. Sample images were photographed at room temperature.

Phase-contrast holographic microscopy

He-Ne (λ = 633 nm) laser was used as a light source of an upright microscope (Eclipse i50, Nikon) equipped with a 20× objective lens (NA = 0.5). Using additional relay optics, we set the overall magnification of the microscope at approximately 30×. Interferograms were recorded using a charge-coupled device (CCD) camera (PCO 2000, 7.4 μm/pixel) of 2 k × 2 k pixel resolution. The transmission-type phase-contrast DHM applied in this study is based on the principle of the Mach-Zehnder interferometer. After passing through a spatial filter, a laser beam was divided into an object beam and reference beam by a beam splitter. The sample was illuminated by the object beam via a condenser lens. A microscope objective (MO) lens caused the wavefront in the object beam to form a curved line, which resulted in the deformation of the object wave phase. The objective beam was collimated using a collimation lens to compensate the phase aberrations caused by MO. The collimated object wave interfered with the collimated reference wave at a slightly tilted incidence angle to generate holograms.

Morphology and surface roughness of EC layer

An angular spectrum algorithm was employed to reconstruct holograms [43,70,71]. This technique does not require a minimum distance between the object plane and hologram plane. The angular spectrum algorithm is flexible and can be effectively used to filter the frequency domain with high accuracy. If $E_0(x_0, y_0; 0)$ is the wave field at plane $z = 0$, the angular spectrum $A(k_x, k_y; 0)$ of the hologram at $z = 0$ is obtained by Fourier transformation:

$$A(k_x, k_y; 0) = \iint E_0(x_0, y_0; 0) \exp\left[-i(k_x x_0 + k_y y_0)\right] dx_0 dy_0 \tag{1}$$

where k_x and k_y are the spatial frequencies of the x and y components, respectively.

The separated zero-order image, virtual image, and real image were clearly observed in the frequency domain. Fourier-domain filtering was applied to the angular spectrum to eliminate unwanted noises, as well as the zero-order and virtual images. A region of interest in the real image was selected using Fourier-domain filtering. The actual image was shifted to the center of the frequency domain. The complex wave field, reconstructed at any plane perpendicular to the propagating z-axis, was then calculated using the following equation:

$$E(x, y; z) = \Im^{-1}\{filter[\Im\{E_0\}] \exp[ik_z z]\}, \tag{2}$$

$$k_z = \sqrt{k^2 - k_x{}^2 - k_y{}^2}$$

where $filter[\Im\{E_0\}]$ denotes the filtered angular spectrum, and $\Im$ and $\Im^{-1}$ are the Fourier transform and inverse Fourier transform, respectively. The reconstructed wave field $E(x,y;z)$ is represented by an array of complex numbers. The amplitude-contrast images were obtained by calculating the intensity distribution as follows:

$$I(m, n) = \mathrm{Re}[E(x, y; z)]^2 + \mathrm{Im}[E(x, y; z)]^2 \tag{3}$$

and the phase-contrast image was derived from the following argument values:

$$\phi(m, n) = \arctan\left\{\frac{\mathrm{Im}[E(x, y; z)]}{\mathrm{Re}[E(x, yl; z)]}\right\} \tag{4}$$

The 2π ambiguity encountered in the wrapped phase image was resolved by the unwrapping process [72]. The physical thickness of cells was calculated using the following equation:

$$d = \lambda(\phi/2\pi)/(n - n_0) \tag{5}$$

where λ is the wavelength, ϕ is the unwrapped phase angle, and $(n\text{-}n_0)$ is the refractive index difference between the cells and surrounding medium.

The hologram images of the field-of-view of approximately 370 μm × 370 μm were captured using a CCD camera with 1600 × 1600 pixels. Considering that the refractive index depends on principal components and different physical states, we discovered that the local refractive index was slightly different in the EC layer. The mean refractive index of ECs tested in this study was approximately 1.375. The stability of the mean refractive index corresponds to the vertical sensitivity of approximately 11 nm [42,73,74]. Assuming that the total refractive index of the cells is constant, we applied Equation (5) to extract information about physical thickness from the original phase image. The obtained 3D phase image clearly shows the physical thickness distribution of the EC layer with high spatial resolution.

The 3D phase images shown in Figures 2D and 2E illustrate the physical thickness distribution of the EC layer with high spatial resolution. Variations in the surface roughness of HUVECs were described mathematically using root mean square (RMS). The surface roughness of cells was calculated using the following equation:

$$x_{rms} = \sqrt{\frac{1}{n}\sum M^2} \tag{6}$$

where n is the number of samples, and M is the thickness of the EC layer.

Dynamic behavior of ECs

To observe the dynamic behavior of cell-to-cell interaction, we visualized HUVECs cultured on collagen or collagenase-treated collagen film using CLSM (OLYMPUS FV-1000). HUVECs' membranes were labeled with a PKH26 red fluorescent cell linker kit (Sigma-Aldrich) and PKH67 green fluorescent cell linker kit (Sigma-Aldrich) according to the manufacturer's protocol.

The green- and red-fluorescence-labeled HUVECs were mixed at a volume ratio of 1:1. These cells were seeded on collagen film, and incubated under humidified 5% CO_2 medium at 37°C. To maintain the standard tissue culture condition for ECs in the flow chamber, we placed all the components, except the syringe pump, in a chamber (Chamlide TC, Live Cell Instrument, Korea) in which the temperature and CO_2 concentration were controlled. The cell culture medium was continuously circulated (1 mL/h) over the cells by a syringe pump (KD Scientific). ECs were cultured on the collagen film or denatured collagenase-treated collagen film for 3 h, and mounted in the flow chamber. Cell movements were observed for 90 min at an interval of 10 min under CLSM.

Growth properties of ECs

The growth properties of HUVECs plated on the treated and untreated collagen films were analyzed using the Apoptosis, DNA damage, and Cell Proliferation (ADDCP) kit (BD Biosciences, USA) and FACS technique. The ADDCP kit contains key markers [bromodeoxyuridine (BrdU), H2AX, and PARP] for the simultaneous determination of important cellular state changes in the cell cycle, such as apoptosis, DNA damage, and cell proliferation. BrdU is an analog of the DNA precursor thymidine. When cells are incubated in the presence of BrdU, the molecule is incorporated into newly synthesized DNA and can be detected with antibodies against BrdU [75]. DNA damage was determined using Phosphorylated H2AX. H2AX is a member of the histone H2A protein family. Phosphorylation of H2AX leads to the recruitment of DNA damage repair proteins at the site of DNA damage [76]. Detection of cleaved PARP is used for the study of apoptosis. During the early phases of apoptosis (programmed cell death), caspase-3 is activated by cleavage [77].

HUVECs were cultured on the collagen film or denatured collagen film for 6, 12, 24, and 36 h. Biological analyses on apoptosis, DNA damage, and cell proliferation were performed according to the manufacturer's protocol. The test samples were then analyzed using a BD FACSCanto II flow cytometer (BD Biosciences, USA). The acquired data were handled with FlowJo (Version 10.0.6) flow cytometry analysis software.

MMP-1 and MMP-2 of ECs

MMP-1 (BC013875.2) and MMP-2 (BC002576.2) in the gateway entry vector were tagged with EGFP or red fluorescent protein (mCherry) by combining the MMPs with gateway destination vector (pDS-XB-EGFP or pDS-XB-mCherry) using LR recombinase (Invitrogen). MMP-1 or MMP-2 vector was mixed with FuGENE HD transfection reagent (Promega). HUVECs were then tagged with MMP-1/EGFP and MMP-2/mCherry. To visualize the spatial distribution of MMP-1 or MMP-2 vector on HUVECs, we cultured HUVECs on collagen or denatured collagenase-treated collagen films for 24 h, and fixed the cells with 4% paraformaldehyde solution for 20 min at room temperature. ECs were then stained with DAPI (Invitrogen) for nuclear staining. The stained EC images were obtained using CLSM (TCS SP5II MP Leica Microscopy Systems, GMBH) with a 20× IR APO water-immersion lens (NA 1, Leica Microscopy Systems, GMBH). The acquired images were analyzed and processed using LAS AF 2.7 software (Leica Microscopy Systems, GMBH). MMP2 has difficulty in discriminating the difference in fluorescence intensity of ECs on the collagen and denatured collagen films because of their weak fluorescence intensities. Therefore, we further improved the contrast and brightness of fluorescence images using Image J software.

Collagen zymography

Collagen zymography assay was performed to detect MMP-1, −2, −9, and −13 [78,79]. To investigate MMP activity on the three different collagen films, samples were centrifuged to dispose cellular debris and then homogenized without a reducing agent to retain the native state of the enzyme. Samples were centrifuged once again at 16,000 g for 20 min at 4°C. The protein concentration of the collected supernatant was measured by a Nanodrop Spectrophotometer (Thermo Scientific, Wilmington, DE, USA). The samples were mixed with an equal volume of 2× non-reducing sample buffer (Santa Cruz Technology, Santa Cruz, CA, USA), and 15–20 μL was loaded in each well. Polyacrylamide gels containing 10 mL of type I collagen (from rat tails; 1 mg/mL, Invitrogen, Carlsbad, CA, USA) were dispersed in a buffered solution consisting of 2.5 mL of gel, 1.5 M Tris–HCl (pH 8.8), 165 μL of 10% SDS, 5.25 mL of 40% polyacrylamide, 165 μL of 50% glycerol, and 4 mL of distilled water. The stacking gel containing 4% polyacrylamide in 1 M Tris–HCl (pH 6.8) was polymerized by adding 100 μL of 10% ammonium persulfate and 10 μL of TEMED. The gel was then electrophoresed at a constant voltage of 90 V. The electrophoresed gel was washed twice in 200 mL of 2.5% Triton X-100 (30 min each) under shaking, and incubated in 100 mM Tris–HCl, 5 mM $CaCl_2$, 0.005% Brij-35, and 0.001% NaN_3 (pH 8.0) for 6–48 h at 37°C. The gel was stained with 0.25% Coomassie brilliant blue G-250 (50% methanol, 10% acetic acid) for 1 h at room temperature and destained (40% methanol, 10% acetic acid). The gels were incubated for 1 h in 5% methanol and 7.5% acetic acid, and kept under cellophane at 4°C. The stained gel images were obtained using a color scanner. The density of MMP bands in the acquired images was analyzed using Quantity One software (Bio-Rad Laboratories, Hercules, CA, USA).

Immunostaining of ECs

HUVECs were cultured on collagen film or denatured collagenase-treated collagen film for 24 h, and fixed with 4% paraformaldehyde solution for 20 min at room temperature to visualize the spatial distribution of adhesion molecules on HUVECs. The antibodies (5 μg/mL) for VCAM, E-selectin, CD146, and VE-cadherin (R&D Systems, USA) were overlaid for 1 h on the EC surface and washed. The secondary antibody (10 μg/mL) conjugated with Alexa Fluor 488 or 594 (Invitrogen) was added again for 1 h and rinsed. ECs were then stained with phalloidin tetramethylrhodamine isothiocyanate (Sigma-Aldrich) or phalloidin-fluorescein (Sigma-Aldrich) for actin staining and with DAPI (Invitrogen) for nuclear staining. The stained EC images were obtained using CLSM (TCS SP5II MP Leica Microscopy Systems, GMBH) with a 20× IR APO water-immersion lens (NA 1, Leica Microscopy Systems, GMBH). The acquired images were analyzed and processed using LAS AF 2.7 software (Leica Microscopy Systems, GMBH).

Bead accumulation on the EC layer

A syringe pump (KDS100: KD Scientific), syringe filled with 1 mL of Medium 200 (microsphere density in medium: 9.1×10^7 particles/mL) mixed with yellow-green fluorescent beads (FluoSpheres Collagen I-Labeled Microspheres, 1.0 μm), and rectangular flow chamber (Chamlide CF, Live Cell Instrument, Korea) were serially connected to silicon tubes (inner diameter/outer diameter = 1.15 mm/ 3.2 mm, Korea Ace) to examine the effect of flow on bead accumulation over HUVEC layers. The test channel in the flow chamber had a height, width, and length of 0.2, 2, and 14 mm, respectively. To maintain the standard tissue-culture conditions (37°C, 5% CO_2) in the flow chamber for ECs, we placed the components, except the syringe pump, in the temperature and CO_2 control chamber (Chamlide TC, Live Cell Instrument, Korea). The ECs cultured on collagen film or denatured collagenase-treated collagen film for 24 h were mounted in the flow chamber. Dynamic motions of the motile beads in the flow chamber were observed using CLSM (OLYMPUS FV-1000) while recording the bead accumulation process on the HUVEC layers. A modified Zeiss Axiovert 200 fluorescence microscope with a 20× (NA = 0.4) phase-contrast objective lens and AxioCam MRc CCD camera was used to monitor the temporal variation in bead accumulation on the HUVEC layers. An X-Cite 120 Q excitation light source (120 W mercury vapor short arc lamp) and shift-free filter set for GFP/RFP were used in the fluorescence imaging experiment. The microscope was operated by Axiovision 4.8.2 (Carl Zeiss). Captured images were analyzed and processed using Image-Pro Plus 7.0 software. The dynamic motions of the motile beads inside the flow chamber were observed by time-lapse microscopy for 10 min. Images were consecutively recorded at an interval of 1 s.

Abbreviations

ADDCP: Apoptosis, DNA damage, and cell proliferation; AFM: Atomic force microscopy; BrdU: Bromodeoxyuridine; CCD: Charge-coupled device; DAPI: 4',6-diamidino-2-phenylindole; DHM: Digital holographic microscopy; EC: Endothelial cell; ECM: Extracellular matrix; EGFP: Enhanced green fluorescent protein; CLSM: Confocal laser scanning microscopy; FACS: Fluorescence-activated cell sorting; HUVECs: Human umbilical vein endothelial cells; MMPs: Matrix-degrading metalloproteinases; MO: Microscope objective; OTS: Octadecyltrichlorosilane; RMS: Root mean square; SEM: Scanning electron microscopy; 3D: Three-dimensional; VCAM: Vascular cell adhesion molecule; VE-cadherin: Vascular endothelial cadherin.

Competing interests

The authors declare that they have no competing interests.

Authors' contributions

ES: design of cell biology experiments, data collection, and data analysis; KWS: experimental set-up and analysis of DHM; JEG: cell biology experiments; YRH: collagen zymography; EY: data analysis; SL: data collection and analysis; SJL: experimental design. All authors read and approved the final manuscript.

Acknowledgements

The authors are grateful for the valuable help in the confocal imaging experiments performed at the POSTECH *in vivo* Microscopy Laboratory of the POSTECH Biotech Center. This study was supported by the National Research Foundation of Korea (NRF) by the Korean government (MSIP; Grant No. 2008–0061991).

Author details

[1]Division of Integrative Biosciences and Biotechnology, Pohang University of Science and Technology, San 31, Hyoja-dong, Nam-Gu, Pohang, Gyeongbuk 790-784, Korea. [2]Center for Biofluid and Biomimic Research, Department of Mechanical Engineering, Pohang University of Science and Technology, San 31, Hyoja-dong, Nam-Gu, Pohang, Gyeongbuk 790-784, South Korea. [3]Department of Biological Sciences, Center for Cognition and Sociality, Institute for Basic Science, 291 Daehak-ro, Yuseong-gu, Daejeon 305-701, South Korea.

References

1. Feldman CL, Coskun AU, Yeghiazarians Y, Kinlay S, Wahle A, Olszewski ME, Rossen JD, Sonka M, Popma JJ, Orav J, Kuntz RE, Stone PH: **Remodeling characteristics of minimally diseased coronary arteries are consistent along the length of the artery.** *Am J Cardiol* 2006, **97:**13–16.
2. Iademarco MF, McQuillan JJ, Rosen GD, Dean DC: **Characterization of the promoter for vascular cell adhesion molecule-1 (VCAM-1).** *J Biol Chem* 1992, **267:**16323–16329.
3. Schindler U, Baichwal VR: **Three NF-kappa B binding sites in the human E-selectin gene required for maximal tumor necrosis factor alpha-induced expression.** *Mol Cell Biol* 1994, **14:**5820–5831.
4. Collins T, Read MA, Neish AS, Whitley MZ, Thanos D, Maniatis T: **Transcriptional regulation of endothelial cell adhesion molecules: NF-kappa B and cytokine-inducible enhancers.** *FASEB J* 1995, **9:**899–909.
5. Read MA, Whitley MZ, Gupta S, Pierce JW, Best J, Davis RJ, Collins T: **Tumor necrosis factor alpha-induced E-selectin expression is activated by the nuclear factor-kappa B and c-JUN N-terminal kinase/p38 mitogen-activated protein kinase pathways.** *J Biol Chem* 1997, **272:**2753–2761.
6. Bardin N, Anfosso F, Masse JM, Cramer E, Sabatier F, Le Bivic A, Sampol J, Dignat-George F: **Identification of CD146 as a component of the endothelial junction involved in the control of cell-cell cohesion.** *Blood* 2001, **98:**3677–3684.
7. Bardin N, Moal V, Anfosso F, Daniel L, Brunet P, Sampol J, Dignat George F: **Soluble CD146, a novel endothelial marker, is increased in physiopathological settings linked to endothelial junctional alteration.** *Thromb Haemost* 2003, **90:**915–920.
8. Zachlederova M, Jarolim P: **Gene expression profiles of microvascular endothelial cells after stimuli implicated in the pathogenesis of vasoocclusion.** *Blood Cells Mol Dis* 2003, **30:**70–81.
9. Zachlederova M, Jarolim P: **The dynamics of gene expression in human lung microvascular endothelial cells after stimulation with inflammatory cytokines.** *Physiol Res* 2006, **55:**39–47.
10. Wojciak-Stothard B, Entwistle A, Garg R, Ridley AJ: **Regulation of TNF-alpha-induced reorganization of the actin cytoskeleton and cell-cell junctions by Rho, Rac, and Cdc42 in human endothelial cells.** *J Cell Physiol* 1998, **176:**150–165.
11. Aepfelbacher M, Essler M: **Disturbance of endothelial barrier function by bacterial toxins and atherogenic mediators a role for Rho/Rho kinase.** *Cell Microbiol* 2001, **3:**649–658.
12. van Buul JD, Voermans C, van den Berg V, Anthony EC, Mul FP, van Wetering S, van der Schoot CE, Hordijk PL: **Migration of human hematopoietic progenitor cells across bone marrow endothelium is regulated by vascular endothelial cadherin.** *J Immunol* 2002, **168:**588–596.
13. Guo M, Wu MH, Granger HJ, Yuan SY: **Focal adhesion kinase in neutrophil-induced microvascular hyperpermeability.** *Microcirculation* 2005, **12:**223–232.

14. Galley HF, Webster NR: **Physiology of the endothelium.** *Br J Anaesth* 2004, **93:**105–113.
15. Halliday I, Atherton M, Care CM, Collins MW, Evans D, Evans PC, Hose DR, Khir AW, Konig CS, Krams R, Lawford PV, Lishchuk SV, Pontrelli G, Ridger V, Spencer TJ, Ventikos Y, Walker DC, Watton PN: **Multi-scale interaction of particulate flow and the artery wall.** *Med Eng Phys* 2011, **33:**840–848.
16. Widlansky ME, Gokce N, Keaney JF Jr, Vita JA: **The clinical implications of endothelial dysfunction.** *J Am Coll Cardiol* 2003, **42:**1149–1160.
17. Cernuda-Morollon E, Ridley AJ: **Rho GTPases and leukocyte adhesion receptor expression and function in endothelial cells.** *Circ Res* 2006, **98:**757–767.
18. Murikipudi S, Methe H, Edelman ER: **The effect of substrate modulus on the growth and function of matrix-embedded endothelial cells.** *Biomaterials* 2013, **34:**677–684.
19. Birkedal-Hansen H, Moore WG, Bodden MK, Windsor LJ, Birkedal-Hansen B, DeCarlo A, Engler JA: **Matrix metalloproteinases: a review.** *Crit Rev Oral Biol Med* 1993, **4:**197–250.
20. Woessner JF Jr: **The family of matrix metalloproteinases.** *Ann N Y Acad Sci* 1994, **732:**11–21.
21. Faia KL, Davis WP, Marone AJ, Foxall TL: **Matrix metalloproteinases and tissue inhibitors of metalloproteinases in hamster aortic atherosclerosis: correlation with in-situ zymography.** *Atherosclerosis* 2002, **160:**325–337.
22. Nagase H: **Activation mechanisms of matrix metalloproteinases.** *Biol Chem* 1997, **378:**151–160.
23. Newby AC: **Dual role of matrix metalloproteinases (matrixins) in intimal thickening and atherosclerotic plaque rupture.** *Physiol Rev* 2005, **85:**1–31.
24. Oda N, Abe M, Sato Y: **ETS-1 converts endothelial cells to the angiogenic phenotype by inducing the expression of matrix metalloproteinases and integrin beta3.** *J Cell Physiol* 1999, **178:**121–132.
25. Nguyen M, Arkell J, Jackson CJ: **Activated protein C directly activates human endothelial gelatinase A.** *J Biol Chem* 2000, **275:**9095–9098.
26. Hummel V, Kallmann BA, Wagner S, Fuller T, Bayas A, Tonn JC, Benveniste EN, Toyka KV, Rieckmann P: **Production of MMPs in human cerebral endothelial cells and their role in shedding adhesion molecules.** *J Neuropathol Exp Neurol* 2001, **60:**320–327.
27. Dobrin PB: **Mechanical properties of arterises.** *Physiol Rev* 1978, **58:**397–460.
28. Glasser SP, Arnett DK, McVeigh GE, Finkelstein SM, Bank AJ, Morgan DJ, Cohn JN: **Vascular compliance and cardiovascular disease: a risk factor or a marker?** *Am J Hypertens* 1997, **10:**1175–1189.
29. Ross R: **The pathogenesis of atherosclerosis: a perspective for the 1990s.** *Nature* 1993, **362:**801–809.
30. Kelly R, Hayward C, Avolio A, O'Rourke M: **Noninvasive determination of age-related changes in the human arterial pulse.** *Circulation* 1989, **80:**1652–1659.
31. McVeigh GE, Morgan DJ, Finkelstein SM, Lemay LA, Cohn JN: **Vascular abnormalities associated with long-term cigarette smoking identified by arterial waveform analysis.** *Am J Med* 1997, **102:**227–231.
32. Lehmann ED, Gosling RG, Sonksen PH: **Arterial wall compliance in diabetes.** *Diabet Med* 1992, **9:**114–119.
33. Heagerty AM, Aalkjaer C, Bund SJ, Korsgaard N, Mulvany MJ: **Small artery structure in hypertension: Dual processes of remodeling and growth.** *Hypertension* 1993, **21:**391–397.
34. Libby P, Theroux P: **Pathophysiology of coronary artery disease.** *Circulation* 2005, **111:**3481–3488.
35. Bank AJ, Wang H, Holte JE, Mullen K, Shammas R, Kubo SH: **Contribution of collagen, elastin, and smooth muscle to in vivo human brachial artery wall stress and elastic modulus.** *Circulation* 1996, **94:**3263–3270.
36. Pandit A, Lu X, Wang C, Kassab GS: **Biaxial elastic material properties of porcine coronary media and adventitia.** *Am J Physiol Heart Circ Physiol* 2005, **288:**H2581–H2587.
37. Stoletov K, Fang L, Choi SH, Hartvigsen K, Hansen LF, Hall C, Pattison J, Juliano J, Miller ER, Almazan F, Crosier P, Witztum JL, Klemke RL, Miller YI: **Vascular lipid accumulation, lipoprotein oxidation, and macrophage lipid uptake in hypercholesterolemic zebrafish.** *Circ Res* 2009, **104:**952–960.
38. Mann CJ, Yu LF, Lo CM, Kim MK: **High-resolution quantitative phase-contrast microscopy by digital holography.** *Opt Express* 2005, **13:**8693–8698.
39. Marquet P, Rappaz B, Magistretti PJ, Cuche E, Emery Y, Colomb T, Depeursinge C: **Digital holographic microscopy: a noninvasive contrast imaging technique allowing quantitative visualization of living cells with subwavelength axial accuracy.** *Opt Lett* 2005, **30:**468–470.
40. Rappaz B, Marquet P, Cuche E, Emery Y, Depeursinge C, Magistretti P: **Measurement of the integral refractive index and dynamic cell morphometry of living cells with digital holographic microscopy.** *Opt Express* 2005, **13:**9361–9373.
41. Kim MK, Yu LF, Mann CJ: **Interference techniques in digital holography.** *J Opt A: Pure Appl Opt* 2006, **8:**S518–S523.
42. Shih LM, Hsu MY, Palazzo JP, Herlyn M: **The cell-cell adhesion receptor Mel-CAM acts as a tumor suppressor in breast carcinoma.** *Am J Pathol* 1997, **151:**745–751.
43. Ross R: **Atherosclerosis is an inflammatory disease.** *Am Heart J* 1999, **138:**S419–S420.
44. Solovey AN, Gui L, Chang L, Enenstein J, Browne PV, Hebbel RP: **Identification and functional assessment of endothelial P1H12.** *J Lab Clin Med* 2001, **138:**322–331.
45. Ramcharan KS, Lip GY, Stonelake PS, Blann AD: **The endotheliome: a new concept in vascular biology.** *Thromb Res* 2011, **128:**1–7.
46. Corada M, Liao F, Lindgren M, Lampugnani MG, Breviario F, Frank R, Muller WA, Hicklin DJ, Bohlen P, Dejana E: **Monoclonal antibodies directed to different regions of vascular endothelial cadherin extracellular domain affect adhesion and clustering of the protein and modulate endothelial permeability.** *Blood* 2001, **97:**1679–1684.
47. Decuzzi P, Godin B, Tanaka T, Lee SY, Chiappini C, Liu X, Ferrari M: **Size and shape effects in the biodistribution of intravascularly injected particles.** *J Control Release* 2010, **141:**320–327.
48. Decuzzi P, Ferrari M: **The adhesive strength of non-spherical particles mediated by specific interactions.** *Biomaterials* 2006, **27:**5307–5314.
49. Saunders WB, Bohnsack BL, Faske JB, Anthis NJ, Bayless KJ, Hirschi KK, Davis GE: **Coregulation of vascular tube stabilization by endothelial cell TIMP-2 and pericyte TIMP-3.** *J Cell Biol* 2006, **175:**179–191.
50. Stratman AN, Saunders WB, Sacharidou A, Koh W, Fisher KE, Zawieja DC, Davis MJ, Davis GE: **Endothelial cell lumen and vascular guidance tunnel formation requires MT1-MMP-dependent proteolysis in 3-dimensional collagen matrices.** *Blood* 2009, **114:**237–247.
51. Sieminski AL, Was AS, Kim G, Gong H, Kamm RD: **The stiffness of three-dimensional ionic self-assembling peptide gels affects the extent of capillary-like network formation.** *Cell Biochem Biophys* 2007, **49:**73–83.
52. Hanjaya-Putra D, Wong KT, Hirotsu K, Khetan S, Burdick JA, Gerecht S: **Spatial control of cell-mediated degradation to regulate vasculogenesis and angiogenesis in hyaluronan hydrogels.** *Biomaterials* 2012, **33:**6123–6131.
53. Galis ZS, Sukhova GK, Lark MW, Libby P: **Increased expression of matrix metalloproteinases and matrix degrading activity in vulnerable regions of human atherosclerotic plaques.** *J Clin Invest* 1994, **94:**2493–2503.
54. Romanic AM, White RF, Arleth AJ, Ohlstein EH, Barone FC: **Matrix metalloproteinase expression increases after cerebral focal ischemia in rats: inhibition of matrix metalloproteinase-9 reduces infarct size.** *Stroke* 1998, **29:**1020–1030.
55. Figarella-Branger D, Schleinitz N, Boutière-Albanèse B, Camoin L, Bardin N, Guis S, Pouget J, Cognet C, Pellissier JF, Dignat-George F: **Platelet-endothelial cell adhesion molecule-1 and CD146: soluble levels and in situ expression of cellular adhesion molecules implicated in the cohesion of endothelial cells in idiopathic inflammatory myopathies.** *J Rheumatol* 2006, **33:**1623–1630.
56. Middleton J, Americh L, Gayon R, Julien D, Mansat M, Mansat P, Anract P, Cantagrel A, Cattan P, Reimund JM, Aguilar L, Amalric F, Girard JP: **A comparative study of endothelial cell markers expressed in chronically inflamed human tissues: MECA-79, Duffy antigen receptor for chemokines, von Willebrand factor, CD31, CD34, CD105 and CD146.** *J Pathol* 2005, **206:**260–268.
57. Weninger W, Rendl M, Mildner M, Mayer C, Ban J, Geusau A, Bayer G, Tanew A, Majdic O, Tschachler E: **Keratinocytes express the CD146 (Muc18/S-endo) antigen in tissue culture and during inflammatory skin diseases.** *J Invest Dermatol* 2000, **115:**219–224.
58. Neidhart M, Wehrli R, Brühlmann P, Michel BA, Gay RE, Gay S: **Synovial fluid CD146 (MUC18), a marker for synovial membrane angiogenesis in rheumatoid arthritis.** *Arthritis Rheum* 1999, **42:**622–630.
59. Bardin N, Reumaux D, Geboes K, Colombel JF, Blot-Chabaud M, Sampol J, Duthilleul P, Dignat-George F: **Increased expression of CD146, a new marker of the endothelial junction in active inflammatory bowel disease.** *Inflamm Bowel Dis* 2006, **12:**16–21.
60. Bardin N, Blot-Chabaud M, Despoix N, Kebir A, Harhouri K, Arsanto JP, Espinosa L, Perrin P, Robert S, Vely F, Sabatier F, Le Bivic A, Kaplanski G, Sampol J,

Dignat-George F: **CD146 and its soluble form regulate monocyte transendothelial migration.** *Arterioscler Thromb Vasc Biol* 2009, **29**:746–753.

61. Schulz C, Wolf K, Harth M, Krätzel K, Kunz-Schughart L, Pfeifer M: **Expression and release of interleukin-8 by human bronchial epithelial cells from patients with chronic obstructive pulmonary disease, smokers, and never-smokers.** *Respiration* 2003, **70**:254–261.
62. Dagur PK, Biancotto A, Wei L, Sen HN, Yao M, Strober W, Nussenblatt RB, McCoy JP Jr: **MCAM-expressing CD4(+) T cells in peripheral blood secrete IL-17A and are significantly elevated in inflammatory autoimmune diseases.** *J Autoimmun* 2011, **37**:319–327.
63. Flanagan K, Fitzgerald K, Baker J, Regnstrom K, Gardai S, Bard F, Mocci S, Seto P, You M, Larochelle C, Prat A, Chow S, Li L, Vandevert C, Zago W, Lorenzana C, Nishioka C, Hoffman J, Botelho R, Willits C, Tanaka K, Johnston J, Yednock T: **Laminin-411 is a vascular ligand for MCAM and facilitates TH17 cell entry into the CNS.** *PLoS One* 2012, **7**:e40443.
64. Anfosso F, Bardin N, Francès V, Vivier E, Camoin-Jau L, Sampol J, Dignat-George F: **Activation of human endothelial cells via S-endo-1 antigen (CD146) stimulates the tyrosine phosphorylation of focal adhesion kinase p125(FAK).** *J Biol Chem* 1998, **273**:26852–26856.
65. Anfosso F, Bardin N, Vivier E, Sabatier F, Sampol J, Dignat-George F: **Outside-in signaling pathway linked to CD146 engagement in human endothelial cells.** *J Biol Chem* 2001, **276**:1564–1569.
66. Wang Z, Yan X: **CD146, a multi-functional molecule beyond adhesion.** *Cancer Lett* 2013, **330**:150–162.
67. Bhagat K, Vallance P: **Inflammatory cytokines impair endothelium-dependent dilatation in human veins in vivo.** *Circulation* 1997, **96**:3042–3047.
68. Raza K, Thambyrajah J, Townend JN, Exley AR, Hortas C, Filer A, Carruthers DM, Bacon PA: **Suppression of inflammation in primary systemic vasculitis restores vascular endothelial function: lessons for atherosclerotic disease?** *Circulation* 2000, **102**:1470–1472.
69. Elliott JT, Tona A, Woodward JT IV, Jones PL, Plant AL: **Thin films of collagen affect smooth muscle cell morphology.** *Langmuir* 2003, **19**:1506–1514.
70. Yu LF, Kim MK: **Wavelength scanning digital interference holography for variable tomographic scanning.** *Opt Express* 2005, **13**:5621–5627.
71. Choi YS, Lee SJ: **Three-dimensional volumetric measurement of red blood cell motion using digital holographic microscopy.** *Appl Opt* 2009, **48**:2983–2990.
72. Cuche E, Marquet P, Depeursinge C: **Simultaneous amplitude-contrast and quantitative phase-contrast microscopy by numerical reconstruction of Fresnel off-axis holograms.** *Appl Opt* 1999, **38**:6994–7001.
73. Dunn A, Richards-Kortum R: **Three-dimensional computation of light scattering from cells.** *IEEE J Sel Top Quant* 1996, **2**:898–905.
74. Farinas J, Verkman AS: **Cell volume and plasma membrane osmotic water permeability in epithelial cell layers measured by interferometry.** *Biophys J* 1996, **71**:3511–3522.
75. Li C: **Specific cell cycle synchronization with butyrate and cell cycle analysis.** *Methods Mol Biol* 2011, **761**:125–136.
76. Kuo LJ, Yang LX: **Gamma-H2AX - a novel biomarker for DNA double strand breaks.** *In Vivo* 2008, **22**:305–309.
77. Krishnakumar R, Kraus WL: **The PARP side of the nucleus: molecular actions, physiological outcomes, and clinical targets.** *Mol Cell* 2010, **39**:8–24.
78. Snoek-van Beurden PA, Von den Hoff JW: **Zymographic techniques for the analysis of matrix metalloproteinases and their inhibitors.** *Biotechniques* 2005, **38**:73–83.
79. Gogly B, Groult N, Hornebeck W, Godeau G, Pellat B: **Collagen zymography as a sensitive and specific technique for the determination of subpicogram levels of interstitial collagenase.** *Anal Biochem* 1998, **255**:211–216.

6

The relationship between mTOR signalling pathway and recombinant antibody productivity in CHO cell lines

Raihana Edros, Susan McDonnell and Mohamed Al-Rubeai*

Abstract

Background: High recombinant protein productivity in mammalian cell lines is often associated with phenotypic changes in protein content, energy metabolism, and cell growth, but the key determinants that regulate productivity are still not clearly understood. The mammalian target of rapamycin (mTOR) signalling pathway has emerged as a central regulator for many cellular processes including cell growth, apoptosis, metabolism, and protein synthesis. This role of this pathway changes in response to diverse environmental cues and allows the upstream proteins that respond directly to extracellular signals (such as nutrient availability, energy status, and physical stresses) to communicate with downstream effectors which, in turn, regulate various essential cellular processes.

Results: In this study, we have performed a transcriptomic analysis using a pathway-focused polymerase chain reaction (PCR) array to compare the expression of 84 target genes related to the mTOR signalling in two recombinant CHO cell lines with a 17.4-fold difference in specific monoclonal antibody productivity (q_p). Eight differentially expressed genes that exhibited more than a 1.5-fold change were identified. *Pik3cd* (encoding the Class 1A catalytic subunit of phosphatidylinositol 3-kinase [PI3K]) was the most differentially expressed gene having a 71.3-fold higher level of expression in the high producer cell line than in the low producer. The difference in the gene's transcription levels was confirmed at the protein level by examining expression of p110δ.

Conclusion: Expression of p110δ correlated with specific productivity (q_p) across six different CHO cell lines, with a range of expression levels from 3 to 51 pg/cell/day, suggesting that p110δ may be a key factor in regulating productivity in recombinant cell lines.

Keywords: CHO, mTOR, Monoclonal antibody, Phosphotidylinositol 3-kinase, Cell engineering

Background

The production of biopharmaceutical products is one of the fastest growing areas in the pharmaceutical industry with sales of $99 billion in 2010 [1,2]. Many of these biopharmaceutical products are used to treat critical diseases, like hepatitis, cancer, heart disease, hemophilia, rheumatoid arthritis, and diabetes. Monoclonal antibodies (mAbs) currently constitute more than 30% of total biopharmaceuticals and the portfolio of products is expected to grow between 7-15% per annum in the coming years. Whilst the future does looks bright for this sector, the emergence of biosimilar competition is ensuring that manufacturers of these products are continuously seeking methods of increasing productivity of cell culture processes to maximise yields.

Numerous attempts have been made to generate cell lines that can produce large amounts of recombinant products. Among the cell lines available for mAb production are myeloma, hybridoma, and Chinese hamster ovary (CHO) cell lines [3,4]. The use of CHO cells in large-scale production is common [3,5-7] because of their ability to express high recombinant protein levels [8,9], grow to high cell densities [10-13], and to grow in serum-free suspension culture [6,14-16]. CHO cells are also suitable for use with expression systems, such as

* Correspondence: m.al-rubeai@ucd.ie
School of Chemical and Bioprocess Engineering and Conway Institute of Biomolecular and Biomedical Research, University College Dublin, Dublin, Ireland

dihydrofolate reductase (DHFR) and glutamine synthetase (GS) [17-19].

High productivity is usually accompanied by slower growth rates [20-25], and generating high-producing cell lines with high growth rates should be advantageous to further improve mAb production. However, maintaining cell-specific productivity at high growth rates is challenging because both processes require high energy. Glutamine and glucose are often consumed in the synthesis of precursors for recombinant proteins rather than in cell proliferation [26], which eventually leads to growth repression.

The correlation between cell volume (size) and productivity has been clearly demonstrated [27] and is reflected in the phenotypic changes in protein content and cell biomass [13,28,29]. This increase in protein content may lead to an increase in cell size [25,29,30]. The integration of both protein synthesis and cell proliferation in response to diverse environmental cues has been reported to be regulated by a signalling pathway known as the mammalian target of rapamycin (mTOR) signalling pathway [31,32].

As shown in Figure 1, upon receiving signals from growth factors and hormones, mTOR is activated by Akt/P13K signalling [33,34]. P13K catalyses the formation of phosphoinositides-3-phosphate (PIP3) from phosphoinositides-2-phosphate (PIP2), which then activates a protein kinase known as Akt [35]. The activated mTOR initiates protein synthesis via two principal pathways; firstly through the inhibition of eukaryotic initiation factor 4E binding protein (4EBP1) and secondly through the activation of ribosomal protein S6 kinase (S6K) [36,37]. These downstream effectors activate protein synthesis by activating translation initiation and elongation factors, including eIF4E, eIF4B, S6, and eEF2 [32,38].

In addition to growth factors and hormones, mTOR also receives signals from phosphatidic acid (PA) [39] and amino acids [38,40,41]. In the presence of serum, protein kinase C (PKC) is activated, and it sequentially activates phospholipase D (PLD) which is responsible for converting PA to phosphatidylcholine (PC) [39]. In contrast, the activation of mTOR by amino acids is unclear [32]. It was reported to be regulated by the Ras protein [41] and/or intracellular amino acids [38,42]. Moreover, adenosine monophosphate kinase (AMPK) can also precisely modulate mTOR activity based on the adenosine monophosphate (AMP) to adenosine triphosphate (ATP) ratio [43]. In the presence of high intracellular ATP levels, AMPK can be deactivated, leading to an increase in protein synthesis [44,45].

A recent study highlighted the role of mTOR in improving viability, cell growth, antibody production, and robustness of the CHO cell line [46]. However, the mechanism underlying these improvements has not yet been elucidated. We believe that using a transcriptomic approach, such as the pathway-focused polymerase chain reaction (PCR) array, which has proven to be useful in a similar study [47], we can elucidate the connections between mTOR, cell growth, and productivity. Thus, this study initially employed two recombinant CHO cell lines with a 17.4-fold difference in specific productivity to

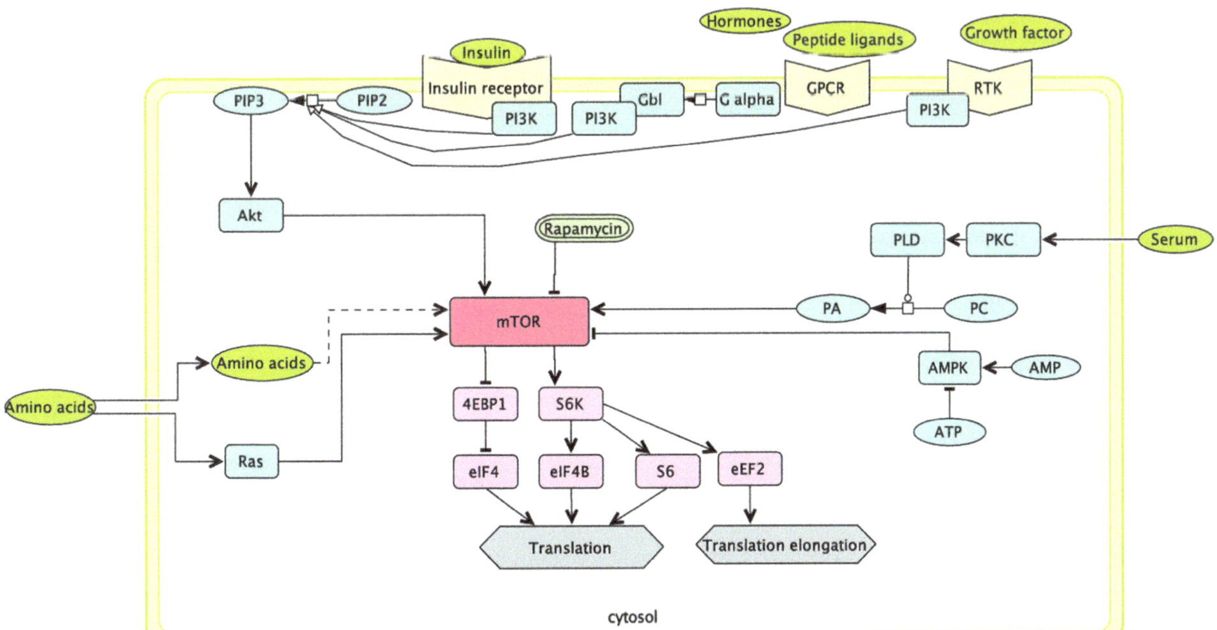

Figure 1 The mTOR pathway signaling pathway. This pathway involves communication between the upstream proteins that respond directly to such extracellular signals as nutrient availability, energy status, and physical stresses.

study the differential expressions of genes related to the mTOR signaling pathway. High productivity has been associated partly with enhanced translation mechanisms [48]. We therefore expected to see upregulation of genes encoding mTOR upstream regulators such as, Akt [49,50], PI3K [51,52], PLD [39,53], and Ras-related GTP binding proteins (RagC) [41,54,55], as well as downstream effectors involved in translation mechanisms such as S6 and S6K [56-59]. Since high gene expression does not necessarily mean that the corresponding protein is also highly upregulated in the cell, we decided to perform Western blot analysis to confirm that the selected potential marker p110δ was expressed in increasing amount in a panel of 6 CHO cell lines with varying levels of productivity.

Methods

Cell lines and maintenance

This study employed a panel of six GS-CHO cell lines (referred to as CL38, CL47, CL76, CL150, CL160 and CL164) producing cB72.3 IgG4 monoclonal antibody with varying levels of productivity. The cell lines were kindly provided by Lonza Biologics (Slough, UK). The cell lines were generated by transfection of the suspension-variant derivatives CHO-K1 (CHOK1SV) with the glutamine synthetase (GS) expression vector pcB72.3 containing light chain (LC) and heavy chain (HC) cDNA, each driven by the hCMV-MIE promoter [60,61]. Cells were maintained in CD-CHO medium supplemented with 25 μM MSX. Using three biological replicates, our experiments were conducted in 125 ml shake flasks (SCHOTT Inc., Elmsford, USA) at 37°C and 140 rpm agitation.

The six cell lines were numbered in descending order according to their specific productivity, with [1] referring to the most productive cell line and [6] referring to the least productive cell line. These cell lines were thus referred to as CL47[1], CL76[2], CL150[3], CL164[4], CL38 [5] and CL160[6] throughout this study . Based on specific productivity, CL47[1] was categorised as the high producer; CL76[2], CL150[3], CL164 [4], and CL38[5], as medium producers; and CL160[6], as the low producer.

Flow cytometric analysis of cell number and viability

Cell suspensions (490 μl) were removed from flasks and placed in a flow cytometry tube and 10 μl propidium iodide (PI) solution added from the stock solution (0.5 μg/ml). Samples were mixed by gentle shaking and analysed immediately for cell numbers and viability, using Cell Lab Quanta SC flow cytometry (Beckman Coulter Inc., CA, USA) equipped with an argon laser (488 nm). Red fluorescence (PI) was collected using a 635 nm band pass filter. Analysis was undertaken by loading an appropriate protocol for the acquired parameters: electronic volume (EV), log side scatter (SS), and PI integral; 10 000 observations were collected for analysis. The evaluation of cell number was achieved by gating areas in the EV versus log SS dot plot in which living cells and dead cells appear.

We used the following equation to calculate the specific growth rate, μ:

$$\mu = \frac{(lnX_{V_1} - lnX_{V_0})}{t_1 - t_0}$$

Where X_{V_1} and X_{V_0} symbolise the viable cell density at time points t_1 and t_0, respectively.

Determination of antibody concentration

Enzyme-linked immunosorbent assay (ELISA) was used to determine the concentration of mAb secreted by the cell lines, as described previously [30]. The antibody was sandwiched between monoclonal anti-human IgG Fc (Sigma-Aldrich, St. Louis, USA) and anti-human kappa light chain horseradish peroxidase (HRP) conjugates (Sigma-Aldrich). The concentration of antibody in the samples was determined using o-phenylenediamine (OPD; Sigma-Aldrich) as a substrate, and the specific productivity, q_p, was calculated using the following equation:

$$q_p = \frac{[MAb]_1 - [MAb]_2}{CVCT}$$

where [MAb] represents the antibody concentration at a particular time, while cumulative viable cell time (CVCT) is the sum of the individual areas given by

$$CVCT = \sum\left(\left(\frac{x_1 + x_0}{2}\right) \times (t_1 - t_0)\right)$$

Where x_0 represents the cell number at the particular time t_0 and x_1 represents the cell number at the elapsed time t_1.

Quantitative real-time PCR (qRT-PCR) analysis

3×10^6 cells were removed from the flask at the mid-exponential phase (day three of the batch culture), centrifuged, and washed twice by PBS. The sample was treated with RNA*later* and stored at −80°C until analysis, at which point it was centrifuged to remove the RNA*later* stabilization reagent. RNA isolation was carried out using the RNeasy Mini Kit (QIAGEN, Valencia, CA, USA) according to the manufacturer's instructions. The concentration of RNA was determined using a NanoDrop ND-1000 UV–vis Spectrophotometer (Nanodrop Technologies, Wilmington, DE, USA), and the integrity of RNA was checked using an Agilent Bioanalyzer (Santa Clara, CA, USA).

The expression levels of mTOR-related genes were quantified using a mouse-mTOR-pathway-focused qRT-PCR array from SA Biosciences (Frederick, Maryland, USA).

The DNA elimination treatment was carried out, and complementary DNA (cDNA) was synthesized from the RNA samples, using the RT^2 First Strand Kit (SA Biosciences) according to the manufacturer's instructions. The cDNA samples were mixed with RT^2 SYBR Green/ROX qRT-PCR Master Mix reagents (SA Biosciences) according to the manufacturer's instructions, and the qRT-PCR was performed on these samples using ABI Prism 7500 FAST sequence detection system (Applied Biosystems, Carlsbad, CA, USA). The C_t values obtained from the qRT-PCR analysis were normalised to five housekeeping genes (beta glucuronidase [*Gusb*], hypoxanthine quinine phosphoribosyl transferase 1 [*Hprt1*], heat shock protein 90 alpha [*Hsp90ab1*], glyceraldehyde-3-phosphate dehydrogenase [*Gapdh*], and beta actin [*Actb*]). The normalization was calculated by using the $2^{-\Delta\Delta CT}$ method, as previously described [62].

Total protein content

The total protein content was determined using the QuantiPro BCA Assay Kit (Sigma-Aldrich). 3×10^6 cells were removed from the flask, centrifuged and then washed once with sterile-filtered PBS. The cells were then centrifuged at 500 g for five minutes and supernatant removed. A mixture of CelLytic M Cell Lysis Reagent (Sigma-Aldrich) and a protease inhibitor cocktail (Sigma-Aldrich) was added to disrupt the cells. The samples were mixed by vortexing for one minute, centrifuged at 15 000 g for 15 minutes at 4°C. The supernatant was transferred to a new microcentrifuge tube and stored at −20°C until analysis.

Western blot analysis

For western blot analysis, 20 μg of total protein was mixed in 2× Laemmli buffer (Sigma-Aldrich) and denatured at 95°C. The samples were then subjected to electrophoresis on an 8% polyacrylamide gel (Thermo Scientific, Waltham, MA, USA) using a Mini Protean 3 electrophoresis system (Bio-Rad Laboratories, Waltham, MA, USA) for 90 minutes. The proteins were transferred electrophoretically to a PVDF membrane (Millipore Corporation, Cheshire, NH, USA) overnight. The membrane was blocked in non-fat milk powder in PBST followed by an overnight incubation with an anti-mouse p110δ PI3K antibody (Santa Cruz Biotechnology, Santa Crus, CA, USA). The membrane was probed with a secondary antibody solution containing anti-goat IgG peroxidase conjugates prior to detection using a SuperSignal West Pico kit (Thermo Scientific) according to the manufacturer's instructions. Upon reaction, the membrane was exposed, developed, and fixed on a film. The band intensity was measured by the AlphaDigiDoc RT2 gel documentation system software (Alpha Innotech Corporation, CA, USA).

Results

Comparing the growth and productivity of the high and low producers

The C47 [1] and CL160 [6] lines, which expressed the recombinant chimeric IgG4 mAb, were grown in batch cultures in three biological replicates. The cell numbers and antibody concentrations, determined by flow cytometric analysis and ELISA, respectively, were determined routinely throughout the culture duration until the cells reached the death phase. The cell growth rates and cumulative viable cell time (CVCT) were determined throughout the exponential phase, from day one to day five of batch culture as shown in Table 1. The viable cell density, viability, and antibody concentration profiles are shown in Figure 2. Following a normal growth curve, CL160[6] outperformed CL47[1] in terms of growth, as demonstrated by the calculated CVCT and growth rate (μ) of 329.66 ± 17.99 (10^9 cell.h/L) and 0.67 ± 0.01 (day^{-1}) in CL160[6], compared to 244.48 ± 4.58 (10^9 cell.h/L) and 0.63 ± 0.01 (day^{-1}) in CL47[1], respectively. However, CL47 [1] sustained high viability towards the decline phase (day 8), while the viability of CL160[6], on the contrary, started to decrease sharply on day seven. In CL 47[1], the specific productivity and total antibody concentration, as indicated in Table 1, were 17.4- and 11-fold higher than CL160[6], respectively.

Differential mTOR-related gene expression between high and low producers

We hypothesized that the differences in CL47 [1] and CL160[6] productivity levels are a consequence of the diverse mechanisms of cellular functions that are controlled by mTOR. Using a commercially available pathway-focused PCR array, the expression of 84 genes related to the mTOR signalling pathway were screened in the template cDNA which was prepared from the samples collected at the mid-exponential phase (day three of the batch culture). Figure 3 demonstrates the relative expression of the mTOR-related genes exhibited by the high producer (CL47[1], treated) in comparison to the low producer (CL160[6], control). Student's t-tests were conducted to determine the significance of the gene expression changes. Out of the 84 mTOR-related genes in this

Table 1 The culture characteristics and cell productivity of high producer (CL47[1]) and low producer (160[6]) cell lines during batch culture

	47[1]	160[6]
CVCT (10^9 cell.hr/L)	244.48 ± 4.58	329.66 ± 17.99
Antibody concentration (μg/ml)	575.24 ± 137.55	52.08 ± 0.77
Growth rate (day^{-1})	0.63 ± 0.01	0.67 ± 0.01
q_p (pg/cell/day;pcd)	50.54 ± 3.18	2.91 ± 1.45

The symbol "±" represents the standard deviation calculated between the three biological replicates.

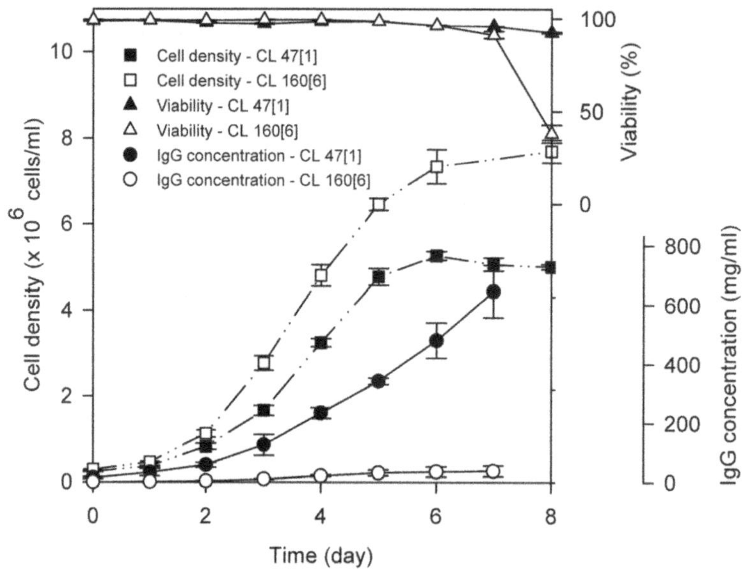

Figure 2 Viable cell density, viability, and antibody concentration profiles of high producer (CL47[1]) and low producer (CL160[6]) cells lines during batch culture. Error bars symbolize the standard deviation as calculated between three biological replicates.

array, eight genes expressed at least a 1.5-fold difference (*Pik3cd, Pik3cg, Pld1, Rragc, Ins2, Telo2, Rps6* and *Prkab1*), with a corresponding statistical significance (P-value < 0.05) between the high and low producers (Figure 4). These genes were all upregulated.

Expression of genes in mTOR signalling pathway

The list of mTOR-related genes with significantly altered expression in CL47[1] relative to CL160[6] is presented in Table 2. Three differentially expressed genes involved in insulin signalling were upregulated. The most upregulated gene, *pik3cd*at 71.3-fold (P-value < 0.05), encodes the catalytic subunit of the delta polypeptide p110δ. The gene encoding a p110δ sibling, the gamma polypeptide (p110γ) subunit, was also significantly expressed but to a lesser extent (6.6-fold; P-value < 0.05). The p110δ and p110γ subunits are classified as Class 1A and 1B, respectively and both receive signals from insulin and various types of growth factors [51,63,64]. Another gene related to insulin signalling that was also significantly expressed was insulin 2 (*Ins2*), with a 3.06-fold change (P-value < 0.05).

Ras-related GTP-binding protein C (*Rragc*) and phospholipase D1 (*Pld1*) were also expressed significantly, with fold changes of 3.08 and 4.63, respectively (P-value < 0.05). Unlike RagC, which has a response specific to amino acid availability [41,54,55], PLD1 is activated by mitogens such as serum [39]. Moreover, the expression of the regulatory

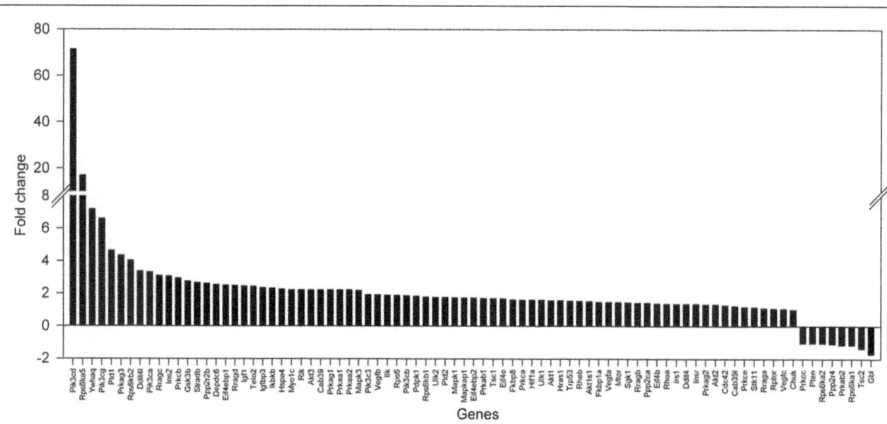

Figure 3 The fold changes of mTOR-related gene expressions of GS-CHO cell lines at the mid-exponential phase of batch culture. Values were calculated based on a relative quantification made between high producer (CL47[1]) and low producer (160[6];control) lines by using the $2^{-\Delta\Delta CT}$ method.

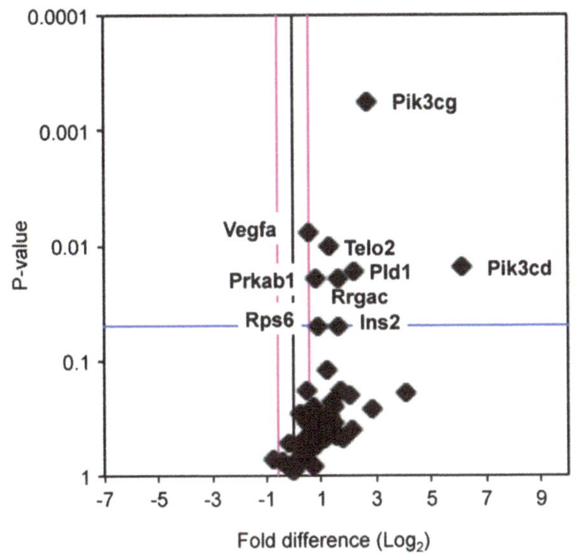

Figure 4 The log$_2$ fold changes in gene expression between high producer (CL47[1]) and low producer (160[6]) lines against t-test P-values. Vertical line represents the threshold of fold changes (i.e., 1.5); horizontal line represents the significant difference of t-test P-values (i.e., 0.05). Significantly expressed genes (P-values < 0.05) are tabulated in Table 2.

gamma subunit of AMPK gene (*Prkag3*) was also observed to be significantly differentially expressed (1.71-fold, P-value < 0.05) though the expression was observed to be minimal compared to other significant differential mTOR-related genes. The activity of AMPK, precisely controlled by the ratio of AMP to adenosine triphosphate (ATP), reflects the cell energy status [43].

Among the genes that encode for downstream effectors of mTOR, only ribosomal protein S6 (*Rps6*) was significantly differentially expressed (1.89-fold, P-value < 0.05) in CL47[1] relative to the CL160[6]. The interactions among the upstream regulators, mTOR, and the downstream effectors are indicated in Figure 1. Although mTOR was not significantly expressed in this study, a gene that could be linked to its presence, known as *Telo2*, was upregulated by 2.44-fold (P-value < 0.05). This gene encodes telomere maintenance 2 (TEL2), which is responsible for mTOR stabilisation and maturation [65,66]. However, the mechanism by which TEL2 stabilizes mTOR remains poorly understood [65,66].

The relationship among p110δ expression and specific productivity, cell growth, and cell size

To investigate whether *Pik3cd* mRNA levels in the high and low producers are predictive of p110δ expression, western blot analysis was performed on samples harvested from the mid-exponential phase (day three) of the batch cultures. Figure 5a shows the expression of the p110δ subunit in GS-CHO cell lines with different q_p levels ranging from 18 to 40 pg/cell/day. (CL47[1] = 50.54 pcd; CL76[2] = 47.23 pcd; CL150[3] = 38.73 pcd; CL164[4] = 21.23 pcd; CL38[5] = 18.34 pcd and CL160[6] = 2.91 pcd). The molecular characteristics of these cell lines have been extensively studied previously [67]. The integrated density values (IDV) corresponding to the band intensities were measured. Pearson correlation coefficient analysis was performed to determine the relationship between p110δ subunit expression and the q_p levels. The analysis revealed that the expression of the p110δ subunit correlated with the overall q_p (ρ = 0.83, P-value < 0.05; Figure 5b). This demonstrates the possibility that the p110δ subunit could be a potential marker describing q_p. Previous studies have demonstrated that PI3K overexpression could influence cell growth and proliferation [68-70]. Thus, we attempted to correlate expression of the p110δ subunit with cell size and CVCT. A Pearson correlation coefficient analysis indicated that neither cell size nor CVCT (data not shown) correlated with the expression of the p110δ subunit (P-value > 0.05).

Discussion

Since Dreesen and Fussenegger [46] recently showed that overexpression of mTOR in a CHO cell line significantly improved recombinant antibody production, we decided to investigate expression of members of the mTOR signalling in a panel of CHO cell lines with different levels of productivity. In order to address this question we used the

Table 2 The mTOR-related genes with at least a 1.5-fold change in expression level (P-values < 0.05)

Gene symbol	Description	Fold change	P-value
Pik3cd	Phosphatidylinositol 3-kinase catalytic delta polypeptide	(+) 71.33	0.02
Pik3cg	Phosphoinositide-3-kinase, catalytic, gamma polypeptide	(+) 6.59	<0.001
Pld1	Phospholipase D1	(+) 4.63	0.02
Rragc	Ras-related GTP binding C	(+) 3.08	0.05
Ins2	Insulin II	(+) 3.06	0.02
Telo2	TEL2, telomere maintenance 2, homolog (S. cerevisiae)	(+) 2.44	0.01
Rps6	Ribosomal protein S6	(+) 1.89	0.05
Prkab1	Protein kinase, AMP-activated, beta 1 non-catalytic subunit	(+) 1.71	0.02

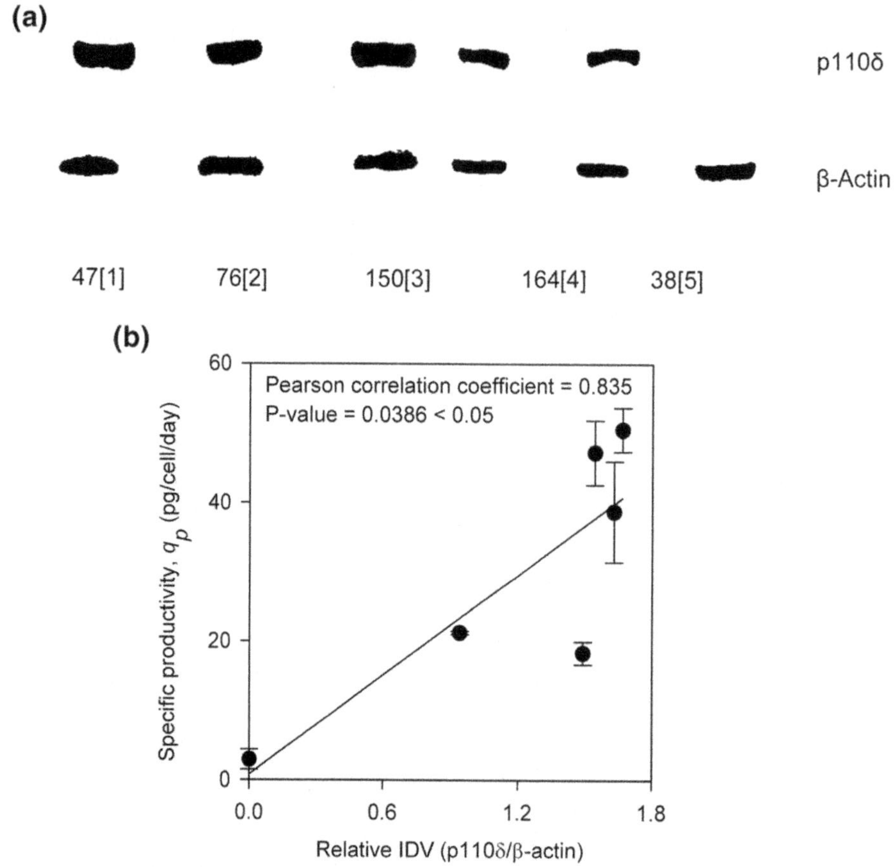

Figure 5 Comparison of p110σ subunit expressions between (a) six GS-CHO cell lines with different productivity levels at the mid-exponential phase of batch culture and (b) the correlation between p110σ expression and q_p. Error bars symbolize the standard deviation as calculated between three biological replicates.

mTOR signalling PCR array for mouse due to the unavailability of the hamster array at that time. Given that the mTOR signalling pathway is conserved from yeast to mammals [71], we used mouse sequences instead of hamster to quantify the levels of mTOR-related gene expressions in this study. Previous transcriptomic studies which relied on cross-hybridization to mouse DNA microarrays have showed considerable success [72] and in general, results from alignment to both mouse and rat reference genomes suggest that CHO genomic sequences are generally more similar to mouse genomic sequences [73]. Of the 84 mTOR-related genes present on the array only eight genes including, Pik3cd, Pik3cg, Ins2, Pld1, Rragc, Prkag3, Rps6 and Telo2 were identified as significantly ($p < 0.05$) upregulated in CL47[1] in comparison to CL160 [6]. This panel of candidate genes could potentially determine high specific productivity in CHO cell lines. Two of the genes that were highly expressed, *Pik3cd* and *Pik3cg*, encode for p110δ and p110γ polypeptides, respectively and are also present on the SA Biosciences hamster mTOR signalling PCR array (PAJJ-098Z). These polypeptides differ at the regulatory subunit structure that is responsible for mediating p110δ and p110γ recruitment to the receptors of interest. The presence of p110δ as a regulatory subunit facilitates the binding of p110γ to the G protein beta subunit-like (Gβl) in response to a stimulated G-protein couple receptor (GPCR). The recruitment of the p110δ subunit to the activated receptor tyrosine kinase (RTK) is, however, mediated by a different regulatory subunit, p85, in response to various extracellular growth factors and insulin signals [35,74,75]. Hence, the different receptors as targets imply that upregulation of *pik3cd* gene could be independent of *pik3cg* (and vice versa), even though these polypeptides share a common role in catalysing phosphorylation of the inositol ring at the D3 position of their downstream effectors, the phosphoinositides.

Although the effects of p110δ in recombinant protein production have yet to be clarified, its associations with growth are better understood. The effects of p110δ overexpression have been correlated to cell growth and cell size in *Drosophila*, where expansion of the wing blade was due to an increase in cell number and cell size. A reverse effect on cell growth was observed in p110δ-deficient cells

[68]. The real cause, which affects the lack of correlation between the CVCT and cell size with p110δ expression, is not clear. The correlation could be weakened by the presence of the medium producer cell line in which other signalling pathways could have compensated for the lack of effects of p110δ expression on cell growth. However, this expression effect is more profound when only two extremes (high and low producers) were considered. CL160 [6] has a smaller cell size (data not shown) than CL47[1], and this could partly result from the defective p110δ expression in CL160[6] [51,68]. There has been some studies published which examined the use of genetic control of the cell cycle to increase q_p. However, these studies cannot confirm that the production of recombinant protein is related to any specific phase of the cell cycle. However, the increase in q_p can be related to cell size rather than the cell cycle phase [17]. This has been shown previously by Bi et al. [30] in a cytostatic system utilising inducible expression of p21^{CIP1} to arrest CHO cells where a fourfold increase of q_p was obtained concomitant to a fourfold increase in cell volume. Regardless of cell cycle phase, the cell size could also be controlled by p110δ expression through protein synthesis, a mechanism that is important for cell growth and proliferation. As cells have to retain a constant cell size during proliferation, a tight coordination between cell growth and cell cycle progression is essential for maintaining cell size and these processes are controlled by p110δ expression as the upstream regulator of mTOR signalling pathway [33,76,77].

While the regulation of p110δ activity is related to cell growth, the activity of p110γ involves the regulation of cell survival [78-80]. An increased resistance to cell death was also observed as a result of the simultaneous inhibition of p53 and activation of nuclear factor κB (NF-κB) signalling pathways by Akt phosphorylation following the overexpression of the p110γ subunit in human cell lines [80].

In addition, *Prkag3, Pld1,* and *Rragc* genes were also significantly expressed in CL47 [1], and these could be related to high specific productivity. These genes encode AMPK, PLD, and Ras-related GTP-binding protein C, which represent upstream regulators of mTOR. The altered expression of these genes may implicate the expression of the *Rsp6* gene, which encodes the S6 protein. The S6 protein regulates the translation of ribosomal protein, elongation factor, and polyA-binding protein, that could lead to ribosome biogenesis [81-83]. This suggests that the improved specific productivity in CL47 [1] could be due to the altered expression of the *Rsp6* gene. Our results were supported by a study conducted by Bi et al. [30]. A significant increase in mAb titre was shown to correlate with higher S6 protein expression in an isopropyl-β-d-thiogalactoside (IPTG)-induced p21^{cip}-arrested CHO cell line [30].

We also found altered expression of the *Ins2* gene in the CL47[1] cell line; this gene is not commonly expressed in non-pancreatic β cells. It is known that all cells contain an insulin gene, but its expression can vary in different cell types. The expression of *Ins2* could be controlled at the transcriptional level. It was shown previously by Kuroda et al. [84] that the *Ins2* gene was fully methylated and becomes demethylated as the cells differentiate into insulin-expressing cells *in vitro* [84]. If CHO cells could produce insulin, this would make the cells less dependent on exogenous insulin, which is certainly an advantageous trait in recombinant protein production.

Conclusion

This study provides a better understanding of the role of the mTOR signalling pathway in GS-CHO cell lines and identifies the importance of this pathway in recombinant protein production. We also highlight the presence of the p110δ subunit of PI3K as a potential key regulator in productivity that may represent a future target for cell-engineering strategies for enhancing productivity. Cell-engineering approaches such as overexpression and/or gene silencing could be used to verify the roles of p110δ and p110γ in cellular functions that may be involved in recombinant protein production.

Competing interests

The authors declare no competing interests.

Authors' contribution

RE carried out the molecular genetic and cell culture studies, and participated in the sequence alignment. All authors wrote, edited and approved the manuscript.

Acknowledgements

The work was funded by grant from HEA, Ireland: PRTLI cycle 4- (Bio) Pharmaceutical and Pharmacological Sciences project and from Majlis Amanah Rakyat, MARA, Malaysia.

References

1. Walsh G: **Biopharmaceutical benchmarks.** *Nat Biotechnol* 2000, **18**(8):831–3.
2. Walsh G: **Biopharmaceutical benchmarks.** *Nat Biotechnol* 2010, **28**(9):917–24.
3. Chu L, Robinson DK: **Industrial choices for protein production by large-scale cell culture.** *Curr Opin Biotechnol* 2001, **12**(2):180–187.
4. Seth G, Charaniya S, Wlaschin KF, Hu W-S: **In pursuit of a super producer—alternative paths to high producing recombinant mammalian cells.** *Curr Opin Biotechnol* 2007, **18**:8.
5. Derouazi M, Girard P, Van Tilborgh F, Iglesias K, Muller N, Bertschinger M, Wurm FM: **Serum-free large-scale transient transfection of CHO cells.** *Biotechnol Bioeng* 2004, **87**(4):537–545.
6. Haldankar R, Li D, Saremi Z, Baikalov C, Deshpande R: **Serum-free suspensin large-scale transient transfection of CHO cells in WAVE bioreactors.** *Mol Biotechnol* 2006, **34**(2):191–199.
7. Xing Z, Kenty BM, Li ZJ, Lee SS: **Scale-up analysis for a CHO cell culture process in large-scale bioreactors.** *Biotechnol Bioeng* 2009, **103**(4):733–746.
8. Lucas BK, Giere LM, DeMarco RA, Shen A, Chisholm V, Crowley CW: **High-level production of recombinant proteins in CHO cells using a dicistronic DHFR intron expression vector.** *Nucleic Acids Res* 1996, **24**(9):1774–1779.

9. Pu H, Cashion LM, Kretschmer PJ, Liu Z: **Rapid establishment of high-producing cell lines using dicistronic vectors with glutamine synthetase as the selection marker.** *Mol Biotechnol* 1998, **10**(1):17–25.
10. Hilal-Alnaqbi A, Hu A, Zhang Z, Al-Rubeai M: **Growth, metabolic activity, and productivity of immobilized and freely suspended CHO cells in perfusion culture.** *Biotechnol Appl Biochem* 2013, **60**:436–445.
11. Kaneko Y, Sato R, Aoyagi H: **Evaluation of Chinese hamster ovary cell stability during repeated batch culture for large-scale antibody production.** *J Biosci Bioeng* 2010, **109**(3):274–280.
12. Konstantinov KB, Tsai Y, Moles D, Matanguihan R: **Control of long-term perfusion chinese hamster ovary cell culture by glucose auxostat.** *Biotechnol Prog* 2008, **12**(1):100–109.
13. Kou TC, Fan L, Ye ZY, Zhou Y, Liu XP, Zhao L, Tan WS: **Process analysis of reduced specific productivity of TNFR-Fc in Chinese hamster ovary cells at high cell density.** *Process Biochem* 2011, **46**(7):1492–1499.
14. Gasser F, Mulsant P, Gillois M: **Long-term multiplication of the Chinese hamster ovary (CHO) cell line in a serum-free medium.** *In Vitro Cell Dev Biol Plant* 1985, **21**(10):588–592.
15. Haldankar R, Kopchick JJ, Ridgway D: **Stable production of a human growth hormone antagonist from CHO cells adapted to serum,ÄêFree suspension culture.** *Biotechnol Prog* 1999, **15**(3):336–346.
16. Sinacore M, Charlebois T, Harrison S, Brennan S, Richards T, Hamilton M, Scott S, Brodeur S, Oakes P, Leonard M: **CHO DUKX cell lineages preadapted to growth in serum-free suspension culture enable rapid development of cell culture processes for the manufacture of recombinant proteins.** *Biotechnol Bioeng* 2000, **52**(4):518–528.
17. Dinnis DM, James DC: **Engineering mammalian cell factories for improved recombinant monoclonal antibody production: lessons from nature?** *Biotechnol Bioeng* 2005, **91**(2):180–189.
18. Page MJ, Sydenham MA: **High level expression of the humanized monoclonal antibody CAMPATH-1H in chinese hamster ovary cells.** *Nat Biotech* 1991, **9**(1):64–68.
19. Wurm FM: **Production of recombinant protein therapeutics in cultivated mammalian cells.** *Nat Biotechnol* 2004, **22**(11):1393–1398.
20. Chusainow J, Yang YS, Yeo JH, Toh PC, Asvadi P, Wong NS, Yap MG: **A study of monoclonal antibody-producing CHO cell lines: what makes a stable high producer?** *Biotechnol Bioeng* 2009, **102**(4):1182–1196.
21. Fann CH, Guarna MM, Kilburn DG, Piret JM: **Relationship between recombinant activated protein C secretion rates and mRNA levels in baby hamster kidney cells.** *Biotechnol Bioeng* 1999, **63**(4):464–472.
22. Gu MB, Todd P, Kompala DS: **Metabolic burden in recombinant CHO cells: effect ofdhfr gene amplification andlacZ expression.** *Cytotechnology* 1995, **18**(3):159–166.
23. Jiang Z, Huang Y, Sharfstein ST: **Regulation of recombinant monoclonal antibody production in chinese hamster ovary cells: a comparative study of gene copy number, mRNA level, and protein expression.** *Biotechnol Prog* 2008, **22**(1):313–318.
24. Kim SJ, Kim NS, Ryu CJ, Hong HJ, Lee GM: **Characterization of chimeric antibody producing CHO cells in the course of dihydrofolate reductase-mediated gene amplification and their stability in the absence of selective pressure.** *Biotechnol Bioeng* 1998, **58**(1):73–84.
25. Pendse GJ, Karkare S, Bailey JE: **Effect of cloned gene dosage on cell growth and hepatitis B surface antigen synthesis and secretion in recombinant CHO cells.** *Biotechnol Bioeng* 1992, **40**(1):119–129.
26. Kidane A, Guan Y, Evans P, Kaderbhai M, Kemp R: **Comparison of heat flux in wild-type and genetically-engineered chinese hamster ovary cells.** *J Therm Anal Calorim* 1997, **49**(2):771–783.
27. Lloyd DR, Holmes P, Jackson LP, Emery AN, Al-Rubeai M: **Relationship between cell size, cell cycle and specific recombinant protein productivity.** *Cytotechnology* 2000, **34**(1):59–70.
28. Carvalhal AV, Marcelino I, Carrondo MJT: **Metabolic changes during cell growth inhibition by p27 overexpression.** *Appl Microbiol Biotechnol* 2003, **63**(2):164–173.
29. Khoo SH, Al-Rubeai M: **Detailed understanding of enhanced specific antibody productivity in NS0 myeloma cells.** *Biotechnol Bioeng* 2009, **102**(1):188–199.
30. Bi JX, Shuttleworth J, Al-Rubeai M: **Uncoupling of cell growth and proliferation results in enhancement of productivity in p21CIP1-arrested CHO cells.** *Biotechnol Bioeng* 2004, **85**(7):741–749.
31. Arsham AM, Neufeld TP: **Thinking globally and acting locally with TOR.** *Curr Opin Cell Biol* 2006, **18**(6):589–97.
32. Fingar DC, Richardson CJ, Tee AR, Cheatham L, Tsou C, Blenis J: **mTOR controls cell cycle progression through its cell growth effectors S6K1 and 4E-BP1/eukaryotic translation initiation factor 4E.** *Mol Cell Biol* 2004, **24**(1):200–216.
33. Ma XM, Blenis J: **Molecular mechanisms of mTOR-mediated translational control.** *Nat Rev Mol Cell Biol* 2009, **10**(5):307–18.
34. Polak P, Hall MN: **mTOR and the control of whole body metabolism.** *Curr Opin Cell Biol* 2009, **21**(2):209–18.
35. Engelman JA, Luo J, Cantley LC: **The evolution of phosphatidylinositol 3-kinases as regulators of growth and metabolism.** *Nat Rev Genet* 2006, **7**(8):606–19.
36. Sarbassov DD, Ali SM, Sabatini DM: **Growing roles for the mTOR pathway.** *Curr Opin Cell Biol* 2005, **17**(6):596–603.
37. Shaw RJ, Cantley LC: **Ras, PI(3)K and mTOR signalling controls tumour cell growth.** *Nature* 2006, **441**(7092):424–30.
38. Proud CG: **mTOR-mediated regulation of translation factors by amino acids.** *Biochem Biophys Res Commun* 2004, **313**(2):429–36.
39. Fang Y, Vilella-Bach M, Bachmann R, Flanigan A, Chen J: **Phosphatidic acid-mediated mitogenic activation of mTOR signaling.** *Sci Signal* 2001, **294**(5548):1942.
40. Peng T, Golub TR, Sabatini DM: **The immunosuppressant rapamycin mimics a starvation-like signal distinct from amino acid and glucose deprivation.** *Mol Cell Biol* 2002, **22**(15):5575–84.
41. Sancak Y, Peterson TR, Shaul YD, Lindquist RA, Thoreen CC, Bar-Peled L, Sabatini DM: **The Rag GTPases bind raptor and mediate amino acid signaling to mTORC1.** *Sci Signal* 2008, **320**(5882):1496.
42. Christie GR, Hajduch E, Hundal HS, Proud CG, Taylor PM: **Intracellular sensing of amino acids in xenopus laevis oocytes stimulates p70 S6 kinase in a target of rapamycin-dependent manner.** *J Biol Chem* 2002, **277**(12):9952–7.
43. Bolster DR, Crozier SJ, Kimball SR, Jefferson LS: **AMP-activated protein kinase suppresses protein synthesis in rat skeletal muscle through down-regulated mammalian target of rapamycin (mTOR) signaling.** *J Biol Chem* 2002, **277**(27):23977–23980.
44. Chong WPK, Sim LC, Wong KTK, Yap MGS: **Enhanced IFNγ production in adenosine-treated CHOCells: a mechanistic study.** *Biotechnol Prog* 2009, **25**(3):866–873.
45. Dennis PB, Jaeschke A, Saitoh M, Fowler B, Kozma SC, Thomas G: **Mammalian TOR: a homeostatic ATP sensor.** *Science* 2001, **294**(5544):1102–5.
46. Dreesen IA, Fussenegger M: **Ectopic expression of human mTOR increases viability, robustness, cell size, proliferation, and antibody production of chinese hamster ovary cells.** *Biotechnol Bioeng* 2011, **108**(4):853–66.
47. Browne SM, Al-Rubeai M: **Analysis of an artificially selected GS-NS0 variant with increased resistance to apoptosis.** *Biotechnol Bioeng* 2011, **108**(4):880–92.
48. O'Callaghan PM, McLeod J, Pybus LP, Lovelady CS, Wilkinson SJ, Racher AJ, Porter A, James DC: **Cell line-specific control of recombinant monoclonal antibody production by CHO cells.** *Biotechnol Bioeng* 2010, **106**(6):938–951.
49. Calvisi DF, Wang C, Ho C, Ladu S, Lee SA, Mattu S, Destefanis G, Delogu S, Zimmermann A, Ericsson J: **Increased lipogenesis, induced by AKT-mTORC1-RPS6 signaling, promotes development of human hepatocellular carcinoma.** *Gastroenterol* 2011, **140**(3):1071–1083. e5.
50. Missiaglia E, Dalai I, Barbi S, Beghelli S, Falconi M, della Peruta M, Piemonti L, Capurso G, Di Florio A, delle Fave G: **Pancreatic endocrine tumors: expression profiling evidences a role for AKT-mTOR pathway.** *J Clin Oncol* 2010, **28**(2):245–255.
51. Boller D, Schramm A, Doepfner KT, Shalaby T, von Bueren AO, Eggert A, Grotzer MA, Arcaro A: **Targeting the phosphoinositide 3-kinase isoform p110δ impairs growth and survival in neuroblastoma cells.** *Clin Cancer Res* 2008, **14**(4):1172–81.
52. Kang S, Bader AG, Vogt PK: **Phosphatidylinositol 3-kinase mutations identified in human cancer are oncogenic.** *Proc Natl Acad Sci USA* 2005, **102**(3):802–807.
53. Toschi A, Lee E, Xu L, Garcia A, Gadir N, Foster DA: **Regulation of mTORC1 and mTORC2 complex assembly by phosphatidic acid: competition with rapamycin.** *Mol Cell Biol* 2009, **29**(6):1411–1420.
54. Kim E, Goraksha-Hicks P, Li L, Neufeld TP, Guan KL: **Regulation of TORC1 by Rag GTPases in nutrient response.** *Nat Cell Biol* 2008, **10**(8):935–945.

55. Sancak Y, Bar-Peled L, Zoncu R, Markhard AL, Nada S, Sabatini DM: **Ragulator-Rag complex targets mTORC1 to the lysosomal surface and is necessary for its activation by amino acids.** *Cell* 2010, **141**(2):290–303.
56. Holz MK, Ballif BA, Gygi SP, Blenis J: **mTOR and S6K1 mediate assembly of the translation preinitiation complex through dynamic protein interchange and ordered phosphorylation events.** *Cell* 2005, **123**(4):569–80.
57. Kimball SR, Shantz LM, Horetsky RL, Jefferson LS: **Leucine regulates translation of specific mRNAs in L6 myoblasts through mTOR-mediated changes in availability of eIF4E and phosphorylation of ribosomal protein S6.** *J Biol Chem* 1999, **274**(17):11647–11652.
58. Neshat MS, Mellinghoff IK, Tran C, Stiles B, Thomas G, Petersen R, Frost P, Gibbons JJ, Wu H, Sawyers CL: **Enhanced sensitivity of PTEN-deficient tumors to inhibition of FRAP/mTOR.** *Proc Natl Acad Sci* 2001, **98**(18):10314–10319.
59. Sahin F, Kannangai R, Adegbola O, Wang J, Su G, Torbenson M: **mTOR and P70 S6 kinase expression in primary liver neoplasms.** *Clin Cancer Res* 2004, **10**(24):8421–8425.
60. Hayward BE, Hussain A, Wilson RH, Lyons A, Woodcock V, McIntosh B, Harris TJR: **The cloning and nucleotide sequence of cDNA for an amplified glutamine synthetase gene from the Chinese hamster.** *Nucl Acids Res* 1986, **14**(2):999–1008.
61. Kalwy S, Rance J, Young R: **Toward more efficient protein expression.** *Mol Biotechnol* 2006, **34**(2):151–156.
62. Livak KJ, Schmittgen TD: **Analysis of relative gene expression data using real-time quantitative PCR and the 2-ΔΔCT method.** *Methods* 2001, **25**(4):402–408.
63. Latres E, Amini AR, Amini AA, Griffiths J, Martin FJ, Wei Y, Lin HC, Yancopoulos GD, Glass DJ: **Insulin-like growth factor-1 (IGF-1) inversely regulates atrophy-induced genes via the phosphatidylinositol 3-kinase/Akt/mammalian target of rapamycin (PI3K/Akt/mTOR) pathway.** *J Biol Chem* 2005, **280**(4):2737–2744.
64. Sanchezmargalet V, Goldfine ID, Vlahos CJ, Sung CK: **Role of phosphatidylinositol-3-kinase in insulin receptor signaling: studies with inhibitor, LY294002.** *Biochem Biophys Res Commun* 1994, **204**(2):446–452.
65. Takai H, Wang RC, Takai KK, Yang H, de Lange T: **Tel2 regulates the stability of PI3K-related protein kinases.** *Cell* 2007, **131**(7):1248–1259.
66. Takai H, Xie Y, de Lange T, Pavletich NP: **Tel2 structure and function in the Hsp90 dependent maturation of mTOR and ATR complexes.** *Genes Dev* 2010, **24**(18):2019–2030.
67. Edros RZ, McDonnell S, Al Rubeai M: **Using molecular markers to characterize productivity in chinese hamster ovary cell lines.** *PLoS One* 2013, **8**(10):e75935.
68. Leevers SJ, Weinkove D, MacDougall LK, Hafen E, Waterfield MD: **The Drosophila phosphoinositide 3-kinase Dp110 promotes cell growth.** *EMBO J* 1996, **15**(23):6584–94.
69. Luo J, McMullen JR, Sobkiw CL, Zhang L, Dorfman AL, Sherwood MC, Logsdon MN, Horner JW, DePinho RA, Izumo S: **Class IA phosphoinositide 3-kinase regulates heart size and physiological cardiac hypertrophy.** *Mol Cell Biol* 2005, **25**(21):9491–9502.
70. Shioi T, Kang PM, Douglas PS, Hampe J, Yballe CM, Lawitts J, Cantley LC, Izumo S: **The conserved phosphoinositide 3-kinase pathway determines heart size in mice.** *EMBO J* 2000, **19**(11):2537–2548.
71. Huang S, Houghton PJ: **Targeting mTOR signaling for cancer therapy.** *Curr Opin Pharmacol* 2003, **3**(4):371–377.
72. Yee JC, Wlaschin KF, Chuah SH, Nissom PM, Hu WS: **Quality assessment of cross-species hybridization of CHO transcriptome on a mouse DNA oligo microarray.** *Biotechnol Bioeng* 2008, **101**:1359–1365.
73. Wlaschin KF, Nissom PM, Gatti Mde L, Ong PF, Arleen S, Tan KS, Rink A, Cham B, Wong K, Yap M, Hu WS: **EST sequencing for gene discovery in Chinese hamster ovary cells.** *Biotechnol Bioeng* 2005, **91**:592–606.
74. Fry MJ: **Structure, regulation and function of phosphoinositide 3-kinases.** *Biochim Biophys Acta* 1994, **1226**(3):237–68.
75. Wymann MP, Pirola L: **Structure and function of phosphoinositide 3-kinases.** *Biochim Biophys Acta* 1998, **1436**(1–2):127–50.
76. Fingar DC, Blenis J: **Target of rapamycin (TOR): an integrator of nutrient and growth factor signals and coordinator of cell growth and cell cycle progression.** *Oncogene* 2004, **23**(18):3151–71.
77. Fingar DC, Salama S, Tsou C, Harlow E, Blenis J: **Mammalian cell size is controlled by mTOR and its downstream targets S6K1 and 4EBP1/eIF4E.** *Genes Dev* 2002, **16**(12):1472–1487.
78. Brazzatti JA, Klingler-Hoffmann M, Haylock-Jacobs S, Harata-Lee Y, Niu M, Higgins MD, Kochetkova M, Hoffmann P, McColl SR: **Differential roles for the p101 and p84 regulatory subunits of PI3Kγ in tumor growth and metastasis.** *Oncogene* 2012, **31**(18):2350–61.
79. Dogra C, Changotra H, Wergedal JE, Kumar A: **Regulation of phosphatidylinositol 3-kinase (PI3K)/Akt and nuclear factor-kappa B signaling pathways in dystrophin-deficient skeletal muscle in response to mechanical stretch.** *J Cell Physiol* 2006, **208**(3):575–585.
80. Guo C, Gasparian AV, Zhuang Z, Bosykh DA, Komar AA, Gudkov AV, Gurova KV: **9-Aminoacridine-based anticancer drugs target the PI3K/AKT/mTOR, NF-kappaB and p53 pathways.** *Oncogene* 2009, **28**(8):1151–61.
81. Avni D, Biberman Y, Meyuhas O: **The 5′ terminal oligopyrimidine tract confers translational control on TOP mRNAs in a cell type-and sequence context-dependent manner.** *Nucleic Acids Res* 1997, **25**(5):995–1001.
82. Jefferies H, Reinhard C, Kozma S, Thomas G: **Rapamycin selectively represses translation of the "polypyrimidine tract" mRNA family.** *Proc Natl Acad Sci* 1994, **91**(10):4441–4445.
83. Terada N, Patel HR, Takase K, Kohno K, Nairn AC, Gelfand EW: **Rapamycin selectively inhibits translation of mRNAs encoding elongation factors and ribosomal proteins.** *Proc Natl Acad Sci* 1994, **91**(24):11477–11481.
84. Kuroda A, Rauch TA, Todorov I, Ku HT, Al-Abdullah IH, Kandeel F, Mullen Y, Pfeifer GP, Ferreri K: **Insulin gene expression is regulated by DNA methylation.** *PLoS One* 2009, **4**(9):e6953.

Optimization of heterologous protein production in Chinese hamster ovary cells under overexpression of spliced form of human X-box binding protein

Galina Gulis*, Kelly Cristina Rodrigues Simi, Renata Rodrigues de Toledo, Andrea Queiroz Maranhao and Marcelo Macedo Brigido

Abstract

Background: The optimization of protein production is a complex and challenging problem in biotechnology. Different techniques for transcription, translation engineering and the optimization of cell culture conditions have been used to improve protein secretion, but there remain many open problems involving post-translational modifications of the secreted protein and cell line stability.

Results: In this work, we focus on the regulation of secreted protein specific productivity (using a recombinant human immunoglobulin G (IgG)) by controlling the expression of the spliced form of human X-box binding protein (XBP-(s)) in Chinese hamster ovary cells (CHO-K1) under doxycycline (DOX) induction at different temperatures. We observed a four-fold increase in specific IgG productivity by CHO cells under elevated concentrations of DOX at 30°C compared to 37°C, without detectable differences in binding activity *in vitro* or changes in the structural integrity of IgG. In addition, we found a correlation between the overexpression of human XBP-1(s) (and, as a consequence, endoplasmic reticulum (ER) size expansion) and the specific IgG productivity under DOX induction.

Conclusions: Our data suggest the T-REx system overexpressing human XBP-1(s) can be successfully used in CHO-K1 cells for human immunoglobulin production.

Keywords: CHO cells, Heterologous protein production, X-box binding protein, T-REx™ system, Doxycycline

Background

The optimization of the production of secreted proteins, such as therapeutic monoclonal antibodies (mAbs), is still a challenging problem in pharmaceutical biotechnology. Although biopharmaceutical products can be produced by many host cell systems, eukaryotic cells are preferred due to their ability to correctly process and modify human proteins. The primary goal is to establish the ideal combination of a rapid accumulation of productive biomass and the maintenance of cell viability for as long as possible. Many different strategies have been considered for improving both cell viability and the productivity of recombinant proteins, including mAbs. These strategies include physiological optimization and genetic and metabolic engineering [1,2].

The most common problem during the optimization of protein production is an error in protein folding in the endoplasmic reticulum (ER). The inhibition of protein folding activates the unfolded protein response (UPR), which is a signal transduction network. Overcoming UPR is one of the many strategies for optimizing protein productivity. For instance, protein production has been tested under the expression of survival proteins that play important roles in UPR, including B-cell lymphoma protein 2 (bcl-2), B-cell lymphoma-extra-large protein (bcl-XL) [3-5], caspase inhibitors [6] and molecular chaperones/heat shock proteins (HSP70) [7]. The role of the spliced form of X-box binding protein (XBP-1(s)) (which plays an important role in regulation processes, such as physical expansion of the ER, increasing the mitochondrial mass and function, increasing the cell size and enhancing total protein synthesis) in optimizing protein production has also been studied [8].

* Correspondence: gulis001@gmail.com
Institute of Biological Sciences, Department of Cell Biology, University of Brasilia, Campus Universitário Darcy Ribeiro, Brasília, DF 70910-900, Brazil

This approach to increasing the secretion capacity of mammalian cells by overexpressing the transcription factor XBP-1(s) was successful in CHO cells; the production of the secreted proteins alkaline phosphatase (SEAP) and alpha-amylase (SAMY) was enhanced upon XBP-1(s) overexpression [9], as was the production of antibody [10]. However, these studies have shown that using overexpression systems without regulation leads to cell apoptosis due to the accumulation of the produced proteins [11].

To overcome accumulation-induced apoptosis, other strategies have been applied to regulate the protein production, such as the use of induction systems. For instance, tetracycline has been used to optimize the overexpression of glycosyltransferases under the control of the Tet on/off system in CHO cells, but unfortunately, a high expression of glycosyltransferases still led to growth inhibition [12]. Furthermore, interesting work using the same expression system has been conducted to control the overexpression of human transferrin (hTf) in human embryonic kidney (HEK-293) cells. That study found favorable concentrations of tetracycline at which the overexpression of hTf was optimal, but again, the high levels of expression limited the cell viability. Such impairment might have been a consequence of the overexpression of the protein of interest, which might have altered the quality of this cell product or even been toxic to the cells [13]. Some studies have attempted to investigate the effect of the expression of an ER-resident molecular chaperone, protein disulfide isomerase (PDI), on the specific production levels of thrombopoietin (TPO) and antibody (Ab) in Chinese hamster ovary cells. Mohan and colleagues used the Tet-off system (in the absence of tetracycline) to regulate PDI, TPO and Ab expression in CHO cells under doxycycline (DOX; a chemical analogue of tetracycline) induction. However, only a small increase in antibody production was observed, and the production of TPO was not affected by PDI expression [14].

Moreover, the optimization of protein production in CHO cells cultured at different temperatures has been addressed. For instance, lowering the temperature from 37°C to 33°C increased the production of erythropoietin (EPO) by approximately four-fold, but at the same time, a low cultivation temperature suppressed cell growth [15]. In addition, a temperature reduction from 37°C to 33°C in the culture of a CHO cell line producing recombinant human granulocyte/macrophage colony-stimulating factor (CHO-K1-hGM-CSF) led to a reduced growth rate, increased cell viability, improved cellular protein production and decreased cell metabolism [16]. One study on the optimization of protein production at 32°C also demonstrated that the specific growth rate of CHO cells producing human mAb decreased by 30–63% at 32°C compared to 37°C. However, the specific antibody productivity of these cells was significantly enhanced at 32°C [17]. Lowering the cultivation temperature even more, from 37°C to 30°C, caused growth arrest associated with a 1.7-fold increase in the specific production of secreted alkaline phosphatase (SEAP) in CHO cells [18].

In this context, we attempted to optimize the specific IgG productivity under different culture temperatures and by regulating the overexpression of apoptotic human protein XBP-1(s) using the T-REx™ system (Invitrogen, Carlsbad, CA, USA). The applied T-REx™ system contains a regulatory plasmid (pcDNA6/TR), which encodes the tetracycline repressor, and an inducible expression plasmid (pcDNA™4/TO/myc-His A) with a tetracycline inductor for expression of the gene of the interest (*xbp-1(s)*). Co-transfected together, these plasmids created a network to regulate XBP-1(s) expression under DOX induction. We cloned *xbp-1(s)* into the T-REx™ system to control its expression with DOX. Then, we transfected the obtained T-REx™-XBP-1(s) system into stably IgG-producing CHO cells and selected stable clones of this system expressing IgG-T-REx-XBP-1(s) to control specific IgG productivity under DOX induction (Figure 1). We determined the optimal concentration of DOX and the temperature at which IgG-T-REx-XBP-1(s) cells produced the maximal amount of IgG without a significant inhibition of cell growth. Moreover, cells treated with DOX for seven days recovered viable cell density to the level of non-treated cells after DOX was washed out from the cell system, and their specific IgG productivity dropped to the basal level. Furthermore, we studied the dependence of specific IgG productivity and viable cell density on the overexpression of XBP-1(s) and ER size expansion.

Methods

Cell lines and media

The CHO-K1 (ATCC®CCL-61™) and Raji (ATCC®CCL-86™) cell lines were purchased from American Type Culture Collection (ATCC, Manassas, VA, USA). CHO-K1 cells were grown and maintained at 37°C or 30°C with 70% humidity and 5% CO_2 in HAM F12 media (Gibco, Big Cabin, OK, USA) supplemented with 2% fetal bovine serum (FBS, Gibco, Big Cabin, OK, USA) and were used in experiments on protein production. Raji cells were grown and maintained at 37°C with70% humidity and 5% CO_2 in RAMP media (Gibco, Big Cabin, OK, USA) supplemented with 10% FBS and were used in FACS direct ligation experiments.

Plasmids and cloning

pCOMIRES HIL anti-CD20 is a tricistronic vector that encodes both the heavy and the light chains of an anti-CD20 antibody along with a neomycin resistance gene under the control of a synthetic CMV promoter. This vector was transfected into CHO-K1 cells to obtain IgG

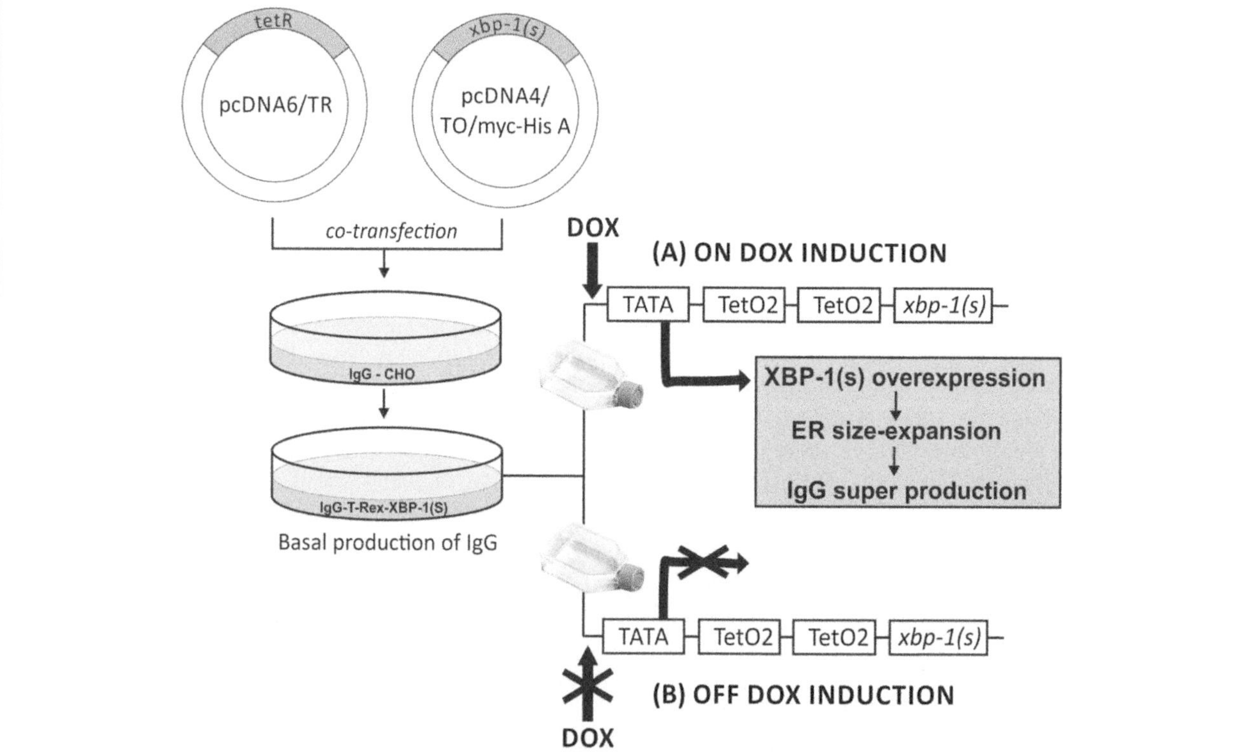

Figure 1 Schematic representation of the DOX-regulated T-Rex™ overexpression XBP-1(s) system. The overproduction of IgG as a result of the XBP-1(s) overexpression and ER size expansion under DOX induction (on DOX induction) **(A)**. The repression of XBP-1(s) overexpression and ER size expansion resulted in the repression of overproduction of IgG in the absence of DOX (off DOX induction) **(B)**.

(anti-CD20)-producing cells. The human *xbp-1(s)* coding sequence was chemically synthesized by GeneScript (Piscataway, NJ, USA). The restriction enzymes *Hind* III and *Bam*H I (Fermentas, Ontario, Canada) were used to obtain the *xbp-1(s)* insert and then clone it into the inducible expression plasmid pcDNA™4/TO/myc-His A from the Invitrogen T-REx™ system (Invitrogen, Carlsbad, CA, USA). This plasmid was used to co-transfect IgG-producing stable clones of CHO cells along with the regulatory plasmid pcDNA6/TR (Invitrogen, Carlsbad, CA, USA). To confirm *xbp-1(s)* cloning, XL1-blue bacterial cells (Stratagene, La Jolla, CA, USA) were transformed with ligated DNA. Ampicillin (Sigma, Ronkonkoma, NY, USA)-selected colonies were isolated and processed for DNA extraction and purification, which was performed using a QIAprep Miniprep Kit (Qiagen, Valencia, CA, USA). Restriction analysis and sequencing (using CMV forward primer 5′-CGCAAATGGGCGGTAGGCGTG-3′ and BGH reverse primer 5′-TAGAAGGCACAGTCG AGG-3′) confirmed the cloning of the *xbp-1(s)* insert.

Transfection with pCOMIRES anti-CD20 DNA (IgG-encoding plasmid) into CHO cells and generation of stable IgG-producing cells

The transfection of pCOMIRES HIL anti-CD20 plasmid (encoding an anti-CD 20 (IgG) antibody, a secretable protein with molecular weight 150 kDa (two light chains, each with molecular weight 25 kDa, and two heavy chains, each with molecular weight 50 kDa)) into CHO cells was performed using a PolyPlus (JetPrime, New York, NY, USA) kit in six-well test plates (TPP, San Diego, CA, USA) according to the manufacturer's instructions. The clones harboring the pCOMIRES HIL anti-CD20 transgene were selected from a mixed population by the single-cell dilution method. Geneticin (Roche, Gaillard, France) was used for selection at 800 μg/mL.

Transfection with the T-REx™ -XBP-1(s) system into stable IgG-producing clones of CHO cells and generation of stable double clones (IgG-T-REx-XBP-1(s) cells)

The co-transfection of T-REx-*xbp-1(s)* plasmid (encoding a spliced form of human apoptotic XBP-1 protein with predicted molecular weight 40 kDa) along with regulatory plasmid pcDNA6/TR into one of the stable IgG-producing clones was performed using a PolyPlus (JetPrime, New York, NY, USA) kit according to the manufacturer's instructions in six-well test plates (TPP, San Diego, CA, USA). Blasticidin (Sigma, Ronkonkoma, NY, USA) and Zeocin (Sigma, Ronkonkoma, NY, USA) were added to a final concentration of 0.5 μg/mL and 50 μg/mL, respectively. The selective markers encoded by regulatory plasmid pcDNA6/TR and expression plasmid pcDNA™4/

TO/myc-His A are against blasticidin and Zeocin, respectively.

Doxycycline induction

Selected IgG-T-REx-XBP-1(s) cells (after the first transfection, IgG clones; after the second, co-transfection, T-REx-XBP-1(s) clones) were induced by DOX at different concentrations: 0 µg/mL for control, 0.1 µg/mL, 0.5 µg/mL and 1 µg/mL. We used these concentrations because we found out that 5 µg/ml and 7.5 µg/ml of doxycycline completely inhibits cells growth for clones and wild type CHO-K1 cells. DOX induction was performed 24 hr after IgG-T-REx-XBP-1(s) cells seeding at a uniform cell density (0.5×10^5 cells/mL) in tissue culture flasks (75 cm^2, TPP, San Diego, CA, USA) and then incubated for seven days at 37°C or 30°C. All cultures reached at least 80% under these conditions. Samples were collected for viable cell density, Ab detection by ELISA, nuclear extract isolation and ER staining. Half of the cells in each group continued to grow for seven more days in DOX-free medium after DOX wash-out. Independently, IgG-T-REx-XBP-1(s) cells were incubated for 42 days at 30°C (150 cm^2 flasks, TPP, San Diego, CA, USA) with or without 1 µg/mL DOX. In all DOX induction experiments, DOX was added (at an appropriate concentration) every three days to the cell culture. Induction experiments were performed twice in duplicate (four independent culture samples per group).

Viability assay

The viable cell density of the IgG-T-REx-XBP-1(s) cells were tested under different DOX concentrations (0 µg/mL, 0.1 µg/mL, 0.5 µg/mL or 1 µg/mL) every day during seven days of cell growth at 37°C and 30°C. Seeding was performed at a uniform cell density (0.06×10^5 cells/mL) in six-well tissue culture plates (TPP, San Diego, CA, USA). At the seventh day all cultures reached at least 80% under these conditions. In addition, the viable cell density of IgG-T-REx-XBP-1(s) cells was tested on the seventh day of growth with DOX and on the seventh day after wash-out in DOX-free medium. In addition, IgG-T-REx-XBP-1(s) cells were tested every seventh day during 42 days of cell growth under 1 µg/mL DOX (or 0 µg/mL as control) at 30°C. The viable cell density was measured using the trypan blue (Sigma, Ronkonkoma, NY, USA) exclusion method with a hemocytometer and light microscope for manual cell counting. Every viable cell density experiment was performed twice in duplicate (single determination from each of two independent culture samples per group in two independent experiments).

ELISA

The supernatants of IgG-T-REx-XBP-1(s) cells in the presence or absence of DOX were collected every seventh day of 37°C or 30°C growth for two weeks or every seventh day for six weeks and processed for analysis by enzyme-linked immunosorbent assay (ELISA) (duplicate determination from each of two independent culture samples per group in two independent experiments). The Lunc/Maxisorp Immunoplate (Thermo Scientific, Waltham, MA, USA) was incubated with primary antibody (goat anti-human IgG (H + L), 1:3000 dilution; Thermo Scientific, Waltham, MA, USA) and blocked with 3% fat-free dehydrated milk solution. After blocking and washing the plate, the supernatants were applied to the plate and incubated for 2 hr. The plate was washed again, and secondary antibody (anti-human IgG Fc-specific, alkaline phosphatase-conjugated, produced in goat, 1:3000 dilution; Sigma, Ronkonkoma, NY, USA) was applied for 1 hr. The plate was washed again, and at the end of the procedure, the signal of absorbance was read at 405 nm by a microplate reader (ELx800 96-well Microplate Reader, MTX Lab Systems, Inc., Vienna, VA, USA) after 4-Nitrophenyl phosphate disodium salt solution (pNPP) (Invitrogen, Carlsbad, CA, USA) addition. In addition, human IgG (whole molecule; Thermo Scientific, Waltham, MA, USA) was used in different concentrations as a control on the same plate.

Isolation and purification of produced proteins

The IgG produced under different temperature conditions by IgG-T-REx-XBP-1(s) cells was purified on the HiTrap™ Protein A HP 1 mL (GE Life Sciences, Pittsburgh, PA, USA) column. The column was first equilibrated with 10 mL Protein A IgG Binding Buffer (Thermo Scientific, Waltham, MA, USA) at a rate of 1 mL/min. Then, the supernatant from IgG-T-REx-XBP-1(s) cells was applied to the equilibrated column. The column was washed with 30 mL Protein A IgG Binding Buffer (Thermo Scientific, Waltham, MA, USA). Then, the protein was eluted with 50 mL IgG Elution Buffer (Thermo Scientific, Waltham, MA, USA), and 2 mL per fraction was collected. Fractions were neutralized with Tris–HCl pH 9.0. The Ab present in the fractions was immunodetected in a dot blot assay. Five microliters of each fraction was directly pipetted onto a nitrocellulose Hybond-C Extra membrane (Amersham® Bioscience, Piscataway, NJ, USA). The membrane was blocked with 3% fat-free milk solution and incubated with anti-human IgG (Fc-specific, alkaline phosphatase-conjugated, produced in goat, 1:3000 dilution) (Sigma, Ronkonkoma, NY, USA), and the proteins were revealed using a BCIP/NBT substrate Kit (Invitrogen, Carlsbad, CA, USA). The Ab-containing fractions were selected for dialysis, which was performed using a Centricon YM-50 (Amicon Bioseparations, Billerica, MA, USA) in PBS buffer (10 mM NaH_2PO_4, 137 mM NaCl, 2.7 mM KCl, pH 7.4).

Western blotting

Anti-CD20 antibody was also detected by western blotting. Five hundred nanograms of IgG sample was loaded in each well of a Bis-Tris gel (NuPAGE® Novex 4-12% Bis-Tris Gel, Invitrogen, Carlsbad, CA, USA) and separated by sodium dodecyl sulfate–polyacrylamide gel electrophoresis (SDS-PAGE) according to the manufacturer's instructions. The proteins were transferred to the Hybond-C Extra nitrocellulose membrane (Amersham® Bioscience, Piscataway, NJ, USA) and blocked in 3% fat-free milk PBS solution. The immunodetection was performed using anti-human IgG (Fc-specific, alkaline phosphatase-conjugated, produced in goat (1:3000 dilution) (Sigma, Ronkonkoma, NY, USA) with a BCIP/NBT substrate Kit™ (Invitrogen, Carlsbad, CA, USA).

XBP-1(s) was also probed by western blotting. The nuclear extracts from the IgG-T-REx-XBP-1(s) cells were prepared as described by Becker and colleagues [10]. Briefly, the nuclear extracts were prepared from 5×10^6 cells/per sample and equal volumes of nuclear extracts were loaded into a Bis-Tris gel (NuPAGE® Novex 4-12% Bis-Tris Gel, Invitrogen, Carlsbad, CA, USA), and SDS-PAGE was performed according to the manufacturer's instructions. Samples were transferred to the Hybond-C Extra nitrocellulose membrane (Amersham® Bioscience, Piscataway, NJ, USA), and after blocking with 3% fat-free milk PBS solution, rabbit anti-human-XBP-1(s) (1:1000 dilution; Sigma, Ronkonkoma, NY, USA) was added, followed by alkaline phosphatase-conjugated anti-rabbit IgG incubation (1:1000 dilution; Sigma, Ronkonkoma, NY, USA). The proteins were revealed using a BCIP/NBT substrate Kit™ (Invitrogen, Carlsbad, CA, USA).

Fluorescence-activated cell sorting (FACS)-ER staining

The IgG-T-REx-XBP-1(s) cells that were grown for seven days under DOX induction and those that were grown for one more week after wash-out were collected at 3×10^5 cells/per staining and washed with HBSS buffer (140 mM NaCl, 4.7 mM KCl, 1 mM $MgCl_2$, 1.5 mM $CaCl_2$, 10 mM glucose, 10 mM HEPES, pH 7.4). After washing with HBSS buffer, the cells were labeled with 250 nM of ER-Tracker™ Green Dye (ER-Tracker™ Green Dye for Live-Cell Endoplasmic Reticulum, Molecular Probes, Invitrogen, Carlsbad, CA, USA) according to the manufacturer's manual. The samples were washed again with HBSS buffer and analyzed using a BD FACS Verse flow cytometer (BD Bioscience, San Jose, CA, USA). Ten thousand events were collected per sample using no gate for acquisition. The dead cells were not excluded in the analysis. We used BD FACSuite to data acquisition. The experiment was performed twice in duplicate.

FACS direct ligation assay

Raji cells were grown for five passages as described above, collected and resuspended in 1 part RAMP media with 10% FBS and 1 part FACS buffer (PBS supplemented with 2% FBS) at 3×10^6 cells/well in a 96-well plate (TPP, San Diego, CA, USA). After centrifugation, the cells were blocked with FcR blocking reagent (MACS, Biotec, Bergisch Gladbach, Germany) on ice for 30 minutes according to the manufacturer's instructions. Purified and dialyzed IgG samples, which were produced by IgG-T-REx-XBP-1(s) cells at 37°C and 30°C, and commercial IgG (rituximab, MabThera, Genetech Inc., South San Francisco, CA, USA) as a positive control were added at 100 ng per well. Samples were incubated on ice for 1 hr and centrifuged after the addition of FACS buffer. The cells were washed twice with FACS buffer and incubated with mouse FITC anti-human IgG (BD Pharmingen™, BD Biosciences, San Jose, CA, USA) according to the manufacturer's manual. The cells were incubated on ice for 30 minutes in the dark, washed again twice with FACS buffer and processed for fluorescence intensity measurements using a BD FACS Verse flow cytometer (BD Bioscience, San Jose, CA, USA). Each experiment was performed twice in duplicate.

Results and discussion

Viability and IgG production under induction with DOX in IgG-T-REx-XBP-1(s) cells cultivated at 37°C and 30°C

To establish a DOX-regulated XBP-1(s) cell line, we first created stably IgG-producing clones of CHO cells by transfecting the pCOMIRES HIL anti-CD20 plasmid into CHO-K1 cells. Then, IgG-CHO clones selected with 800 μg/mL of Geneticin were co-transfected with the T-REx-XBP-1(s) system and processed for second selection using 0.5 μg/mL of blasticidin and 50 μg/mL of Zeocin. From these double clones harboring both pCOMIRES HIL anti-CD20 and T-REx-XBP-1(s) system plasmids, we chose one out of 20 for DOX induction at different concentrations (0 μg/mL (control), 0.1 μg/mL, 0.5 μg/mL or 1 μg/mL) and grew them at 37°C or 30°C. Every day, cells were collected to monitor viable cell density. We found that the viable cell density of the IgG-T-REx-XBP-1(s) cells grown at 37°C under 1 μg/mL or 0.5 μg/mL DOX was slightly lower compared to the control and to the cells with 0.1 μg/mL DOX induction (Figure 2A). Moreover, the viable cell density of IgG-T-REx-XBP-1(s) cells grown at 37°C without DOX or with 0.1 μg/mL DOX was 1.25-fold higher compared to the cells grown at 30°C (Figure 2A and 2B). These data agree with several previous studies [15-18]. Moreover, the viable cell density of IgG-T-REx-XBP-1(s) cells grown at 37°C under 1 μg/mL DOX induction was increased by 20.5% compared to cells incubated at 30°C under the same DOX concentration (Figure 2A and 2B). Thus,

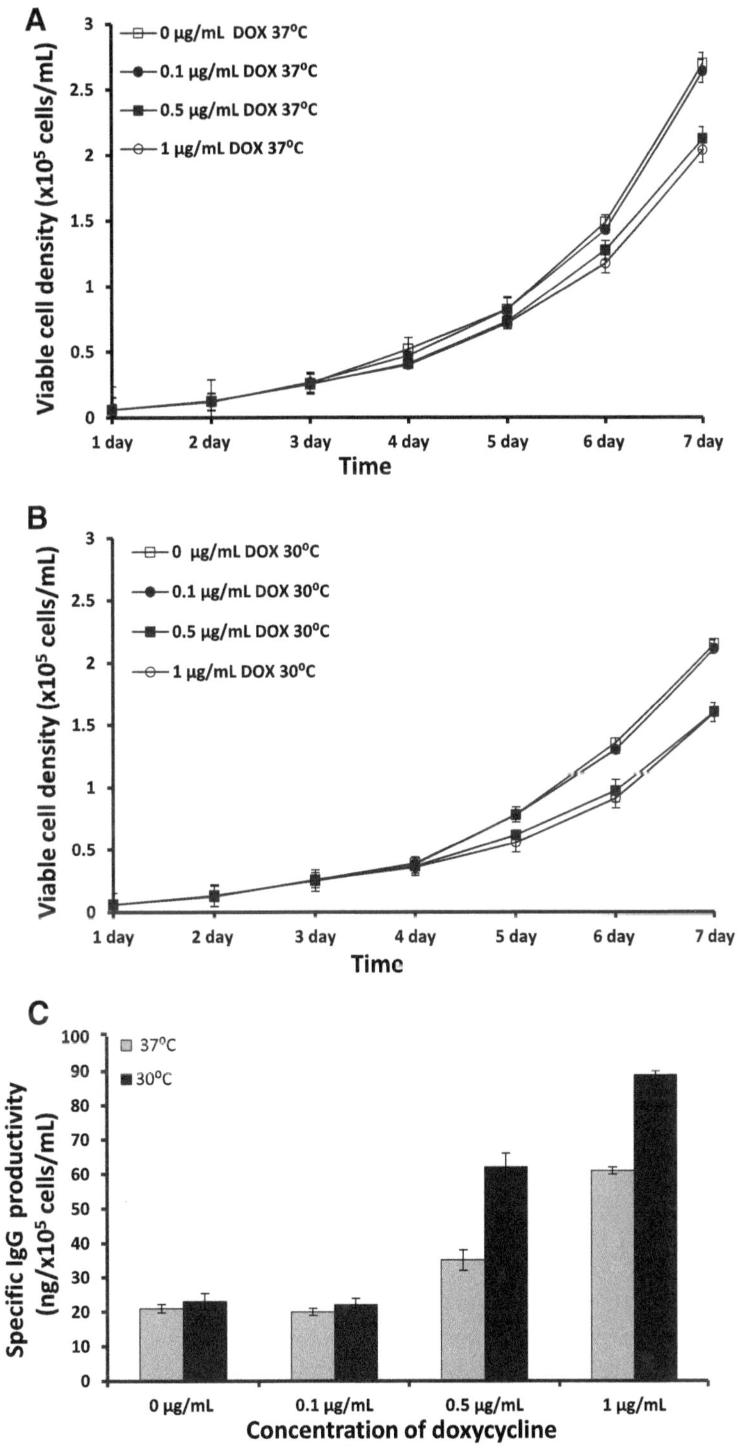

Figure 2 Viable cell density of IgG-T-Rex-XBP-1(s) cells cultured at 37°C (A) or 30°C (B) and their specific IgG productivity at 37°C and 30°C (C) under induction with 0 µg/mL (control), 0.1 µg/mL, 0.5 µg/mL and 1 µg/mL DOX. Error bars represent the standard deviation of the mean of two readings from each of two independent culture samples per group in two independent experiments, n = 4.

IgG-T-REx-XBP-1(s) cells grew more slowly at 30°C with or without DOX compared to those at 37°C.

The supernatants from all cells were collected after seven days of induction and tested by ELISA to determine their IgG yields. The specific IgG productivity depended on the concentration of DOX: under 0.5 µg/mL and 1 µg/mL DOX, the increase of specific IgG productivity was 40% and 66%, respectively, compared to the basal level of specific IgG productivity (0 or 0.1 μg/mL DOX) at 37°C (Figure 2C). These data demonstrate that IgG-T-REx-XBP-1(s) cells produced three-fold more IgG compared to untreated cells, even at low viable cell density. At 0.1 μg/mL DOX, there was no improvement in specific IgG productivity at 37°C. Moreover, the data from ELISA indicate that protein production in the cells incubated at 30°C increased four-fold and three-fold under 1 μg/mL and 0.5 μg/mL DOX, respectively. Once more, induction at a low concentration of DOX (0.1 μg/mL) did not increase specific IgG productivity at 30°C, as above at 37°C. In contrast, the specific IgG productivity by IgG-T-REx-XBP-1(s) cells at 30°C increased by 31.5% and 43.5% compared to induction at 37°C under 1 μg/mL and 0.5 μg/mL DOX concentrations, respectively (Figure 2C). However, we did not detect any effect of low temperature on specific IgG productivity *per se* (without the induction of DOX). Tigges and Fussenegger [9] reported the same lack of effect in CHO cells expressing SEAP, whereas other authors reported an increase in the production of different proteins at low temperature and with no inductor [15-18]. These deviations in experimental results may be due to differences in the proteins and cell lines used in these studies. In conclusion, our data demonstrate a successful improvement of specific IgG productivity using 1 μg/mL DOX in IgG-T-REx-XBP-1(s) cells at 30°C.

Effect of XBP-1(s) expression and ER size expansion on protein production in IgG-T-REx-XBP-1(s) cells

To test the hypothesis that XBP-1(s) expression and ER size expansion indirectly regulate protein production, IgG-T-REx-XBP-1(s) cells were incubated under different concentrations of DOX for seven days at 30°C, and then the same cells were washed with DOX-free medium and incubated for seven more days at the same temperature in fresh DOX-free medium. The supernatant was collected before wash-out and after seven days of incubation in DOX-free medium and processed for ELISA. ELISA showed that specific IgG productivity by cells with DOX (first seven days) increased in a DOX concentration-dependent manner. The specific IgG productivity by cells incubated with 1 μg/mL DOX and 0.5 μg/mL DOX reached four-fold and three-fold that of the untreated cells (treated (t), Figure 3A). In addition, ELISA demonstrated that after DOX removal, the specific IgG productivity returned to the basal level (washed, (w) Figure 3A). Moreover, viability analysis indicated that the growth of the IgG-T-REx-XBP-1(s) cells under different concentrations of DOX was slightly inhibited (first seven days, treated (t), Figure 3B) and then restored to the same level (washed (w), Figure 3B) as cells that had never been exposed to DOX (0 μg/mL (t or w), Figure 3B). In addition, IgG-T-REx-XBP-1(s) cells were used to prepare nuclear extracts, which were analyzed by western blotting for the immune detection of XBP (s). Human XBP-1 (s) was overexpressed in a DOX concentration-dependent manner (first seven days, on DOX induction, treated (t) Figure 3C), but it was absent in cells that were washed and incubated in DOX-free medium for seven days (last seven days, off DOX induction, washed (w), Figure 3C).

ER expansion was also observed by flow cytometry. The fluorescence change of ER-Tracker™ was used as a measure of ER size expansion. In this analysis, cell staining was more intense in samples treated with 1 μg/mL or 0.5 μg/mL DOX for seven days (on DOX induction, treated (t), Figure 4A) than in those treated with 0.1 μg/mL (on DOX induction, treated (t), Figure 4A) or 0 μg/mL or those that were washed out (w), Figure 4B). Moreover, the signal from the washed out cells was equal among different conditions (off DOX induction, washed (w), Figure 4B), which indicated that DOX was responsible for a cascade of processes leading to ER size expansion. In addition, measurements of the median fluorescence intensity (MFI, Figure 4C) obtained from FACS analysis showed that the MFI of the cells treated with 1 μg/mL or 0.5 μg/mL DOX was 2.7-fold or 1.85-fold higher, respectively, than the MFI of non-treated cells or washed out cells (Figure 4C). Thus, cells under DOX induction and low temperature grew more slowly but, at the same time, exhibited a greater increase in specific IgG productivity. Our data also demonstrate that the wash-out of DOX from the cells restored their viable cell density but reduced their specific IgG productivity to basal levels. Taken together, our results indicate the optimal conditions for specific IgG productivity under DOX induction. XBP-1(s) was overexpressed under induction with DOX, which led to the ER size expansion, and this resulted in an increase of specific IgG productivity/secretion. Our findings corroborate the data obtained by Tigges and Fussenegger, who reported an increased production of SEAP and SAMY under the expression of XBP-1(s) and the expansion of ER and Golgi [9].

Binding activity of the recombinant proteins produced at different temperatures

To assess its binding activity, the recombinant IgG produced by cells at different temperatures (37°C and 30°C) was purified and tested for its ability to recognize CD20 at the cell surface of Raji cells that were subjected to flow cytometry analysis. FACS analysis indicated no significant difference in the binding activity of IgG produced by IgG-T-REx-XBP-1(s) cells at different temperatures under DOX induction from commercial IgG (rituximab) (Figure 5A). Moreover, protein samples obtained from the IgG-T-REx-XBP-1(s) cells at different temperatures under DOX induction were submitted to western blotting analysis, and the results did not suggest any differences in structural integrity

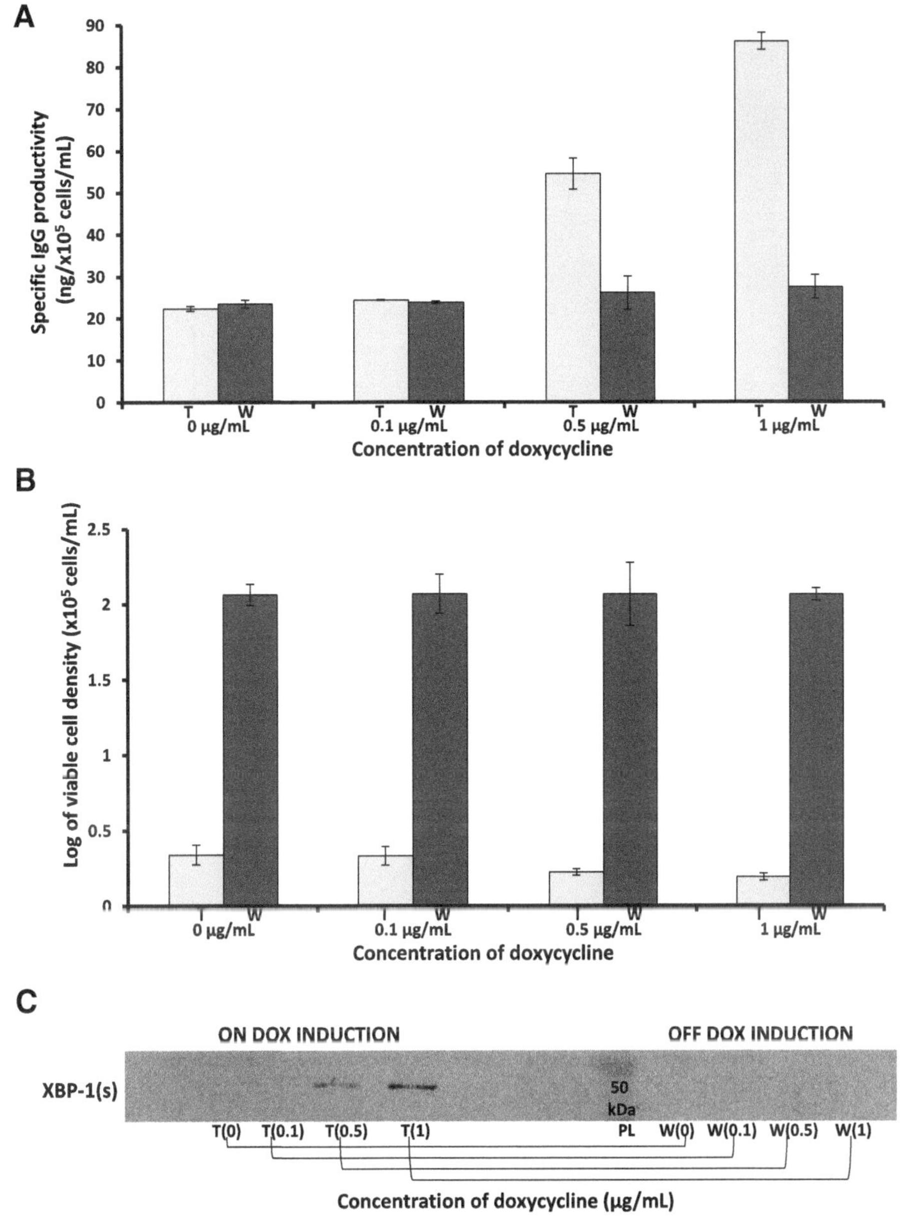

Figure 3 The specific IgG productivity (A), log of viable cell density (B) and western blot analysis of nuclear extracts (C) from the IgG-T-Rex-XBP-1(s) cells grown at 30°C under 0 μg/mL, 0.1 μg/mL, 0.5 μg/mL or 1 μg/mL DOX for seven days (treated, t) and from the same cells seven days after DOX wash-out (washed, w). PL, protein ladder. Error bars represent the standard deviation of the mean of double determination from each of two independent culture samples per group in two independent experiments, n = 4.

of IgG produced at different temperatures (Figure 5B). These data support the use of low-temperature culture conditions under induction by DOX to increase protein production without eliminating the binding activity and structural integrity of the protein of interest.

Establishing a stable protein-producing cell line

To establish a stable cell line, IgG-T-REx-XBP-1(s) cells were seeded and grown at 30°C under 1 μg/mL DOX (or 0 μg/mL DOX as control) for 42 days. The supernatant was collected every seventh day and submitted to ELISA, which showed stable specific IgG productivity around an average of 80.2 ng/10^5 cells/mL by cells incubated with DOX compared to the basal level (25.7 ng/10^5 cells/mL in cells without DOX) (Figure 6A). Moreover, the viable cell density of IgG-T-REx-XBP-1(s) cells under 1 μg/mL DOX was lower than cells without DOX for 42 days (Figure 6B), which allowed cells to be kept in the same

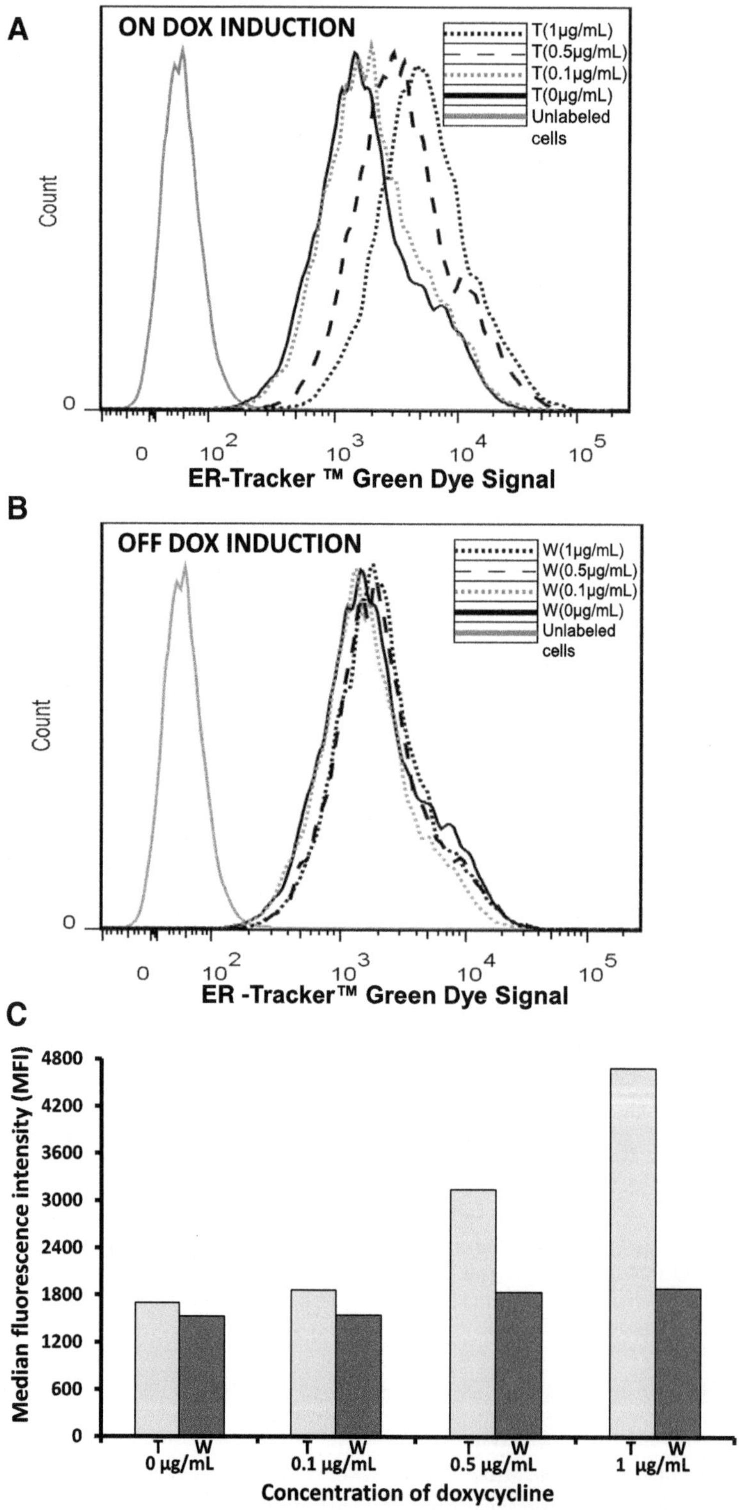

Figure 4 FACS analysis. The samples were run through the flow cytometer until 10.000 events were collected using no gate. The IgG-T-Rex-XBP-1(s) cells were grown under induction with 0 μg/mL, 0.1 μg/mL, 0.5 μg/mL or 1 μg/mL DOX (on DOX induction, treated (t)) for seven days at 30°C. Then, cells from each group were washed with DOX-free media and grown in DOX-free media for seven more days at 30°C (off DOX induction, washed (w)). The cells from each group were stained with ER-tracker™ green dye, and cell counts vs. ER-Tracker™ signal from IgG-T-REx-XBP-1(s) cells were measured (on DOX induction, treated (t) **(A)** and off DOX induction, washed (w) **(B)**). Median fluorescence intensity (MFI) of ER-Tracker™ Green Dye from IgG-T-REx-XBP-1(s) cells (on DOX induction, treated (t) and off DOX induction, washed (w) **(C)**).

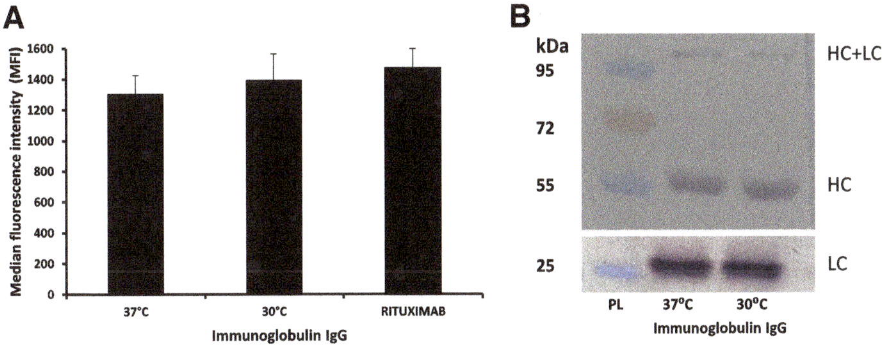

Figure 5 FACS and Western analyses. Median fluorescence intensity measurements of mouse FITC anti-human IgG ligated to the tested IgG (produced by IgG-T-REx-XBP-1(s) cells under DOX induction at 37°C and 30°C (and rituximab, control) that were previously incubated with Raji cells) **(A)**. Western blot analysis of IgG produced by IgG-T-REx-XBP-1(s) cells at 37°C and 30°C under DOX induction (PL, protein ladder) **(B)**.

culture flask without passaging to new flasks. The described approach might be useful for the production of recombinant secreted proteins, as cells growing under the combination of DOX induction and low temperature multiply more slowly but are more productive.

Conclusion

Many studies have been published on improving recombinant protein production. In general, the published data suggest that the optimization of the production of specific target proteins requires specifically adjusted growth

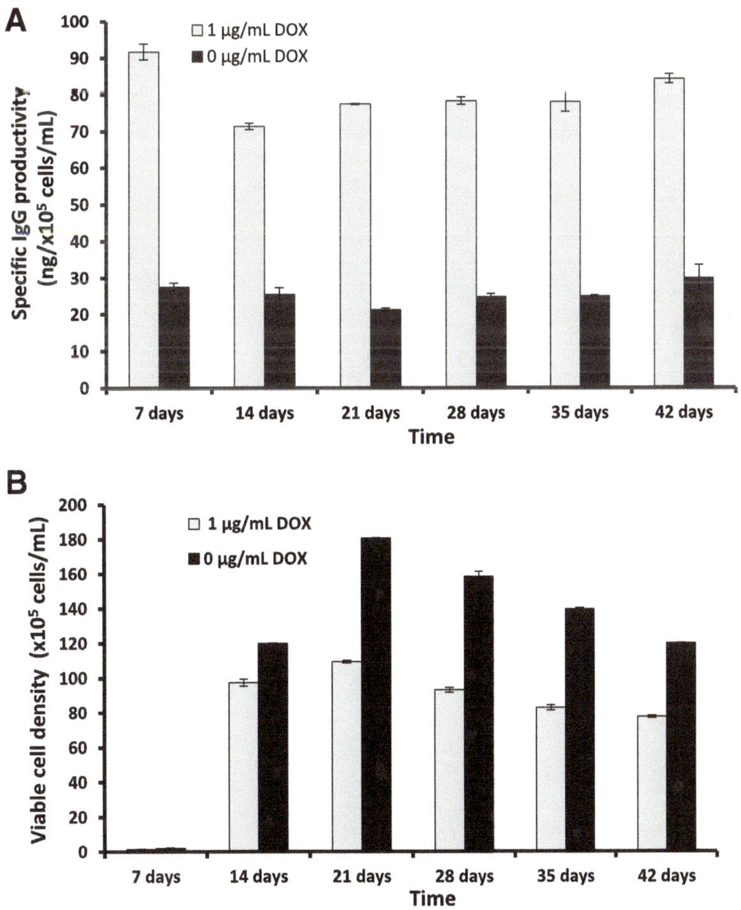

Figure 6 Specific IgG productivity (A) and viable cell density (B) of the IgG-T-REx-XBP-1(s) cells under 1 µg/mL DOX induction (0 µg/mL DOX, control) for 42 days at 30°C. Error bars represent the standard deviation of the mean of double determination from each of two independent culture samples per group in two independent experiments, n = 4.

conditions and a carefully chosen cell line. In the present study, we optimized the conditions for IgG (human anti-CD20) specific productivity in CHO-K1 cells. We showed that the combination of low temperature (30°C) and XBP-1(s) overexpression regulated by DOX induction significantly improved anti-CD20 specific productivity: under 1 μg/mL DOX treatment, specific IgG productivity was increased by 32% compared to the cells grown under the same concentration at 37°C and 74% compared to the cells grown without DOX induction at 37°C or 30°C. Moreover, the results of our study indicate the direct dependence of specific IgG productivity on the concentration of DOX (under 0.5 μg/mL, the increase was 2.7-fold, and under 1 μg/mL, the increase was 3.9-fold), which allows for the precise regulation of specific IgG productivity in CHO-K1 cells. In addition to the concentration dependence, we demonstrated the possibility of returning the specific IgG over productivity to the basal level of specific productivity by removing DOX. This step also restored the viable cell density, which permitted the cells to overcome the problem of accumulation of the target protein. In the production of proteins, it may be possible to use the T-Rex-XBP-1(s) system to turn up and down the production of protein, repeating this cycle several times to accumulate higher amounts of target protein without a loss of cell viability. We also observed a DOX concentration-dependent relationship involving XBP-1(s) overexpression (western analysis), ER size expansion (FACS measurements) and specific IgG productivity (ELISA). Finally, our data demonstrate that it is possible, under DOX induction at low temperature, to produce a target protein for an extended period of time. Taken together, our data suggest the T-REx-XBP-1(s) system can be used in CHO-K1 cells for human immunoglobulin production.

Abbreviations

Ab: Antibody; bcl-2: B-cell lymphoma protein 2; bcl-XL: B-cell lymphoma-extra-large protein; CHO-K1: Chinese hamster ovary cells; CHO-K1-hGM-CSF: CHO cell line producing recombinant human granulocyte/macrophage colony-stimulating factor; DOX: Doxycycline; ELISA: Enzyme-linked immunosorbent assay; EPO: Erythropoietin; ER: Endoplasmic reticulum; FACS: Fluorescence-activated cell sorting; HEK-293: Human embryonic kidney cells 293; HSP70: Heat shock proteins 70; hTf: Human transferrin; IgG: Human immunoglobulin G; mAbs: Monoclonal antibodies; MFI: Median fluorescence intensity; PDI: Protein disulfide isomerase; pNPP: 4-Nitrophenyl phosphate disodium salt solution; SAMY: alpha-amylase; SEAP: Secreted alkaline phosphatase proteins; SDS-PAGE: Sodium dodecyl sulfate–polyacrylamide gel electrophoresis; Tet: Tetracycline; TPO: Thrombopoietin; UPR: Unfolded protein response; XBP-(s): Spliced form of human X-box binding protein.

Competing interests

The authors declare that they have no competing interests.

Authors' contributions

GG, AQM and MMB designed the study, interpreted the results and wrote the manuscript. GG, KCRS and RRT performed the experiments and interpreted the results. All authors read and approved the final manuscript.

Acknowledgments

The work was funded by grant from BNDES, Brazil. GG is a fellow of PNPD postdoctoral program from CAPES. KCRS is a fellow of the CAPES graduate program.

References

1. Costa AR, Rodrigues ME, Henriques M, Azeredo J, Oliveira R: **Guidelines to cell engineering for monoclonal antibody production.** *Eur J Pharm Biopharm* 2010, **74**:127–138.
2. Frenzel A, Hust M, Schirrmann T: **Expression of recombinant antibodies.** *Front Immunol* 2013, **4**(217):1–20.
3. Fussenegger M, Fassnacht D, Schwartz R, Zanghi JA, Graf M, Bailey JE, Pörtner R: **Regulated overexpression of the survival factor bcl-2 in CHO cells increases viable cell density in batch culture and decreases DNA release in extended fixed-bed cultivation.** *Cytotechnology* 2000, **32**(1):45–61.
4. Itoh Y, Ueda H, Suzuki E: **Overexpression of bcl-2, apoptosis suppressing gene: Prolonged viable culture period of hybridoma and enhanced antibody production.** *Biotechnol Bioeng* 1995, **48**(2):118–122.
5. Meents H, Enenkel B, Eppenberger HM, Werner RG, Fussenegger M: **Impact of coexpression and coamplification of sICAM and antiapoptosis determinants bcl-2/bcl-x (L) on productivity, cell survival, and mitochondria number in CHO-DG44 grown in suspension and serum-free media.** *Biotechnol Bioeng* 2002, **80**(6):706–716.
6. Sauerwald TM, Betenbaugh MJ, Oyler GA: **Inhibiting apoptosis in mammalian cell culture using the caspase inhibitor XIAP and deletion mutants.** *Biotechnol Bioeng* 2002, **77**(6):704–716.
7. Lasunskaia EB, Fridlianskaia II, Darieva ZA, da Silva MS, Kanashiro MM, Margulis BA: **Transfection of NS0 myeloma fusion partner cells with HSP70 gene results in higher hybridoma yield by improving cellular resistance to apoptosis.** *Biotechnol Bioeng* 2003, **81**(4):496–504.
8. Shaffer AL, Shapiro-Shelef M, Iwakoshi NN, Lee AH, Qian SB, Zhao H, Yu X, Yang L, Tan BK, Rosenwald A, Hurt EM, Petroulakis E, Sonenberg N, Yewdell JW, Calame K, Glimcher LH, Staudt LM: **XBP1, downstream of Blimp-1, expands the secretory apparatus and other organelles, and increases protein synthesis in plasma cell differentiation.** *Immunity* 2004, **21**(1):81–93.
9. Tigges M, Fussenegger M: **Xbp1-based engineering of secretory capacity enhances the productivity of Chinese hamster ovary cells.** *Metab Eng* 2006, **8**(3):264–272.
10. Becker E, Florin L, Pfizenmaier K, Kaufmann H: **An XBP-1(s) dependent bottle-neck in production of IgG subtype antibodies in chemically defined serum-free Chinese hamster ovary (CHO) fed-batch processes.** *J Biotechnol* 2008, **135**(2):217–223.
11. Becker E, Florin L, Pfizenmaier K, Kaufmann H: **Evaluation of a combinatorial cell engineering approach to overcome apoptotic effects in XBP-1(s) expressing cells.** *J Biotechnol* 2010, **146**(4):198–206.
12. Umaña P, Jean-Mairet J, Bailey JE: **Tetracycline-regulated overexpression of glycosyltransferases in Chinese hamster ovary cells.** *Biotechnol Bioeng* 1999, **65**(5):542–549.
13. Jones J, Nivitchanyong T, Giblin C, Ciccarone V, Judd D, Gorfien S, Krag SS, Betenbaugh MJ: **Optimization of tetracycline-responsive recombinant protein production and effect on cell growth and ER stress in mammalian cells.** *Biotechnol Bioeng* 2005, **91**(6):722–732.
14. Mohan C, Park SH, Chung JY, Lee GM: **Effect of doxycycline-regulated protein disulfide isomerase expression on the specific productivity of recombinant CHO cells: thrombopoietin and antibody.** *Biotechnol Bioeng* 2007, **98**(3):611–615.
15. Yoon SK, Song JY, Lee GM: **Effect of low culture temperature on specific productivity, transcription level, and heterogeneity of erythropoietin in Chinese hamster ovary cells.** *Biotechnol Bioeng* 2003, **82**(3):289–298.
16. Bollati-Fogolín M, Forno G, Nimtz M, Conradt H, Etcheverrigaray M, Kratje R: **Temperature reduction in cultures of hGM-CSF-expressing CHO cells: effect on productivity and product quality.** *Biotechnol Prog* 2005, **21**(1):17–21.

17. Yoon SK, Hwang SO, Lee GM: **Enhancing effect of low culture temperature on specific antibody productivity of recombinant Chinese hamster ovary cells: clonal variation.** *Biotechnol Prog* 2004, **20**(6):1683–1688.
18. Kaufmann H, Mazur X, Fussenegger M, Bailey J: **Influence of low temperature on productivity, proteome and protein phosphorylation of CHO cells.** *Biotechnol Bioeng* 1999, **63**(5):573–582.

Rapid and simple detection of methicillin-resistance *staphylococcus aureus* by *orfX* loop-mediated isothermal amplification assay

Jianyu Su[1], Xiaochen Liu[1], Hemiao Cui[1], Yanyan Li[2], Dingqiang Chen[3*], Yanmei Li[4,5*] and Guangchao Yu[6]

Abstract

Background: Methicillin-resistant *Staphylococcus aureus* (MRSA) has become one of the most prevalent pathogens responsible for nosocomial infections throughout the world. As clinical MRSA diagnosis is concerned, current diagnostic methodologies are restricted by significant drawbacks and novel methods are required for MRSA detection. This study aimed at developing a simple loop-mediated isothermal amplification (LAMP) assay targeting on *orfX* for the rapid detection of methicillin-resistance *Staphylococcus aureus* (MRSA).

Results: The protocol was designed by targeting *orfX*, a highly conserved open reading frame in *S. aureus*. One hundred and sixteen reference strains, including 52 Gram-positive and 64 Gram-negative isolates, were included for evaluation and optimization of the *orfX*-LAMP assay. This assay had been further performed on 667 *Staphylococcus* (566 MRSA, 25 MSSA, 53 MRCNS and 23 MSCNS) strains and were comparatively validated by PCR assay using primers F3 and B3, with rapid template DNA processing, simple equipments (water bath) and direct result determination (both naked eye and under UV light) applied. The indispensability of each primer had been confirmed, and the optimal amplification was obtained under 65°C for 45 min. The 25 μl reactant was found to be the most cost-efficient volume, and the detection limit was determined to be 10 DNA copies and 10 CFU/reaction. High specificity was observed when *orfX*-LAMP assay was subjected to 116 reference strains. For application, 557 (98.4%, 557/566) and 519 (91.7%, 519/566) tested strains had been detected positive by LAMP and PCR assays. The detection rate, positive predictive value (PPV) and negative predictive value (NPV) of *orfX*-LAMP were 98.4%, 100% and 92.7% respectively.

Conclusions: The established *orfX*-LAMP assay had been demonstrated to be a valid and rapid detection method on MRSA.

Keywords: Loop-mediated isothermal amplification (LAMP), MRSA, *OrfX*

Background

As a group of Gram-positive bacteria, staphylococci strains are responsible for various tissues infection and a wide range of diseases, including skin infections, pneumonia, endocarditis, osteomyelitis, gastroenteritis, scalded skin syndrome and toxic shock syndrome [1-4]. Within species of staphylococci, *Staphylococcus aureus* has been considered to be a major concern in both medicine and food safety [1-3,5,6]. For other clinically significant *Staphylococcus* strains, coagulase-negative staphylococci (CoNS) strains have been reported as the most frequently isolated pathogens in intravascular catheter related infections (CRI) (accounting for approximately 28% of nosocomial bloodstream infection), thus become a frequent cause of nosocomial infection and bacteremia, especially in patients with indwelling medical devices [7-9]. Since the first report in 1961, methicillin-resistant *Staphylococcus aureus* (MRSA) has become one of the most prevalent pathogens responsible for nosocomial infections throughout the world, which has raised a global challenge for clinicians, hospital epidemiologists and administrators [10-16]. The *mecA* gene, encoding the PBP2a protein and causing methicillin resistance in staphylococci, is located on staphylococcal cassette chromosome *mec* (SCC*mec*).

* Correspondence: jyksys@126.com; yanmeili@live.cn
[3]Department of Laboratory Medicine, First Affiliated Hospital of Guangzhou Medical College, Guangzhou 510120, China
[4]Guangzhou Women and Children's Medical Center, Guangzhou 510623, China
Full list of author information is available at the end of the article

SCC*mec* contains the *mec* gene complex (the *mecA* gene and its regulators) and the *ccr* gene complex (encoding site-specific recombinases responsible for the mobility of SCC*mec*) [17-23]. In recent years, MRSA strains have been considered to be a major example of the leading "Super Bugs", with broad resistance to practically all β-lactam antibiotics and usually other multiple drugs due to the associated resistance genes carried by SCC*mec* [1-3,17-20]. As consequence, the increasing awareness for the risk and hazard of MRSA strains raised the demands of an early, cost-effective, timely, specific and sensitive detection assay [5,6,10,20].

As clinical MRSA diagnosis is concerned, *Staphylococcus* strains have commonly been identified via routine standard procedures including colony morphology, Gram staining, testing of catalase, hyaluronidase and coagulase, as well as the Vitek 2 automated system and the API-Staph commercial kit, which makes the detection of 16SrRNA somehow irrelevant. However, despite the advantages of this assay, the further development and broad application of multiplex-PCR has been restricted by many factors, such as cross-reaction of different sets of primers, self-inhibition due to formation of dimmers, reduced amplification efficiency caused by the simultaneous and parallel amplification, undetectable influence of different targets, as well as the requirement for standard full use of external and internal quality control (both positive and negative control for each targets) for each assay. As a novel analytical assay since the past decade, mass spectrometry has been applied for detection and diagnostics of various clinical microorganisms, with advantages on high-throughput process, sensitivity and specificity [24]. However, this methodology requires trained personnel, operating space, complicated sample preparation procedure, as well as expensive equipment.

In the last decade, loop-mediated isothermal amplification (LAMP) had been reported as a novel nucleic acid amplification method [25-27] (Figure 1) and applied to the detection of various pathogenic organisms, including *Escherichia coli* O157 with the associated toxins *stx1* and *stx2* [28], *Yersinia pseudotuberculosis* [29], *Salmonella* [30-33], *Vibrio parahaemolyticus* with the associated toxins *tdh* and *trh* [34-39], *Psuedomonas aeruginosa* [40] and *Listeria monocytogenes* [41]. The LAMP methodology relies on an auto-cycling strand displacement DNA synthesis performed by the Bst DNA polymerase large fragment under the isothermal conditions between 60-65°C, and the amplicons are mixtures of many different sizes of stem-loop DNAs with several inverted repeats of the target sequence and cauliflower-like structures with multiple loops, which significantly simplify the reaction itself and the result determination. The LAMP assay includes 4 or 6 primers (targeting on 6 or 8 specific regions) to perform the amplification, and requires 45-60 min and 30-45 min without and with loop primers, respectively. In comparision with PCR assay, the positive points of LAMP method were simplicity, rapidity, specificity and sensitivity (10-1000 times higher than PCR).

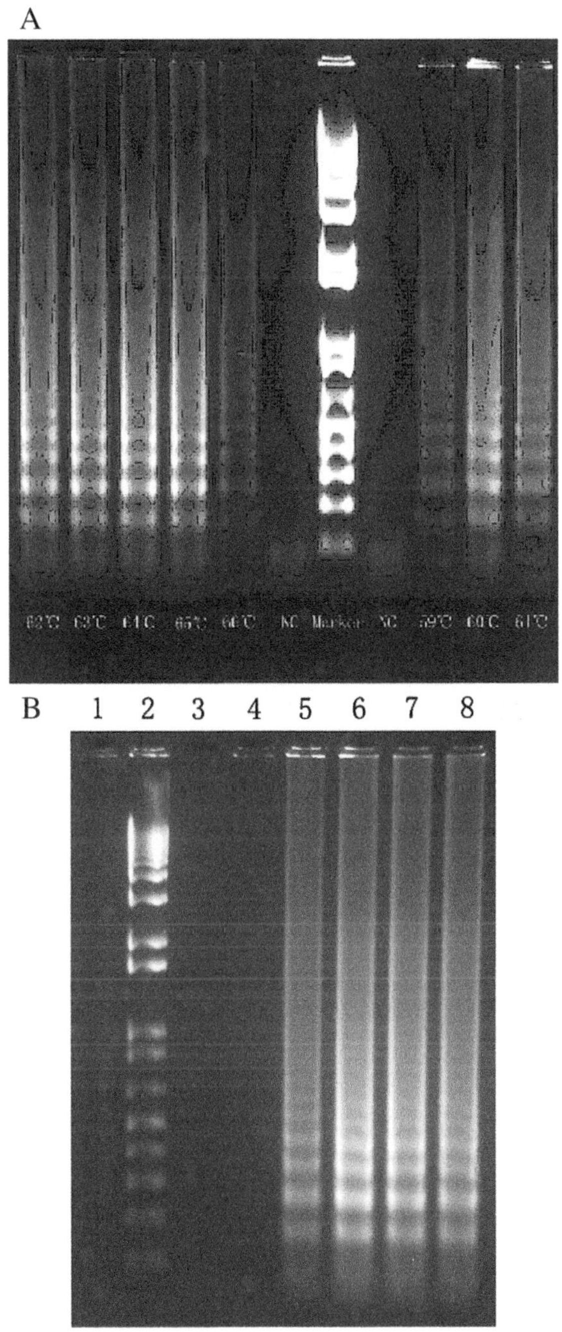

Figure 1 Electrophoresis of *orfX*-LAMP. A: Under different temperatures. **B**: Detection limit, lane 1-8: negative control (distilled water), DNA Marker, negative contro1 (template DNA sample of other bacteria), 1 copies, 10 copies, 10^2 copies, 10^3 copies and 10^4 copies.

In the current study, a simple *orfX*-LAMP assay had been developed, evaluated, optimized and further applied to detect a large scale of clinical MRSA strains, requiring approximately 60 mins for the whole process.

Results

Optimization of *orfX*-LAMP assay

The *orfX*-LAMP amplification had generated a large number of ladder-like pattern bands on agarose gel due to the characteristic structure up to the loading wells, with a bottom band of 212-bp size amplicon obtained by sequencing. *OrfX*-LAMP assay had been performed under isothermal condition between 59°C and 66°C. Despite none of significant difference observed, the electrophoresis of LAMP products which were amplified under 64°C exhibited slightly larger amount of DNA amplicons when compared to other temperatures (Figure 1), which was consistent with studies previously. As reaction length was concerned, various time points had been studied under 65°C and with 10 ng template DNA, including 15 min, 30 min, 45 min, 60 min, 75 min, 90 min, 105 min and 120 min. For the LAMP assay with loop primers, amplification had been initially detected at 30 min, and reached the maximal at 45 min (Figure 2). Nevertheless, without loop primers, amplification products could not be observed until 90 min. All amplicons of *orfX*-LAMP assay had been determined by electrophoresis, with observation directly by naked eye, as well as under UV light with Sybr Green dyed (Figure 3).

Figure 2 Evaluation of *orfX*-LAMP assay under different time points. A: Electrophoresis of *orfX*-LAMP reaction under different time points: 1-10: DNA Marker, 15, 30, 45, 60, 75, 90, 105 and 120 min, negative control. **B**: Sybr Green of *orfX*-LAMP reaction under different time points.

Sensitivity and specificity of *orfX*-LAMP assays

The detection limit of *orfX*-LAMP assay had been studied by both minimal CFU of bacterial and template DNA amount, which was found to be 10 copies of DNA/tube and 10 CFU/reaction (LAMP was positive for sample containing $1X10^4$ CFU/ml, with 1 μl was included in the reaction system) (Figure 1), in comparision with the detection limit of PCR which was 10^2 copies of DNA/tube and 10^3 CFU/reaction, indicating that the sensitive of LAMP was 10-100 times sensitive than PCR. As specificity was concerned, for primers, amplification could not be obtained in the absence of FIP, BIP, F3 or B3, due to the indispensable role each of the primers plays in auto-cycling strand placement. The LAMP was only performed in the existence of both inner primers and outer primers. In addition, for the *orfX*-LAMP assay, high specificity was acquired when this assay was subjected to 116 reference strains, without any false positive amplification for non-MRSA strains (Table 1).

Application of LAMP assays on clinical MRSA strains

After establishment and optimization, *orfX*-LAMP assay were applied to the detection of 667 clinical *Staphylococcus* strains, including 566 MRSA, 25 MSSA, 53 MRCNS and 23 MSCNS strains (Table 2), with comparative validation by standard PCR assay. For application, rapid DNA preparation process, simple heating equipments and results determination by observation directly by naked eye and under UV light had been applied. As shown by the results, for total of 566 MRSA strains, 557 (98.4%, 557/566) and 519 (91.7%, 519/566) had been detected positive by LAMP and PCR assays, respectively. However, for other 101 non MRSA isolates, none and 3 yielded positive amplification. The detection rate, positive predictive value (PPV) and negative predictive value (NPV) of *orfX*-LAMP were 98.4%, 100% and 92.7%, respectively. All reaction had been performed in triplicate to ensure reproducibility.

Discussion

As one of the globally widespread clinical pathogens, MRSA strains have attained a heightened concern for its

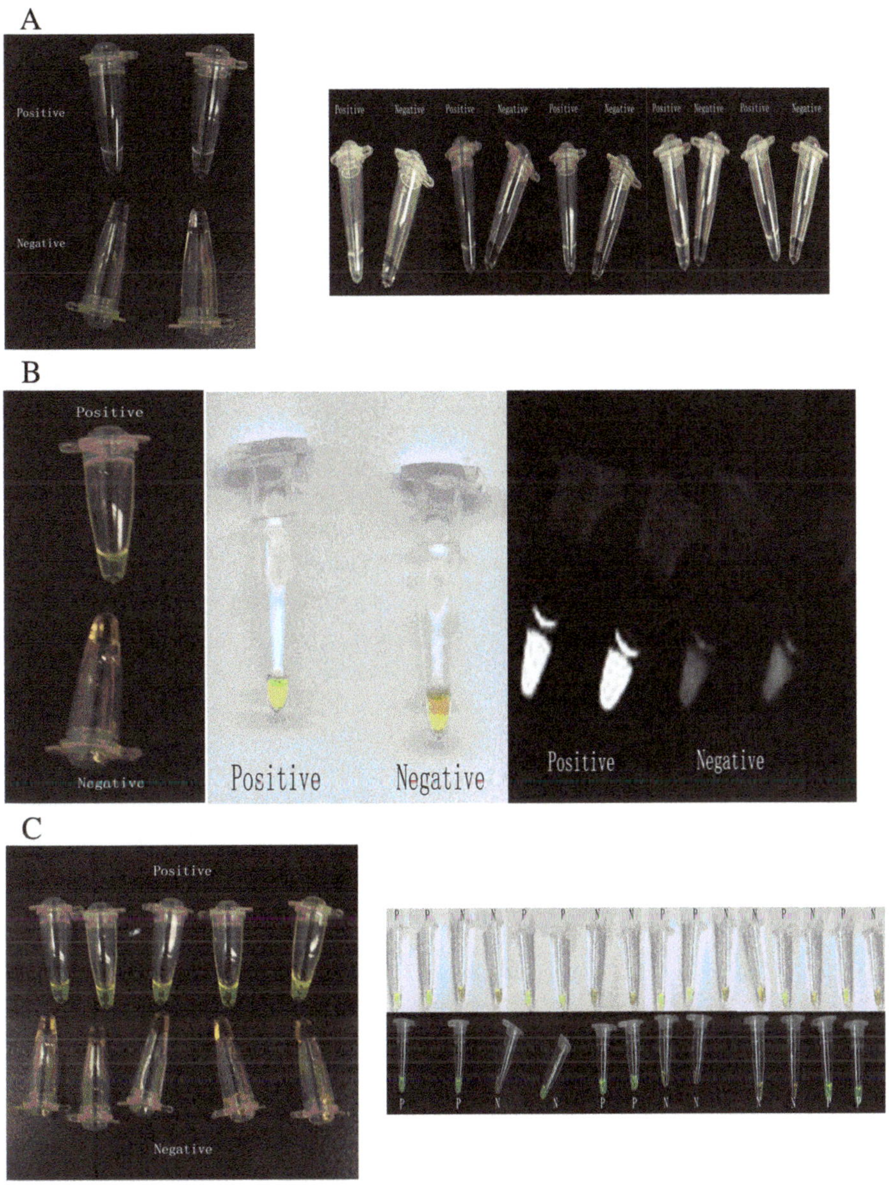

Figure 3 Result determination of *orfX*-LAMP assays by color change. A: Results determination of *orfX*-LAMP by naked eyes. **B**: Results determination of *orfX*-LAMP by Sybr Green with dark background, light backgrounds and under UV light. **C**: Results determination of *orfX*-LAMP in application by Sybr Green.

rapid and accurate detection, which has become a common issue in the diagnostics of MRSA. Current MRSA diagnosis methodologies include routine standard procedures (including colony morphology, Gram staining, testing of catalase, hyaluronidase and coagulase), Vitek 2 automated system, API-Staph kit, immunological assays, mass spectrometry, PCR (regular PCR and quantitative PCR), etc. MRSA strains exhibit resistance to practically all β-lactam antibiotics and commonly other drugs due to the *mecA* and other resistance genes carried by SCC*mec*. However, for type I (1B) and IV (2B) SCC*mec* (with class B mec complex), the *mecI*-encoded repressor function may lead to the low-level methicillin resistance. In the case of type II (2A) and III (3A) SCCmec (with class A mec complex), the IS*431*-mediated deletion of *mecI* and further derepression of *mecA* transcription may cause the expression of methicillin resistance. As consequence, differing from biochemical methodologies, the accurate diagnostic identification of MRSA relies on the detection of highly specific and conserved targets within SCC*mec*. As *orfX* was considered to be highly specific target for both *S. aureus* and SCC*mec*, a distinctive MRSA detection methodology based on *orfX*-LAMP was developed in the present study. In addition, a

Table 1 Reference strains included in the evaluation of *orfX*-LAMP

Reference strains	No.	*orfX*
Gram-positive microorganisms		
Methicillin-resistant *Staphylococcus aureus* 10442, COL, N315, 85/2082, CA05, JCSC 1978, JCSC 4469, MR108, M03-68, WIS	10	+
Methicillin-sensitive *Staphylococcus aureus* ATCC6358, ATCC12598, ATCC12600, ATCC13150, ATCC13565, ATCC14458, ATCC19095, ATCC23235, ATCC27664, ATCC25923, ATCC29213	11	-
Methicillin-resistant coagulase-negative staphylococci: *S. epidermidis* ATCC29887, ATCC700586	2	-
Methicillin-sensitive coagulase-negative staphylococci: *S. epidermidis* ATCC12228, 6508, 9142, 9142-M10; *S. haemolyticus* 1922; *S. hominis* ATCC27844; *S. capitis* ATCC27840; *S. saprophiticus* ATCC35552; *S. sciuri* ATCC29062; *S. schleiferi* ATCC43808; *S. intermedius* ATCC49052	11	-
Listeria monocytogenes ATCC15313ATCC19111, ATCC19114, ATCC19115, ATCC19116, ATCC19118, WF 03213, WF 06410, WF 08211	8	-
Listeria invanovii ATCC19119, WF 06319	2	-
Listeria welshimeri WF 05086	1	-
Listeria seeligeri WF 04426	1	-
Bacillus cereus ATCC11778, ATCC12480, ATCC13061, ATCC14579, ATCC25621, ATCC53522	6	-
Gram-negative organisms		
Escherichia coli O157: H7 WF01201 ATCC43889, NCTC12900	3	-
Escherichia coli O157: H7 WF07803, WF06544, WF05395	3	-
Escherichia coli O157: H7 WF04402, WF05311, WF06837	3	-
Escherichia coli O157: H7 WF04587, WF08385, WF06349	3	-
Escherichia coli O26: H11 WF054489	1	-
Escherichia coli O127: H6 WF073522	1	-
Escherichia coli O148: H28 WF063224	1	-
Escherichia coli C600 ATCC25922, ATCC43984, ATCC43985, ATCC8739, C600, DH5α	6	-
Vibrio parahaemolyticus O1: K25 WF 04213	1	-
Vibrio parahaemolyticus O1: K56 WF 02314	1	-
Vibrio parahaemolyticus O3: K6 WF 04232	1	-
Vibrio parahaemolyticus O3: K6 WF 01031, WF 04506, WF 06215	3	-
Vibrio parahaemolyticus O3: K6 WF 02321, WF 03218, WF 05612	3	-
Vibrio parahaemolyticus O3: K12 WF 02108	1	-
Vibrio parahaemolyticus O3: K72 WF 02471	1	-
Vibrio parahaemolyticus O4: K37 WF 01309, WF 04238	2	-
Vibrio parahaemolyticus O4: K55 WF 03256, WF 07521	2	-
Vibrio parahaemolyticus O4: K68 WF 07926, WF 03422, WF 02613	3	-
Vibrio parahaemolyticus O10: K28 WF 06438	1	-
Vibrio parahaemolyticus ATCC17802, ATCC27969	2	-
Vibrio vulnificus ATCC27562	1	-
Vibrio mimicus ATCC33653	1	-

Table 1 Reference strains included in the evaluation of *orfX*-LAMP *(Continued)*

Psuedomonas aeruginosa ATCC 27853	1	-
Salmonella enterica ATCC13076, ATCC14028, ATCC19585, ATCC29629	4	-
Salmonella typhimurium ATCC 14028, WF 04313	2	-
Salmonella choleraesuis ATCC 13312	1	-
Salmonella enteritidis WF 05148, WF 07086	2	-
Salmonella typhi WF 03201, WF 05026, WF 08138	3	-
Salmonella paratyphi WF 06426	1	-
Salmonella aberdeen WF 04542	1	-
Salmonella gallinarum WF 05938	1	-
Klebsiella pneumoniae ATCC 13883	1	-
Enterobacter cloacae ATCC 23355	1	-
Yersinia enterocolitica ATCC9610, ATCC27729	2	-
Total	116	

number of attributes, including rapidity, simplicity in operation, specificity, sensitivity and expense, were also evaluated to verify the application of the *orfX*-LAMP assay for MRSA detection. Targeting on six distinct regions, inner and outer primers of the LAMP assay were highly specific in comparison with conventional PCR techniques. In this study, the specificity of the *orfX*-LAMP assay was verified by indispensibility of each primer, without any false positive test obtained from reference strains and 100% PPV for application. As sensitivity was concerned, in comparison

Table 2 Application of the *orfX*-LAMP on clinical *Staphylococcus* strains

Samples	LAMP	PCR
MRSA		
Bloodstream	100% (36/36)	94.4% (34/36)
Respiratory tract	98.6% (145147)	91.8% (135/147)
Skin and soft tissue	98.0% (238/243)	92.6% (225/243)
Urinary tract	98.6% (73/74)	91.9% (68/74)
Other	98.5% (65/66)	86.4% (57/66)
Subtotal	98.4% (557/566)	91.7% (519/566)
MSSA	0% (0/25)	0% (0/25)
MRCNS	0% (0/53)	0.0% (2/53)
MSCNS	0% (0/23)	0.0% (1/23)
Subtotal	0% (0/101)	0.0% (3/101)

Samples, Blood culture specimens included blood inoculated into aerobic and/or anaerobic blood culture medium. Tracheal secretion specimens included secretions from the trachea and bronchia and bronchoalveolar lavage fluid specimens. Wound specimens included surgical wound specimens but also specimens from ulcers, fistulae, abscesses, drainage fluids, catheters sites, percutaneous endoscopic gastrostomy insertion sites, and tracheostomas. Urine contained native urine from urinary catheters or bladder puncture and inoculated culture systems. Other specimens consisted of skin swabs specimens (from the nares, axilla, groin, or perianal area), as well as swab specimens of the throat, tonsils, eye, ear, and vagina; body liquids; puncture exudates.

with regular PCR (ranging from 10^3-10^5 CFU/reaction) [6] and previous LAMP detection methodoloy (ranging from 10^2-10^3 copies of DNA) [42], the detection limit of *orfX*-LAMP was found to be 10 copies DNA/tube and 10 CFU/reaction. For rapidity, approximate 60 min was required in the application of *orfX*-LAMP, including DNA process, isothermal reaction and result determination.

Conclusions

The first study on detection of MRSA by LAMP assay was reported lately [6], however, 3 individual LAMP assays with targets including 16SrRNA, *femA* and *mecA* were laborious and time and expense demanding. Also, in another preliminary *Staphylococcus* LAMP identification assay, 120 min and 2 targets (*spaA* and *mecA*) were required for the procedure [42]. In this current study, a simple LAMP assay targeting on *orfX* for rapid MRSA detection had been developed, evaluated and optimized, which required approximately 60 min for each test and used the highly conserved target *orfX*. This *orfX*-LAMP had been further applied to the detection of 667 clinical *Staphylococcus* strains, and the sensitivity, specificity, PPV and NPV were found to be 98.4%, 100%, 100% and 92.7%, respectively. In comparison with conventional PCR, this established LAMP assay exhibited advantages on detection limit, sensitivity, simplicity and rapidity. In conclusion, this *orfX*-LAMP assay had been demonstrated to be a valid and rapid detection method on MRSA, which will undoubtedly aid in the broad application of bacteriological MRSA detection.

Methods

Bacterial strains

For development and evaluation of the *orfX*-LAMP assay, 116 reference strains were included, with various species of gram-negative and -positive isolates (Table 1). As application was concerned, the optimized LAMP and PCR assays was performed on a total of 667 *Staphylococcus* strains, including 566 MRSA, 25 MSSA, 53 MRCNS and 23 MSCNS strains (Table 2), which had been previously isolated from various clinical samples during 2001-2006 and preliminarily identified [6,9,15]. All strains were identified as *S. aureus* using standard procedures: colony morphology, Gram staining, testing of catalase, hyaluronidase and coagulase, the Vitek 2 automated system and the API-Staph commercial kit. Methicillin resistance was determined by susceptibility testing on oxacillin-screening agar, confirmed by latex agglutination for PBP2a and *mecA* detection by PCR [43].

Primer design

The protocol of this LAMP assay for rapid MRSA detection was designed to target on the specific *orfX*, which located on the site of SCC*mec* and had been considered to be a highly conserved open reading frame in *S. aureus*. The sequences of orfX had been acquired on GenBank, including SCCmec type I (NCTC 10442, AB033763), SCCmec type II (N315, D86934), SCCmec type III (Mu50, gi:57634611), CA05, MR108, 86-3P, M03-6, 43000, SCCmec type V (WIS). A set of LAMP primers (Table 3) targeting 8 distinct regions on *orfX* was designed via PrimerExplorer (V4), including forward inner primer (FIP) with the complementary sequence of F1 (F1c), a T–T–T–T linker and F2, backward inner primer (BIP) with the complementary sequence of B1 (B1c), a T–T–T–T linker and B2, the outer primers F3 and B3 located outside of the F2 and B2 regions, loop primers LF and LB located between F2 and F1 or B1 and B2, respectively.

Establishment and optimization of *orfX*-Lamp assay

To evaluate and optimize the *orfX*-LAMP assay, 52 Gram-positive and 64 -negative isolates were employed as reference strains. Cultural conditions and DNA extraction of gram-negative and gram-positive strains were performed as described previously [41,44-47]. In brief, template DNA from *S. aureus* strains were prepared from overnight Tryptic Soy Broth (TSB) cultures at 37°C with shaking, and the collected culture was then diluted

Table 3 The sequences and information of LAMP primers

Label	5'pos	3'pos	len	Tm	5'dG	3'dG	GCrate	Sequence
F3	204	222	19	56.03	−5.56	−4.06	0.42	ACCACAATCMACAGTCATT
B3	398	415	18	55.58	−7.53	−4.56	0.50	CCCGCATCATTTGATGTG
FIP			39					CAAAGTCGCTTTGCCCTTGG-GATGCTATCTTCCGAAGGA
BIP			41					GATCAAACGGCCTGCACAAG-GRAATGTCATTTTGCT RAATG
F2	243	261	19	55.33	−5.14	−4.71	0.47	GATGCTATCTTCCGAAGGA
F1c	291	310	20	61.82	−4.16	−5.45	0.55	CAAAGTCGCTTTGCCCTTGG
B2	377	397	21	55.47	−4.51	−4.07	0.38	GR(G or A)AATGTCATTTTGCTRAATG
B1c	326	345	20	61.55	−3.92	−4.66	0.55	GATCAAACGGCCTGCACAAG
LF	266	286	21	61.30	−6.24	−5.45	0.48	TGCGTTGGTTCAATTCTTGGG
LB	346	367	22	60.11	−5.26	−3.73	0.45	GACGTCTTACAACGCAGTAACT

10-fold in 10 mM Tris-HCl (pH 8.0) containing 1 mM EDTA. The suspension was boiled for 10 min and further kept on ice. After centrifugation at 12,000 *g* for 3 min, the resulting supernatant was used as templates for LAMP and PCR assays. Evaluation and optimization of this *orfX*-LAMP assay included the study of mixture volumn (4 volumns, with 12.5 μl, 25 μl and 50 μl), reaction temperature (8 temperatures, with 59°C, 60°C, 61°C, 62°C, 63°C, 64°C, 65°C and 66°C), reaction time (8 time points, with15 min, 30 min, 45 min, 60 min, 75 min, 90 min, 105 min and 120 min), specificity (including the primers and strains) and detection limit. LAMP assays was carried out in 3 different reaction mixture volumns, containing 1.6 μM (each) of the primers FIP and BIP, 0.2 μM (each) of the primers F3 and B3, 0.8 μM (each) of primers LF and LB, 1.6 mM of deoxynucleoside triphosphates, 6 mM $MgSO_4$, 1 M betain (Sigma, St. Louis, MO, USA), 1 X thermopol buffer (New England Biolabs, Ipswich, MA, USA), and specified amounts of target genomic DNA. The reaction was initiated by heating at 95°C for 3 min, then chilled on ice for 30 sec, with 1 μl (8 U) of Bst DNA polymerase (New England Biolabs, Ipswich, MA, USA) further added. After incubation at various temperatures ranging from 59°C to 66°C for 15 min-120 min, the reaction was terminated by heating at 80°C for 2 min. As PCR assay was concerned, the amplification was carried out in 50 μl reactant, using the outer primers F3 and B3. The thermal profile was 94°C for 5 min, followed by 30 cycles of 94°C for 30 sec, 50°C for 30 sec, and 72°C for 30 sec and a final extension cycle at 72°C for 7 min. The amplified products (5 μl/well) were analyzed by gel electrophoresis in 2% agarose gels and stained with ethidium bromide for 10 min. Template DNA from MRSA 85/2082 was diluted for serial 10-fold and the detection limit of LAMP and PCR assays were ascertained by both minimal CFU of bacterial and plasmid DNA (with *orfX* recombinated in a T vector). For LAMP assays, the lowest bands from amplicons were sequenced by an ABI PRISM 310 genetic analyzer (PE Biosystems, Foster City, CA, USA).

Application of LAMP assays on clinical MRSA strains

After optimization, the *orfX*-LAMP assay had been further performed on 667 *Staphylococcus* (566 MRSA, 25 MSSA, 53 MRCNS and 23 MSCNS) strains, and was comparatively validated by standard PCR assay. Template DNA was prepared through a rapid procedure as aforementioned, with the required time as less than 30 min. Heating and isothermal amplification were separately performed on simple equipments including water bath and heating block. Amplification products of LAMP assay were dyed with Sybr Green, positive or negative were determined through both visually observation of the color change by naked eye and a fluorescence assay under UV [6,28,33]. This experiment was performed in triplicate to ensure reproducibility.

Abbreviations

LAMP: Loop-mediated isothermal amplification; MRSA: Methicillin-resistant *Staphylococcus aureus*; MSSA: Methicillin-sensitive *Staphylococcus aureus*; MRCNS: Methicillin-resistant coagulase-negative staphylococci; MSCNS: Methicillin-sensitive coagulase-negative staphylococci.

Competing interests

The authors declare that they have no competing interests.

Authors' contributions

JS carried out the LAMP detection assays and drafted the manuscript. XL carried out the DNA preparation and participated in the bacterial processing. HC performed the statistical analysis. YL carried out the sample collection and bacterial processing. DC participated in the design of the study and revised the manuscript. YL conceived of the study and participated in its design and coordination. GY participate in the design of the study and manuscript revision. All authors read and approved the final manuscript.

Acknowledgements

We sincerely thank Dr. Zhenbo Xu from University of Maryland for his excellent support for this study. We are grateful for Dr. Xihong Zhao from Taiwan National University for his review for this manuscript. This work was supported by the National 973-Plan (2012CB720800), the National Natural Science Foundation of China (31201362 and 31101278), the national science and technology support program (2012BAD37B01), the Fundamental Research Funds for the Central Universities (2014ZM0060) and the Funding of Guangdong Outstanding Dissertation (Dr. Zhenbo Xu).

Author details

[1]College of Light Industry and Food Sciences, South China University of Technology, Guangzhou 510640, China. [2]State Key Laboratory of Food Science and Technology, Jiangnan University, Wuxi 214122, China. [3]Department of Laboratory Medicine, First Affiliated Hospital of Guangzhou Medical College, Guangzhou 510120, China. [4]Guangzhou Women and Children's Medical Center, Guangzhou 510623, China. [5]Department of Microbial Pathogenesis, University of Maryland, Baltimore 21201, USA. [6]First Affiliated Hospital of Jinan University, Guangzhou 510620, China.

References

1. Schito GC: **The importance of the development of antibiotic resistance in *Staphylococcus aureus*.** *Clin Microbiol Infect* 2006, **12**(Suppl. 1):3–8.
2. Tacconelli E: **Methicillin-resistant *Staphylococcus aureus*: risk assessment and infection control policies.** *Clin Microbiol Infect* 2008, **14**:407–410.
3. Harbarth S: **Control of endemic methicillin-resistant *Staphylococcus aureus* –recent advances and future challenges.** *Clin Microbiol Infect* 2006, **12**:1154–1162.
4. Xu Z, Li L, Shi L, Shirtliff M: **Class 1 integron in staphylococci.** *Mol Biol Rep* 2011, **38**:5261–5279.
5. Chambers HF: **The changing epidemiology of *Staphylococcus aureus*?** *Emerg Infect Dis* 2001, **7**:178–182.
6. Xu Z, Li L, Chu J, Peters B, Harris M, Li B, Shi B, Shirtliff M: **Development and application of loop-mediated isothermal amplification assays on rapid detection of various types of staphylococci strains.** *Food Res Int* 2012, **47**:166–173.
7. Ben-Ami R, Navon-Venezia S, Chwartz D, Carmeli Y: **Infection of a ventriculoatrial shunt with phenotypically variable *Staphylococcus epidermidis* masquerading as polymicrobial bacteremia due to various Coagulase-Negative Staphylococci and kocuria varians.** *J Clin Microbiol* 2003, **41**:2444–2447.
8. Nouwen JL, Van Belkum A, De Marie S, Sluijs J, Wielenga JJ, Kluytmans JW, Verbrugh HA: **Clonal expansion of *Staphylococcus epidermidis* strains causing human catheter-related infections in a hemato-oncologic department.** *J Clin Microbiol* 1998, **36**:2696–2702.

9. Xu Z, Shi L, Alam M, Li L, Yamasaki S: **Integron-bearing methicillin-resistant coagulase-negative staphylococci in South China, 2001-2004.** *FEMS Microbiol Lett* 2008, **278:**223–230.
10. Kluytmans-VandenBergh MFQ, Kluytmans JAJW: **Community-acquired methicillin-resistant *Staphylococcus aureus*: current perspectives.** *Clin Microbiol Infect* 2006, **12**(Suppl. 1)**:**9–15.
11. Chang S, Sun C, Yang L, Luh K, Hsieh W: **Increasing nosocomial infections of methicillin-resistant Staphylococcus aureus at a teaching hospital in Taiwan.** *Int J Antimicrob Agents* 1997, **8:**109–114.
12. Zetola N, Francis JS, Nuermberger EL, Bishai WR: **Community-acquired methicillin-resistant *Staphylococcus aureus*: an emerging threat.** *Lancet Infect Dis* 2008, **5:**275–286.
13. Okuma K, Iwakawa K, Turnidge JD, Grubb WB, Bell JM, O'Brien FG, Coombs GW, Pearman JW, Tenover FC, Kapi M, Tiensasitorn C, Ito T, Hiramatsu K: **Dissemination of new methicillin-resistant *Staphylococcus aureus* clones in the community.** *J Clin Microbiol* 2002, **40:**4289–4294.
14. Vandenesch F, Naimi T, Enright MC, Lina G, Nimmo GR, Heffernan H: **Community-acquired methicillin-resistant *Staphylococcus aureus* carrying Panton-Valentine leukocidin genes: worldwide emergence.** *Emerg Infect Dis* 2003, **9:**978–984.
15. Xu Z, Shi L, Zhang C, Zhang L, Li X, Cao Y, Li L, Yamasaki S: **Nosocomial infection caused by class 1 integron-carrying *Staphylococcus aureus* in a hospital in South China.** *Clin Microbiol Infect* 2007, **13:**980–984.
16. Xu Z, Li L, Alam M, Yamasaki S, Shi L: **First confirmation of integron-bearing methicillin-resistant *Staphylococcus aureus*.** *Curr Microbiol* 2008, **57:**264–268.
17. Ito T, Katayama Y, Asada K, Mori N, Tsutsumimoto K, Tiensasitorn C, Hiramatsu K: **Structural comparison of three types of staphylococcal cassette chromosome *mec* integrated in the chromosome in methicillin-resistant *Staphylococcus aureus*.** *Antimicrob Agents Chemother* 2001, **45:**1323–1336.
18. Katayama Y, Ito T, Hiramatsu K: **A new class of genetic element, staphylococcus cassette chromosome *mec*, encodes methicillin resistance in *Staphylococcus aureus*.** *Antimicrob Agents Chemother* 2000, **44:**1549–1555.
19. Hanssen A, Kjeldsen G, Sollid JUE: **Local variants of staphylococcal cassette chromosome *mec* in sporadic methicillin-resistant *Staphylococcus aureus* and methicillin-resistant coagulase-negative Staphylococci: evidence of horizontal gene transfer?** *Antimicrob Agents Chemother* 2004, **48:**285–296.
20. Oliveira DC, Lencastre H: **Multiplex PCR strategy for rapid identification of structural types and variants of the *mec* element in methicillin-resistant *Staphylococcus aureus*.** *Antimicrob Agents Chemother* 2002, **46:**2155–2161.
21. Qi W, Ender M, O'Brien F, Imhof A, Ruef C, McCallum N, Berger-Bächi B: **Molecular epidemiology of methicillin-resistant *staphylococcus aureus* in Zürich, Switzerland (2003): prevalence of type IV SCC*mec* and a New SCC*mec* element associated with isolates from intravenous drug users.** *J Clin Microbiol* 2005, **43:**5164–5170.
22. You R, Gui Z, Xu Z, Shirtliff M, Yu G, Zhao X, Shi L, Li B, Su J, Li L: **Methicillin-resistance *staphylococcus aureus* detection by an improved rapid PCR assay.** *Afr J Microbiol Res* 2012, **6:**7131–7133.
23. Xu Z, Lin L, Zhao X, Chu J, Li B, Shi L, Su J, Shirtliff M: **Development and application of a novel multiplex polymerase chain reaction (PCR) assay for rapid detection of various types of staphylococci strains.** *Afr J Microbiol Res* 2011, **5:**1869–1873.
24. Sauer S, Kliem M: **Mass spectrometry tools for the classification and identification of bacteria.** *Nat Rev Microbiol* 2010, **8:**74–82.
25. Notomi T, Okayama H, Masubuchi H, Yonekawa T, Watanabe K, Amino N, Hase T: **Loop-mediated isothermal amplification of DNA.** *Nucleic Acids Res* 2000, **28:**e63.
26. Mori Y, Nagamine K, Tomita N: **Detection of loop-mediated isothermal amplification reaction by turbidity derived from magnesium pyrophosphate formation.** *Biochem Bioph Res Co* 2001, **289:**150–154.
27. Nagamine K, Watanabe K, Ohtsuka K: **Loop-mediated isothermal amplification reaction using a nondenatured template.** *Clin Chem* 2001, **47:**1742–1743.
28. Zhao X, Li Y, Wang L, You L, Xu Z, Li L, He X, Liu Y, Wang J, Yang L: **Development and application of a loop-mediated isothermal amplification method on rapid detection Escherichia coli O157 strains from food samples.** *Mol Biol Rep* 2010, **37:**2183–2188.
29. Horisaka T, Fujita K, Iwata T, Nakadai A, Okatani AT, Horikita T, Taniguchi T, Honda E, Yokomizo Y, Hayashidani H: **Sensitive and specific detection of Yersinia pseudotuberculosis by loop-mediated isothermal amplification.** *J Clin Microbiol* 2004, **42:**5349–5352.
30. Ohtsuka M, Yanagawa K, Takatori K: **Detection of *Salmonella enterica* in naturally contaminated liquid eggs by loop-mediated isothermal amplification, and characterization of *Salmonella* isolates.** *Appl Environ Microbiol* 2005, **71:**6730–6735.
31. Okamura M, Ohba Y, Kikuchi S, Suzuki A, Tachizaki H, Takehara K, Ikedo M, Kojima T, Nakamura M: **Loop-mediated isothermal amplification for the rapid, sensitive, and specific detection of the O9 group of *Salmonella* in chickens.** *Vet Microbiol* 2008, **132:**197–204.
32. Ueda S, Kuwabara Y: **The rapid detection of *Salmonella* from food samples by Loop-mediated isothermal amplification (LAMP).** *Biocontrol Sci* 2009, **14:**73–76.
33. Zhao X, Wang L, Chu J, Li Y, Li Y, Xu Z, Li L, Shirtliff M, He X, Liu Y, Wang J, Yang L: **Development and application of a rapid and simple loop-mediated isothermal amplification method for food-borne *Salmonella* detection.** *Food Sci Biotechnol* 2010, **19:**1655–1659.
34. Chen S, Ge B: **Development of a *toxR*-based loop-mediated isothermal amplification assay for detecting Vibrio parahaemolyticus.** *BMC Microbiol* 2010, **10:**41–45.
35. Nemoto J, Sugawara C, Akahane K, Hashimoto K, Kojima T, Ikedo M, Konuma H, Hara-Kudo Y: **Rapid and specific detection of the thermostable direct hemolysin gene in *Vibrio parahaemolyticus* by loop-mediated isothermal amplification.** *J Food Prot* 2009, **72:**748–754.
36. Yamazaki W, Ishibashi M, Kawahara R, Inoue K: **Development of a loop-mediated isothermal amplification assay for sensitive and rapid detection of *Vibrio parahaemolyticus*.** *BMC Microbiol* 2008, **8:**163–167.
37. Yamazaki W, Kumeda Y, Misawa N, Nakaguchi Y, Nishibuchi M: **Development of a loop-mediated isothermal amplification assay for sensitive and rapid detection of the *tdh* and *trh* genes of *Vibrio parahaemolyticus* and related *Vibrio* species.** *Appl Environ Microbiol* 2010, **76:**820–828.
38. Zhao X, Wang L, Chu J, Li Y, Li Y, Xu Z, Li L, Shirtliff M, He X, Liu Y, Wang J, Yang L: **Rapid detection of *Vibrio parahaemolyticus* strains and virulent factors by loop-mediated isothermal amplification assays.** *Food Sci Biotechnol* 2010, **19:**1191–1197.
39. Wang L, Zhao X, Chu J, Li Y, Li Y, Li C, Xu Z, Zhong Q: **Application of an improved loop-mediated isothermal amplification detection of *Vibrio parahaemolyticus* from various seafood samples.** *Afr J Microbiol Res* 2011, **5:**5765–5771.
40. Zhao X, Wang L, Li Y, Xu Z, Li L, He X, Liu Y, Wang J, Yang L: **Development and application of a loop-mediated isothermal amplification method on rapid detection of *Pseudomonas aeruginosa* strains.** *World J Microbiol Biotechnol* 2011, **27:**181–184.
41. Wang L, Li Y, Xu Z, Zhong Q: **Development and application of a simple loop-mediated isothermal amplification method on rapid detection of *Listeria monocytogenes* strains.** *Mol Biol Rep* 2012, **39:**445–449.
42. Misawa Y, Yoshida A, Saito R, Yoshida H, Okuzumi K, Ito N, Okada M, Moriya K, Koike K: **Application of loop-mediated isothermal amplification technique to rapid and direction detection of methicillin-resistant *Staphylococcus aureus* (MRSA) in blood cultures.** *J Infect Chemother* 2007, **13:**134–140.
43. Xu Z, Li L, Shirtliff M, Peters B, Li B, Peng Y, Alam M, Yamasaki S, Shi L: **Resistance class 1 integron in clinical methicillin-resistant *Staphylococcus aureus* strains in Southern China, 2001-2006.** *Clin Microbiol Infect* 2011, **17:**714–717.
44. Parida M, Horioke K, Ishida H: **Rapid detection and differentiation of dengue virus serotypes by a real-time reverse transcription-loop-mediated isothermal amplification assay.** *J Clin Microbiol* 2005, **43:**2895–2903.
45. Kaneko H, Kawana T, Fukushima E, Suzutani T: **Tolerance of loop-mediated isothermal amplification to a culture medium and biological substances.** *J Biochem Bioph Meth* 2007, **70:**499–501.
46. Xu Z, Li L, Shirtliff M, Alam M, Yamasaki S, Shi L: **Occurrence and characteristics of class 1 and 2 integrons in *Pseudomonas aeruginosa* isolates from patients in Southern China.** *J Clin Microbiol* 2009, **47:**230–234.
47. Xu Z, Li L, Shirtliff M, Peters B, Peng Y, Alam M, Yamasaki S, Shi L: **First report of class 2 integron in clinical *Enterococcus faecalis* and class 1 integron in *Enterococcus faecium* in South China.** *Diag Microbiol Infect Dis* 2010, **68:**315–317.

Expression and characterization of a GH43 endo-arabinanase from *Thermotoga thermarum*

Hao Shi[1,2], Huaihai Ding[1,2], Yingjuan Huang[1,2], Liangliang Wang[1,2], Yu Zhang[1,2], Xun Li[1,2] and Fei Wang[1,2*]

Abstract

Background: Arabinan is an important plant polysaccharide degraded mainly by two hydrolytic enzymes, endo-arabinanase and α-L-arabinofuranosidase. In this study, the characterization and application in arabinan degradation of an endo-arabinanase from *Thermotoga thermarum* were investigated.

Results: The recombinant endo-arabinanase was expressed in *Escherichia coli* BL21 (DE3) and purified by heat treatment followed by purification on a nickel affinity column chromatography. The purified endo-arabinanase exhibited optimal activity at pH 6.5 and 75°C and its residual activity retained more than 80% of its initial activity after being incubated at 80°C for 2 h. The results showed that the endo-arabinanase was very effective for arabinan degradation at higher temperature. When linear arabinan was used as the substrate, the apparent K_m and V_{max} values were determined to be 12.3 ± 0.15 mg ml^{-1} and $1{,}052.1 \pm 12.7$ μmol ml^{-1} min^{-1}, respectively (at pH 6.5, 75°C), and the calculated k_{cat} value was 349.3 ± 4.2 s^{-1}.

Conclusions: This work provides a useful endo-arabinanase with high thermostability andcatalytic efficiency, and these characteristics exhibit a great potential for enzymatic conversion of arabinan.

Keywords: Arabinan, Arabinose, Endo-arabinanase, *Thermotoga thermarum*

Background

Lignocellulosic biomass, the most abundant renewable carbon resources in the biosphere, has exhibited valuable industrial applications in many fields, such as pulp and paper, food processing, detergent, textile, synthetic biology, organic synthesis, pharmaceuticals and bioenergy production [1,2]. Arabinan, a polysaccharide constituent of hemicellulose, is composed of α-1,2- and/or α-1,3-linked to a α-1,5-linked L-arabinofuranosyl backbone residues [3-5]. Arabinan mainly consists of L-arabinose residues, existing in rhamnogalacturonan regions of pectins in the cell walls of several plants [6]. Microbial enzymes including hydrolases have represented an important part in industry due to their various favorable properties [7]. To generate L-arabinose from arabinan, two major hydrolytic enzymes, endo-arabinanase and α-L-arabinofuranosidase, are essential to the synergetic action. Endo-arabinanase (EC 3.2.1.99) can hydrolyze the arabinan α-1,5-linked L-arabinofuranosyl backbone to produce arabino-oligosaccharides or L-arabinoses, whereas α-L-arabinofuranosidase (EC 3.2.1.55) is capable of thoroughly hydrolyzing arabino-oligosaccharides and arabinan side chain substitutions to produce L-arabinoses [3]. In addition, exo-arabinanase (EC 3.2.1.-) is able to release terminal arabinose (non-reducing end with net retention of the anomeric configuration) or arabino-oligosaccharides from arabinan [8,9]. Based on the amino acid sequence similarities, endo-arabinanases and exo-arabinanases have been classified into glycoside hydrolase (GH) family 43 and GH93 family, respectively, whereas α-L-arabinofuranosidases belong to five GH families (GH3, 43, 51, 54, 62) (http://www.cazy.org/). These arabinan-degrading enzymes from different GH families have presented various applications in industrial processes such as food technology, nutritional medical research and organic synthesis [4].

A few reports about the endo-arabinanases from bacteria and fungi have been published so far [5,6,8,10-19]. Among all the studies above, only a novel hyperthermophilic endo-arabinanase from *Thermotoga petrophila* was biophysically characterized and the three-dimensional (3D) structure was released [1,17]. Although the genome

* Correspondence: hgwf@njfu.edu.cn
[1]College of Chemical Engineering, Nanjing Forestry University, Nanjing 210037, China
[2]Jiangsu Key Lab of Biomass-Based Green Fuels and Chemicals, Nanjing 210037, China

of hyperthermophilic *Thermotoga thermarum* DSM 5069 has been sequenced, no information is currently available on the endo-arabinanase. In this study, a putative GH43 β-xylosidase from *Thermotoga thermarum,* Tth Abn, was cloned and expressed. Its biochemical characterization and application in arabinan degradation were also investigated. Upon the evaluation of substrate specificity, it was interesting to find that the recombinant enzyme was an endo-arabinanase, which has not yet been reported.

Results

Amino acid sequence of Tth Abn

The *Tth abn* gene isolated from the *T. thermarum* genome is 1,437 bp in length coding for 479 amino acids. There is a signal peptide sequence (32 amino acids, 96 bp) in Tth Abn when analyzed by SignalP 4.0 (http://www.cbs.dtu.dk/services/SignalP/). Therefore, only the DNA fragment of 1,341 bp was amplified. Amino acid sequence of Tth Abn in GenBank is described as a putative β-xylosidase. However, it contains putative domains of α-L-arabinofuranosidase (EC 3.2.1.55), endo-arabinanase (EC 3.2.1.99) and β-xylosidase (EC 3.2.1.37) through the BLAST. In this study, the provisional β-xylosidase was finally confirmed as an endo-arabinanase through the biochemical analysis. Tth Abn exhibited 64% identity to the β-xylosidase from *Thermotoga* sp. RQ2 (GenBank accession number YP_001738698) and *Thermotoga petrophila* RKU-1 (GenBank accession number YP_001244233), 59% identity to putative β-xylosidase *Dictyoglomus thermophilum* H-6-12 (GenBank accession number YP_002251442), and 50% identity to the endo-arabinanase from *B. subtilis* Abn2 H318A mutant (GenBank accession number YP_002251442 2X8T). To gain insights into the evolutionary relationships among endo-arabinanases, the phylogenetic trees with 13 candidate sequences were constructed using the NJ and MP methods, respectively, which both supported the same topological structures as shown in Figure 1. Phylogenetic analysis indicated that Tth Abn from *T. thermarum* had a close relationship with an endo-arabinanase from *T. petrophila* and a putative β-xylosidase from the same genus *T.* sp.

Homology modeling revealed that Tth Abn had a similar 5-fold β-propeller structure as the *B. subtilis* endo-arabinanase Abn2 H318A mutant [11] (Figure 2). Both sequence comparison and homology modeling indicated that Asp^{146} and Glu^{199} residues were the catalytic nucleophile and proton donor, respectively [11]. Furthermore, there are other conserved amino acid residues adjacent to the catalytic center, His^{9}, Asp^{10}, Pro^{11}, Trp^{38}, Ala^{73} and Asn^{143}, which may play some roles in catalytic activity.

Expression and purification of Tth Abn

Mature protein coding 447 amino acids was successfully expressed in the cytoplasmatic fraction of *E. coli* BL21 (DE3). Then, the protein in the cell-free extract was purified by more than 95% homogeneity after heat treatment and affinity chromatography (Additional file 1: Table S1). Finally, the purified fusion enzyme showed a single band on a SDS-PAGE gel with an estimated molecular weight of 36 kDa (Figure 3). To examine the oligomerization state of the enzyme, size exclusion

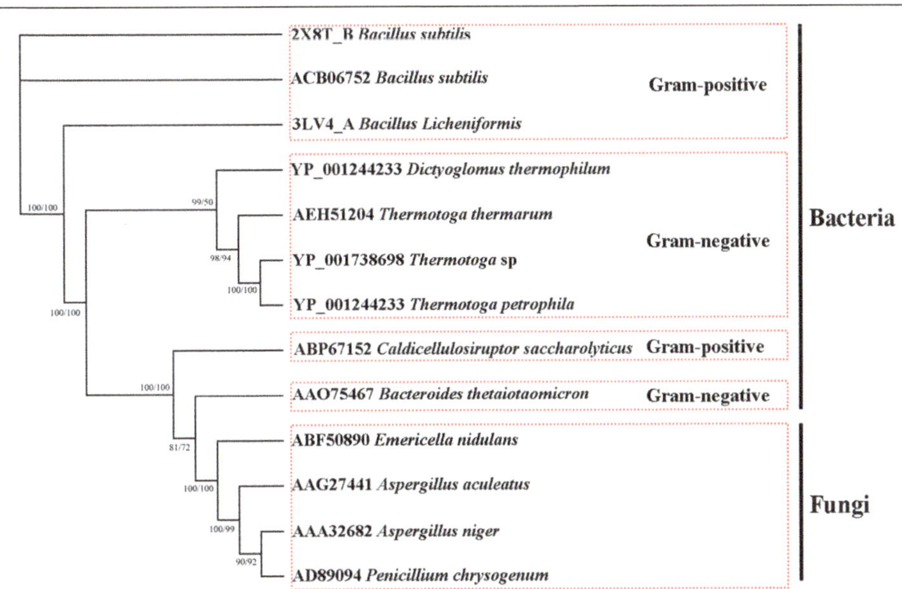

Figure 1 Neighbor-Joining (NJ) and Maximum-Parsimony (MP) trees resulted from analysis of endo-arabinanase with 13 amino acid sequences. Numbers on nodes correspond to percentage bootstrap values for 1000 replicates. The former confidence number in the Figure 1 is for NJ tree and the latter is for MP tree.

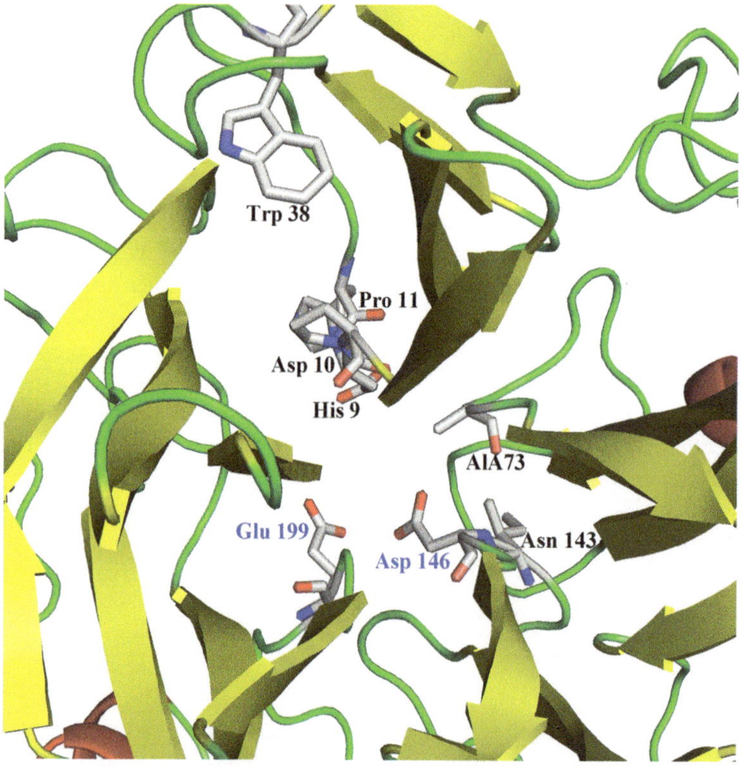

Figure 2 Predicted three-dimensional structure of Tth Abn. The potential catalytic amino acid residues are Asp[146] and Glu[199].

chromatography was performed. The native protein formed monomer in solution with a calculated MW 35,622 Da according to the calibration curve of the gel filtration column.

Biochemical properties of Tth Abn

Enzymatic characteristics of purified recombinant Tth Abn were determined and summarized in Tables 1, 2 and 3. Substrate specificity was assayed with different substrates (Table 1). It was interesting to find that Tth Abn was active towards linear arabinan, debranched arabinan and sugar beet arabinan, but there was no activity towards 1,4-β-D-mannan, galactan, *p*-Nitrophenyl-β-D-xylopyranoside (*p*NPX), and *p*-nitrophenyl-α-L-arabinofuranoside (*p*NPAF). These results indicated that the enzyme only exhibited arabinanase activity.

Effect of pH on Tth Abn activity was determined in 50 mM imidazole-potassium buffer ranging from pH 5.5 to 8.5 (Figure 4a). Tth Abn was found to exhibit an optimum activity at pH 6.5 and was able to retain more than 95% of its initial activity at 70°C for 1 h (Figure 4b). As a function of temperature, the enzyme displayed the highest activity at 75°C and retained more than 90% of its maximum activity even at 85°C after 10 min incubation (Figure 4c). Incubations at different temperatures to determine the enzyme's thermal stability were also carried out. As shown in the result, the enzyme kept nearly its initial activity at 75°C after 2 h incubation, and could still retain more than 80% of its original activity at 80°C for 2 h (Figure 4d).

Effects of metal ions and chemicals on Tth Abn activity were shown in Table 2. In various assays, the enzyme activity was apparently stimulated by 1 mM Mn^{2+}, Ca^{2+}, Ba^{2+}, or chemical reagents 0.05% Tween 60 and 0.1% SDS. On the contrast, Al^{3+}, Cu^{2+} and Zn^{2+} could inhibit the enzyme activity evidently. Kinetic studies in the presence of three arabinans as the substrates at optimum temperature and pH allowed the determination of the Michaelis-Menten parameters as shown in Table 3.

Degradation of arabinan by Tth Abn

Catalytic ability of Tth Abn on arabinan was investigated by analyzing the digestion products of linear , debranched and sugar beet arabinans (Figure 5a,b,c). Clearly from the result, the end products for hydrolysis of linear and debranched arabinans were arabinose, arabinobiose and arabinotriose after degradation for 3 h. However, it displayed very weak catalytic ability towards the hydrolysis of sugar beet arabinan.

Discussion

Enzymatic hydrolysis of arabinose-linked hemicellulose in plant cell wall requires endo-arabinanase, arabinofuranosidase and arabinoxylan arabinofuranohydrolase, which

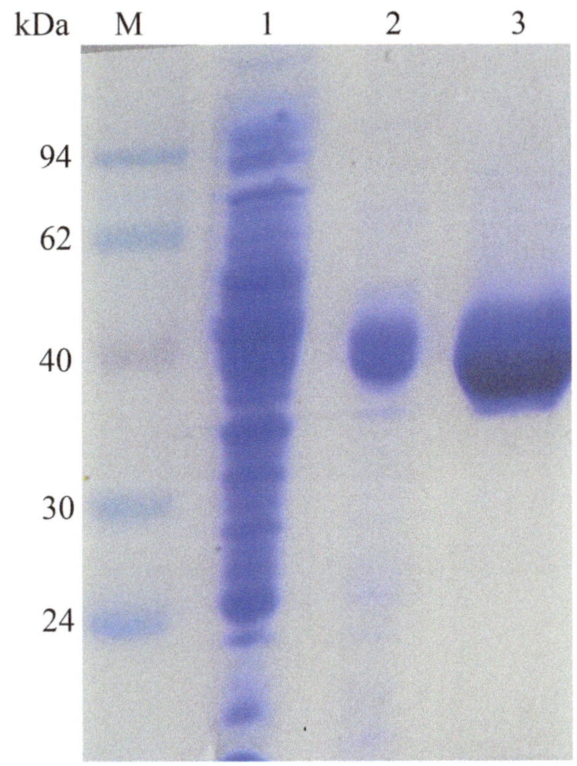

Figure 3 SDS-PAGE analysis of recombinant Tth Abn in *E. coli* BL21 (DE3). Lane M: protein marker, lane 1: cell-free extract of *E. coli* BL21 (DE3) harboring recombinant plasmids, lane 2: the purified Tth Abn eluted with 0.2 M imidazole buffer, lane 3: the purified Tth Abn eluted with 0.4 M imidazole buffer.

Table 2 Effects of cations and chemical reagents on the activity of purified Tth Abn

Cations[a]/Chemical reagents[b]	Residual activity (%)
Control	100 ± 0.6
Mg^{2+}	99 ± 1.2
Zn^{2+}	14 ± 0.3
Mn^{2+}	128 ± 2.9
Ba^{2+}	113 ± 3.6
Ca^{2+}	113 ± 1.7
Al^{3+}	0
Cu^{2+}	0
Co^{2+}	82 ± 1.7
Ni^{2+}	31 ± 0.5
Tween 60	133 ± 2.5
Tris	41 ± 0.4
SDS	111 ± 1.8

[a]Final concentration, 1 mM. [b]Final concentration, 0.05% Tween 60 and Tris, 0.1% SDS. Values shown were the mean of triplicate experiments, and the variation about the mean was below 5%.

are capable of releasing arabinosyl oligomers and L-arabinose from arabinose-containing polysaccharides [20].

We reported an endo-arabinanase (Tth Abn) from the hyperthermophilic bacterium *T. thermarum* here which was originally predicted as a β-xylosidase yet possessed the capability of arabinan depolymerization. Similar results have been published that endo-arabinanase from *Thermotoga petrophila* RKU-1 and *B. subtilis* were confirmed as endo-arabinanases through biophysical characterizations although once were predicted as β-xylosidases [1,20]. It is well known that hydrolases from GH43 family utilize a single displacement mechanism, which results in an inverted anomeric configuration for hydrolysis products [8]. Amino acid sequence alignment and homology modeling indicated that Asp^{146} and Glu^{199} in Tth Abn were the catalytic nucleophile (base) and proton donor, respectively. The result is also supported by the fact that aspartic acid and glutamic acid residues are the catalytic nucleophile and proton donor in all members of GH43 family (http://www.cazy.org). Residue Asp^{10} may act as a pKa modulator and maintain the correct alignment of the general acid residue relative to the substrate [21]. Residues Asp^{10}, Asp^{146} and Glu^{199} located in the deep cavity at the centre of the β-propeller are also found in all members of GH32, 43, 62 and 68 [22]. As the consequence of the flexibility reduction of the polypeptide chain, residue Pro^{11} may contribute to the high thermal stability of the enzyme [23].

Phylogenetic analysis and enzymatic properties showed that Tth Abn was distant with the endo-arabinanases from *Bacillus subtilis*, *Bacillus licheniformis* and *Aspergillus aculeatus* [4,5,19]. However, it revealed a close relationship with *T. thermarum* and *T. petrophila*, indicating that they share the similar properties [1,17]. Other endo-arabinanases from genus *Thermotoga* including *T. thermarum* endo-arabinanase have not yet been studied. Since there is 64% amino acid sequences similarity between endo-arabinanases *T. thermarum* and *T. petrophila* and is also inferred by the experimental data shown in Table 3, it is confirmed that Tth Abn could be an endo-arabinanase with some specific properties.

Table 1 Specific activity of Tth Abn on various substrates

Substrate	Specific activity (μmol arabinose min^{-1} mg^{-1})
Linear arabinan	237.7 ± 10.0
Debranched arabinan	199.7 ± 6.8
Sugar beet arabinan	42.5 ± 1.5
1,4-β-D-Mannan	ND
Galactan	ND
p-Nitrophenyl-α-L-arabinofuranoside	ND
p-Nitrophenyl-β-D-xylopyranoside	ND

ND: not detected. Values shown were the mean of triplicate experiments, and the variation about the mean was below 5%.

Table 3 Characteristics of endo-arabinanases from *T. thermarum* DSM 5069 and other microorganisms

Strain	K_m (mg ml^{-1})	V_{max} (μmol ml^{-1} min^{-1})	k_{cat} (s^{-1})	k_{cat}/K_m (ml mg^{-1} s^{-1})	Optimal temp (°C)	Reference
Thermotoga thermarum	[a]12.3 ± 0.15 [b]3.3 ± 0.12 [c]28.6 ± 0.88	[a]1,052.1 ± 12.7 [b]510.9 ± 17.5 [c]164.0 ± 4.6	[a]349.3 ± 4.2 [b]169.6 ± 5.9 [c]54.4 ± 1.6	[a]28.4 ± 0.4 [b]51.4 ± 0.1 [c]1.9 ± 0.0	75	This study
Caldicellulosiruptor saccharolyticus	[b]18.1	ND	[b]49.8	[b]2.8	75	[3]
Bacillus licheniformis	[b]19 ± 0.2	ND	[b]161 ± 1.2	[b]8.47 ± 0.16	35	[5]
Bacillus subtilis	[a]2.0 ± 0.24	[a,d]250 ± 1.2	ND	ND	50	[20]
Thermotoga petrophila	[b]19.9 ± 4.7	[b]478.2 ± 63.6	ND	ND	73	[1]
Aspergillus aculeatus	[b]66	11	ND	ND	50	[16]

ND: not determined. Data were carried out with linear arabinan (a), debranched arabinan (b), or sugar beet arabinan (c) as substrate. Values shown were the mean of triplicate experiments, and the variation about the mean was below 5%. d: μmol mg^{-1} min^{-1}.

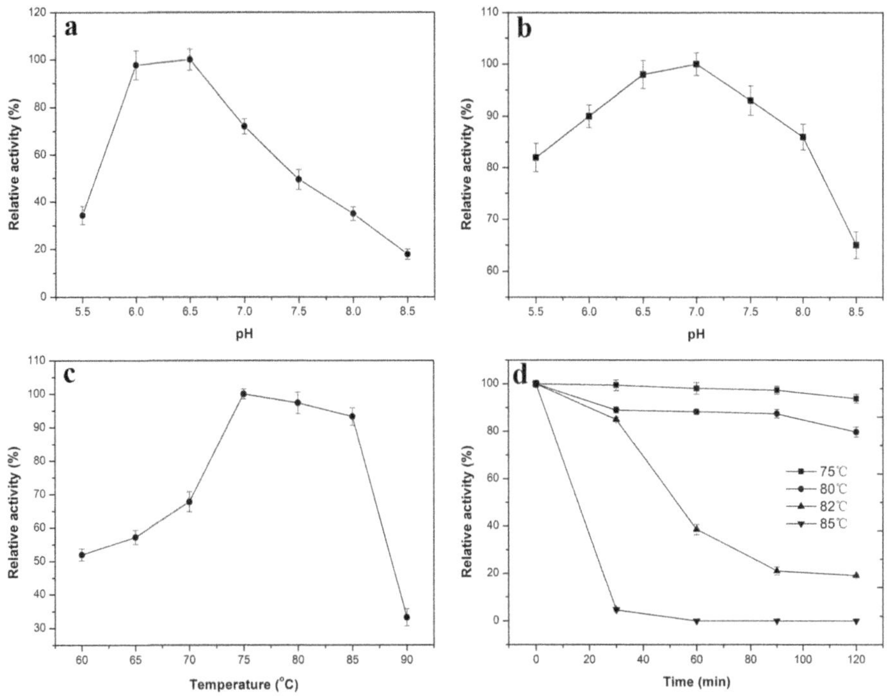

Figure 4 Effects of pH and temperature on the activity and stability of the recombinant Tth Abn endo-arabinanase. a. Optimal pH of the Tth Abn. **b**. pH stability of the Tth Abn. **c**. Effect of temperature on Tth Abn activity. **d**. The thermostability of the Tth Abn. The residual activity was monitored, and the maximum activity was defined as 100% **(a, c)** or initial activity was defined as 100% **(b, d)**. Values shown were the mean of triplicate experiments, and the variation about the mean was below 5%.

Tth Abn exhibited good thermostability when incubated at 75° for 2 h. Endo-arabinanase from Gram-negative bacterium *T. petrophila* RKU-1 has been reported to have a pH optimum at 6.5 and retained 95% of initial activity at 90°C for 10 h [1]. While Gram-positive bacteria *B. subtilis* and *B. licheniformis* endo-arabinanases show optimum activity at near-neutral pH at 60°C and 35°C, respectively [4]. Endo-arabinanase *Aspergillus aculeatus* from fungus displays optimum activity at pH 5.5 and 50°C [16]. In general, endo-arabinanases from bacteria exhibit the near-neutral pH optima, while those from fungi display the acid pH optima. The enzyme activities regarding pH from bacteria and fungi are also consistent with their inherent properties.

Cations and chemical reagents exhibited different effects on the activity of the endo-arabinanase Tth Abn. With the evidence that one Ca^{2+} ion was observed in the catalytic cavity of *Bacillus subtilis* about 5 Å below the catalytic carboxylates and the presence of ions in an equivalent location in other arabinanases [22].

Compared to other endo-arabinanases, Tth Abn demonstrated the highest V_{max} value among all the known

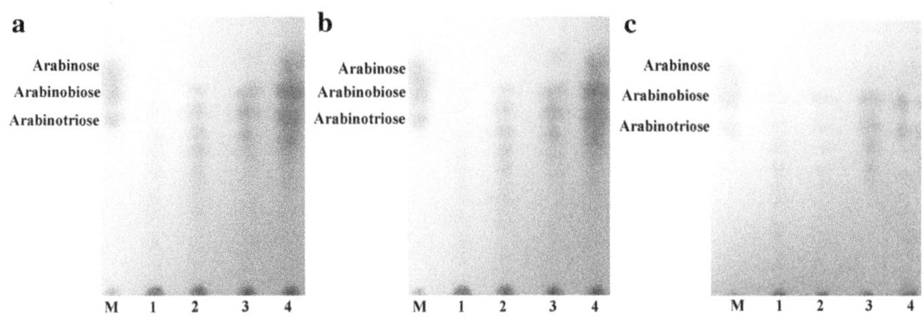

Figure 5 Analysis of linear arabinan and debranched arabinan hydrolyzed by Tth Abn endo-arabinanase. The products of the reaction were examined with TLC. M: arabinose, arabinobiose and arabinotriose. **a**. Lane 1, 2, 3, 4: linear arabinan samples (2%, wt/vol) incubated with 1 μg Tth Abn endo-arabinanase for 0.5 h, 1 h, 2 h and 3 h. respectively. **b**. Lane 1, 2, 3, 4: debranched arabinan samples (2%, wt/vol) incubated with 1 μg Tth Abn endo-arabinanase for 0.5 h, 1 h, 2 h and 3 h, respectively. **c**. Lane 1, 2, 3, 4: sugar beet arabinan samples (2%, wt/vol) incubated with 1 μg Tth Abn endo-arabinanase for 0.5 h, 1 h, 2 h and 3 h, respectively.

endo-arabinanases from thermophile and hyperthermophile when linear arabinan was used as the substrate. The catalytic efficiency (k_{cat}/K_m) of endo-arabinanase Tth Abn on debranched arabinanwas approximately 6-fold and 18-fold higher than *B. licheniformis* and *Caldicellulosiruptor saccharolyticus*, respectively (Table 3). In addition, catalytic efficiency for linear, debranched and sugar beet arabinans existed obvious differences (linear > debranched > sugar beet). This may be caused by substrate's composition and structure. As we know, a typical sugar composition of linear arabinan is: arabinose: galactose: rhamnose: galacturonic acid = 97.5: 0.4: 0.1: 2.0. For debranched and sugar beet arabinans, the contents of arabinose are lower than linear arabinan, and sugar beet arabinan has more substitutions attached to backbone while 1,2- and 1,3-α-L-arabinofuranosyl branch units of debranched arabinan have been all removed.

The fact that Tth Abn could hydrolyze linear and debranched arabinan into oligomers arabinose, arabinobiose and arabinotriose implied that the Tth Abn had endo-arabinanase activity but no exo activity. Only a few endo-arabinanases especially that from hyperthermophile have been reported to possess the function in generating arabino-oligosaccharides [1,17]. Although it was insensitive to the branched arabinan (eg. sugar beet arabinan), Tth Abn could be used together with other arabinan-degrading enzymes in synergistic reaction for the hydrolysis of branched ones.

Conclusions

In this study, a useful endo-arabinanase (Tth Abn) from *T. thermarum* DSM 5069 was expressed in *E. coli* with desirable features, such as high thermostability, catalytic efficiency, and hydrolytic reaction at high temperature. This is easily envisioned that Tth Abn endo-arabinanase exhibits a great potential for enzymatic conversion of arabinan through synergetic action with other arabinan-degrading enzymes.

Methods

Substrates

Linear arabinan, debranched arabinan, sugar beet arabinan, 1,4-β-D-mannan and galactan were purchased from Megazyme (Wicklow, Ireland). *p*NPX and *p*NPAF were purchased from Sigma Chemical Co., Ltd. (St. louis, USA).

Bacterial strains and growth conditions

E. coli Top10 (Novagen, San Diego, USA) was used for routine molecular cloning work and *E. coli* BL21 (DE3) (Novagen) was employed as the host for the expression of the recombinant Tth Abn. *E. coli* strains, both were grown in Luria-Bertani (LB) medium. Ampicillin (100 μg ml^{-1}, Shanghai, China) and isopropyl-β-D-thiogalactopyranoside (IPTG, 0.5 mM, Dalian, China) were added when necessary.

Construction of plasmids and strains

DNA manipulations were carried out according to standard methods [24]. Restriction enzymes and high-fidelity Ex-Taq DNA polymerase were purchased from Takara Biotechnology Co. Ltd. (Dalian, China) and used according to the manufacturer's instructions. DNA was extracted from agarose gels with BIOMIGA Gel Extraction Kit (Shanghai, China). Sequencing was performed using ABI 3730 DNA analyzer (Applied Biosystems, Foster City, USA). PCR amplifications were conducted using high-fidelity Ex-Taq DNA polymerase, and the resulting products were purified with BIOMIGA PCR Purification Kit (Shanghai, China).

The coding sequence of *Tth abn* gene was amplified from *T. thermarum* genomic DNA using primers 5'-GGAATTCCATATGGTATTCAACTGGGCAACTGTA-CAC-3' (forward) and 5'- CCGCTCGAGATCTTCGA TCCGAACTCCCCAG-3' (reverse). The primers contained restrictions sites *Nde*I and *Xho*I (underlined) for forward and reverse primers, respectively. The amplified DNA fragment was digested with *Nde*I and *Xho*I, and inserted into the corresponding sites in plasmid pET-20b (Novagen) to produce pET-20b-*Tth abn*. The correctness of the insert was confirmed by DNA sequencing.

Expression and purification of recombinant Tth Abn

E. coli BL21 (DE3) cells harboring recombinant plasmids were grown (200 rpm, 37°C) in LB medium (200 ml) with appropriate antibiotic selection. When the OD_{600} reached 0.6 ~ 0.8, the expression of Tth Abn was induced by the addition of 0.5 mM IPTG and the culture was further incubated (150 rpm, 25°C, 12 h) . Cells were harvested by centrifugation (10000 rpm, 4°C, 5 min). The pellet was washed twice with 20 mM Tris–HCl buffer (pH 8.0), and re-suspended in 5 ml of 5 mM imidazole, 0.5 mM NaCl, and 20 mM Tris–HCl buffer (pH 7.9). All steps were carried out at 4°C. The cell extracts after sonication were heat treated at 70°C for 30 min, and subsequently cooled in an ice bath. After centrifugation (15000 g, 4°C, 20 min), the resulting supernatant was loaded onto a 1 ml nickle affinity column (Novagen) and the bounded proteins were eluted by discontinuous imidazole gradient (30–1000 mM). The fractions that contained Tth Abn were dialyzed overnight against storage buffer (20 mM Na-phosphate, 50 mM NaCl, 10% glycerol, pH 7.0) and then were kept at −80°C until further use. The analysis of production and purity of the enzymes were determined by a 12% SDS-PAGE gel [25] using broad range molecular weight markers (MBI Fermemtas, Burlington, Canada) as standards. The protein concentration was determined by Bradford method using bovine serum albumin (BSA) as a standard

[26]. Oligomerization state of Tth Abn was analyzed by size exclusion chromatography on a AKTA*FPLC*™ (GE Healthcare Life Sciences, New Jersey, USA) system with a Superdex 200 10/30 GL column as described by Zhang et al. [27].

Biochemical characterization

Unless otherwise stated, all the enzyme-catalyzed reactions were performed at 75°C for 10 min in 50 mM imidazole-potassium buffer (pH 6.5) containing 0.1 μg enzyme and 0.4% (w/v) linear arabinan. The total reaction system was 0.2 ml. The reducing sugar content after hydrolysis of the polysaccharides was determined by the 3,5-dinitrosalicylic acid method [28] with L-arabinose as the standard. One enzyme activity unit was defined as the amount of enzyme producing 1 μmol of arabinose per minute. All assays were performed in triplicate.

Optimum temperature and pH for enzymatic activity of the Tth Abn were examined with linear arabinan 0.4% (w/v) as substrate for 10 min. The effect of pH on the activity was tested at pH 5.5 to 8.5 in 50 mM imidazole-potassium buffer at 75°C. The effect of temperature was assayed at temperatures ranging from 60°C to 95°C in 50 mM imidazole-potassium buffer at pH 6.5. The pH stability was determined from pH 5.5 to 8.5 at 75°C for 1 h. Thermostability of the enzyme was estimated by incubating the diluted solution of enzyme at different temperatures (75°C, 80°C, 82°C and 85°C). Samples were removed after different time intervals (30 min, 60 min, 90 min, and 120 min) and kept in ice for 10 min.

Enzymatic activity was also examined in the presence of metal ions and chemical reagents. Magnesium chloride ($MgCl_2$), zinc sulfate ($ZnSO_4$), barium chloride ($BaCl_2$), nickel sulfate ($NiSO_4$), manganese chloride ($MnCl_2$), calcium chloride ($CaCl_2$), cobalt chloride ($CoCl_2$), copper sulfate ($CuSO_4$) or aluminum sulfate ($Al_2(SO_4)_3$) were used at a concentration of 1 mM, and the chemical reagents Tween 60, Tris or sodium dodecyl sulfate (SDS) were added at 0.05%, 0.05% and 0.1%, respectively in the reaction mixture. The enzyme was incubated with each cation or reagent at 80°C for 2 h before the addition of linear arabinan 0.4% (w/v) to initiate the enzyme reaction. The activity of the enzyme without adding chemical reagent or metal ion was defined as 100%. The kinetic parameters, apparent K_m and V_{max} values were calculated according to the Lineweaver-Burk method at optimum pH and temperature using linear arabinan as substrate at concentrations ranging from 1.25 to 30 mg ml^{-1}.

Substrate specificity and degradation

The abilities of the enzyme to hydrolyze the natural and modified substrates, including linear arabinan, debranched arabinan, sugar beet arabinan, 1,4-β-D-mannan, galactan, *p*NPX and *p*NPAF, were measured at 75°C for various time (0.5 h, 1 h, 2 h and 3 h). The reaction mixture (600 μL) contained 2% substrate (wt/vol) and 1 μg enzyme in 50 mM imidazole-potassium buffer (pH 6.5). The reaction was stopped by cooling the mixture at 4°C in a water bath. Thin-layer chromatography (TLC) was applied to monitor the degradation of linear arabinan using arabinose, arabinobiose and arabinotriose as the standard markers. TLC was performed using G plates (Qingdao, China) with a mobile phase containing *n*-butanol/acetic acid/water (3:1:1). The sugars that degraded were detected using the orcinol/concentrated sulfuric acid reagent [29].

Bioinformatics analysis

Multiple sequence alignment was conducted using Clustal X2 [30]. Phylogenetic relationships were deduced using the Neighbor-Joining (NJ) and Maximum-Parsimony (MP) methods as performed in PAUP 4.0 [31]. Homology modeling was performed using Swiss Model [32-34]. The crystal structure of a GH43 endo-arabinanase from *Bacillus subtilis* (PDB entry code: 2X8T) was used as three-dimensional template for restraint-based modeling as implemented in SPDBV_4.04_PC.

Nucleotide sequence accession number

GenBank accession number of the *T. thermarum* DSM5069 endoarabinanase is AEH51204.

Additional file

Additional file 1: Supplementary Information.

Competing interests

The authors declare that they have no competing interests.

Authors' contributions

HS carried out the cloning and expression and drafted the manuscript. HD, YH, LW, and XL helped to purify and characterize the Tth Abn. YZ helped to revise the manuscript. FW directed the overall study and revised the manuscript. All authors have read and approved the final manuscript.

Acknowledgements

This work was financially supported by the National Natural Science Foundation of China (No. 31170537, 31370572), Jiangsu Provincial Government (CXZZ11_0526), and the Doctorate Fellowship Foundation of Nanjing Forestry University as well as the Priority Academic Program Development of Jiangsu Higher Education Institutions (PAPD).

References

1. Squina FM, Santos CR, Ribeiro DA, Cota J, de Oliveira RR, Ruller R, Mort A, Murakami MT, Prade RA: **Substrate cleavage pattern, biophysical characterization and low-resolution structure of a novel hyperthermostable arabinanase from Thermotoga petrophila.** *Biochem Biophys Res Commun* 2010, **399:**505–511.
2. Himmel ME, Ding SY, Johnson DK, Adney WS, Nimlos MR, Brady JW, Foust TD: **Biomass recalcitrance: Engineering plants and enzymes for biofuels production.** *Science* 2007, **315:**804–807.

3. Hong MR, Park CS, Oh DK: **Characterization of a thermostable endo-1,5-α-l-arabinanase from Caldicellulorsiruptor saccharolyticus.** *Biotechnol Lett* 2009, **31**:1439–1443.
4. Leal TF, Sá-Nogueira I: **Purification, characterization and functional analysis of an endo-arabinanase (AbnA) from Bacillus subtilis.** *FEMS Microbiol Lett* 2004, **241**:41–48.
5. Seo ES, Lim YR, Kim YS, Park CS, Oh DK: **Characterization of a recombinant endo-1,5-α-l-arabinanase from the isolated bacterium Bacillus licheniformis.** *Biotechnol Bioprocess Eng* 2010, **15**:590–594.
6. Sakamoto T, Ihara H, Kozaki S, Kawasaki H: **A cold-adapted endo-arabinanase from Penicillium chrysogenum.** *Biochim Biophys Acta Gen Subj* 2003, **1624**:70–75.
7. Sanchez S, Demain AL: **Enzymes and Bioconversions of Industrial, Pharmaceutical, and Biotechnological Significance.** *Org Process Res Dev* 2011, **15**:224–230.
8. Wong DWS, Chan VJ, Batt SB: **Cloning and characterization of a novel exo-α-1,5-L-arabinanase gene and the enzyme.** *Appl Microbiol Biotechnol* 2008, **79**:941–949.
9. Carapito R, Imberty A, Jeltsch JM, Byrns SC, Tam PH, Lowary TL, Varrot A, Phalip V: **Molecular basis of arabinobio-hydrolase activity in phytopathogenic fungi crystal structure and catalytic mechanism of Fusarium graminearum GH93 exo-alpha-L-arabinanase.** *J Biol Chem* 2009, **284**:12285–12296.
10. Kaji A, Saheki T: **Endo-arabinanase from Bacillus subtilis F-11.** *Biochim Biophys Acta* 1975, **410**:354–360.
11. de Sanctis D, Bento I, Inacio JM, Custodio S, de Sá-Nogueira I, Carrondo MA: **Overproduction, crystallization and preliminary X-ray characterization of Abn2, an endo-1,5-alpha-arabinanase from Bacillus subtilis.** *Acta Crystallographica Section F-Structural Biology and Crystallization Communications* 2008, **64**:636–638.
12. Dunkel MP, Amado R: **Analysis of endo-(1->5)-alpha-L-arabinanase degradation patterns of linear (1->5)-alpha-L-arabino-oligosaccharides by high-performance anion-exchange chromatography with pulsed amperometric detection.** *Carbohydr Res* 1995, **268**:151–158.
13. McKie VA, Black GW, Millward-Sadler SJ, Hazlewood GP, Laurie JI, Gilbert HJ: **Arabinanase A from Pseudomonas fluorescens subsp cellulosa exhibits both an endo- and an exo- mode of action.** *Biochem J* 1997, **323**(Pt 2):547–555.
14. Nurizzo D, Turkenburg JP, Charnock SJ, Roberts SM, Dodson EJ, McKie VA, Taylor EJ, Gilbert HJ, Davies GJ: **Cellvibrio japonicus alpha-L-arabinanase 43A has a novel five-blade beta-propeller fold.** *Nat Struct Biol* 2002, **9**:665–668.
15. Scott M, Pickersgill RW, Hazlewood GP, Gilbert HJ, Harris GW: **Crystallization and preliminary X-ray analysis of arabinanase A from Pseudomonas fluorescens subspecies cellulosa.** *Acta crystallographica Section D, Biological crystallography* 1999, **55**:544–546.
16. Skjot M, Kauppinen S, Kofod LV, Fuglsang C, Pauly M, Dalboge H, Andersen LN: **Functional cloning of an endo-arabinanase from Aspergillus aculeatus and its heterologous expression in A. oryzae and tobacco.** *Mol Gen Genomics* 2001, **265**:913–921.
17. Squina FM, Prade RA, Wang HL, Murakami MT: **Expression, purification, crystallization and preliminary crystallographic analysis of an endo-1,5-alpha-L-arabinanase from hyperthermophilic Thermotoga petrophila.** *Acta Crystallographica Section F-Structural Biology and Crystallization Communications* 2009, **65**:902–905.
18. Yamaguchi A, Tada T, Nakaniwa T, Kitatani T, Takao M, Sakai T, Nishimura K: **Crystallization and preliminary X-ray diffraction analysis of a thermostable endo-1,5-alpha-L-arabinanase from Bacillus thermodenitrificans TS-3.** *Acta Crystallographica Section D-Biological Crystallography* 2004, **60**:1149–1151.
19. Yamaguchi A, Tada T, Wada K, Nakaniwa T, Kitatani T, Sogabe Y, Takao M, Sakai T, Nishimura K: **Structural basis for thermostability of endo-1,5-alpha-L-arabinanase from Bacillus thermodenitrificans TS-3.** *J Biochem* 2005, **137**:587–592.
20. Inacio JM, de Sá-Nogueira I: **Characterization of abn2 (yxiA), Encoding a Bacillus subtilis GH43 Arabinanase, Abn2, and Its Role in Arabino-Polysaccharide Degradation.** *J Bacteriol* 2008, **190**:4272–4280.
21. de Sanctis D, Inacio JM, Lindley PF, de Sá-Nogueira I, Bento I: **New evidence for the role of calcium in the glycosidase reaction of GH43 arabinanases.** *FEBS J* 2010, **277**:4562–4574.
22. Pons T, Naumoff DG, Martinez-Fleites C, Hernandez L: **Three acidic residues are at the active site of a beta-propeller architecture in glycoside hydrolase families 32, 43, 62, and 68.** *Proteins-Structure Function and Bioinformatics* 2004, **54**:424–432.
23. Watanabe K, Kitamura K, Suzuki Y: **Analysis of the critical sites for protein thermostabilization by proline substitution in oligo-1,6-glucosidase from Bacillus coagulans ATCC 7050 and the evolutionary consideration of proline residues.** *Appl Environ Microbiol* 1996, **62**:2066–2073.
24. Mahuku GS: **A simple extraction method suitable for PCR-based analysis of plant, fungal, and bacterial DNA.** *Plant Mol Biol Report* 2004, **22**:71–81.
25. Lammirato C, Miltner A, Kaestner M: **Effects of wood char and activated carbon on the hydrolysis of cellobiose by β-glucosidase from Aspergillus niger.** *Soil Biol Biochem* 2011, **43**:1936–1942.
26. Laemmli UK: **Cleavage of structural proteins during the assembly of the head of bacteriophage T4.** *Nature* 1970, **227**:680–685.
27. Zhang Y, Raudah S, Teo H, Teo GWS, Fan R, Sun X, Orner BP: **Alanine-shaving mutagenesis to determine key interfacial residues governing the assembly of a nano-cage maxi-ferritin.** *J Biol Chem* 2010, **285**:12078–12086.
28. Miller GL: **Use of dinitrosalicylic acid reagent for determination of ruducing sugar.** *Anal Chem* 1959, **31**:426–428.
29. Okuyama M, Okuno A, Shimizu N, Mori H, Kimura A, Chiba S: **Carboxyl group of residue Asp647 as possible proton donor in catalytic reaction of alpha-glucosidase from Schizosaccharomyces pombe.** *Eur J Biochem* 2001, **268**:2270–2280.
30. Larkin MA, Blackshields G, Brown NP, Chenna R, McGettigan PA, McWilliam H, Valentin F, Wallace IM, Wilm A, Lopez R, Thompson JD, Gibson TJ, Higgins DG: **Clustal W and clustal X version 2.0.** *Bioinformatics* 2007, **23**:2947–2948.
31. Wilgenbusch JC, Swofford D: **Inferring evolutionary trees with PAUP*.** *Curr Protoc Bioinformatics* 2003. Chaper 6, unit 6.4. http://www.currentprotocols.com/protocol/bi0604.
32. Schwede T, Kopp J, Guex N, Peitsch MC: **SWISS-MODEL: an automated protein homology-modeling server.** *Nucleic Acids Res* 2003, **31**:3381–3385.
33. Guex N, Peitsch MC: **SWISS-MODEL and the Swiss-PdbViewer: an environment for comparative protein modeling.** *Electrophoresis* 1997, **18**:2714–2723.
34. Arnold K, Bordoli L, Kopp J, Schwede T: **The SWISS-MODEL workspace: a web-based environment for protein structure homology modelling.** *Bioinformatics* 2006, **22**:195–201.

The avian cell line AGE1.CR.pIX characterized by metabolic flux analysis

Verena Lohr[1,2*†], Oliver Hädicke[1†], Yvonne Genzel[1], Ingo Jordan[2], Heino Büntemeyer[3], Steffen Klamt[1] and Udo Reichl[1,4]

Abstract

Background: In human vaccine manufacturing some pathogens such as Modified Vaccinia Virus Ankara, measles, mumps virus as well as influenza viruses are still produced on primary material derived from embryonated chicken eggs. Processes depending on primary cell culture, however, are difficult to adapt to modern vaccine production. Therefore, we derived previously a continuous suspension cell line, AGE1.CR.pIX, from muscovy duck and established chemically-defined media for virus propagation.

Results: To better understand vaccine production processes, we developed a stoichiometric model of the central metabolism of AGE1.CR.pIX cells and applied flux variability and metabolic flux analysis. Results were compared to literature dealing with mammalian and insect cell culture metabolism focusing on the question whether cultured avian cells differ in metabolism. Qualitatively, the observed flux distribution of this avian cell line was similar to distributions found for mammalian cell lines (e.g. CHO, MDCK cells). In particular, glucose was catabolized inefficiently and glycolysis and TCA cycle seem to be only weakly connected.

Conclusions: A distinguishing feature of the avian cell line is that glutaminolysis plays only a minor role in energy generation and production of precursors, resulting in low extracellular ammonia concentrations. This metabolic flux study is the first for a continuous avian cell line. It provides a basis for further metabolic analyses to exploit the biotechnological potential of avian and vertebrate cell lines and to develop specific optimized cell culture processes, e.g. vaccine production processes.

Keywords: Avian cell line AGE1.CR.pIX, Biomass composition, Flux variability analysis, Metabolic network modeling, Glutaminolysis

Background

Novel therapeutic options have been made possible with production of animal cell culture-derived biopharmaceuticals. An enormous amount of insight has been published on production of recombinant proteins (including antibodies) since the first products have been licensed for application in humans 1986 and 1987. CHO cells are the main substrate for recombinant proteins and therefore most metabolic studies describe these cells [1,2]. Regarding vaccine manufacturing, literature on properties and metabolism of host cells during growth and virus replication as well as process design and optimization is less abundant and not as focused. One reason is that safety considerations and execution of pre-clinical and clinical trials are paramount for vaccines because this medication usually is administered to healthy recipients. Another reason is the multitude of process options and host cell systems used for propagation of pathogens. For historic reasons, primary cultures (today usually embryonated chicken eggs or chicken embryo fibroblasts), diploid cell strains (human embryonic WI-38 and MRC-5), and few continuous cell lines (i.e. macaque-derived Vero cells or MDCK cells) are associated with the lowest possible risks to the vaccine recipients. Production with such cellular substrates often is performed in the presence of calf serum and adherent culture with little room for optimization. Furthermore, standardization of experiments is difficult with primary chicken material.

* Correspondence: verena.lohr@probiogen.de
†Equal contributors
[1]Max Planck Institute for Dynamics of Complex Technical Systems, Sandtorstr. 1, 39106 Magdeburg, Germany
[2]Current address: ProBioGen AG, Goethestr. 54, 13086 Berlin, Germany
Full list of author information is available at the end of the article

Several virus strains and viral vectors, e.g. Modified Vaccinia Virus Ankara (MVA), only replicate efficiently in avian cells [3]. Accordingly, primary chicken fibroblasts are still a commonly accepted production substrate. As vectored vaccines are gaining increasing importance [4], the potential of continuous suspension cells that proliferate in chemically-defined media are attracting more and more attention. Therefore, continuous avian cell lines are being investigated as a viable option to replace primary material. One possible new cell candidate that meets these criteria is the avian designer cell line AGE1.CR.pIX (in the following: CR.pIX) that proliferates in fully scalable suspension culture and is adapted to growth in a chemically-defined medium [5,6]. The latter property is an advantage for metabolic flux analysis as no unknown or complex components such as animal sera or hydrolysates that complicate carbon and nitrogen balance closure are present.

Metabolic models have been developed for a variety of cell lines to study physiological states or changes in metabolism as a response to different cultivation strategies, medium composition or stimuli like virus infection or accumulation of toxins. For example, the human cell lines HEK293 [7,8] and AGE1.HN [9,10], other mammalian cells like CHO [11], BHK [12], MDCK [13,14] and hybridoma cells [15,16] or insect cell lines like Sf9 [17,18] have been studied because of their relevance for the production of biopharmaceuticals. For all these cell lines, the two main pathways for energy generation and precursor supply were found to be glucose catabolism via glycolysis and glutamine catabolism via the TCA cycle, referred to as glutaminolysis [19-21]. Therefore, cell culture media usually contain substantial amounts of both substrates. Associated with an overflow metabolism based on glucose and a high glutaminolysis activity is the accumulation of lactate and ammonia in the cultivation broth with a negative impact on cell growth and product formation [22,23].

To describe cell metabolism, various mathematical methods have been developed and comprehensive reviews on applied approaches and their benefits for mammalian cell culture research are available [24,25]. Mass balancing techniques like metabolic flux analysis (MFA) can be performed for analysis of cellular growth, requiring only a comparatively small set of experimental data. During the exponential growth phase of batch cultivations or continuous cultivations this method can be used to evaluate possible intracellular flux distributions based on measurements of exchange fluxes. However, if the system is underdetermined, no unique flux distribution can be calculated. One approach to circumvent this limitation is to perform ^{13}C labelling experiments. The higher experimental effort of such studies comes with the advantage of an increased number of constraints and uniquely resolvable flux distributions [26]. Another possibility is to apply flux variability analysis (FVA), a method related to MFA that calculates a flux range instead of a distinct flux value [27]. The advantage of this method is that it is experimentally less demanding as it requires (as MFA) only extracellular metabolite and cell concentrations as well as the biomass composition. Intracellular flux ranges from measured extracellular rates (uptake and release of metabolites) can then be computed with the help of software tools such as the *CellNetAnalyzer* toolbox [28].

As a starting point for studying metabolism of avian CR.pIX cells during growth, we developed a network model for the central metabolism oriented in size and scope on other published models of the central metabolism of mammalian and insect cell lines (see Figure 1). Pathways were selected based on entries from avian species in the Kyoto Encyclopedia of Genes and Genomes (KEGG) database. As it was not known whether biomass composition of avian cells is comparable to other examined cells, we first determined relative proportions of biomass components experimentally. FVA and MFA were then applied to assess the metabolic behaviour of CR.pIX cells during exponential growth in a 1 L stirred tank reactor (STR). Finally, to verify emerging hypotheses regarding metabolic pathways, especially glutaminolysis, enzyme activity measurements and additional growth experiments were performed.

Results and discussion

Biomass composition

To allow metabolic flux analysis, some cell characteristics need to be known, i.e. specific dry cell weight and biomass composition. As it was not known whether similar results to other eukaryotic cells could be expected for an avian cell, most characteristics were determined experimentally rather than presumed.

The specific cell dry weight of CR.pIX cells was measured with 314 pg/cell. Amounts of DNA (2.3 ± 0.5% of biomass) and RNA (3.1 ± 0.2%) per CR.pIX cell were comparable to published values obtained for other eukaryotic cells. The protein content (55.2 ± 8.4%) was lower in CR.pIX cells than the typical range of 70–75% determined for other cells [7,16,29]. However, there are reports supporting a low protein content such as Zupke et al. who found 60% protein in mouse hybridoma cells and Carnicer et al. who measured a protein content of only 37% for yeast cells [30,31]. The relative amounts of amino acids of whole cell protein determined from CR.pIX cells are in general similar to published data on yeast [31], mammalian cells [9,29,32] and insect cell lines [33] (see Additional file 1: Table S1). The remaining fraction of biomass was assigned to lipids and carbohydrates in a ratio of 1:2. The estimated lipid content (13.1%) agrees with previous reports that vary between 9 and 20%, whereas the assumed carbohydrate

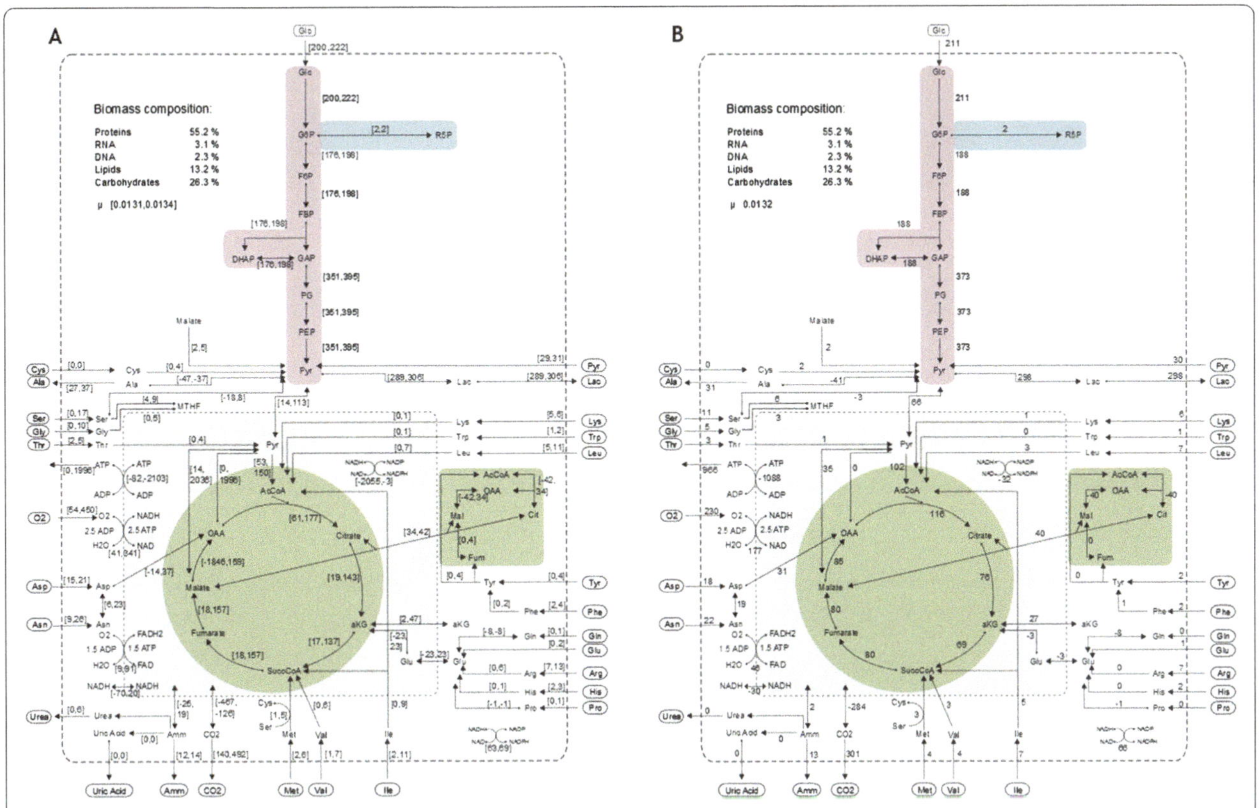

Figure 1 Metabolic network model of the central metabolism of avian CR.pIX cells. Main parts are glycolysis (purple), amino acid catabolism and the TCA cycle (green) which takes place in the mitochondria (grey). Reaction reversibility indicated by arrow heads. All fluxes are given in [μmol/gDW/h]. **A**: shows calculated flux ranges from scenario 1 (by FVA) in square brackets, **B**: shows determined rate values from scenario 2 (by MFA). The values are also given in Additional file 1: Table S2.

content (26.3%) is at the upper limit of the wide range of 3.5–25.0% reported in literature [7,16,29,30,34]. However, a sensitivity analysis showed that absolute fractions of lipids and carbohydrates have only a minor impact on rate values of the calculated flux distributions (data not shown).

Growth phases of CR.pIX cells in STR

The growth of CR.pIX cells can be divided into distinct phases. First, an initial lag phase was observed lasting for about 24 h with only slightly increasing viable cell concentrations. Thereafter, cells grew exponentially until 172 h. Following the exponential growth phase, a reduced growth was observed until 230 h, but viability was still above 90% in this intermediate growth phase (Figure 2A). After having reached the maximum viable cell concentration of 1.3×10^7 cells/mL, cell death started.

The region chosen for MFA and FVA was 24–97 h, where a growth rate of 0.0132 h^{-1} was observed. In the following, only this time frame is considered when describing and discussing extracellular rates and intracellular flux distributions as it is very likely that the quasi steady state assumption holds. Furthermore, it is also the typical time frame in which cells are grown before infection in virus production processes (MVA or influenza A and B virus) [6,35]. Studies in this time frame are therefore most relevant in terms of vaccine manufacturing using CR.pIX cells.

Exchange rates from measured extracellular metabolites

Determined exchange rates of CR.pIX were compared to rates given in other publications for mammalian cell lines (Table 1). We chose 6 data sets where equal units were given or could be calculated for comparison, thereof three publications on the metabolism of human AGE1.HN cells [9,10,36], one publication on HEK293 cells [7], and one on MDCK cells [13]. The majority of concentration profiles of CR.pIX cultivations are similar to concentration profiles obtained for other transformed cell lines studied in the past. For most of the metabolites, the measured exchange rates during CR.pIX cultivation have the same direction (uptake or release) and are in the same range compared to those found for other cell lines. Most notably, as observed for other cell lines, almost all of the consumed glucose is converted to lactate and excreted in the medium (Figure 2B). Key differences in the rates

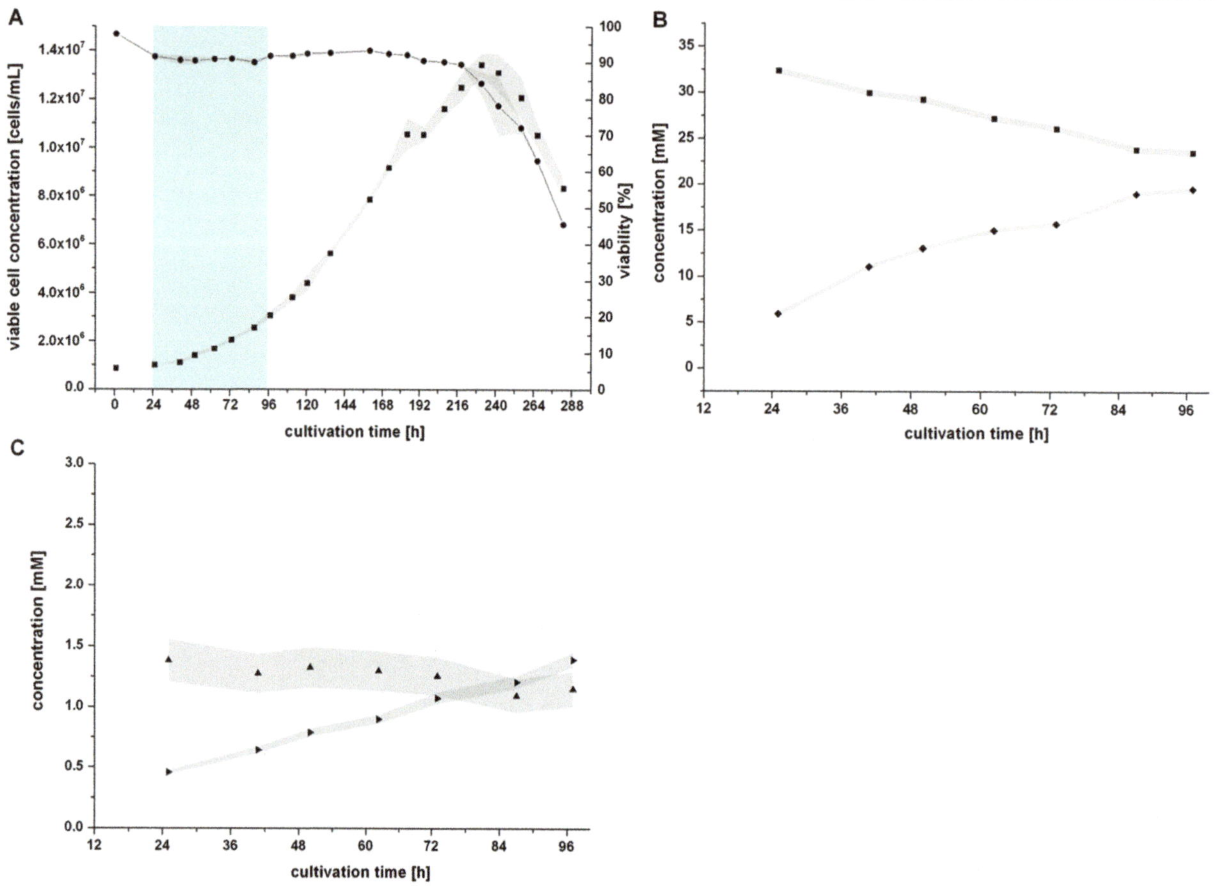

Figure 2 Cultivation of CR.pIX cells in 1 L stirred tank reactor. A: viable cell concentration (■) and viability (●) with the time frame used for MFA indicated by a pale blue box. Bands around data points indicate standard deviation from triplicate measurements. **B**: extracellular concentrations glucose (■) and lactate (♦); **C**: extracellular concentrations of glutamine (▲) and ammonia (▶). Bands around data points indicate measurement errors from assay validations (see Additional file 1: Table S3).

of CR.pIX cell cultivations were observed for the following metabolites:

First, glutamine uptake is zero with CR.pIX cells (Figure 2C) as the observed glutamine concentration decrease is solely due to glutamine hydrolysis. Associated with this, only low amounts of ammonia accumulated in the medium (1.4 mM after 97 h: maximal 2.8 mM at the end of the batch cultivation at 210 h). For production of cell culture-derived products this is a beneficial characteristic as higher ammonia concentrations are known to be toxic with negative impact on cell growth and product formation [22]. The observed ammonia concentration is at the lower limit compared to other cells, where concentrations between 3 and 8 mM were reported to lead to impaired growth [37-39].

The second difference is observable for glycine which is taken up by CR.pIX cells, but released by other cell lines. This observation can be explained by the fact that glycine is reported to be an essential amino acid for birds and therefore needs to be provided for CR.pIX cells by the medium [40]. The third finding is that aspartate and asparagine are consumed by CR.pIX cells in higher amounts or at least at higher rates compared to other cell lines. Both amino acids can be converted to oxaloacetate and thus channeled into the TCA cycle. One could hypothesize that CR.pIX cells compensate in that way for the negligible glutamine uptake to fill the TCA cycle intermediate pools. We also observed that the measured uptake rates for the essential amino acids histidine, lysine and tryptophan were slightly below their minimal requirements for biomass formation with the determined growth rate. For the following analyses we therefore adjusted their upper boundaries slightly to fit the minimal stoichiometric demands for the measured specific growth rate.

Intracellular flux distribution

Using a metabolic network model of the central metabolism of avian cells that we developed according to literature data and database entries, the metabolism of CR.pIX cells during cultivation in 1 L STR was studied using FVA (details see section "Flux variability and

Table 1 Determined extracellular fluxes for cultivation of CR.pIX cells in comparison with literature data

	CR.pIX STR	AGE1.hn Shaker[a]	AGE1.hn Tube[b]	AGE1.hn Tube[c]	HEK293 STR[d,f]	MDCK STR[e,f]
Time frame	24–97 h	18–90 h	72–169 h	0–74 h	0–72 h	16–80 h
Initial glc conc. [mM]	33	20–25	30	30	15	30
Initial gln conc. [mM]	1.5	5	2	4	1	2
Glc	-211.2 ± 10.8[g]	-123.0	-126.9	-429.2	-395.5	-1108.7
Lac	297.6 ± 8.8	263.3	87.6	639.8	439.0	2108.0
Pyr	-30.1 ± 0.8	-28.7	-2.9	-76.5	-	0.0
Gln	-0.1 ± 0.1	-37.1	-41.2	-95.5	-66.5	-70.9
Amm	12.6 ± 0.9	-	-	91.7	48.5	51.8
Glu	-1.4 ± 1.1	4.0	0.9	17.0	-1.4	-42.1
Ala	31.6 ± 4.9	17.5	8.0	20.6	11.8	-49.4
Arg	-9.9 ± 2.8	-8.2	-4.4	-12.4	-15.3	-11.7
Asn	-17.1 ± 8.5	-2.9	-2.0	-7.3	-4.6	-1.9
Asp	-17.8 ± 3.2	-14.1	-8.9	-16.2	-0.2	-2.0
Cys	0.0	0.2	-1.1	-0.2	-	-2.4
Gly	-4.8 ± 4.8	2.6	-0.8	6.8	17.0	2.3
His	-2.3 ± 0.4	-1.7	-0.7	-3.2	-3.5	-2.8
Ile	-6.2 ± 4.4	-8.1	-1.4	-14.7	-21.8	-12.3
Leu	-7.9 ± 3.2	-11.5	-3.2	-23.8	-24.7	-23.5
Lys	-5.5 ± 0.3	-6.9	-3.5	-16.6	-12.6	-22.9
Met	-4.1 ± 2.0	-3.1	-1.2	-6.6	-4.9	-6.4
Phe	-2.5 ± 1.0	-3.3	-1.5	-6.3	-5.4	-8.9
Pro	-0.1 ± 0.1	7.1	13.1	13.4	0.8	6.7
Ser	-8.3 ± 8.3	-12.2	-5.7	-33.4	-17.7	-16.8
Thr	-3.4 ± 1.9	-4.9	-2.0	-8.0	-9.5	-14.9
Trp	-1.5 ± 0.2	-0.9	-0.4	-0.8	-	-5.2
Tyr	-1.8 ± 1.8	-2.8	-1.4	-5.5	-4.6	-5.0
Val	-4.3 ± 2.8	-8.4	-1.8	-14.2	-19.1	-17.0
Uric acid	0.0	-	-	-	-	-

All rates are given in µmol/gDW/h.
[a]from [9]; [b]from [36]; [c]from [10]; [d]from [7]; [e]from [13,14].
[f]these fluxes were re-calculated to [µmol/gDW/h] with the given fluxes and cell dry weights.
[g]standard deviation from Monte-Carlo sampling using validated standard deviations of the methods as boundaries.

metabolic flux analysis"). The time span of 24 h and 97 h was chosen for analysis, as within this time interval during exponential growth, the quasi steady state assumption holds. FVA uses as few constraints for the metabolic network as possible in order to get an unbiased look at the metabolism. This generated a distribution of flux ranges that is named scenario 1 in this manuscript. However, flux ranges are sometimes difficult to interpret. Therefore, after having performed FVA, we set few reasonable constraints, e.g. uptake and excretion rates were set as fixed and the pyruvate carboxylase activity was assumed to be inactive. With these constraints a unique flux distribution could be calculated which is discussed as scenario 2 in the following.

All calculated flux ranges (scenario 1; determined by FVA) and fluxes (for scenario 2, determined by MFA) are given in the Additional file (Additional file 1: Table S2). Together with the network, main fluxes are also depicted in Figure 1. Although only flux ranges can be calculated with the constraints of scenario 1 (Figure 1A), some conclusions can already be drawn from them as we will discuss below. However, most of the following discussion will refer to the unique flux distribution calculated for scenario 2 (Figure 1B). With the applied constraints, the linear equation system had two degrees of redundancy. Therefore, the sum of the variance weighted squared residuals (h) need to be below 5.99 to exclude significant measurement errors (with an applied significance level of $\alpha = 0.05$).

Since we obtained an *h*-value of 0.63 we concluded that the model fitted the data sufficiently well.

Glycolysis and cytosolic pyruvate metabolism

Glucose is taken up with a high rate (211 μmol/gDW/h) and then processed via glycolytic reactions to pyruvate. Only 1% of the glucose is fed into the pentose phosphate pathway for nucleotide synthesis. Such low pentose phosphate influxes were also observed in other studies with transformed cells, e.g. 1.5% and 2.9% in human AGE1.HN cells [9,36]. The calculated ranges for the rates of glycolysis are relatively narrow as they mainly depend on the glucose uptake rate that was given by the measurement of extracellular glucose. Cytosolic pyruvate is mainly derived from glycolysis (92%); the rest of the cytoplasmic pyruvate is taken up by the cells and in small parts obtained from cysteine degradation and the malic enzyme reaction (together 1%). Similar to other transformed cells, cytosolic pyruvate is mainly converted to lactate (73%) or alanine (10%). Only a small fraction (16%) is transported to the mitochondria where it is converted to acetyl-CoA and further oxidized. However, if a complete pentose phosphate pathway (PPP) is included in the model, the exact ratio between glycolysis and PPP is unresolvable and the corresponding flux ranges would be notably wider. Since CO_2 is produced via the PPP, a high PPP flux would reduce the amount of carbon that re-enters the lower part of glycolysis, and therefore even less pyruvate would enter the TCA cycle.

In general, the flux distribution at the cytosolic pyruvate node shows that the majority of the glucose-associated metabolism of CR.pIX cells is used to generate ATP and reducing equivalents via glycolysis. The known Warburg effect of enhanced aerobic glycolysis, leading mostly to lactate, can thus also be observed in transformed cells from avian origin. Furthermore, there seems to be a weak coupling to the mitochondrial TCA cycle as only a minority of the cytosolic pyruvate is channeled into the mitochondria. Other studies have shown that several transformed cell lines show low activities of the three pyruvate dependent enzymes pyruvate dehydrogenase (PDH), pyruvate decarboxylase (PC) and phosphoenolpyruvate carboxykinase (PEPCK) that can serve as connectors between glycolysis and TCA cycle [41,42]. For our cells, we calculated that the PDH (r33) must carry some flux (since the lower boundary of this reaction is strictly positive), but the non-resolvable PC reaction (r46) is not obligatory as its flux range also includes zero. Measurements on activities of anaplerotic enzymes will be discussed in a later subsection.

Mitochondrial pyruvate metabolism and TCA cycle

The mitochondrial pyruvate pool is fed from the cytosolic pyruvate pool (64%) and the anaplerotic malate conversion (34%). 100% of the efflux is catalyzed by the pyruvate dehydrogenase thereby generating 88% of the mitochondrial acetyl-CoA. Other influxes towards this TCA precursor come from amino acids, namely isoleucine (4%) and leucine (7%), tryptophan, and lysine (together 1%). The subsequent reactions of the TCA cycle are fed from other catabolic reactions: α-ketoglutarate from cytosolic influx and succinyl-CoA from isoleucine, valine, serine and methionine degradation. This shows the dependency of the TCA cycle activity on relatively high amino acid uptake and catabolic rates in CR.pIX cells.

Oxidative phosphorylation & ATP balance

We calculated high fluxes for the oxidation of NADH (177 μmol/gDW/h) and $FADH_2$ (46 μmol/gDW/h). Since a transport reaction for NADH between the cytosol and mitochondria was included in the model, the pools of reducing equivalents for both compartments are balanced simultaneously. A minor flux for the corresponding transport reaction shifting reducing equivalents from the mitochondria to the cytosol was calculated with 30 μmol/gDW/h. By setting this transport reaction to zero, cytosolic and mitochondrial reducing equivalents have to be balanced separately. Besides an increased oxygen uptake and ATP synthesis rate, this leads to no significant differences in the flux distributions (data not shown).

ATP generation via the TCA cycle and the electron transport chain makes up to 74% of the total ATP generated. A smaller fraction (26%) is derived from glycolysis. This is similar to results from studies on mouse hybridoma and CHO cells [43,44] where values of 60–80% of ATP production via the TCA cycle have been reported. Although considerable ATP amounts are needed for lipid synthesis, amino acid transporters, upper glycolysis and biomass, the high (unspecific) ATP consumption through reaction r77 (966 μmol/gDW/h) indicates that there seems to be an overload with ATP which is either used for certain maintenance processes or passed through futile cycles.

Amino acid and nitrogen metabolism

Degradation rates of some amino acids, e.g. tryptophan and histidine, are low, indicating that these amino acids are taken up to match the anabolic demand for synthesizing biomass rather than to further metabolize them via catabolic routes. The calculated rate ranges for most fluxes of the amino acid catabolism are constrained to relatively small intervals (e.g. for degradation of valine and histidine). There is no uric acid synthesis (r68) as no uric acid could be measured experimentally. Although this reaction is known to take place in avian cells, this is not a fully unexpected finding. Uric acid was detected in livers and kidneys of chicken and turkey, but not in all bird tissues [40]. CR.pIX cells, as other transformed cells, use ammonia as a sink and release the excess cytosolic ammonia (that is produced by deamination reactions of amino acids) into the medium where it accumulates. However, due to low glutamine exploitation, the ammonium concentration stays at a low level so that toxic concentrations are not reached.

Interestingly, the fluxes through the glutamine synthetase (r48) (i.e. in the direction of glutamate synthesis) are negative as glutamine uptake was not sufficient for biomass synthesis and glutamine therefore has to be synthesized from the glutamate pool. Glutamate itself is likely to be mainly derived from degradation reactions of isoleucine, valine and aspartate. Especially aspartate is taken up in large parts (18 μmol/gDW/h) and can be converted to glutamate. Other amino acids that could compensate for glutamine conversion to glutamate are arginine, histidine and proline. However, only arginine is taken up in considerable amounts by the cells so that one could speculate that the path from glutamate to α-ketoglutarate is generally not used extensively by the CR.pIX cells and thus, the classical dependency on glutamine is not given for CR.pIX cells. To elucidate unambiguously the source of carbon for glutamate synthesis, ^{13}C labelling experiments would be required, which are beyond the scope of the current contribution.

In vitro enzyme activities

To validate the two hypotheses from the previous section that i) glycolysis is only weakly connected to the TCA, and that ii) glutaminolysis plays a minor role in energy supply and precursor generation, we measured the maximum *in vitro* enzyme activities of the related enzymes.

For the first hypothesis, we measured maximum activities of PC and PDH. For both reactions, low enzyme activities of 0.2 nmol/10^6cells/min (i.e. 39 μmol/gDW/h) were measured as for MDCK cells [45]. These values supported the applied constraint of a missing carboxylating anaplerotic reaction from pyruvate towards oxaloacetate. However, the measured low PDH activity seems to contradict the calculated flux of 102 μmol/gDW/h. One possible explanation is given by the significantly smaller calculated lower boundary of this flux when applying fewer and less stringent assumptions with scenario 1. However, at least a basal activity is required as the lower boundary is strictly positive (53 μmol/gDW/h).

The second hypothesis, stating that glutaminolysis is inversed, was validated via measurements of glutaminase (converting glutamine to glutamate) and glutamine synthetase (converting glutamate to glutamine). Glutaminase showed a maximal activity of 0.1 nmol/10^6cells/min (i.e. 19 μmol/gDW/h), which is an even lower value than the measured 1.5–7.4 nmol/10^6cells/min for MDCK cells [45,46]. The flux from glutamate to glutamine was predicted by flux analysis to be active towards glutamine in order to account for the unmet demand of glutamine for biomass formation. The measured enzyme activity of the catalyzing enzyme glutamine synthetase showed a maximal activity of 5.9 nmol/10^6cells/min (i.e. 1135 μmol/gDW/h). This value is on the one hand significantly higher than the value for the reverse reaction and on the other hand at the upper limit of reported values for MDCK cells (0.6–6.5 nmol/10^6cells/min, [45,46]). Thus, although values from *in vitro* enzyme activity measurement and flux analysis are not directly convertible, the measured enzyme activities support the calculated flux distribution towards glutamine and the low level of glutamine usage.

Growth in glutamine-free medium

Since the uptake flux for glutamine was calculated as zero and the enzyme measurements supported the low glutaminolytic activity, we speculated whether glutamine is even dispensable for growth of CR.pIX cells. To validate this hypothesis, we performed shaker flask cultivations of CR.pIX cells in CD-U2 medium with 2 mM glutamine and in CD-U2 medium without glutamine addition. As no serum is present in the medium, other glutamine sources can be excluded. Cultures with or without glutamine were run in parallel over 5 passages. For passage 5, a growth curve was recorded. Figure 3 shows that cells without glutamine did not grow as well as in the presence of glutamine during the first three passages (Figure 3A). Viability also dropped below 90% in these cultures (Figure 3B). After passage 3, cell growth performance in glutamine-free medium increased significantly. Finally, cells in the 5th passage in glutamine-free medium did grow comparably fast as in the presence of glutamine. Maximal viable cell concentrations of 1.4×10^7 cells/mL were achieved with viabilities above 90%. This experiment demonstrated that CR.pIX cells are capable of glutamine-independent growth. With CHO cells, this was achieved after several selection rounds [47], but could not be achieved with other cells [48,49]. Considering that we did not measure glutamine uptake fluxes, we were surprised at the intermittent decrease in proliferation rates and viabilities of CR.pIX cultures after glutamine removal. However, extracellular and intracellular glutamine mediate also non-nutritional effects that can impact apoptosis and heat shock responses [50,51]. We hypothesize that the short adaptation period after glutamine removal may be due to shifts in glutamine-responsive signaling pathways. Infection experiments with poxvirus MVA showed that productivity of CR.pIX cells is not negatively affected by glutamine absence (data not shown). The feature of glutamine-free growth of CR.pIX cells may be advantageous in particular for high cell density cultivations were ammonia accumulates to very high concentrations in the culture supernatant and prevents further increases in cell numbers and productivity [38,52].

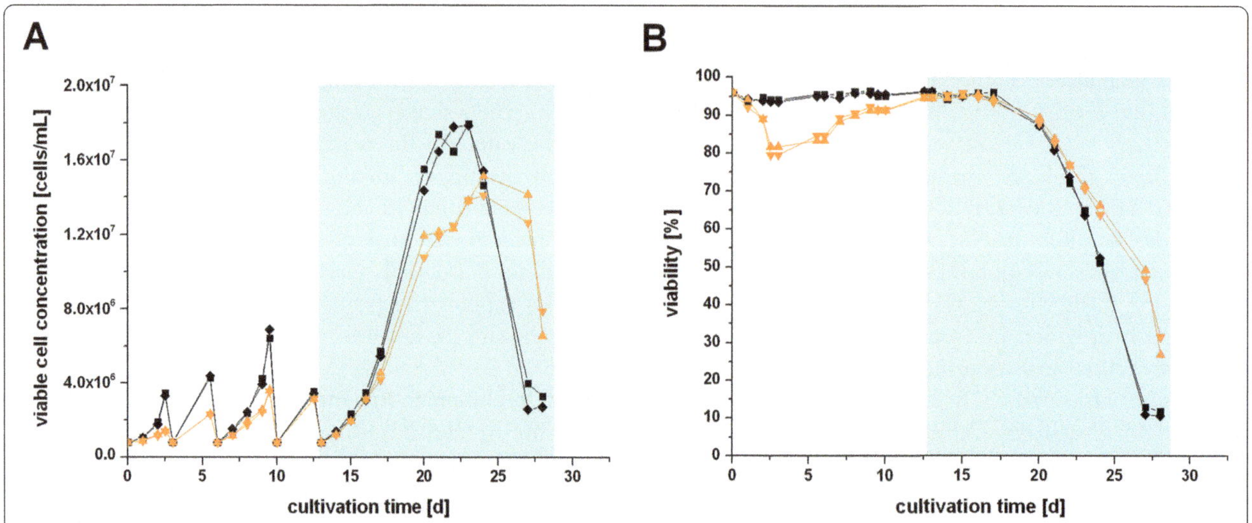

Figure 3 Cultivation of CR.pIX cells with (black squares) or without glutamine (orange triangles) in shaker flasks (two parallel cultures each). Viable cell concentrations **(A)** and viabilities **(B)** are shown for 5 subsequent passages of CR.pIX cells in the respective medium. The pale blue box highlights the 5th passage for which a growth curve was monitored.

Conclusions

Avian cells are important substrates for producing a number of licensed vaccines and various viral vector-based processes under development. The only avian substrates that are currently approved for production are embryonated chicken eggs or primary chicken embryo fibroblasts. Because processes using primary cultures are difficult to optimize and standardize, access to well characterized continuous avian cell lines, such as the CR.pIX cell line derived from a muscovy duck embryo, would be a significant step forward.

Here, we determined metabolic properties of the CR.pIX cell line to complement previous developments on scalable vaccine production processes in chemically defined media. The current study comprised three parts: i) analysis of biomass composition and measurements of cell and extracellular metabolite concentrations, ii) construction of a stoichiometric network for the central metabolism of avian cells and subsequent analysis with flux variability and metabolic flux analysis, and iii) experiments and analytics to validate hypotheses derived from the stoichiometric analysis.

The first part expands the available data base on the metabolism of immortalized cell lines. Results on biomass composition and growth properties of the avian CR.pIX cell line agree with data determined for other transformed mammalian and insect cell lines used in biopharmaceutical production.

Flux variability and metabolic flux analyses in the second part reveal an overflow metabolism similar to mammalian cells with high uptake of glucose followed by conversion to and release of lactate and alanine. The TCA cycle was only weakly connected to glycolysis and appeared to depend on the influx of amino acids, mainly isoleucine, valine and aspartate. Finally, our measurements together with flux (variability) analysis indicated that glutamine supplementation is not an essential requirement for CR.pIX cell proliferation.

The derived hypothesis on low glutaminolysis activity could be confirmed by enzyme activity measurements. However, the hypothesis of a weak connection between the cytosolic and mitochondrial pyruvate pools could not be confirmed entirely. To elucidate whether cytosolic pyruvate is transported in significant amounts from the cytosol to the mitochondria, e.g. ^{13}C labelling experiments could be helpful, which was beyond the scope of the current contribution. Low dependence on glutamine was confirmed by passaging cells in glutamine-free medium and agrees well with the observation that ammonia is accumulating to very low concentrations in the medium. This is a very beneficial property of CR.pIX cells especially for fed-batch cultivation strategies where ammonia often accumulates to toxic levels.

Overall, the present study provides a basis for a more focused metabolic analysis of avian cell lines and thus helps to further optimize medium and feed requirements. Acute infection and associated take-over of the metabolism by a virus may shift the requirements for certain nutrients. The derived model may therefore also provide a context for future studies on the metabolism of CR.pIX cells during the virus replication phase.

Methods

Cell cultivations

CR.pIX suspension cells were routinely cultivated in 250 mL shaker flasks (working volume 150 mL, with membrane cap for aeration) in a chemically-defined medium (CD-U2, PAA) supplemented with glutamine (2 mM final concentration, Sigma), alanine (2 mM final concentration, Sigma) and a recombinant insulin-like growth factor (Long-R3 IGF, 10 ng/mL final concentration, Sigma). Flasks were incubated at 185 rpm with 5 cm amplitude, 37°C and 5% CO_2. Pre-cultures for bioreactor experiments, cultures for determination of biomass composition as well as the comparative cultivations in glutamine-containing and glutamine-free medium were performed in shaker flasks. For inoculation of a 1 L stirred tank bioreactor (STR, DasGip AG, vessel: Spinner Type BS, cellferm-pro® system), 8×10^5 cells/mL in 1 L CD-U2 medium were transferred into the bioreactor. The bioreactor was equipped with a pitched-blade stirrer that was operated at 120 rpm. The temperature was controlled at 37°C; pO_2 was set to 50% by pulsed aeration with air enriched with variable contents of O_2 and CO_2. During the first 60 h, pH was slowly reduced from 7.4 (initial value of the medium) to 7.0 and controlled at this value subsequently by addition of CO_2 or 1 M Na_2CO_3.

Measurement of biomass composition

All assays and analytics were performed with cells in the exponential growth phase. Therefore, samples were taken from shaker flask cultures at 72 h post inoculation. For each analysis, 3 biological replicates were measured with at least 3 technical replicates each.

For determination of dry cell weight, cell suspension containing about 1×10^8 cells/mL was centrifuged (500 × g, 10 min) and re-suspended in 10 mL of fresh medium. Samples were dried at 60°C until weight remained constant. The specific dry cell weight was calculated against weight of empty tubes and against a medium control (10 mL of fresh medium dried in tubes).

The BCA assay (Pierce/ThermoScientific) was used to determine the cellular protein content. Samples containing 4×10^6 cells were centrifuged (500 × g, 10 min) and re-suspended in 1 mL lysis buffer (from CytoTox 96® Non-Radioactive Cytotoxicity Assay, Promega). After 15 min incubation at room temperature, samples were sonicated for 3 min in a water bath. Standard curves were prepared using BSA and analyzed together with samples by measuring the absorption at 562 nm (inifinite200 plate reader, TecanGroup).

For DNA purification, a culture volume containing 5×10^6 cells was taken and purified with the QIAamp DNA Blood Mini kit (Qiagen) according to the procedure "protocol for cultured cells". After purification, 2 μL of assay buffer (in which the samples were dissolved later) were pipetted onto 16 spots on a NanoQuantPlate (TecanGroup) as blank. After blanking, 2 μL of samples were measured. DNA concentration was then determined by multiplication of the measured absorption coefficient with 50 as a dsDNA specific factor. The same method as for DNA content analysis was employed for RNA content analysis. Here, purification was done with the NucleoSpin RNAII kit (Macherey-Nagel) according to the procedure "total RNA purification from cultured cells and tissue". RNA concentration was determined from absorption measurements at 260 nm, but by multiplication with a factor of 40. Measurement procedure was analogous to the DNA analysis.

Amino acid contents were determined as follows: about 1×10^8 cells were washed with 0.9% NaCl solution and re-suspended in 2 M HCl. Cells were incubated in acid at 100°C for 24 h. After this hydrolyzation, the solution was neutralized by adding 2 M NaOH and subsequently analysed by HPLC. Aspartate and glutamate percentages were corrected for a hydrolysis rate of 47.5% as described in literature [7].

A HPLC method was used to measure the amino acid composition of cell protein [53]. Due to harsh hydrolytic conditions, some amino acid concentrations could not be measured, but were estimated from literature values (see Additional file 1: Table S1).

Cell and metabolite concentration measurements

Samples were taken once or twice a day. Cell concentration, viability and average cell diameter were determined using the Vi-CELL™ XR device (Beckman Coulter). Samples for measuring extracellular metabolite concentrations were centrifuged after sampling and the supernatant was stored at -80°C until analysis. Measurement of glucose, lactate and ammonia concentrations was done with a Bioprofile 100 Plus (Nova Biomedical). A HPLC method was used to measure the concentration of pyruvate. Extracellular amino acid concentrations for metabolic flux analysis were determined using a derivatization method and subsequent measurement in a HPLC system [54]. Uric acid concentration was measured with an enzymatic reaction assay kit (Uric Acid Assay Kit, BioVision) including standard curves and controls as described in the manual. Uric acid was only measured at culture start and end. Standard deviations for all concentration measurements (determined by method validations) can be found in Additional file 1: Table S3.

In vitro determination of maximum enzyme activities

Maximum activities of pyruvate carboxylase (PC), pyruvate dehydrogenase (PDH), glutamine synthetase (GS) and glutaminase (GLNase) in exponentially growing cells were measured. Therefore, cells were harvested from shaker flasks 72 h after inoculation. After a washing step in 0.9% NaCl, dry cell pellets were frozen at -80°C and stored until analysis. Enzyme extraction and assays were performed as described for MDCK cells [46]. Three assay parameters had to be optimized before measuring

CR.pIX cell extracts: pH, amount of cells that are used and – for GLNase and GS activity assay – incubation time. After these pre-tests, CR.pIX cell extracts were measured with the following modifications: a two-fold higher cell concentration as for MDCK cells was necessary for the GLNase and GS assay and the incubation time of the GLNase assay was set to 30 min instead of 60 min. The assay plates were analyzed in a 96 well plate reader (Tecan) and enzyme activities were calculated from absorption values.

Metabolic network

A stoichiometric network model of the central metabolism of avian cells (Figure 1, Table 2) was constructed based on KEGG database entries from *Gallus gallus* (chicken), *Meleagris gallopavo* (turkey) and *Taeniopygia guttata* (zebra finch) as well as on information available from previously published networks and from studies on avian metabolism [40]. Overall, the network consists of 97 reactions and 72 metabolites, resulting in a stoichiometric matrix with dimensions 72 × 97 and 28 degrees of freedom. All implemented reactions are listed in the Table 2.

Generally, the network includes

- Two compartments: cytosol and mitochondria
- Main pathways of energy metabolism: glycolysis, TCA cycle and oxidative phosphorylation
- Catabolism of amino acids
- Transport of metabolites and corresponding energy demands (as described before [14])
- Synthesis of macromolecules (biomass)
- Synthesis of urea and uric acid
- A simplified/lumped pentose-phosphate pathway (as in [9])

Calculation of extracellular rates

During the exponential growth phase, metabolism is considered to be in a pseudo steady-state with constant reaction rates [55]. The growth rate μ was calculated via curve fitting of the viable cell concentration measurements (X(t)) to:

$$X(t) = X(t_0) \cdot e^{\mu \cdot (t - t_0)} \tag{1}$$

Uptake or release rates v_i (in unit [μmol/gDW/h] of external metabolites c_i were computed via curve fitting of the concentration data to:

$$c_i(t) = c_i(t_o) + v_i \cdot \frac{DW(t_0)}{\mu} \cdot \left(e^{\mu \cdot (t - t_0)} - 1\right), \tag{2}$$

where we applied the experimentally determined cell dry weight and the calculated μ from Equation (1).

For glutamine, a first order degradation process to ammonia and 5-oxopyrrolidine-2-carboxylic acid was assumed. The rate constant $k_{Gln} = 0.0032\ h^{-1}$ of this process was determined experimentally (n = 3) using medium without cells under standard cultivation conditions in a bioreactor. Based on this degradation rate, glutamine and ammonia rates (Equation (2)) were corrected as described elsewhere [56]. To compute standard deviations of the fluxes, we performed a Monte-Carlo simulation and added a normally distributed error with mean zero and corresponding standard deviation (derived from assay validations) at each time point and for each measurement (metabolites and cell concentration). With this, we generated 100,000 random sample data sets and recalculated for each data set the rates of uptake, release and growth. From these simulated data sets, the empirical variances and standard deviations σ_{r_i} for each of these rates were computed.

Flux variability and metabolic flux analysis

We restrict our analysis to the time span of exponential growth from 24 h to 97 h. Within this time interval the quasi steady state assumption holds and, together with the stoichiometric matrix **N** and the vector **r** of reaction rates (fluxes), the reaction rates can be determined by solving a system of linear equations. Given the measured and known rates, **r** and **N** can be partitioned into the known ($\mathbf{r_k}$, $\mathbf{N_k}$) and unknown part ($\mathbf{r_u}$, $\mathbf{N_u}$) we can write the steady state equation as

$$\frac{dc}{dt} = \mathbf{0} = \mathbf{N} \cdot \mathbf{r} = \mathbf{N_u r_u} + \mathbf{N_k r_k}$$

yielding after reformulation the central equation for metabolic flux analysis

$$\mathbf{N_u r_u} = -\mathbf{N_k r_k}. \tag{3}$$

A detailed description of how to further proceed with Equation (3) is given elsewhere [57-59]. If the linear equation system is over-determined, a variance-weighted least squares estimation can be applied to calculate uniquely determined reaction rates. However, Equation (3) usually defines an underdetermined system and, hence, the flux distribution cannot be calculated uniquely. For identifying at least feasible ranges of the fluxes we use flux variability analysis (FVA), an approach that is based on linear programming [27]. Here, the minimal and maximal possible flux that a reaction can have in any flux distribution that is consistent with Equation (3) is calculated separately for each unknown reaction $r_{u,i}$. Further constraints arise by the irreversibility of some reactions (constrained to non-negative values) or because some reaction rates are fixed to specific (measured or known) values reflecting the given environmental conditions. To avoid unbounded fluxes and

Table 2 Reactions included in the metabolic network model of CR.pIX cells

	Uptake rates
r1, Glc	$Glc \rightarrow Glc_{cyt}$
r2, Pyr	$Pyr + 0.33\ ATP_{cyt} \rightarrow Pyr_{cyt}$
r3, O2	$O_2 \rightarrow O_{2,cyt}$
r4, Gln	$Gln + 0.33\ ATP_{cyt} \rightarrow Gln_{cyt}$
r5, Glu	$Glu + ATP_{cyt} \rightarrow Glu_{cyt}$
r6, Ala	$Ala + 0.33\ ATP_{cyt} \rightarrow Ala_{cyt}$
r7, Asp	$Asp + ATP_{cyt} \rightarrow Asp_{cyt}$
r8, Arg	$Arg + 0.33\ ATP_{cyt} \rightarrow Arg_{cyt}$
r9, Asn	$Asn + 0.33\ ATP_{cyt} \rightarrow Asn_{cyt}$
r10, Cys	$Cys + 0.33\ ATP_{cyt} \rightarrow Cys_{cyt}$
r11, Gly	$Gly + 0.33\ ATP_{cyt} \rightarrow Gly_{cyt}$
r12, His	$His + 0.33\ ATP_{cyt} \rightarrow His_{cyt}$
r13, Ile	$Ile + 0.33\ ATP_{cyt} \rightarrow Ile_{cyt}$
r14, Leu	$Leu + 0.33\ ATP_{cyt} \rightarrow Leu_{cyt}$
r15, Lys	$Lys + 0.33\ ATP_{cyt} \rightarrow Lys_{cyt}$
r16, Val	$Val + 0.33\ ATP_{cyt} \rightarrow Val_{cyt}$
r17, Met	$Met + 0.33\ ATP_{cyt} \rightarrow Met_{cyt}$
r18, Phe	$Phe + 0.33\ ATP_{cyt} \rightarrow Phe_{cyt}$
r19, Pro	$Pro + 0.33\ ATP_{cyt} \rightarrow Pro_{cyt}$
r20, Ser	$Ser + 0.33\ ATP_{cyt} \rightarrow Ser_{cyt}$
r21, Thr	$Thr + 0.33\ ATP_{cyt} \rightarrow Thr_{cyt}$
r22, Trp	$Trp + 0.33\ ATP_{cyt} \rightarrow Trp_{cyt}$
r23, Tyr	$Tyr + 0.33\ ATP_{cyt} \rightarrow Tyr_{cyt}$
	Glycolysis
r24, G6P	$Glc_{cyt} + ATP_{cyt} \rightarrow G6P_{cyt}$
r25, F6P	$G6P_{cyt} \leftrightarrow F6P_{cyt}$
r26, FBP	$F6P_{cyt} + ATP_{cyt} \leftrightarrow FBP_{cyt}$
r27, DHAP	$FBP_{cyt} \leftrightarrow GAP_{cyt} + DHAP_{cyt}$
r28, GAP	$DHAP_{cyt} \leftrightarrow GAP_{cyt}$
r29, PG	$GAP_{cyt} \leftrightarrow NADH_{cyt} + ATP_{cyt} + PG_{cyt}$
r30, PEP	$PG_{cyt} \leftrightarrow PEP_{cyt}$
r31, PEP_Pyr	$PEP_{cyt} \rightarrow Pyr_{cyt} + ATP_{cyt}$
r32, Pyr_Lac	$Pyr_{cyt} + NADH_{cyt} \leftrightarrow Lac_{cyt}$
r33, PDH	$Pyr_{mit} + CoA_{mit} \rightarrow AcCoA_{mit} + CO_{2,mit} + NADH_{mit}$
	Pentose phosphate pathway
r34, R5P	$G6P_{cyt} \rightarrow R5P_{cyt} + CO_{2,cyt} + 2\ NADPH_{cyt}$
	TCA cycle
r35, OAA	$Mal_{cyt} \leftrightarrow OAA_{cyt} + NADH_{cyt}$
r36, Cit	$AcCoA_{cyt} + OAA_{cyt} \rightarrow Cit_{cyt} + CoA_{cyt}$
r37, Fum_Mal	$Fum_{cyt} \leftrightarrow Mal_{cyt}$
r38, Cit_{mito}	$AcCoA_{mit} + OAA_{mit} \rightarrow Cit_{mit} + CoA_{mit}$
r39, OAA_{mito}	$Mal_{mit} \leftrightarrow OAA_{mit} + NADH_{mit}$
r40, Fum_Mal_{mito}	$Fum_{mit} \leftrightarrow Mal_{mit}$
r41, $SCoA_{mito}$	$aKG_{mit} + CoA_{mit} \rightarrow SuccCoA_{mit} + CO_{2,mit} + NADH_{mit}$
r42, Fum_{mito}	$SuccCoA_{mit} \leftrightarrow Fum_{mit} + CoA_{mit} + ATP_{mit} + FADH_{2,mit}$
r43, aKG_{mito}	$Cit_{mit} \leftrightarrow aKG_{mit} + CO_{2,mit} + NADH_{mit}$
	Anaplerosis
r44, Ana_PyrI	$Mal_{cyt} \rightarrow Pyr_{cyt} + CO_{2,cyt} + NADPH_{cyt}$
r45, Ana_PyrII	$Mal_{mit} \leftrightarrow Pyr_{mit} + CO_{2,mit} + NADPH_{mit}$
r46, PC	$Pyr_{mit} + CO_{2,mit} + ATP_{mit} \rightarrow OAA_{mit}$
	Amino acid catabolism
r47, GDH	$Glu_{mit} \leftrightarrow aKG_{mit} + Amm_{mit} + NADPH_{mit}$
r48, GS	$Gln_{cyt} \leftrightarrow Glu_{cyt} + Amm_{cyt} + ATP_{cyt}$
r49, Ala_{cat}	$Ala_{cyt} + aKG_{cyt} \leftrightarrow Pyr_{cyt} + Glu_{cyt}$
r50, Asn_Asp	$Asn_{cyt} \leftrightarrow Asp_{cyt} + Amm_{cyt}$
r51, His_{cat}	$His_{cyt} \rightarrow Glu_{cyt} + 2\ Amm_{cyt} + CO_{2,cyt}$
r52, Ile_{cat}	$Ile_{cyt} + aKG_{cyt} + 2\ CoA_{mit} + ATP_{mit} \rightarrow SuccCoA_{mit} + AcCoA_{mit} + Glu_{cyt} + NADH_{mit} + FADH_{2,mit}$
r53, Leu_{cat}	$Leu_{cyt} + aKG_{cyt} + 3\ CoA_{mit} + 2\ ATP_{mit} \rightarrow 3\ AcCoA_{mit} + Glu_{cyt} + NADH_{mit} + FADH_{2,mit}$
r54, Lys_{cat}	$Lys_{cyt} + 2\ aKG_{cyt} + 2\ CoA_{mit} + NADPH_{cyt} \rightarrow 2\ AcCoA_{mit} + 2\ Glu_{cyt} + 2\ CO_{2,mit} + 2\ NADH_{cyt} + 2\ NADH_{mit} + FADH_{2,mit}$
r55, Met_{cat}	$Met_{cyt} + Ser_{cyt} + CoA_{mit} + 3\ ATP_{cyt} + ATP_{mit} \rightarrow Cys_{cyt} + SuccCoA_{mit} + Amm_{cyt} + NADH_{mit} + CO_{2,cyt}$
r56, Phe_{cat}	$Phe_{cyt} + O_{2,cyt} + NADPH_{cyt} \rightarrow Tyr_{cyt}$
r57, Pro_{cat}	$Pro_{cyt} \leftrightarrow Glu_{cyt} + 2\ NADH_{cyt}$
r58, Thr_{cat}	$Thr_{cyt} \rightarrow Pyr_{mit} + Amm_{mit} + CO_{2,mit} + NADH_{cyt} + NADH_{mit} + FADH_{2,mit}$
r59, Trp_{cat}	$Trp_{cyt} + 2\ CoA_{mit} + 3\ O_{2,cyt} + NADPH_{cyt} \rightarrow Ala_{cyt} + 2\ AcCoA_{mit} + Amm_{cyt} + 2\ CO_{2,mit} + 2\ CO_{2,cyt} + NADH_{cyt} + 2\ NADH_{mit} + FADH_{2,mit}$
r60, Val_{cat}	$Val_{cyt} + aKG_{cyt} + CoA_{mit} + ATP_{mit} \rightarrow SuccCoA_{mit} + Glu_{cyt} + CO_{2,cyt} + 3\ NADH_{mit} + FADH_{2,mit}$
r61, Tyr_{cat}	$Tyr_{cyt} + aKG_{cyt} + 2\ CoA_{cyt} + 2\ O_{2,cyt} \rightarrow Fum_{cyt} + 2\ AcCoA_{cyt} + Glu_{cyt} + CO_{2,cyt}$
r62, Ser_{cat}	$Ser_{cyt} \leftrightarrow Pyr_{cyt} + Amm_{cyt}$
r63, Cys_{cat}	$Cys_{cyt} + aKG_{cyt} + O_{2,cyt} \rightarrow Pyr_{cyt} + Glu_{cyt}$
r64, Asp_{cat}	$Asp_{cyt} + aKG_{mit} \leftrightarrow Glu_{cyt} + OAA_{mit}$
r65, Arg_{cat}	$Arg_{cyt} + aKG_{cyt} \rightarrow 2\ Glu_{cyt} + Urea_{cyt} + NADH_{cyt}$
	MTHF & uric acid synthesis
r66, MTHF_I	$Ser_{cyt} + THF_{cyt} \rightarrow Gly_{cyt} + MTHF_{cyt}$
r67, MTHF_II	$Gly_{cyt} + THF_{cyt} \rightarrow NADH_{cyt} + CO_{2,cyt} + Amm_{cyt} + MTHF_{cyt}$
r68, UricAcid	$Asp_{cyt} + 2\ Gln_{cyt} + Gly_{cyt} + 2\ MTHF_{cyt} + 7\ ATP_{cyt} + CO_{2,cyt} \rightarrow UricAcid_{cyt} + Fum_{cyt} + 2\ Glu_{cyt} + 2\ THF_{cyt}$
	Lipid synthesis
r69, CH_{Lip}	$18\ AcCoA_{cyt} + 18\ ATP_{cyt} + 11\ O_{2,cyt} + 27\ NADPH_{cyt} \rightarrow CH + 18\ CoA_{cyt} + 9\ CO_{2,cyt}$
r70, PC_{Lip}	$GAP_{cyt} + Ser_{cyt} + 27.6\ ATP_{cyt} + 17.6\ AcCoA_{cyt} + 4\ MTHF_{cyt} + 2\ NADH_{cyt} + 31.2\ NADPH_{cyt} \rightarrow PC + 17.6\ CoA_{cyt} + 4\ THF_{cyt}$
r71, PE_{Lip}	$GAP_{cyt} + Ser_{cyt} + 18.6\ ATP_{cyt} + 17.6\ AcCoA_{cyt} + MTHF_{cyt} + 2\ NADH_{cyt} + 31.2\ NADPH_{cyt} \rightarrow PE + 17.6\ CoA_{cyt} + THF_{cyt}$

Table 2 Reactions included in the metabolic network model of CR.pIX cells *(Continued)*

r72, PS_{Lip}	GAP_{cyt} + Ser_{cyt} + 18.6 ATP_{cyt} + 17.6 $AcCoA_{cyt}$ + 2 $MTHF_{cyt}$ + 2 $NADH_{cyt}$ + 31.2 $NADPH_{cyt}$ → PS + 17.6 CoA_{cyt} + 2 THF_{cyt}
r73, PGL_{Lip}	2 GAP_{cyt} + 17.6 ATP_{cyt} + 17.6 $AcCoA_{cyt}$ + 4 $NADH_{cyt}$ + 31.2 $NADPH_{cyt}$ → PGL + 17.6 CoA_{cyt}
r74, PI_{Lip}	GAP_{cyt} + $G6P_{cyt}$ + 17.6 ATP_{cyt} + 17.6 $AcCoA_{cyt}$ + 2 $NADH_{cyt}$ + 31.2 $NADPH_{cyt}$ → PI + 17.6 CoA_{cyt}
r75, SM_{Lip}	2 Ser_{cyt} + 27.8 ATP_{cyt} + 16.8 $AcCoA_{cyt}$ + 3 $MTHF_{cyt}$ + 2 $NADH_{cyt}$ + 29.6 $NADPH_{cyt}$ → SM + 16.8 CoA_{cyt} + 3 THF_{cyt}
r76, DPG_{Lip}	3 GAP_{cyt} + 35.2 ATP_{cyt} + 35.2 $AcCoA_{cyt}$ + 6 $NADH_{cyt}$ + 62.4 $NADPH_{cyt}$ → DPG + 35.2 CoA_{cyt}
	Release rates
r77, ATP_{main}	ATP_{cyt} → maintenance
r78, Lac_{out}	Lac_{cyt} → Lac
r79, Ala_{out}	Ala_{cyt} → Ala
r80, $UricAcid_{out}$	Uric $acid_{cyt}$ → Uric acid
r81, $Urea_{out}$	$Urea_{cyt}$ → Urea
r82, Amm_{out}	Amm_{cyt} → Amm
r83, $CO_{2\ out}$	$CO_{2,cyt}$ → CO_2
r84, Pyr_{out}	Pyr_{cyt} → Pyr
	Transport reactions, oxidative phosphorylation
r85, $NADH_{cyt,trans}$	$NADH_{cyt}$ ↔ $NADH_{mit}$
r86, ATP_{trans}	ATP_{cyt} ↔ ATP_{mitt}
r87, $CO_{2\ trans}$	$CO_{2,cyt}$ ↔ $CO_{2,mit}$
r88, MAL_{trans}	Mal_{cyt} + Cit_{mit} ↔ Mal_{mit} + Cit_{cyt}
r89, Glu_{trans}	Glu_{cyt} ↔ Glu_{mit}
r90, Pyr_{trans}	Pyr_{cyt} ↔ Pyr_{mit}
r91, aKG_{trans}	aKG_{cyt} ↔ aKG_{mit}
r92, Amm_{trans}	Amm_{cyt} ↔ Amm_{mit}
r93, $FADH_{ox}$	$O_{2,cyt}$ + 2 $FADH_{2,mit}$ → 3 ATP_{mit}
r94, $NADH_{mit,trans}$	$NADH_{mit}$ ↔ $NADPH_{mit}$
r95, $NADH_{ox}$	$O_{2,cyt}$ + 2 $NADH_{mit}$ → 5 ATP_{mit}
r96, $NADH_{cyt,trans}$	$NADH_{cyt}$ ↔ $NADPH_{cyt}$
	Synthesis of macromolecules and biomass
r97, μ	0.552 proteins + 0.263 carbohydrates + 0.131 lipids + 0.023 DNA + 0.031 RNA → biomass

Proteins [1 g] = 955.79 Asp_{cyt} + 1344.29 Ala_{cyt} + 543.16 Gln_{cyt} + + 817.99 Glu_{cyt} + 1024.72 Arg_{cyt} + 362.10 Asn_{cyt} + 19.61 Cys_{cyt} + 913.16 Gly_{cyt} + 271.58 His_{cyt} + 240.54 Ile_{cyt} + 588.41 Leu_{cyt} + 724.21 Lys_{cyt} + 191.97 Val_{cyt} + 114.57 Met_{cyt} + 201.91 Phe_{cyt} + 205.97 Pro_{cyt} + 247.44 Ser_{cyt} + 202.07 Thr_{cyt} + 21.17 Trp_{cyt} + 61.87 Tyr_{cyt} + 24046.3 ATP_{cyt}

Lipids [1 g] = 181 CH + 661.4 PC + 250.3 PE + 90.9 PI + 24.9 PS + 12.6 PGL + 81.4 SM + 26.8 DPG

DNA [1 g] = 3009 $R5P_{cyt}$ + 3912 Asp_{cyt} + 5717 Gln_{cyt} + -2106 $NADH_{cyt}$ + 1505 Gly_{cyt} + 22569 ATP_{cyt} + 903 $NADH_{mit}$ + -2407 Mal_{cyt} + 903 $NADPH_{cyt}$ + -5717 Glu_{cyt} + 5417 $MTHF_{cyt}$ + -5417 THF_{cyt}

RNA [1 g] = 3020 $R5P_{cyt}$ + 3606 Asp_{cyt} + 6316 Gln_{cyt} + 293 $O_{2,cyt}$ + -2435 $NADH_{cyt}$ + 1477 Gly_{cyt} + 22614 ATP_{cyt} + 586 $NADH_{mit}$ + -2069 Mal_{cyt} + -2954 $NADPH_{cyt}$ + -6316 Glu_{cyt} + 4431 $MTHF_{cyt}$ + -4431 THF_{cyt}

Carbohydrates [1 g] = 6172.8 $G6P_{cyt}$ + 21605 ATP_{cyt}

to reflect thermodynamic limits under physiological conditions, upper bounds of unknown reaction rates were set to 10000 μmol/gDW/h. The optimization problems studied by FVA then read

$$
\begin{gathered}
r_{u,i}^{\min} = \min_{r} r_{u,i} \quad \text{OR} \quad r_{u,i}^{\max} = \max_{r} r_{u,i} \\
s.t. \\
\mathbf{N_u r_u} = -\mathbf{N_k r_k} \\
0 \le r_i \le 10000 \quad \forall i \quad \text{reaction} \quad i \quad \text{irreversible} \\
-10000 \le r_i \le 10000 \quad \forall i \quad \text{reaction} \quad i \quad \text{reversible}
\end{gathered}
\tag{4}
$$

As a result from the $2{*}u$ optimizations (u = number of unknown fluxes), we obtained the physiologically feasible flux range for the unknown reactions. Moreover, if the computed minimal and maximal rate of a reaction coincide ($r_{u,i}^{\min} = r_{u,i}^{\max}$), the reaction rate follows to be uniquely determined. Note that FVA as described above does not make any assumption about biological objectives in contrast to flux balance analysis [59]. The objective function in equation (4) only serves as a tool to identify the feasible flux ranges. This approach has also been used to estimate flux distributions in CHO cells and was introduced under the term flux-spectrum [60].

We define two scenarios where we consider different sets of constraints. In scenario 1, we applied the measured rate values plus/minus their corresponding standard deviations as upper/lower boundaries for the corresponding rates. In case that the lower boundary of an uptake rate becomes negative, we set it to zero. Rates that were measured not to be active (e.g. uric acid excretion and alanine uptake) were also fixed to zero. No further constraints were used so that the results obtained with this scenario are robust against measurement errors and assumptions on reaction activities. However, since the system is underdetermined with these constraints, only flux ranges can be calculated via FVA.

In scenario 2, we set the measured uptake/excretion rates as fixed constraints. To resolve the anaplerotic fluxes and since the corresponding flux range contains zero (see Results), we assumed the pyruvate-carboxylase (r46) to be inactive. This assumption is biologically reasonable as it is often described and assumed to be negligible in transformed cell cultures (e.g. in [25,61]) and supported by our enzyme measurements. With these constraints the network is over-determined with two degrees of redundancy. By applying MFA with a variance-weighted least squares estimation [57], we can calculate a unique flux distribution. To avoid a significant deviation of the estimated against the measured growth rate, we set the measurement variance of the growth rate to 1×10^{-9} in scenario 2.

All computations presented in this study were performed with our software *CellNetAnalyzer*, a MATLAB toolbox with graphical user interface facilitating metabolic network analysis [28]. It can be downloaded from www.mpi-magdeburg.mpg.de/projects/cna/cna.html, and the network project files will be made available on this site within *CellNetAnalyzer*'s model repository.

Additional file

Additional file 1: Table S1. Amino acid composition of cell protein from CR.pIX cells and other cell lines. **Table S2.** Calculated rate ranges (scenario 1; by FVA) or rates (scenario 2; by MFA) for CR.pIX cells cultured in stirred tank reactors. **Table S3.** Measured concentrations, standard deviations of validated assays, and measurement devices.

Competing interests
Patent applications covering avian cell lines (including AGE1.CR.pIX) have been filed by I. Jordan.

Authors' contributions
VL and OH conceived the experiments or calculations, interpreted the data and drafted the manuscript. IJ and YG edited the manuscript and participated in planning the experiments. HB helped in data analyses and manuscript drafting. SK contributed to the conception of the study and edited the manuscript. UR helped in drafting the manuscript and generally supervised the project. All authors read and approved the final manuscript.

Acknowledgements
The authors thank C. Best, N. Wynserski, I. Behrendt and S. König for their excellent technical assistance and A. Laqoda for her advice in RNA, DNA and protein quantification. Also, advice and assistance by Anne-Kareen Blechert in enzyme activity assays was highly appreciated.

Author details
[1]Max Planck Institute for Dynamics of Complex Technical Systems, Sandtorstr. 1, 39106 Magdeburg, Germany. [2]Current address: ProBioGen AG, Goethestr. 54, 13086 Berlin, Germany. [3]Cell Culture Technology, Bielefeld University, Universitätsstr. 25, 33615 Bielefeld, Germany. [4]Bioprocess Engineering, Otto von Guericke University Magdeburg, Universitätsplatz 2, 39106 Magdeburg, Germany.

References
1. Butler M: **Optimisation of the cellular metabolism of glycosylation for recombinant proteins produced by Mammalian cell systems.** *Cytotechnology* 2006, **50**(1–3):57–76.
2. Lim Y, Wong NS, Lee YY, Ku SC, Wong DC, Yap MG: **Engineering mammalian cells in bioprocessing - current achievements and future perspectives.** *Biotechnol Appl Biochem* 2010, **55**(4):175–189.
3. Blanchard TJ, Alcami A, Andrea P, Smith GL: **Modified vaccinia virus Ankara undergoes limited replication in human cells and lacks several immunomodulatory proteins: implications for use as a human vaccine.** *J Gen Virol* 1998, **79**(Pt 5):1159–1167.
4. Draper SJ, Heeney JL: **Viruses as vaccine vectors for infectious diseases and cancer.** *Nat Rev Microbiol* 2010, **8**(1):62–73.
5. Jordan I, Vos A, Beilfuss S, Neubert A, Breul S, Sandig V: **An avian cell line designed for production of highly attenuated viruses.** *Vaccine* 2009, **27**(5):748–756.
6. Lohr V, Rath A, Genzel Y, Jordan I, Sandig V, Reichl U: **New avian suspension cell lines provide production of influenza virus and MVA in serum-free media: studies on growth, metabolism and virus propagation.** *Vaccine* 2009, **27**(36):4975–4982.
7. Nadeau I, Sabatie J, Koehl M, Perrier M, Kamen A: **Human 293 cell metabolism in low glutamine-supplied culture: interpretation of metabolic changes through metabolic flux analysis.** *Metab Eng* 2000, **2**(4):277–292.
8. Nadeau I, Jacob D, Perrier M, Kamen A: **293SF metabolic flux analysis during cell growth and infection with an adenoviral vector.** *Biotechnol Prog* 2000, **16**(5):872–884.
9. Niklas J, Schrader E, Sandig V, Noll T, Heinzle E: **Quantitative characterization of metabolism and metabolic shifts during growth of the new human cell line AGE1.HN using time resolved metabolic flux analysis.** *Bioprocess Biosyst Eng* 2011, **34**(5):533–545.
10. Priesnitz C, Niklas J, Rose T, Sandig V, Heinzle E: **Metabolic flux rearrangement in the amino acid metabolism reduces ammonia stress in the alpha1-antitrypsin producing human AGE1.HN cell line.** *Metab Eng* 2012, **14**(2):128–137.
11. Altamirano C, Illanes A, Casablancas A, Gamez X, Cairo JJ, Godia C: **Analysis of CHO cells metabolic redistribution in a glutamate-based defined medium in continuous culture.** *Biotechnol Prog* 2001, **17**(6):1032–1041.
12. Teixeira A, Cunha AE, Clemente JJ, Moreira JL, Cruz HJ, Alves PM, Carrondo MJ, Oliveira R: **Modelling and optimization of a recombinant BHK-21 cultivation process using hybrid grey-box systems.** *J Biotechnol* 2005, **118**(3):290–303.
13. Sidorenko Y, Wahl A, Dauner M, Genzel Y, Reichl U: **Comparison of metabolic flux distributions for MDCK cell growth in glutamine- and pyruvate-containing media.** *Biotechnol Prog* 2008, **24**(2):311–320.
14. Wahl A, Sidorenko Y, Dauner M, Genzel Y, Reichl U: **Metabolic flux model for an anchorage-dependent MDCK cell line: characteristic growth phases and minimum substrate consumption flux distribution.** *Biotechnol Bioeng* 2008, **101**(1):135–152.
15. Bonarius HP, Houtman JH, Schmid G, de Gooijer CD, Tramper J: **Metabolic-flux analysis of hybridoma cells under oxidative and reductive stress using mass balances.** *Cytotechnology* 2000, **32**(2):97–107.
16. Bonarius HP, Hatzimanikatis V, Meesters KP, de Gooijer CD, Schmid G, Tramper J: **Metabolic flux analysis of hybridoma cells in different culture media using mass balances.** *Biotechnol Bioeng* 1996, **50**(3):299–318.
17. Bernal V, Carinhas N, Yokomizo AY, Carrondo MJ, Alves PM: **Cell density effect in the baculovirus-insect cells system: a quantitative analysis of energetic metabolism.** *Biotechnol Bioeng* 2009, **104**(1):162–180.
18. Carinhas N, Bernal V, Yokomizo AY, Carrondo MJ, Oliveira R, Alves PM: **Baculovirus production for gene therapy: the role of cell density, multiplicity of infection and medium exchange.** *Appl Microbiol Biotechnol* 2009, **81**(6):1041–1049.
19. Kovacevic Z, Brkljac O, Bajin K: **Control and function of the transamination pathways of glutamine oxidation in tumour cells.** *Biochem J* 1991, **273**(Pt 2):271–275.
20. Moreadith RW, Lehninger AL: **The pathways of glutamate and glutamine oxidation by tumor cell mitochondria. Role of mitochondrial NAD(P)+-dependent malic enzyme.** *J Biol Chem* 1984, **259**(10):6215–6221.
21. Reitzer LJ, Wice BM, Kennell D: **Evidence that glutamine, not sugar, is the major energy source for cultured HeLa cells.** *J Biol Chem* 1979, **254**(8):2669–2676.
22. Glacken MW, Fleischaker RJ, Sinskey AJ: **Reduction of waste product excretion via nutrient control: possible strategies for maximizing product and cell yields on serum in cultures of mammalian cells.** *Biotechnol Bioeng* 1986, **28**(9):1376–1389.
23. Miller WM, Blanch HW, Wilke CR: **A kinetic analysis of hybridoma growth and metabolism in batch and continuous suspension culture: effect of nutrient concentration, dilution rate, and pH.** *Biotechnol Bioeng* 1988, **32**(8):947–965.
24. Boghigian BA, Seth G, Kiss R, Pfeifer BA: **Metabolic flux analysis and pharmaceutical production.** *Metab Eng* 2010, **12**(2):81–95.
25. Quek LE, Dietmair S, Kromer JO, Nielsen LK: **Metabolic flux analysis in mammalian cell culture.** *Metab Eng* 2010, **12**(2):161–171.
26. Ahn WS, Antoniewicz MR: **Metabolic flux analysis of CHO cells at growth and non-growth phases using isotopic tracers and mass spectrometry.** *Metab Eng* 2011, **13**(5):598–609.
27. Mahadevan R, Schilling CH: **The effects of alternate optimal solutions in constraint-based genome-scale metabolic models.** *Metab Eng* 2003, **5**(4):264–276.
28. Klamt S, Saez-Rodriguez J, Gilles ED: **Structural and functional analysis of cellular networks with Cell NetAnalyzer.** *BMC Syst Biol* 2007, **1**:2.
29. Xie L, Wang DI: **Applications of improved stoichiometric model in medium design and fed-batch cultivation of animal cells in bioreactor.** *Cytotechnology* 1994, **15**(1–3):17–29.
30. Zupke C, Stephanopoulos G: **Intracellular flux analysis in hybridomas using mass balances and in vitro (13)C nmr.** *Biotechnol Bioeng* 1995, **45**(4):292–303.

31. Carnicer M, Baumann K, Toplitz I, Sanchez-Ferrando F, Mattanovich D, Ferrer P, Albiol J: **Macromolecular and elemental composition analysis and extracellular metabolite balances of Pichia pastoris growing at different oxygen levels.** *Microb Cell Fact* 2009, **8**:65.
32. Selvarasu S, Ho YS, Chong WP, Wong NS, Yusufi FN, Lee YY, Yap MG, Lee DY: **Combined in silico modeling and metabolomics analysis to characterize fed-batch CHO cell culture.** *Biotechnol Bioeng* 2012, **109**(6):1415–1429.
33. Ferrance JP, Goel A, Ataai MM: **Utilization of glucose and amino acids in insect cell cultures: quantifying the metabolic flows within the primary pathways and medium development.** *Biotechnol Bioeng* 1993, **42**(6):697–707.
34. Xie L, Wang DI: **Fed-batch cultivation of animal cells using different medium design concepts and feeding strategies.** *Biotechnol Bioeng* 1994, **43**(11):1175–1189.
35. Lohr V, Genzel Y, Jordan I, Katinger D, Mahr S, Sandig V, Reichl U: **Live attenuated influenza viruses produced in a suspension process with avian AGE1.CR.pIX cells.** *BMC Biotechnol* 2012, **12**:79.
36. Niklas J, Priesnitz C, Rose T, Sandig V, Heinzle E: **Primary metabolism in the new human cell line AGE1.HN at various substrate levels: increased metabolic efficiency and alpha1-antitrypsin production at reduced pyruvate load.** *Appl Microbiol Biotechnol* 2012, **93**(4):1637–1650.
37. Schneider M, Marison IW, von Stockar U: **The importance of ammonia in mammalian cell culture.** *J Biotechnol* 1996, **46**(3):161–185.
38. Yang M, Butler M: **Effect of ammonia on the glycosylation of human recombinant erythropoietin in culture.** *Biotechnol Prog* 2000, **16**(5):751–759.
39. Slivac I, Blajic V, Radosevic K, Kniewald Z, Gaurina Srcek V: **Influence of different ammonium, lactate and glutamine concentrations on CCO cell growth.** *Cytotechnology* 2010, **62**(6):585–594.
40. Stevens L: *Avian Biochemistry and Molecular Biology.* Cambridge: Cambridge University Press; 1996.
41. Fitzpatrick L, Jenkins HA, Butler M: **Glucose and glutamine metabolism of a murine B-lymphocyte hybridoma grown in batch culture.** *Appl Biochem Biotechnol* 1993, **43**(2):93–116.
42. Neermann J, Wagner R: **Comparative analysis of glucose and glutamine metabolism in transformed mammalian cell lines, insect and primary liver cells.** *J Cell Physiol* 1996, **166**(1):152–169.
43. Xie L, Wang DI: **Energy metabolism and ATP balance in animal cell cultivation using a stoichiometrically based reaction network.** *Biotechnol Bioeng* 1996, **52**(5):591–601.
44. Zamorano F, Wouwer AV, Bastin G: **A detailed metabolic flux analysis of an underdetermined network of CHO cells.** *J Biotechnol* 2010, **150**(4):497–508.
45. Janke R, Genzel Y, Handel N, Wahl A, Reichl U: **Metabolic adaptation of MDCK cells to different growth conditions: effects on catalytic activities of central metabolic enzymes.** *Biotechnology and bioengineering* 2011, **108**(11):2691–2704.
46. Janke R, Genzel Y, Wahl A, Reichl U: **Measurement of key enzyme activities in mammalian cell lines using rapid and sensitive microplate-based assays.** *Biotechnol Bioeng* 2010, **107**(3):566–81.
47. Taschwer M, Hackl M, Hernandez Bort JA, Leitner C, Kumar N, Puc U, Grass J, Papst M, Kunert R, Altmann F, Borth N: **Growth, productivity and protein glycosylation in a CHO EpoFc producer cell line adapted to glutamine-free growth.** *J Biotechnol* 2012, **157**(2):295–303.
48. Capiaumont J, Legrand C, Carbonell D, Dousset B, Belleville F, Nabet P: **Methods for reducing the ammonia in hybridoma cell cultures.** *J Biotechnol* 1995, **39**(1):49–58.
49. Hassell T, Gleave S, Butler M: **Growth inhibition in animal cell culture. The effect of lactate and ammonia.** *Appl Biochem Biotechnol* 1991, **30**(1):29–41.
50. Fuchs BC, Bode BP: **Stressing out over survival: glutamine as an apoptotic modulator.** *J Surg Res* 2006, **131**(1):26–40.
51. Niederlechner S, Klawitter J, Baird C, Kallweit AR, Christians U, Wischmeyer PE: **Fibronectin integrin signaling is required for L-glutamine's protection against gut injury.** *PLoS One* 2012, **7**(11):e50185.
52. Chen P, Harcum SW: **Effects of amino acid additions on ammonium stressed CHO cells.** *J Biotechnol* 2005, **117**(3):277–286.
53. Genzel Y, Ritter JB, König S, Alt R, Reichl U: **Substitution of glutamine by pyruvate to reduce ammonia formation and growth inhibition of mammalian cells.** *Biotechnol Prog* 2005, **21**(1):58–69.
54. Buentemeyer H: **Methods for off-line analysis of nutrients and products in mammalian cell culture.** In *Methods in Biotechnology. Volume 24.* 2nd edition. Edited by Poertner R. New York: Humana Press Inc; 2007:253–268.
55. Stephanopoulos GN, Aristidou AA, Nielsen J: *Metabolic Engineering - Principles and Methodologies.* Orlando: Academic Press; 1998.
56. Ozturk SS, Palsson BO: **Chemical decomposition of glutamine in cell culture media: effect of media type, pH, and serum concentration.** *Biotechnol Prog* 1990, **6**(2):121–128.
57. van der Heijden RT, Heijnen JJ, Hellinga C, Romein B, Luyben KC: **Linear constraint relations in biochemical reaction systems: I. Classification of the calculability and the balanceability of conversion rates.** *Biotechnol Bioeng* 1994, **43**(1):3–10.
58. Klamt S, Schuster S, Gilles ED: **Calculability analysis in underdetermined metabolic networks illustrated by a model of the central metabolism in purple nonsulfur bacteria.** *Biotechnol Bioeng* 2002, **77**(7):734–751.
59. Orth JD, Thiele I, Palsson BO: **What is flux balance analysis?** *Nat Biotechnol* 2010, **28**(3):245–248.
60. Llaneras F, Pico J: **An interval approach for dealing with flux distributions and elementary modes activity patterns.** *J Theor Biol* 2007, **246**(2):290–308.
61. Mancuso A, Sharfstein ST, Tucker SN, Clark DS, Blanch HW: **Examination of primary metabolic pathways in a murine hybridoma with carbon-13 nuclear magnetic resonance spectroscopy.** *Biotechnol Bioeng* 1994, **44**(5):563–585.

Ectopic expression of human acidic fibroblast growth factor 1 in the medicinal plant, *Salvia miltiorrhiza*, accelerates the healing of burn wounds

YaQing Tan[1†], Kevin Yueju Wang[2†], Nan Wang[1], GangQiang Li[1] and DeHu Liu[1*]

Abstract

Background: Healing of burns is a complex process and very few effective treatments exist to facilitate the burn recovery process. Human acidic fibroblast growth factor 1 (FGF-1) plays an important role in a variety of biological processes, including angiogenesis, and tissue repair. *Salvia miltiorrhiza* is widely used in traditional Chinese medicine as an herb for the treatment of various diseases, including cardiovascular and cerebrovascular diseases, and traumatic injuries. We present that expression of FGF-1 in *S. miltiorrhiza* significantly accelerates the healing of burn wounds.

Results: The human *fgf-1* gene was fused with a barley α-amylase signal peptide DNA sequence and driven by a 35S promoter for constitutive expression in transgenic *S. miltiorrhiza* plants. The highest yield of recombinant FGF-1 obtained from leaves of transgenic *S. miltiorrhiza* lines was 272 ng/fresh weight. Aqueous extracts from transgenic *S. miltiorrhiza* exhibited FGF-1 activity approximately 19.2-fold greater than that of the standard FGF-1. Compared to the standard FGF-1 or the extracts obtained from non-transgenic plants, it stimulated proliferation of Balb/c 3 T3 mouse fibroblast cells assessed with the standard MTT assay and promoted angiogenesis in the chicken embryo chorioallantoic membrane (CAM) assay. Topical application of the extract significantly accelerated the burn wound healing process.

Conclusions: The product appears to retain the biological activity of both FGF-1 as well as the medicinal properties of the plant. The extracts from transgenic *S. miltiorrhiza* combines the therapeutic functions of FGF-1 and the medicinal plant, *S. miltiorrhiza*. Topical application of the product can reduce the costs associated with extraction, purification, and recovery.

Keywords: Human acidic fibroblast growth factor 1, Medicinal plant, *Salvia miltiorrhiza*, Burn wound healing

Background

Various factors, such as heat, electricity, UV-light, and chemicals, can cause burns. Although most burns are not fatal, over 300,000 deaths occur globally each year as a direct result of burns, with 90% of burn deaths occurring in developing countries [1]. Healing of burns is a complex process and very few effective treatments exist to facilitate the burn recovery process. Antibiotics are often used to prevent or treat infections and skin or synthetic grafts are used to facilitate the healing process in cases of third-degree burns [2].

Fibroblast growth factors (FGFs) comprise a large family genes involved in growth and differentiation. They are found in both invertebrates and vertebrates [3,4]. FGFs have been demonstrated to play essential functions in development, metabolism, and repair of various tissues and organs. Human FGFs contain 22 family members. Human FGF-1 is one of the most studied and characterized members of the superfamily. FGF-1 has various biological functions and has been associated with various

* Correspondence: liudehu2006@126.com
†Equal contributors
[1]Biotechnology Research Institute, Chinese Academy of Agricultural Sciences, 12 Southern Zhong Guan Cun Road, Beijing 100081, China
Full list of author information is available at the end of the article

stages of morphogenesis, tissue repair, angiogenesis and wound healing [3,4].

FGF-1 is present in various tissues, especially within the extracellular matrix, however, isolating and purifying this protein is a complicated and tedious process. Recombinant DNA technology has enabled FGF-1 to be produced in *E.coli* [5], yeast [6], mammalian cells [7] and plants [8] for potential medical applications. Most studies designed to evaluate its effect on wound healing and tissue regeneration have administered recombinant FGF-1 directly to animal [9,10] or human wound sites [10]. Like other FGFs, however, FGF-1 has low stability and is very sensitive to degradation by proteases [3,4]. Unfortunately, recombinant FGF-1 exhibits a very short half-life *in vivo* which limits its medical application.

Salvia miltiorrhiza, also known as Danshen, native to China and Japan, is a traditional Chinese medicinal (TCM) herb. The extract of its root has been widely used clinically to treat and prevent diseases, such as cardiovascular disease, hyperlipidemia, cancer, liver fibrosis and cirrhosis, etc. [11]. Leaf extract of *S. miltiorrhiza* also contains considerable amounts of pharmacological compounds even it is discarded as waste during root harvest. More than seventy compounds from *S. miltiorrhiza* have been investigated. Three major bioactive compounds, tanshinone-I, tanshinone-IIA, and cryptotanshinone have been extensively studied and reported to have a variety pharmacological activities, including anti-inflammatory [12], antioxidant [13], antitumor [14] and anti-platelet aggregation [15]. TCM compendiums indicate that *S. miltiorrhiza* reduces blood stasis by promoting blood circulation and repairing damaged tissues. The use of *S. miltiorrhiza* to treat coronary heart disease and hypertension, as well as for stroke patients, is in clinical phase IV trials (clinicaltrials.gov). However, like many TCM herbs, the precise mechanism underlying the biological activity of *S. miltiorrhiza* still remains to be completely elucidated.

In the present study, FGF-1 was constitutively expressed in the medicinal plant, *S. miltiorrhiza*. Presence and levels of FGF-1 were analyzed by ELISA and Western blot. Aqueous extracts of transgenic *S. miltiorrhiza* were applied directly to fertilized chicken egg embryo chorioallantoic membranes (CAM) and to second degree burn wounds on Sprague Dawley (SD) rats. Results indicated that the transgenic plants exhibited a both the biological function of FGF-1 and the medicinal properties of *S. miltiorrhiza*. Data indicated that the extract obtained from the transgenic plants could promote cell proliferation, speed up the growth of new blood vessels and significantly accelerate the burn wound healing process compared to the use of either FGF-1 or *S. miltiorrhiza* alone. The use of the transgenic plant extract also avoids the complicated process of FGF-1 extraction and purification.

Results

Gene cloning and plant transformation

Human FGF-1 was cloned into the expression vector, pBI121, with a Kozak consensus sequence inserted before the start codon and a barley alpha amylase signal sequence at the 5′ end, driven by a cauliflower mosaic virus 35S (CaMV35S) promoter (Figure 1) for strong constitutive expression. At least 120 independent putative transgenic events were obtained. All putative transformants were analyzed by PCR amplification using P1 and P2 primers. Results indicated that fifty-six of the putative transformants had the predicted 0.5 kb band.

ELISA and western blot analysis

The level of rFGF-1 produced in the confirmed transgenic lines was determined by ELISA. Results revealed that the yield of rFGF-1 in lines T65, T97 and T117 were131, 146 and 272 ng/g fresh leaf weight, respectively, as calculated by comparison with a standard curve constructed using commercial FGF-1 (see Additional file 1: Table S1).

Recombinant human acidic fibroblast growth factor 1 (rFGF-1) protein expression in transgenic lines T65, T97 and T117 was determined by the Western blot analysis (Figure 2). A positive response using FGF-1 monoclonal antibody demonstrated that transgenic lines (T65, T97, and T117) had a 16 kDa band corresponding in size to standard FGF-1. No band was observed in protein samples extracted from WT plants. The results of the Western blot also indicated that T117 had a higher level of expression than T65. It corresponded with the results obtained in the ELISA assay. Since T117 had the highest level of rFGF-1, it was examined further. Transgene copy number in line T117 was determined by Southern blot analyses (see Additional file 2: Figure S1). Results indicated that there was a single T-DNA copy integrated in line T117.

Proliferative effects of transgenic *S. miltiorrhiza* (T-SM) on cell proliferation *in vitro*

After 72 h of culture, MTT assay revealed that wild-type *S. miltiorrhiza* (WT), standard FGF-1 (S-FGF) and T-SM

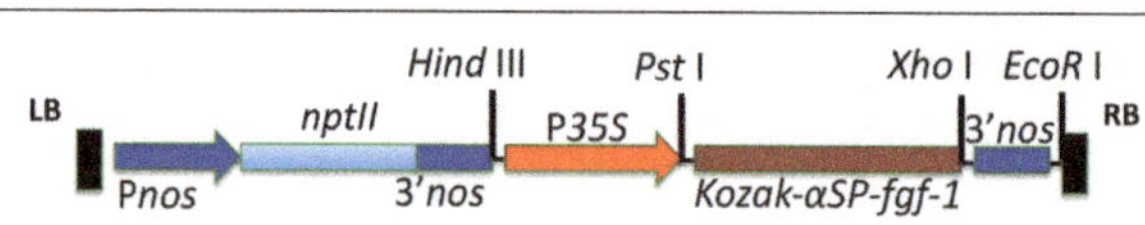

Figure 1 Plant expression construct pBI121-*α*FGF. The construct was made in the T-DNA vector, pBI121. The coding region of *fgf-1* was fused with an alpha amylase signal (*αSP*). A Kozak sequence was inserted before the start codon. *Pnos*, promoter from nopaline synthase gene; nptII, neomycin phosphotransferase II gene; *3′nos*, nopaline synthase terminator; *p35S*, *CaMV 35S* promoter. LB and RB, left and right border T-DNA sequence. *Hind* III, *Pst* I, *Xho* I and *EcoR* I were restriction enzymes used for cloning FGF-1.

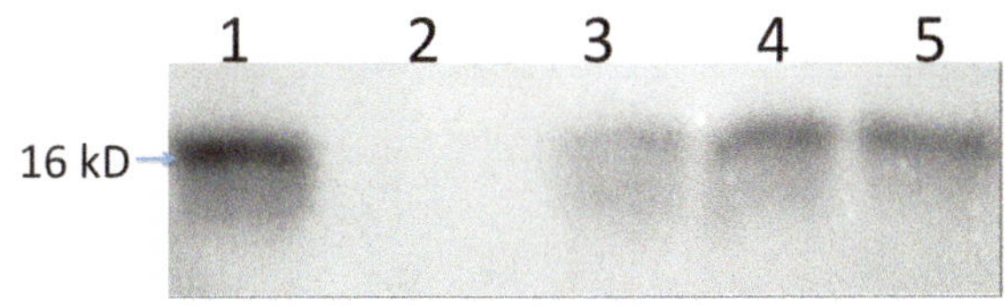

Figure 2 Western immunoblotting assay. Standard FGF-1 (Lane 1), rFGF-1 expression in wild type plants (Lane 2), and transgenic plant lines, T65 (Lane 3), T97 (Lane 4) and T117 (Lane 5). 10 μL of supernatant from each sample was used in the immunoblotting assay.

Table 1 Enzymatic activity of the recombinant FGF-1 from transgenic *S. miltiorrhiza*

	FGF-1 content	ED50 (dilution)	AU/mL	U/mg
S-FGF	200 ng/mL	$4^{1.5}$	50	2.5×10^5
T-SM	10 ng/mL	$4^{1.47}$	48	4.8×10^6

Effective dose 50 (ED50) is determined by MTT assay curve using Balb/c 3 T3 cells. The dilution fold is used to determine that compounds are active at that concentration to proliferate 50% of the cell growth. The enzymatic activity of recombinant FGF-1 from transgenic *S. miltiorrhiza* (T-SM) is determined in terms of AU/mL in relation to a standard human acid FGF-1 (S-FGF) with a known activity (50 AU/mL). The specific enzymatic activity Units of FGF in T-SM was calculated as 48 AU/mL ($4^{1.47}/4^{1.50} \times 50$ AU/mL), with a conversion of $(1000 \times 1000/10 \text{ ng}) \times 48 = 4.8 \times 10^6$ U/mg.

significantly promoted Balb/c 3 T3 cell proliferation compared to the Saline control (Figure 3). The viability of the cells decreased when the cells were treated with diluted concentrations of samples. After the second (4^2 fold) dilution, WT showed had no effect on cellular viability. After the third (4^3 fold) dilution, the absorbance of T-SM (0.076 ± 0.0034) was still significantly higher than that of S-FGF (0.065 ± 0.0026). The results indicated that treatment with T-SM greatly increased the cell proliferation compared with that of the other groups. The estimated effective dose (ED50) stimulating 50% of the maximal cell number response was determined from MTT proliferation curves. The dilution fold of ED50 for S-FGF and T-SM were $4^{1.5}$ and $4^{1.47}$, respectively. The converted enzyme activity of T-SM (4.8×10^6 IU/mg) was approximately 19.2-fold greater than that of the standard FGF-1 (2.5×10^5 IU/mg) (Table 1).

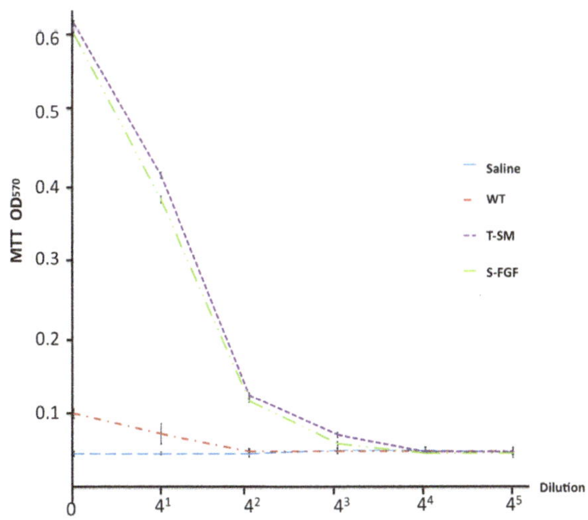

Figure 3 MTT assay for Balb/c 3 T3 cell proliferation analysis. After 72 h treatment, the proliferation rate in the extract of transgenic *S. miltiorrhiza* plant (T-SM) was significantly higher compared to the standard FGF-1 (S-FGF), wild-type plant (WT) and saline solution ($p \leq 0.001$). Data for 4 replicates and 3 repeats were averaged, and error bars represent SD.

T-SM stimulates angiogenesis *in vivo*

The CAM assay is widely used to study angiogenesis *in vivo*. Our study demonstrated that treatment with extracts from WT, S-FGF and T-SM all stimulated angiogenesis (Figure 4). At day 11, a very few large vessels with only few branching were observed with in the samples treated with just saline. In comparison to the saline treatment, a greater number of large vessels and branches were formed in the WT treatment. T-SM treatment induced a significantly greater number of large vessels and branches compared to all of the other treatments. Even though the concentration of T-SM rFGF-1 used in the CAM assay was 1/18.5 of the S-FGF treatment (≈2.7 ng vs. 50 ng, respectively), it stimulated more vessels. The greatest amount of branch formation was also observed with T-SM-induced angiogenesis in the CAM assay.

T-SM Accelerates wound healing process

Images of burn wounds treated with saline, WT, S-FGF (FGF-1 1000 ng/wound), and T-SM (rFGF-1 ≈ 2.7 ng/wound) at 0, 3, 7, 14, and 21 d after administering the burn treatment are shown in Figure 5. On day 3, shrinkage of the wounds was observed in all groups. WT and saline groups had begun to form dark granulation tissue. In contrast, T-FGF-1 treated wounds appeared smooth and whitish. FGF-1 treated wounds were also smooth but exhibited some granulation tissue. On day 7, wounds in all groups continued to shrink and develop thick granulation tissue. On day 14, while considerable contraction of wounds was observed in all treatments, T-SM treated wounds began rebuilding healthy granulation tissue. On day 21, the S-FGF and T-SM treated wounds were almost healed, while WT and saline treated wounds were still very visible and whitish in appearance. Wound surface area tracing measurements (Figure 6) did not identify significant difference among the four treatments on day 0, 7 and 14 and also revealed that the healing process appeared normal. On day 21, however, both T-SM and S-FGF treated wounds were significantly reduced ($P < 0.01$) in wound surface area (0.14 ± 0.06 and 0.21 ± 0.05, respectively) compared to the saline (0.64 ± 0.10)

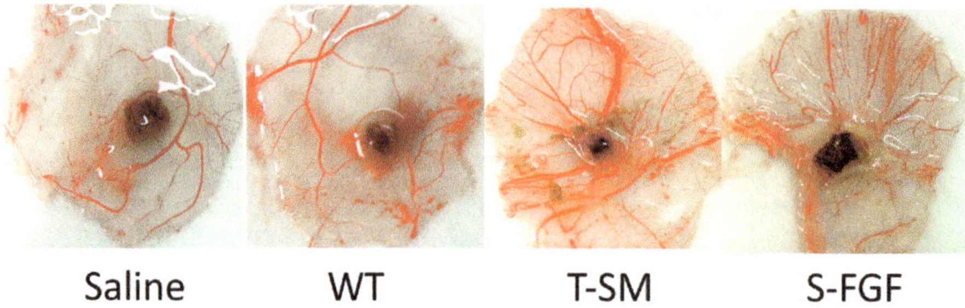

Figure 4 Extracts from transgenic plant promotes angiogenesis in the CAM assay. A: Saline solution was used as negative control. 5 μL of aqueous extract from wild-type *S. miltiorrhiza* plants (WT) and transgenic *S. miltiorrhiza* (T-SM). S-FGF treatment contained 50 ng of FGF-1 produced in *E. coli* and obtained commercially.

and WT (0.58 ± 0.11) treated wounds. S-FGF and T-SM treated wounds were almost completely healed. No difference in wound surface area was observed between saline and WT treated burns. Importantly, while the concentration of FGF-1 in the S-FGF treatment was 1000 ng FGF-1/wound, it was only ≈ 2.7 ng rFGF-1 in the T-SM treatment. However, they exhibited a similar ability to accelerate the wound healing process.

Histopathological assessments of burn wound healing process

Histopathological analysis of normal tissues and the burn wound healing process was conducted by sectioning of tissues and staining with hematoxylin and eosin staining (H & E). A standard second degree burn wound was established (see Additional file 3: Figure S2). Normal skin had an intact cuticle, hair follicles, sebaceous glands, collagen and a prickle cell layer (A and B). Epidermal necrolysis was observed 24 h after burn treatment (C and D). In the dermis, dead cells disintegrated and collagen fibers became disorganized. Hair follicles were also damaged and vasodilation and congestion of the blood vessels were evident.

On day 7, the saline treated burn wounds exhibited inflammatory cells in epidermal necrolysis tissue. An inflammatory response in the wound area was not obvious (Figure 7). In other treatments, homogeneous necrosis was evident in the epidermal exudate of burn wounds but with less inflammatory cells. The inflammatory response in the wound area was obvious with an increased number of inflammatory cells (B, C, D). On day 14, in contrast to saline treated wounds, the burn wounds in all other treatments exhibited rapidly growing granulation tissue, including fibroblasts, capillaries, and inflammatory cells. In S-FGF and T-SM treated burn wounds, a greater amount of newly formed epidermis began to migrate into the wound area (C and D). On day 21, saline treated burn wounds still exhibited a thick surface layer of dead cells. A large number of inflammatory cells were also observed with some bleeding near the epidermis. An incomplete monolayer of epidermal cells was also observed. Burn wounds in all the other treatments exhibited good re-epithelialization. A thicker prickle cell layer was observed in S-FGF treated wounds than in T-SM treated wounds. Tissue directly in the wound area began to form hair follicles and other skin appendages. T-SM treated wounds exhibited less necrotic tissue compared to all the other treatments.

Discussion

Our studies demonstrated that the medicinal plant, *S. miltiorrhiza*, can be used to produce functional human

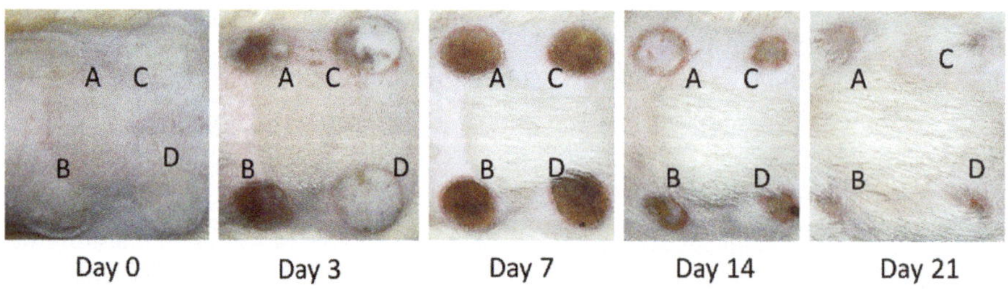

Figure 5 Extracts from transgenic plants accelerate the burn wound healing process. Second degree burn wounds were administered on the dorsal skin of SD rats. The area of the burn wounds was treated with **(A)**, 100 μL of saline **(B)** aqueous extract from wild type plants; **(C)**, 1000 ng of commercial FGF-1, and; **(D)** 100 μL of aqueous extract from transgenic *S. miltiorrhiza*. Photographs were taken on Day 0, 3, 7, 14 and 21 after burn injury was administered.

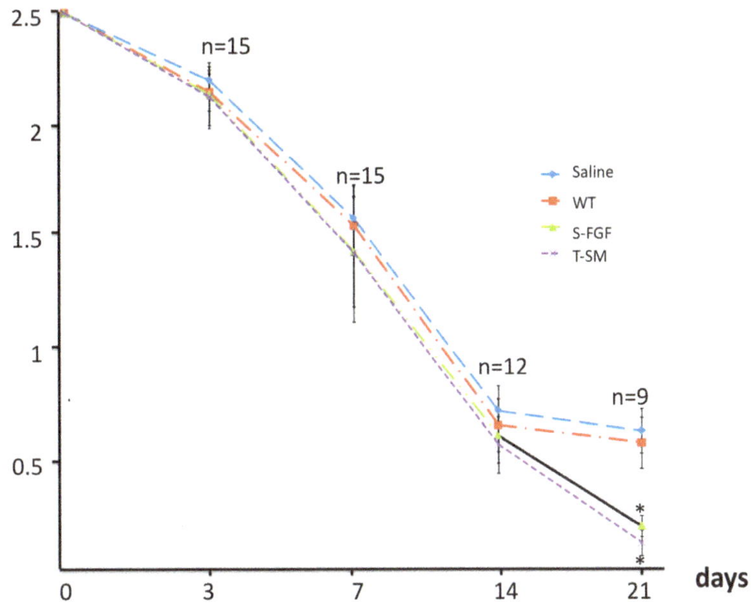

Figure 6 Burn wound surface assay. Area as evaluated using the transparent tracing method described by Bohannon and Pfaller (1983). Wounds were treated with either 100 μL of saline, 1000 ng commercial FGF-1, or 100 μL aqueous extract from wild type or transgenic *S. miltiorrhiza*. Burn wound area was recorded on day 3 (n = 15), 7 (n = 15), 14 (n = 12) and 21 (n = 9). The data represent the Means ± se. n: rats number for each treatment. *indicates significant difference (P < 0.01) between the specified treatment and the saline treatment.

FGF-1. In addition to acquiring the therapeutic function of FGF-1, the transgenic plants also contained natural, endogeneous, medicinal compounds, such as tanshinone-I, tanshinone-IIA and cryptotanshinone. The therapeutic effect of the transgenic plants was significant and greater than using either FGF-1 or the medicinal plant alone. Western blot (Figure 2) assays indicated that FGF-1 was expressed in the transgenic *S. miltiorrhiza*.

The highest accumulation of recombinant FGF-1 level in transgenic *S. miltiorrhiza* was 272 ng/g fresh leave weight in line T117. We found that topical application of crude extracts from transgenic *S. miltiorrhiza* could enhance angiogenesis and accelerate the burn wound healing process.

The drawback in using plants to produce recombinant proteins is that plant cells often recognize the proteins as foreign and target them for degradation. Since only the expected single band was observed in the western blot assay, this indicated that the rFGF-1 in *S. miltiorrhiza* had not been subjected to proteolysis. Reports have suggested that foreign gene expression, and protein yield in transgenic plants could increase after producing several generations (T1-T4) of plants from self-pollinated primary (T0) transgenic lines [16,17]. We selected the line T117, which has a single transgene integration event as determined by Southern blot (Additional file 2: Figure S1), to produce subsequent generations in order to determine if the yield of rFG-1 in transgenic *S. miltiorrhiza* can be increased.

Extracts of non-transgenic *S. miltiorrhiza* can inhibit the stickiness of blood platelets and decrease fibrin [11,15]. The extract from this plant is widely used in

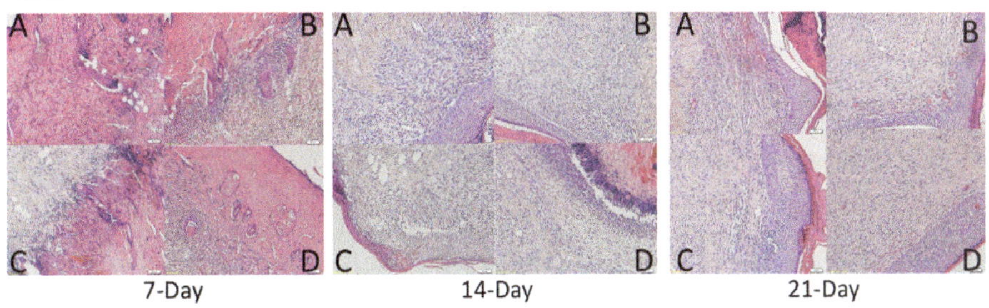

Figure 7 H&E staining. Cryosections (2 μm thick) stained with hematoxylin (H) and eosin (E) to examine the burn wound healing process after application of different treatments. Samples were taken on day 7, 14 and 21. **A**: 100 μL of saline; **B**: 100 μL saline aqueous extract from wild type (WT) *S. miltiorrhiza;* **C**: 1000 ng FGF-1; and **D**: 100 μL saline aqueous extract from transgenic *S. miltiorrhiza*. Micrographs were taken at 200x.

traditional Chinese medicine to treat cardiovascular and cerebrovascular diseases [11]. One of its compounds, salvianolic acid, was reported to enhance *in vitro* angiogenesis in rat endothelial cells through the up-regulation of VEGF and VEGF receptors genes that promote cell growth and differentiation [18]. In the CAM assays, we also observed that the *S. miltiorrhiza* treatment produced more blood vessels (Figure 4). These results indicate that *S. miltiorrhiza* may facilitate angiogenesis. Very few studies have reported that *S. miltiorrhiza* has an effect on the burn wound healing process. One study indicated that injection of *S. miltiorrhiza* and ligustrazine can effectively reduce myocardial damage in patients with severe burns [19]. Animal studies also indicate that *S. miltiorrhiza* enhances the wound healing process [20,21]. Randomized clinical trial of *S. miltiorrhiza* suggested that it is effective in reducing skin flap ischaemia necrosis after mastectomy [20]. *S. miltiorrhiza* can stimulate the cell proliferation and it has the potential to be used for wounding healing or cosmetic treatment [22]. In this study, the *S. miltiorrhiza* treatment stimulated fibroblast cell proliferation (Figure 3). The burn wound healing process is complex. We did not observe that *S. miltiorrhiza* treatment of burn wounds reduced burn wound area compared to the saline treatment (Figures 5 and 6). H&E staining, however, indicated that *S. miltiorrhiza* treatment can promote the formation of new cells and tissues to accelerate wound re-epithelialization. *S. miltiorrhiza* can promote the circulation system in wounded tissue (Figure 7). It may improve the burn wound healing process by removing necrotic cells, attenuating the inflammatory response and apoptosis [12,21], reducing oxidative stress [12], and possibly facilitate the repair of tubular epithelia structures [23].

The function of FGFs in tissue repair, angiogenesis and wound healing is well documented [3,4]. FGF-1 and FGF-2 promote cell proliferation and migration, as well as induce the physical organization of endothelia cells into blood vessel structures [24,25]. These properties can facilitate the burn wound healing process. The MTT, CAM assays, and burn wound healing test used in the present study also demonstrated that the application of FGF-1 enhanced cell proliferation and induced a greater number of new blood vessels and reduced the burn wound surface area more than treatments in which FGF-1 was absent (Figures 3, 4 and 5). These results were further supported by the histological analysis. The healing rate in the S-FGF treated wounds was more rapid compared to either the saline or WT treated wounds. The healing rate in S-FGF and T-SM treated burn wounds was similar. H&E staining revealed less necrosis in T-SM treated wounds than in S-FGF treated wounds. Approximately 2.7 ng of the rFGF-1 was used in the T-SM treatment while 1000 ng FGF-1 was used in the S-FGF treatment. This indicates that the T-SM treatment was significantly more efficient in its effect on angiogenesis and the burn wound healing process. The FGF-1 used in the S-FGF treatment was produced in *E.coli.* Common microbial hosts such as *E. coli* can produce high yields of recombinant protein but lack the requisite machinery for post-translational modification required for protein stability and bioactivity [26]. In contrast, the protein synthesis process in plants is very similar to animal cells. The structure and function of rFGF-1 derived from transgenic *S. miltiorrhiza* may be more similar to native, human FGF-1 and thus have higher activity than FGF-1 obtained from *E. coli. S. miltiorrhiza* exhibits anti-inflammatory [12], anti-oxidant [13] and anti-bacterial activity [27]. The medicinal properties of *S. miltiorrhiza* combined with FGF-1 functions appears to accelerate burn wound healing by improving blood circulation and providing more nutrients and oxygen to the wound area.

FGFs are readily degradable *in vivo*, which results in a loss of biological activity and function [3,4,28,29]. FGF-1 biological activity was maintained in the extracts obtained from transgenic *S. miltiorrhiza* and applied topically to burn wounds. *S. miltiorrhiza* is a medicinal plant and contains natural compounds that also have various pharmaceutical functions. Overexpression of *fgf-1* in *S. miltiorrhiza* provides the benefits of both FGF-1 and the medicinal properties of the plant, which may enhance its therapeutic effects. Plant specific glycans can harm humans and injection of recombinant plant therapeutic proteins may result in an immunogenic response in humans [30]. Topical application of transgenic *S. miltiorrhiza* extracts avoids plant-specific glycosylated immunogenic concerns and reduces the complex and costly purification and recovery process associated with the production of therapeutic compounds in microorganisms.

Conclusions

We successfully overexpressed *fgf-1* in transgenic *S. miltiorrhiza.* Supernatant from homogenated transgenic *S. miltiorrhiza* plants promoted angiogenesis and accelerated the burn wound healing process. The product system combines the therapeutic functions of FGF-1 and the medicinal plant, *S. miltiorrhiza*. Topical application of the product can reduce the costs associated with extraction, purification, and recovery. The yield of rFGF-1 could be further increased by optimizing the gene codon, targeting the protein to the specific subcellular compartments, or selfing of high-expressing lines to obtain homozygous plants. The mechanism associated with the enhanced biological activity of *S. miltiorrhiza* and FGF-1 combined in the same extract will require further study.

Methods

Materials

The plasmids, pUC-haFGF (containing *fgf-1* gene), pTΩ4A, the plant expression construct pBI121, and *Agrobacterium tumefaciens* LBA4404 are materials maintained in our lab. Seeds of *S. miltiorrhiza* were purchased from Shangluo Northwestern Medical Plant Company of Shaanxi (Shaanxi, China). Balb/c 3 T3 mouse fibroblast cells were bought from the Cell Culture Center, Institute of Basic Medical Sciences of Chinese Academy of Medical Sciences and School of Basic Medicine of Peking Union Medical College (Beijing, China). Fertilized eggs were obtained from the Institute of Animal Sciences (IAS), Chinese Academy of Agricultural Science (CAAS), Beijing, China. Male Sprague–Dawley (SD) rats (aged 6 to 7 weeks) were purchased from Vital River Laboratories Company (Beijing, China). All animal protocols were approved by the CAAS Institutional Ethics Committee.

Plasmid construction

In order to enhance recombinant FGF-1 expression level, a Kozak consensus sequence was inserted before the start codon and a barley alpha amylase signal peptide [31,32] was fused to the FGF-1 sequence for extracellular targeting of the recombinant FGF-1. Three rounds of PCR were performed. The first round of PCR used P1 and P2 primers. P1 (forward): 5′-*CTTTCTGCCAGCTT GGCCTCCGGACAAGTT*TACAAGAAGCCAAAGTT GCTTTACT-3′ (Italicized sequences indicate the partial 3′- end alpha amylase signal peptide sequence obtained from barely (*Hordeum vulgare*) and P2 (FGF-1 reverse); 5′-AA*CTCGAG*TTAATCAGAAGAAACTGGCAAT-3′ (*Xho* I site is italicized). PCR product resulting from the use of the P1 and P2 primers was used as the template for the second round PCR with P3 and P2 primers. P3 (forward):5′-*TCCCTCTCCCTCTTCCTCGTCCTCCTT GGC*CTTTCTGCCAGCTTGGCCTCCGGAC-3′ (Italicized sequences indicate the middle region of the alpha amylase signal peptide sequence). The second round PCR product resulting from the use of the P3 and P2 primers was used as the template for the third round PCR with P4 and P2 primers. P4 (forward): 5′- AA*CTGCAG* **AACA**ATG*GCGAACAAACATTTG*TCCCTCTCCCTC TTCCTCGTCCTCC (Italicized sequences indicate the *Pst* I site and 5′-end partial alpha amylase signal peptide sequence, respectively. The bold indicates the Kozak sequence). The third round PCR product was analyzed on a 1% agarose gel and subcloned into a TA cloning vector. After sequence verification (TaKaRa, Dalian, China), the product was digested using *Pst* I/*Xho* I and inserted between the *35S* promoter and *Nos* terminator of pTΩ4A. The resulting *35S-(fgf-1)-Nos* cassette was digested using *Hind* III/*EcoR* I and then inserted into a plant binary vector, pBI121 using the same sites, to generate the plant expression construct, pBI121-aFGF.

Plant transformation and selection

Transformation of *S. miltiorrhiza* utilized the method described by Horschet et al. [33] with modification. The binary vector pBI121-aFGF described above was introduced into *A.tumefaciens* strain LBA4404 using the triparental mating method. The resulting culture was plated on YEP (Yeast extract 10 g/L, Peptone 10 g/L, NaCl 5 g/L, pH 7.0) plates containing selective antibiotics (kanamycin 50 mg/L and rifampicin 50 mg/L). A single colony was analyzed by PCR using P1 and P2 primers, inoculated into 50 mL of YEB medium containing kanamycin (50 mg/L) and rifampicin 50 mg/L, and grown at 28°C, 240 rpm until the OD_{600} reached 1.0. The culture was pelleted by centrifugation and re-suspended in MS liquid medium to obtain an approximate OD_{600} of 0.5. Explants (0.5 cm × 0.5 cm) were excised from 2- to 3-week-old sterile *S. miltiorrhiza* seedlings and immersed for 5 to 8 min in the *Agrobacterium* suspension described above for 5 to 8 min. The explants were then blotted on sterile filter paper and plated on a co-cultivation medium (MS, 6-BA 2.0 mg/L) in the dark for 2 days at 25°C. After co-culture, the explants were transferred onto selection medium (MS, 6-BA 1.0 mg/L, kanamycin 50 mg/L, cefotaxime 400 mg/L). Cultures were incubated at 25°C/23°C (day/night temperature) with a 16-hr photoperiod. Explants were transferred to fresh selection medium every 2 weeks to generate shoots. Shoots were then transferred to a rooting medium (1/2 MS, sucrose 2.0%, kanamycin 30 mg/L) to obtain roots. Plants were acclimated for two weeks in vermiculite with half strength MS liquid medium and then transferred to soil.

Southern blot analysis of putative transgenic plants

Putative transgenic plants were analyzed by DNA hybridization. Total leaf genomic DNA was extracted according to Minas et al. [34]. Thirty μg DNA was digested overnight with single site *EcoR* I or *Hind* III restriction enzymes. The digested product was run on a 0.8%agarose gel. Gel denaturation and neutralization was followed by a routine DNA Southern Blot [35]. The gel was blotted on positively charged nylon membranes by capillary transfer with 20 × SSC following the protocol described in the instruction manual for the DIG High Prime DNA labeling and detection kit (Roche Diagnostics, Indianapolis, USA). DNA fixation, hybridization and wash procedures also followed the manufacturer's instructions. DIG-labeled probe was amplified with *fgf-1* specific primers P1 and P4. Immunological detection was performed according to the manufacturer's instructions.

ELISA and immunoblot analysis

The ELISA procedure was performed according to the method described in the instruction manual of the Human FGF-acidic ELISA Construction Kit (Antigenix America, Huntington Station, USA). Standard FGF-1 produced in *E.coli* (Wanxing-Bio, Shanghai, China) was used to create a series of dilutions (0, 3.125, 6.25, 12.5, 25, 50 and 100 ng/mL) in 1XPBS (phosphate buffered saline) buffer to construct a standard curve. Total soluble protein extracts were made by grinding 20 mg of fresh leaf material in 200 μL1XPBS buffer. The mixture was centrifuged at 12, 000 g for 10 min at 4°C. Recombinant FGF-1 (rFGF-1) protein present in 100 μL supernatant was loaded in duplicate wells of a microtiter plate and the OD_{650} was obtained. The concentration of rFGF-1 was determined by comparison with the constructed FGF-1 standard curve.

For immunoblot analysis, 50 mg of leaf disks were ground in 200 μL of deionizer water. The extract was centrifuged at 12, 000 g for 10 min at 4°C. 100 μL of supernatant was collected and concentrated to 20 μL at 4°C for 5 min using a vacuum concentrator (SPD1010 SpeedVac Syetem, Thermo Scientific, USA). 10 μL of the concentrated supernatant was separated by 15% SDS-PAGE, and transferred to a nitrocellulose membrane with an iBlot® blotting system (Invitrogen). Western blot analysis of rFGF-1 was performed as described by Li et al. (2012). FGF-1 antibody (Sino Biological Inc. Beijing, China) was used as the primary antibody. Nonspecific binding was blocked by incubation in 3% bovine serum albumin (BSA) in TBST (25 mM Tris–HCl, pH 7.4, 0.14 mM NaCl, and 0.05% Tween 20) for 2 h at 4°C. The membrane was washed three times for 10 min each in TBST and then incubated in streptavidin-HRP for 1 h. The membrane was then wash three times in TBST and then stained with 4-Chloro-1-naphthol (4CN) to visualize the results.

Cellular proliferation

The effect of wild-type *S. miltiorrhiza* (WT), transgenic *S. miltiorrhiza* (T-SM) or standard FGF-1 (S-FGF) on cell viability and proliferation was examined by thiazolyl blue tetrazolium bromide (MTT) assay described by Mosman [36]. Leaf tissue (0.25 g fresh weight) of WT and T-SM was ground in 0.75 mL of 1XPBS buffer respectively, and centrifuged at 12,000 g for 10 min at 4°C. The supernatants were diluted 5 times in RPMI 1640 medium supplemented with 0.4% fetal bovine serum (FBS). The yield of rFGF-1 in diluted aqueous extract from T-SM was 10 ng/mL determined by ELISA assay. S-FGF (200 ng/mL, enzyme activity unit: 50 AU/mL) was used as positive control. RPMI 1640 medium with 0.4% FBS was used as blank control. All samples (WT, S-FGF and S-FGF) were 4-fold serially (4°- to 4^{5}- fold dilution) diluted in RPMI 1640 medium with 0.4% FBS.

A concentration of 6.5×10^3 BALB/c 3 T3 cells per well were seeded in a 96-well microplate containing 100 μL of RPMI 1640 medium with 10% FBS. After 24 h of incubation in a humidified incubator with 5% CO_2 at 37°C, all the medium was refreshed in 100 μL of RPMI 1640 with 0.4% FBS. The cultures were maintained for another 24 h. And then, the cells were exposed to 100 μL of sample solutions prepared previously with 100U/mL penicillin and 100 μg/mL streptomycin. After 72 h incubation, 25 μL of 5 mg/mL MTT (AMRESCO, USA) was added to each well. After 5 h treatment with MTT, 120 μL of 50% DMSO was added to the well and mixed thoroughly with pipette. The plate was shaken at room temperature 15 min. The absorbance of each well was determined at 570 nm using a microplate reader (Bio-rad, USA). Four replicates from each group were analyzed. The experiment was repeated 3 times. Results are expressed as mean ± standard error. Student's *t* test was used to evaluate significance differences between groups at $p < 0.05$.

CAM assay

The effect of various extracts on angiogenesis (blood vessel growth) was analyzed by a CAM assay using methycellulose disks described by Larger et al. [37]. Each disk was wetted with 5 μL of aqueous extract from either non-transgenic, wild-type *S. miltiorrhiza* (WT) or transgenic *S. miltiorrhiza* (T-SM), plants. 5 μL of a saline solution and 50 ng of standard FGF-1 (S-FGF) were used as negative and positive controls. 10 μL of 0.5% methylcellulose was dried in a well, 2 mm diameter and 0.6 mm deep, of culture plate. Forty five chick eggs were cleaned with a formalin solution and then incubated at 37°C and 75% RH. Abnormal embryo development was verified and the embryos with malformations or dead embryos were discarded. On day 9 of the incubation period, previously prepared saline, S-FGF, T-SM, or WT methylcellulose disks were placed on the CAM. The treated area was carefully excised after 48 h of incubation and washed with deionized water. The excised area was spread on 3MM filter paper and photographed with a Sony DSLR digital camera.

Burn wound surface area assay on SD rats

Male SD rats were anesthetized using diethyl ether. Their dorsal hair was shaved as clean as possible and the skin sterilized using 70% ethanol. A 9-cm by 8-cm square of dorsal skin was further depilated using 8% sodium sulfide. The bare skin was sterilized using 1% povidone iodine solution and cleaned with 0.9% NaCl. Four circular (2.5 cm^2) second degree burns were produced with YLS-5Q (Biowill Co, LTD. Shanghai, China) on the dorsal skin of each SD rat. The burn device was set at 80°C and applied for 8 seconds. After surgery, each wound was treated for 5 min with either 100 μL of saline, S-FGF

(100 ng FGF-1), or 100 μL of T-SM or WT saline aqueous extract for 5 min. The wound was then covered with sterile gauze. After treatment, each rat was housed in an individual cage, 16 × 16 × 9 cm. The burn area was treated once for 5 minutes every other day for 21 days. During the experimental period, all the rats had free access to fresh water and pelleted feed. The wound surface area was determined using the transparent tracing method described by Bohannon and Pfaller [38]. Wound area = (7.5 cm × 7.5 cm × paper weight)/0.1604 g. The area of each wound was recorded on day 3 (n = 15), 7 (n = 15), 14 (n = 12) and 21 (n = 9). Results are expressed as mean ± standard error. Student's t test was used to evaluate significance differences between groups at $p < 0.05$.

Histological staining

Two rats were sacrificed on the day or one day after the burn was administered to determine the severity of the burn. Three rats were euthanized after 7-, 14-, and 21 d of treatment with the different solutions (Saline, S-FGF, T-SM and WT). Tissues including and surrounding the burn wounds were removed and frozen in liquid nitrogen. Wound tissues were cut and processed into 2-μm thick cryocut sections. The effect of different treatments on the wound healing process were determined histologically using a hematoxylin (H) and eosin (E) staining kit (Huaxia Bio Co., Beijing, China). The status of the wound was determined by the presence of inflammatory cells, necrotic cells and tissue, epidermal appendages, granulation tissue, and angiogenesis. Images were taken in the center of each histological section at a 200× magnification.

Additional files

Additional file 1: Table S1. Analysis of recombinant FGF-1 accumulation in the selected transgenic *S. miltiorrhiza* lines. Recombinant FGF-1 accumulation levels in aqueous extract from transgenic lines were analyzed by ELISA. The yield was calculated by the standard FGF-1 curve (Y = 0.0042X + 0.1418). Y is the protein measurement of OD_{650} in 100 μL of 1XPBS buffer. FW: fresh weight of the leaf. The detailed method was described in material and method. Data represents mean of duplicates.

Additional file 2: Figure S1. Southern blot. DNA hybrization assay using *fgf-1* as a probe. Genomic DNA (30 μg) from wild-type (WT) and the transgenic (T117) plants was digested with either, *HindIII* or *EcoRI* restriction enzymes. Lane 1, wild type. Lane 2–9, transgenic lines. Lane 1, *fgf-1* PCR product; Lane 2, WT/*HindIII*; Lane 3, WT/*EcoRI*, Lane 4, T117/*HindIII*; Lane 5, T117/*EcoRI*.

Additional file 3: Figure S2. Deep-second burn degree determined by H&E analysis. Normal skin had an intact cuticle, hair follicles, sebaceous glands, collagen and a prickle cell layer (A and B). Epidermal necrolysis was observed 24 h after burn treatment (C and D). Micrographs were taken at 200×.

Abbreviations

FGF: Fibroblast growth factor; CAM: Chicken embryo chorioallantoic membrane assay; MTT: Thiazolyl blue tetrazolium bromide colorimetric assay; TCM: Traditional Chinese medicine.

Competing interests

The authors declare that they have no competing interests.

Authors' contributions

YT carried out the gene cloning, participated in the plant transformation, experiments and drafted the manuscript. KW designed and analyzed data, and helped to write the paper. NW and GL participated in experiments. DL conceived of the study, and participated in its design and coordination and helped to draft the manuscript. All authors read and approved the final manuscript.

Acknowledgements

This work was supported by the Chinese National Biotechnology Development Plan (2013ZX08005-004), The Researcher Foundation of the Chinese Academy of Agricultural Sciences, and the National Institute of General Medical Sciences of the National Institutes of Health (8P20GM103447).

Author details

[1]Biotechnology Research Institute, Chinese Academy of Agricultural Sciences, 12 Southern Zhong Guan Cun Road, Beijing 100081, China. [2]Department of Natural Sciences, Northeastern State University, Broken Arrow, Oklahoma 74014, USA.

References

1. Peck MD: **Epidemiology of burns throughout the world. Part I: distribution and risk factors.** *Burns* 2011, **37**:1087–1100. doi:10.1016/j.burns.2011.06.005.
2. National Institute of General Medical Sciences. http://www.nigms.nih.gov/Education/Factsheet_Burns.htm.
3. Yun YR, Won JE, Jeon E, Lee S, Kang W, Jo H, Jang JH, Shin US, Kim HW: **Fibroblast growth factors: biology, function, and application for tissue regeneration.** *J Tiss Eng* 2010, **148142**:. doi:10.4061/2010/218142.
4. Zakrzewska M, Marcinkowska E, Wiedlocha A: **FGF-1: from biology through engineering to potential medical applications.** *Crit Rev Clin Lab Sci* 2008, **45**:91–135. doi:10.1080/10408360701713120.
5. Zazo M, Lozano RM, Ortega S, Varela J, Díaz-Orejas R, Ramírez JM, Giménez-Gallego G: **High-level synthesis in Escherichia coli of shortened and full-length human acidic fibroblast growth factor and purification in a form stable in aqueous solutions.** *Gene* 1992, **113**:231–238.
6. Fantoni A, Bill RM, Gustafsson L, Hedfalk K: **Improved yields of full-length functional human FGF1 can be achieved using the methylotrophic yeast *Pichia pastoris*.** *Protein Expr Purif* 2007, **52**:31–39.
7. Zhou Y, Ren LZ, Zhu JG, Yan S, Wang H, Song N, Li L, Ouyang H, Pang D: **Construction of a recombinant human FGF1 expression vector for mammary gland-specific expression in human breast cancer cells.** *Mol Cell Biochem* 2011, **354**:39–46.
8. Liu J, Ma P, Sun Y, Yang M, Li Y, Wu Y, Zhu X, Wang X: **Expression of human acidic fibroblast growth factor in Nicotiana benthamiana with a potato-virus-X-based binary vector.** *Biotechnol Appl Biochem* 2007, **48**:143–147.
9. Pandit AS, Wilson DJ, Feldman DS: **Fibrin scaffold as an effective vehicle for the delivery of acidic fibroblast growth factor (FGF-1).** *Biomater Appl* 2000, **14**:229–242.
10. Tan Y, Xiao J, Huang Z, Xiao Y, Lin S, Jin L, Feng W, Cai L, Li X: **Comparison of the therapeutic effects recombinant human acidic and basic fibroblast growth factors in wound healing in diabetic patients.** *J Health Sci* 2008, **54**:432–440. doi:10.1248/jhs.54.432.
11. Wang B-Q: ***Salvia miltiorrhiza*: chemical and pharmacological review of a medicinal plant.** *J Med Plants Res* 2010, **4**:2813–2820.
12. Yin HQ, Choi YJ, Kim YC, Sohn DH, Ryu SY, Lee BH: ***Salvia miltiorrhiza Bunge* and its active component cryptotanshinone protects primary cultured rat hepatocytes from acute ethanol-induced cytotoxicity and fatty infiltration.** *Food Chem Toxicol* 2009, **47**:98–103.
13. Liu CS, Cheng Y, Hu JF, Zhang W, Chen NH, Zhang JT: **Comparison of antioxidant activities between salvianolic acid B and Ginkgo biloba extract (EGb 761).** *Acta Pharmacol Sin* 2006, **27**:1137–1145.

14. Zhou ZT, Yang Y, Ge JP: **The preventive effect of salvianolic acid B on malignant transformation of DMBA-induced oral premalignant lesion in hamsters.** *Carcinogenesis* 2006, **27:**826–832.
15. Park JW, Lee SH, Yang MK, Lee JJ, Song MJ, Ryu SY, Chung HJ, Won HS, Lee CS, Kwon SH, Yun YP, Choi WS, Shin HS: **15,16-Dihydrotanshinone I, a major component from Salvia miltiorrhiza Bunge (Dansham), inhibits rabbit platelet aggregation by suppressing intracellular calcium mobilization.** *Arch Pharm Res* 2008, **31:**47–53.
16. De Jaeger G, Scheffer S, Jacobs A, Zambre M, Zobell O, Goossens A, Depicker A, Angenon G: **Boosting heterologous protein production in transgenic dicotyledonous seeds using Phaseolus vulgaris regulatory sequences.** *Nature Biotechnol* 2002, **20:**1265–1268.
17. Hood EE, Woodard SL, Horn ME: **Antibody manufacturing in transgenic plants: myths and realities.** *Curr Opin Biotechnol* 2002, **13:**630–635.
18. Lay IS, Chiu JH, Shiao MS, Liu WY, Wu CW: **Crude extract of *Salvia miltiorrhiza* and salvianolic acid B enhance in vitro angiogenesis in murine SVR endothelial cell line.** *Planta Med* 2003, **69:**26–32.
19. Ding LT, Zhu YG, Gu ZQ, Cai LL, Peng JL, Lü GZ: **Effect of *Salvia miltiorrhiza* and Ligustrazine injection on the early myocardial damage of patients with severe burn.** *Zhonghua Shao Shang Za Zhi (in Chinese)* 2012, **28:**170–172.
20. Chen J, Lv Q, Yu M, Zhang X, Gou J: **Randomized clinical trial of Chinese herbal medications to reduce wound complications after mastectomy for breast carcinoma.** *Br J Surg* 2010, **97:**1798–1804.
21. Yin X, Yin Y, Cao FL, Chen YF, Peng Y, Hou W-G, Sun SK, Luo ZJ: **Tanshinone IIA attenuates the inflammatory response and apoptosis after traumatic injury of the spinal cord in adult rats.** *PLoS One* 2012, **7**(6):e38381. doi:10.1371/journal.pone.0038381.
22. Chen YS, Lee SM, Lin YJ, Chiang SH, Lin CC: **Effects of danshensu and salvianolic Acid B from *Salvia miltiorrhiza* Bunge (Lamiaceae) on cell proliferation and collagen and melanin production.** *Molecules* 2014, **19:**2029–2041. doi:10.3390/molecules19022029.
23. Pan RH, Xie FY, Chen HM, Xu LZ, Wu XC, Xu LL, Yao G: **Salvianolic acid B reverses the epithelial-to-mesenchymal transition of HK-2 cells that is induced by transforming growth factor-β.** *Arch Pharm Res* 2011, **34:**477–483.
24. Caron A, Michelet S, Caron A, Sordello S, Ivanov MA, Delaère P, Branellec D, Schwartz B, Emmanuel F: **Human FGF-1 gene transfer promotes the formation of collateral vessels and arterioles in ischemic muscles of hypercholesterolemic hamsters.** *J Gene Med* 2004, **6:**1033–1045.
25. Cao RH, Brakenhielm E, Pawliuk R, Wariaro D, Post MJ, Wahlberg E, Leboulch P, Cao Y: **Angiogenic synergism, vascular stability and improvement of hind-limb ischemia by a combination of PDGF-BB and FGF-2.** *Nature Med* 2003, **9:**604–613.
26. Faye L, Gomord V: **Post translational modification of therapeutic proteins in plants.** *Curr Opin Plant Biol* 2004, **7:**171–181.
27. Feng H, Xiang H, Zhang J, Liu G, Guo N, Wang X, Wu X, Deng X, Yu L: **Genome-wide transcriptional profiling of the response of Staphylococcus aureus to cryptotanshinone.** *J Biomed Biotechnol* 2009, **2009:**617509. doi:10.1155/2009/617509.
28. Andreopoulos FM, Persaud I: **Delivery of basic fibroblast growth factor (bFGF) from photoresponsive hydrogel scaffolds.** *Biomaterials* 2006, **27:**2468–2476.
29. Cai S, Liu Y, Xiao ZS, Prestwich GD: **Injectable glycosaminoglycan hydrogels for controlled release of human basic fibroblast growth factor.** *Biomaterials* 2005, **26:**6054–6067.
30. Bardor M, Faveeuw C, Fitchette A-C, Gilbert D, Galas L, Trottein F, Faye L, Lerouge P: **Immunoreactivity in mammals of two typical plant glyco-epitopes, core a(1,3)-fucose and core xylose.** *Glycobiology* 2003, **13:**427–434.
31. During K, Hippe S, Kreuzaler F, Schell J: **Synthesis and self-assembly of a functional monoclonal antibody in transgenic *Nicotiana tabacum*.** *Plant Mol Biol* 1990, **15:**281–293.
32. Chen MH, Huang LF, Li HM, Chen YR, Yu SM: **Signal peptide-dependent targeting of a rice α-amylase and cargo proteins to plastids and extracellular compartments of plant cells.** *Plant Physiol* 2004, **135:**1367–1377.
33. Horsch RB, Fry JE, Hofimann NL, Eichholtz D, Rogers SG, Fraley RT: **A simple and general method for transferring genes into plants.** *Science* 1985, **227:**1229–1231.
34. Minas K, McEwan NR, Newbold CJ, Scott KP: **Optimization of a high-throughput CTAB-based protocol for the extraction of qPCR-grade DNA from rumen fluid, plant and bacterial pure cultures.** *FEMS Microbiol Lett* 2011, **25:**162–169. doi:10.1111/j.1574-6968.2011.02424.x.
35. Li G, Wang KY, Li D, Wang N, Liu D: **Cloning, expression and characterization of a gene from earthworm *Eisenia foetida* encoding a blood-clot dissolving protein.** *PLoS One* 2012, **7:**e53110. doi:10.1371/journal.pone.0053110.
36. Mosmann T: **Rapid colorimetric assay for cellular growth and survival: application to proliferation and cytotoxicity assays.** *J Immunol Methods* 1983, **65:**55–63.
37. Larger E, Marre M, Corvol P, Gasc JM: **Hyperglycemia-induced defects in angiogenesis in the chicken chorioallantoic membrane model.** *Diabetes* 2004, **53:**752–761.
38. Bohannon RW, Pfaller BA: **Documentation of wound surface area from tracings of wound perimeters.** *Pbys Tkr* 1983, **63:**1622–1624.

12

Improved elongation factor-1 alpha-based vectors for stable high-level expression of heterologous proteins in Chinese hamster ovary cells

Nadezhda A Orlova[1,2], Sergey V Kovnir[1,2], Julia A Hodak[1,2], Ivan I Vorobiev[1,2*], Alexandre G Gabibov[2,3] and Konstantin G Skryabin[1]

Abstract

Background: Establishing highly productive clonal cell lines with constant productivity over 2–3 months of continuous culture remains a tedious task requiring the screening of tens of thousands of clonal colonies. In addition, long-term cultivation of many candidate lines derived in the absence of drug selection pressure is necessary. Expression vectors based on the elongation factor-1 alpha (EEF1A) gene and the dihydrofolate reductase (DHFR) selection marker (with separate promoters) can be used to obtain highly productive populations of stably transfected cells in the selection medium, but they have not been tested for their ability to support target gene amplification under gradually increasing methotrexate pressure.

Results: We have modified EEF1A-based vectors by linking the DHFR selection marker to the target gene in the bicistronic RNA, shortening the overall plasmid size, and adding an Epstein-Barr virus terminal repeat fragment (EBVTR) element. Presence of the EBVTR element increased the rate of stable transfection by the plasmid by 24 times that of the EBVTR-minus control and improved the rate of methotrexate-driven gene amplification. The mean expression level of the enhanced green fluorescent protein (eGFP) used herein as a model protein, increased up to eight-fold using a single round of amplification in the case of adherent colonies formation and up to 4.5-fold in the case of suspension polyclonal cultures. Several eGFP-expressing cell populations produced using vectors with antibiotic resistance markers instead of the DHFR marker were compared with each other. Stable transfection of Chinese hamster ovary (CHO) DG44 cells by the p1.2-Hygro-eGFP plasmid (containing a hygromycin resistance marker) generated highest eGFP expression levels of up to 8.9% of the total cytoplasmic protein, with less than 5% of the cell population being eGFP-negative.

Conclusions: The p1.1 vector was very effective for stable transfection of CHO cells and capable of rapid MTX-driven target gene amplification, while p1.2-Hygro achieved similar eGFP expression levels as p1.1. The set of vectors we have developed should speed-up the process of generating highly productive clonal cell lines while substantially decreasing the associated experimental effort.

Keywords: CHO cells, High level expression, Stable cell line generation, Molecular cloning

* Correspondence: ptichman@gmail.com
[1]Laboratory of Mammalian Cell Bioengineering, Centre "Bioengineering", Russian Academy of Sciences, 60-letija Oktyabrya 7, Moscow 117312, Russia
[2]Laboratory of Biocatalysis, Institute of Bioorganic Chemistry, Russian Academy of Sciences, Miklukho-Maklaya 16/10, Moscow 119971, Russia
Full list of author information is available at the end of the article

Background

Most of the proteins currently employed for therapeutic use are produced by stably transfected mammalian cells, of which the most popular is the Chinese hamster ovary (CHO) cell line. Establishing highly productive clonal cell lines that exhibit constant productivity over a 2–3 month period of continuous culture remains a tedious task, requiring tens of thousands of clonal colonies to be screened, followed by the long-term cultivation of candidate lines in the absence of an appropriate selection pressure. Generally, the expression levels of a target gene may be increased by its amplification in the genome [1], which is usually achieved by linking the target gene to the murine dihydrofolate reductase (DHFR) gene with stepwise increases in the concentration of the DHFR inhibitor, methotrexate (MTX), in the selection medium.

Target gene amplification is a time-consuming process, resulting in cell populations that often contain unstable clones, and in the absence of an appropriate selection pressure, reduced production levels. The probability of obtaining a highly productive clonal cell line can be increased significantly by utilizing plasmids based on non-coding parts of the elongation factor-1 alpha gene (EEF1A) from Chinese hamster, as described by Running Deer and Allison [2]. Expression vector pDEF38, introduced by these authors, differs significantly from the widely employed vectors with the core promoter of the human ortholog elongation factor 1 alpha gene (EF1a). EEF1A-based expression vector includes 4.1 kb upstream and 4.2 kb downstream flanking areas of the EEF1A gene, so the ORF of the of the target gene replaces the coding exons of the elongation factor 1 alpha protein in the natural EEF1A gene, mimicking with all possible accuracy the structure of the natural gene in the resulting expression plasmid. It was shown that presence of both flanking areas in the EEF1A-based vectors results in the 6- to 35- fold increase of the average expression level comparing to commercial vectors with CMV or EF1alpha promoters. Removal of the downstream flanking area from the expression vector resulted in the 4-fold drop in the expression level.

Original expression vector pDEF38 contained the DHFR selection marker with a separate SV40 promoter and was not tested for its ability to support target gene amplification under gradually increasing MTX pressure. DHFR-compatible vectors, bearing the neomycin resistance gene instead of the DHFR gene, were also described in the same work. Existing EEF1A-based vectors, despite their high promoter strength and their long-term production level stability, do not accommodate very large plasmid sizes. Consequently, this can result in low-level genome integration and inability to maintain the target gene amplification step, possibly as a result of vector fragmentation and autonomous amplification of the DHFR-coding area.

Since EEF1A-based vectors are much longer than CMV-based vectors, they are expected to have lower transfection efficiency and, subsequently, lower numbers of stably transfected cells. It was shown, that the insertion the concatemer fragment of the Epstein-Barr virus terminal repeats (EBVTR) [3,4] in the expression vectors increase the rate of stably transfected colonies formation by five to ten fold [5]. The molecular mechanism of this effect is poorly understood. It is known that G-rich repeats in the EBVTR bind to the cellular protein terminal repeat binding protein (TRBP) [3] and at least two binding sites of TRBP were identified in the repetitive cellular DNA [6]. EBVTR areas are involved in the integration of the Epstein-Barr virus into the chromosomal DNA [7]. EBV-infected cells may harbour the virus in the chromosome-integrated form, as the independently replicating episome or the mixture of both forms [8]. Region of the EBV, called oriP, maintains the episomal replication of the EBV genome, interacts with the EBV-encoded nuclear antigen-1 (EBNA-1) and allows EBV plasmids to separate in mitosis through binding to chromosomes [9]. EBVTR concatemer used for enhancement of expression plasmids, however, contains no sequences from the oriP region and no DNA fragments with significant homology toward oriP region, so the EBNA-1 – mediated persistence of the EBVTR-containing plasmid as the episome in the transfected cells is highly unlikely.

We hypothesized that significant improvements to EEF1A-based vectors might be achieved by: 1) inserting the EBVTR element outside of the EEF1A flanking DNA; 2) linking the DHFR open reading frame to the target gene by the internal ribosome entry site (IRES) thereby preventing the possibility of separate amplification of the selection marker; 3) reducing of the length of the backbone DNA, which is necessary for maintaining the plasmid in the bacterial host. Similar improvements might be applied to DHFR-compatible EEF1A-based vectors used for monocistronic expression of a target gene; in this case by placing the antibiotic resistance genes outside the context of the non-coding parts of the elongation factor 1 alpha gene, which might decrease genetic linkage between the selection marker and the target gene.

Here, we report on the functional properties of EEF1A-based plasmid vectors for bicistronic and monocistronic expression. We also describe the corresponding methods for obtaining highly productive and stable cell lines that maintain constant productivity levels after genome amplification of the integrated plasmid, utilizing the DHFR-negative cell line CHO DG44 [10,11]. In addition, we used the enhanced green fluorescent protein (eGFP) as a model target protein, and show eGFP accumulation inside CHO cells.

Methods

Molecular cloning

The sequences of the primers used for cloning expression plasmids are shown in Additional file 1: Table S1. The backbone vector, pBL-2, was obtained by two stages of inverted PCR using long adapter primers and the pUC18 plasmid as a template. Non-functional parts of the plasmid including the pLac promoter and the LacZ gene were removed. Inverted PCR was performed as described previously [12]. Oligonucleotides and PCR reagents were from Evrogen, JSC (Moscow, Russia). PCR products were purified from 1% agarose gels by the Wizard SV Gel and PCR Clean-Up System (Promega, Madison, WI). T4 polynucleotide kinase and *Mal*I restriction endonuclease (Sibenzyme, Novosibirsk, Russia) were used. T4 DNA ligase (Fermentas, Vilnius, Lithuania) was used for inverted PCR product circularization. The *Escherichia coli* TOP10 strain (Invitrogen, Carlsbad, CA) was used for cloning. Plasmids were isolated with a GeneJet Plasmid Purification Kit (Fermentas, Vilnius, Lithuania).

The pAL-EBV plasmid, containing a fragment of a concatemer of EBV terminal repeats, as described previously [5] and nearly undistinguishable from the human herpes virus 4 strain K4123-MiEBV sequence [GenBank: KC440852.1] was made from synthetic oligonucleotides cloned into a pAL-TA (Evrogen) vector. The ORF encoding mouse DHFR was obtained by PCR using pOptivec-TOPO linearized vector (Invitrogen) as a template. The fragment encoding the attenuated encephalomyocarditis virus (EMCV) IRES was obtained from the pOptivec-circ plasmid (self-ligated pOptivec-TOPO) by restriction. These fragments were cloned into the pBL-2 plasmid via assembly of two different intermediate constructs, PBL-2-ID and PBL-2-ID-EBV. DNA modification enzymes for routine molecular cloning were obtained from Fermentas or Sibenzyme.

Construction of p1.1 vectors

Fragments corresponding to the upstream and downstream flanking regions (8532–12603 and 14545–18794 sequences of [GenBank:AY188393]) of the CHO elongation factor 1 gene were obtained by PCR using CHO DG44 cell (Invitrogen) genomic DNA as a template. The modular assembly cloning technique used herein is described in detail elsewhere [13].

Assembled CHO genomic regions were cloned into the intermediate plasmids, PBL-2-ID and PBL-2-ID-EBV, resulting in p1.1(EBVTR-) and p1.1 expression vectors, respectively (Figure 1).

Construction of p1.2 vectors

p1.2-Mono, the intermediate backbone plasmid for expression vectors bearing antibiotic resistance genes was obtained by removal of the region containing the EMCV IRES and the DHFR ORF from the p1.1 expression vector. Plasmid pAL-3CH123, containing first three modules of the downstream flanking region of the *EEF1A* was used as the source of the donor DNA insert fragment, replacing the deleted IRES and DHFR area, so both flanking regions of the *EEF1A* remained unaltered (Figure 2). Antibiotic resistance genes and the SV40 promoter and terminator regions were obtained by amplification with adaptor primers, using pcDNA3.1/Hygro, pcDNA3.1(+), and self-ligated pcDNA4/HisMax-TOPO (Invitrogen) as PCR templates. Antibiotic resistance cassettes were sub-cloned into T-vectors and then transferred into the p1.2-Mono backbone by restriction-ligation resulting in p1.2-Hygro, p1.2-Neo and p1.2-Zeo. A DNA fragment encoding eGFP and a consensus Kozak sequence (GCCGCCATGG) [14] was obtained by PCR with adaptor primers and the pEGFP-C2 plasmid (Clontech, Mountain View, CA) as a template and then cloned into the polylinker area of p1.1 and p1.2 vectors, thereby resulting in p1.1(EBVTR-)eGFP, p1.1eGFP, p1.2-HygroeGFP, p1.2-NeoeGFP and p1.2-ZeoeGFP expression plasmids.

Purified plasmids for transfection and the control plasmid pEGFP-N2 (Clontech) were prepared using an EndoFree Plasmid Maxi Kit (Qiagen, Valencia, CA). For stable cell line generation all plasmids except p1.2-HygroeGFP were linearized by restriction with *Pvu*I by cutting inside the ampicillin resistance gene *bla* sequence. The plasmid p1.2-HygroeGFP was restricted with *BspH*I, which introduced two breaks near the *bla* gene.

Cell culture

A DHFR-negative CHO DG44 cell line (Invitrogen) was cultured in shaking flasks in the chemically defined medium, CD DG44 (Invitrogen), supplemented with 0.18% Pluronic F-68 (Invitrogen) and 4 mM L-glutamine (Invitrogen). The cells were passaged 24 h before transfection. For direct colony generation in 96-well culture plates, transfection was performed using Fugene HD reagent (Promega), containing 60 μg of DNA and 180 μl of the reagent per 15 millions of cells in 30 ml of the above medium. Plasmids p1.2 were transfected by electroporation in Gene Pulser Electroporation Buffer (Bio-Rad, Hercules, CA) using a cuvette with a 4 mm gap with 7.5 million cells and 15 μg of linearized DNA for each transfection. Cells were counted by trypan blue exclusion and fluorescence microscopy at 48 h post-transfection. For direct generation of colonies, transiently transfected cell cultures were transferred into CHO-A culture medium (Invitrogen) lacking hypoxanthine and thymidine (HT), and seeded at 5000 cells/well in the culture plates. Cells were grown undisturbed for 14 days and inspected by fluorescence microscopy. The medium was changed and the plates were

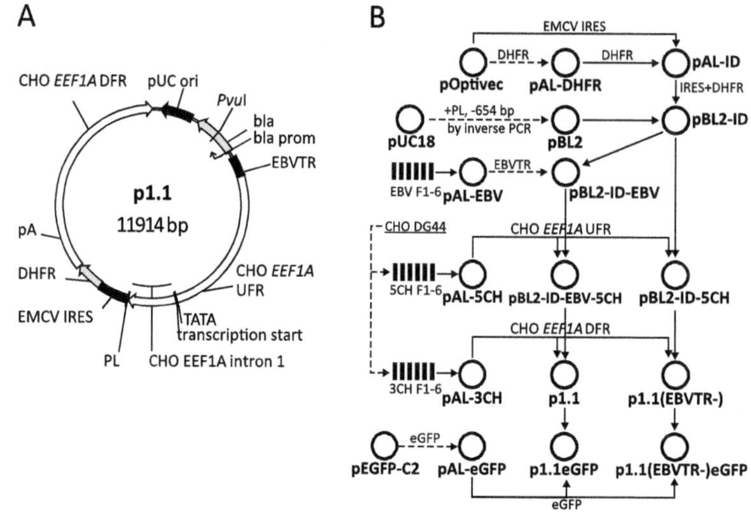

Figure 1 Map of the p1.1 plasmid vector and the cloning scheme for p1.1-based plasmids. A. Plasmid map. UFR: upstream flanking region of the EEF1A gene; DFR: downstream flanking region; PL: polylinker region; pA: polyadenylation signal; bla – ampicillin resistance gene; bla prom – promoter of the ampicillin resistance gene. **B.** Cloning scheme for p1.1-based plasmids. Generation of cloning inserts by PCR with adaptor primers is depicted by dashed lines; generation of cloning inserts by restriction is depicted by solid lines. EBV F1-6: corresponding synthetic fragments of the EBVTR element. 5CH F1-6: corresponding fragments of the upstream flanking region of the EEF1A gene; 3CH F1-6: corresponding fragments of the downstream flanking region of the EEF1A gene.

cultivated for 5–10 additional days until the first 10% of the wells containing colonies became confluent.

To generate stably transfected cell populations using p1.1eGFP and p1.1(EBVTR-)eGFP plasmids, transiently transfected cultures were transferred to OptiCHO medium (Invitrogen) lacking HT, and thereafter cultivated in shaking flasks with medium exchange every 3 days until the cell viability increased to 85% (approximately 22–27 days of cultivation).

MTX-driven target gene amplification in culture plates was performed by seeding the cells from stably transfected cell populations into 96-well culture plates at a density of 5000 cells/well in the CHO-A culture medium, supplemented with 0, 50, 100, 200, 400 or 800 nM MTX. Three plates were used for each concentration of MTX. The cells were grown undisturbed for 14 days, after which the plates were inspected by microscopy and the culture medium was changed every 4 days until the first 10% of wells in each plate became confluent. Plates were screened again by fluorescence microscopy, and cells from the 16 brightest wells from each plate were transferred into a 48-well plate, grown to confluence, and then transferred into 24-well

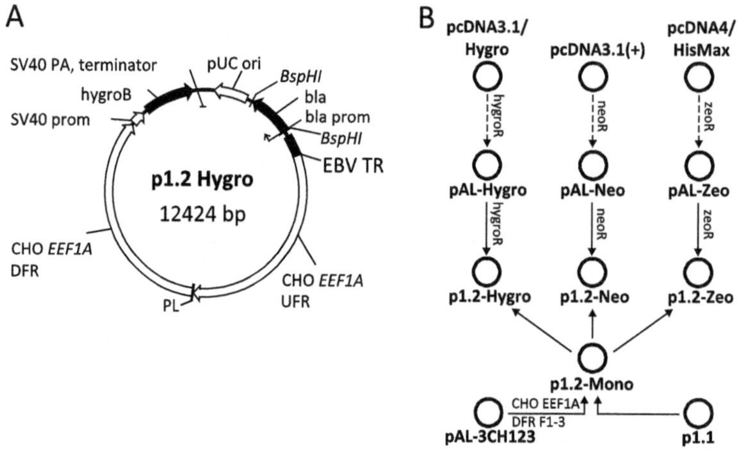

Figure 2 Map of the p1.2-Hygro plasmid vector and the cloning scheme for p1.2-based plasmids. A. Plasmid map. UFR: upstream flanking region of the EEF1A gene; DFR: downstream flanking region; Pl: polylinker region; SV40 prom and SV40 PA: promoter and polyadenylation signal from the SV40 virus. **B.** Cloning scheme for p1.2-based plasmids.

plates. Colonies lacking normal proliferation speeds or attached to the surface of the plates too tightly for dislodging by pipetting were discarded. Cells from the eight brightest wells for each MTX concentration were dislodged from their plates, lysed as described below, and then used to determine eGFP levels. Six randomly picked colonies, obtained in the presence of 400 and 800 nM MTX, were transferred into a 6-well plate and grown with shaking in OptiCHO medium with passages made every three days for 60 days. Samples for eGFP level determination were collected every second passage.

Target gene amplification for polyclonal cell populations was performed for the suspension culture of CHO DG44 cells, stably transfected by the p1.1eGFP plasmid in presence of 50 nM MTX, as described above. Concentration of the MTX in the culture medium was increased by two-fold steps, each after two consecutive passages, until the cell viability decreased below 85%. Resulting culture, obtained in presence of 0.8 μM MTX, was split into four flasks, supplemented by 0.8; 1.6; 3.2; 6.4 μM MTX and cultured until the cell viability returned to at least 85% (7–12 days).

Generation of polyclonal cell populations involving transfected p1.2 plasmids were performed by seeding transiently transfected cells in 6-well culture plates, using 1 million of viable cells per well in 5 ml of DG44 medium, supplemented with the corresponding antibiotic, or 5 ml of OptiCHO medium with 200 nM MTX for control transfections using p1.1 plasmids. The concentrations of the antibiotics used are shown in Figure 3. Plates were cultivated with shaking until the cell viability returned to at least 85% (20 days), after which the medium was changed every 4 days.

Determination of eGFP concentrations in cell lysates

Cell culture samples containing approximately one million of cells were centrifuged and the cell pellets were resuspended in phosphate buffered saline and recentrifuged. The washed cell pellets were resuspended in 0.1 ml of lysis solution containing 150 mM NaCl, 50 mM Tris–HCl at pH 7.5, 1% Triton X-100, a protease inhibitor cocktail (Sigma, St. Louis, MO), and then incubated for 30 min on ice with stirring. Cell debris was removed by centrifugation. The concentration of eGFP in the lysate from the H-4 cell population (Figure 3) was measured by spectrophotometry at a wavelength of 488 nm using a molar extinction coefficient of 55,000 $M^{-1} \cdot cm^{-1}$ and an eGFP molecular weight of 32.7 kDa [15]. The fluorescence intensity of eGFP in all of the lysates was measured along with the serially diluted calibration samples, which were prepared from the H-4 lysate containing a known concentration of eGFP. Total protein concentration in the lysates was measured by the Bradford method with bovine serum albumin as a standard.

FACS analysis and quantitative PCR

Undiluted cell culture samples were subject to FACS FC 500 (Beckman Coulter, Krefeld, Germany) analysis at an emission at 488 nm and detection through a 530/40-nm bandpass filter. At least 10,000 individual cells were counted for each sample analysed. Quantitative PCR analysis of the expression plasmid copy numbers in the genomes of stably transfected cells was performed using an iCycler iQ thermocycler (Bio-Rad) and qPCRmix-HS SYBR (Evrogen) reaction mixture with the primers shown in Additional file 1: Table S2. The highly purified p1.1eGFP plasmid was used as a quantity calibrator using five different concentrations for each determination performed in triplicate. PCR was performed three times with three to four replicates for each sample. Genomic DNA was extracted from cells with a Genomic DNA Purification Kit (Fermentas) and quantified using a Qubit fluorometer (Invitrogen) and the dsDNA HS kit (Invitrogen). Quantity calibrator plasmid was used as the external standard for the quantification of genomic DNA samples by fluorometry.

Results and discussion

Construction of expression plasmids

Since the transfection efficiency and, probably, the genome integration rate of an expression plasmid is inversely proportional to its size [16], we made a minimal backbone plasmid by eliminating most of the unnecessary elements from the pUC18 plasmid. The resulting plasmid, pBL-2, lacks the f1 origin of replication, and the bacterial promoter of the LacZ gene along with the LacZ ORF itself and some flanking DNA regions. Overall, the resulting plasmid length decreased some 600 bp from 2686 to 2032 bp.

The upstream and downstream regions of the EEF1A gene were obtained from CHO DG44 cell genomic DNA using the modular assembly cloning technique described previously [13]. A concatemer of terminal repeats from the Epstein-Barr virus (EBVTR) [3,4] was assembled from synthetic oligonucleotides using the same technique and was inserted along with the IRES from the encephalomyocarditis virus and the murine DHFR open reading frame into the pBL-2 vector. Cloning the upstream and downstream flanking areas of the EEF1A gene into the pBL-2-ID-EBV plasmid resulted in the expression vector p1.1 (Figure 1). A control vector, lacking the EBVTR fragment, was assembled similarly and is denoted here as p1.1(EBVTR-). The p1.1 plasmid was approximately 1.5 kbp shorter than the original EEF1A-based plasmid, pDEF38, despite addition of the EBVTR fragment. The eGFP ORF with the synthetic consensus Kozak sequence [14] was cloned into both vectors and the resulting plasmids p1.1eGFP and p1.1(EBVTR-)eGFP were used for CHO cell transfections.

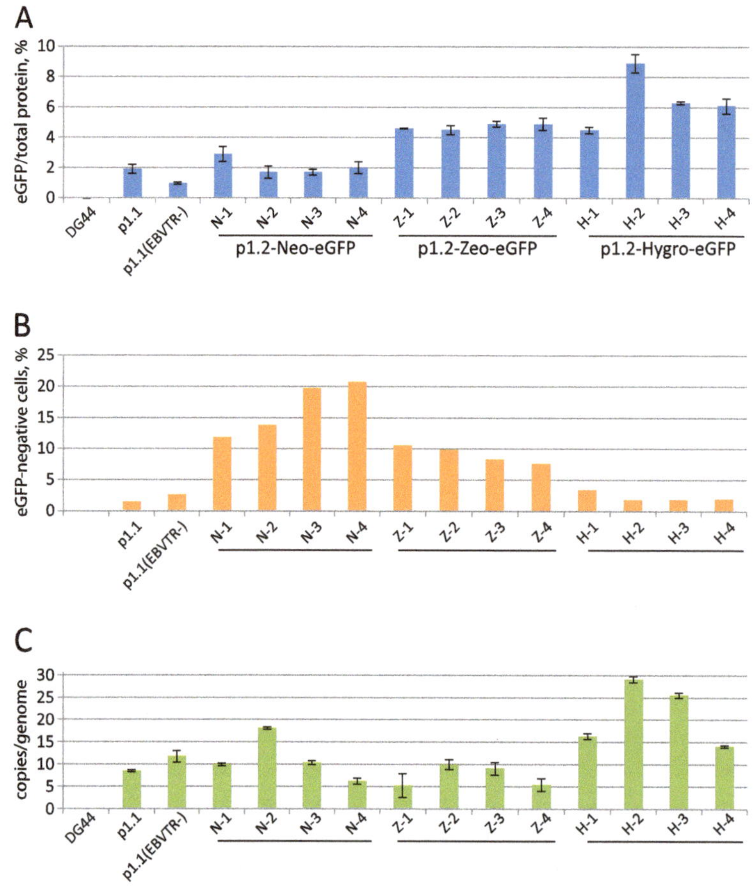

D

Antibiotic	G-418	Zeocin	Hygromycin
Code of cell population	N-1 N-2 N-3 N-4	Z-1 Z-2 Z-3 Z-4	H-1 H-2 H-3 H-4
Concentration,	0,75 1 1,25 1,5	0,5 0,75 1 1,25	0,25 0,5 0,75 1

Figure 3 Properties of the cell populations stably transfected by p1.2-based plasmids under various drug selection stringencies. DG44: untransfected CHO DG44 cells; p1.1: cells stably transfected by the p1.1eGFP plasmid and selected in the presence of 200 nM MTX; p1.1(EBVTR-): transfection by the p1.1(EBVTR-)eGFP plasmid using the same conditions. **A.** Level of intracellular eGFP in cell populations. Error bars indicate the standard deviation, n = 2. **B.** Proportion of eGFP-negative cell populations measured by FACS. **C.** Number of copies of genome-integrated plasmids measured by Q-PCR. Amplicons are located inside the eGFP ORF and one representative value experiment from three independent measurements is shown. Error bars represents standard deviations, n = 3-4. The apparent level of the eGFP ORF DNA for the untransfected CHO DG44 cells is below 0.1 copies per one haploid genome. **D.** Codes for the different cell populations and the concentrations of antibiotics employed.

Generation of stably transfected colonies using p1.1-based plasmids

Transient transfection of the DHFR-deficient CHO DG44 cells resulted in significantly decreased transfection efficiencies for both of the EEF1A-based plasmids relative to the cytomegalovirus (CMV)- promoter-based 4700 bp pEGFP-N2 plasmid, and approximately the same transfection efficiencies and eGFP expression levels for plasmids with or without the EBVTR element (Table 1).

At the same time, stable integration rate (or rate of establishment of stable episomal maintenance) of the p1.1eGFP plasmid was 24 times higher than that of the p1.1(EBVTR-)eGFP control plasmid in the selection medium lacking both HT and MTX (Table 2), clearly indicating that the EBVTR element was active in the very large expression plasmid. Transfection and selection of stably transformed CHO DG44 cells by the p1.1eGFP plasmid was repeated with the selection medium supplemented with 50 nM MTX. In this case, the eGFP expression level increased twice for the ten most productive wells (Figure 4A). Thus, the p1.1 plasmid is suitable for creation of stably transfected cell clones or populations under variable selection stringencies.

Table 1 Properties of the transiently transfected cells used in this study

Plasmid name	pEGFP-N2	p1.1eGFP	p1.1(EBVTR-)eGFP
eGFP-expressing cells	22.8%	5.8%	6.0%
Fluorescence intensity, RFU/50 cells	114	35.3	32.0
Viable cells	86%	83%	84%

MTX-driven target gene amplification

Since the EBVTR element was effective at increasing the incidence of stable transfection, we tested its ability to speed up the transgene amplification process. Polyclonal populations of CHO DG44 cells, transfected by p1.1eGFP and p1.1(EBVTR-)eGFP plasmids and selected for stable integration by suspension cultivation in the absence of MTX and HT, were seeded in the 96-well culture plates in the presence of 0–800 nM MTX and grown undisturbed until visible colonies developed. For the p1.1(EBVTR-)eGFP plasmid lacking the EBVTR element, no viable cell colonies were obtained in the presence of 200, 400 and 800 nM MTX. The eGFP expression levels of the highest expressing colonies, obtained in the presence of 0 and 100 nM MTX, was in the same range as the colonies obtained by direct plating of transiently transfected cells in the absence of MTX (data not shown). In the case of the p1.1-eGFP plasmid, multiple colonies were obtained for all of the concentrations of MTX tested. After visual screening by fluorescence microscopy and expansion, the eight brightest colonies for each concentration of MTX employed were grown to confluency in 24-well culture plates. The relative eGFP expression levels for these colonies (Figure 4B) was approximately 8 times higher when cultivated with 800 nM MTX, and approximately 2 times higher when cultivated with 400 nM MTX, compared with cultivation without MTX. Six randomly chosen colonies, obtained in the presence of 400 and 800 nM MTX, were scaled up, re-adapted to suspension culture and cultivated for 60 days. No substantial decay in the eGFP expression level was detected for each colony (data not shown). Thus, the p1.1 vector is suitable for target gene amplification in the presence of MTX. The resulting cell clones are sufficiently stable for cell bank generation and subsequent large-scale cultivation.

Table 2 Colony formation efficiency for p1.1eGFP and p1.1(EBVTR-)eGFP plasmids

Plasmid name	p1.1eGFP	p1.1(EBVTR-)eGFP
Total number of colonies in 10 culture plates	2342	95
eGFP-expressing colonies in 10 culture plates and their proportions	2093 (89.4%)	52 (54.7%)
Fluorescence intensity of the brightest well, RFU/50 cells	210	45.1

Target gene amplification process was also tested for polyclonal cell population, obtained by the primary selection in the presence of 50 nM MTX. Sequential addition of MTX from 100 nM to 400 nM gave no decrease in cell viability, eGFP level was also constant (data not shown). Further addition of 0.8 – 6.4 μM of MTX, performed in one step for multiple culture flasks, resulted in the concentration-dependent increase of eGFP content (Figure 4C), peaking at 9% of the total protein in the case of 6.4 μM of MTX. Analysis of the copy numbers of the integrated plasmids using quantitative PCR (Figure 4D) showed that higher MTX level and higher eGFP content correspond to higher copy number of the integrated plasmid. Hereby, the vector p1.1 is suitable for obtaining highly productive cell populations or clones by direct clone selection in culture plates in the presence of MTX or by the multi-step target gene amplification in the suspension culture.

Polyclonal cell populations stably transfected by p1.2 plasmids

Heterologous expression of a functionally active target protein often requires co-expression of a small number of protein processing enzymes. For example, the blood clotting factor IX expression systems used with CHO or BHK cells rely on co-expression of the signal protease PACE/furin [17] and the vitamin-K recharging enzyme, VKORC1 [18]. Generally, the expression levels of such "helper" proteins should be lower than that of the target protein, but of comparable magnitude. If the target protein is coded by a plasmid bearing a DHFR selection marker, helper proteins may be coded by plasmids with the same structure, but bearing antibiotic resistance markers. We have tested resistance markers for three widely used antibiotics, G418 (a neomycin analogue), zeocin, and hygromycin, in the EEF1A-based expression vector, which was modified by removing the IRES fragment and the DHFR open reading frame from the p1.1 plasmid, and insertion of the corresponding antibiotic resistance genes outside of the EEF1A gene flanking areas and controlled by a separate SV40 promoter. The resulting vectors, denoted p1.2-Neo, p1.2-Zeo, and p1.2-Hygro, were used for insertion of the eGFP protein ORF. All three resulting plasmids showed similar transient transfection efficiencies in CHO DG44 cells (19–24% by electroporation), and the resulting cell populations were used to generate stably transfected cell populations in the suspension culture under variable selection pressures for each antibiotic used. The control plasmids p1.1eGFP and p1.1(EBVTR-)eGFP were transfected using the same procedure and stably transfected cell populations were generated by suspension cultivation in the presence of

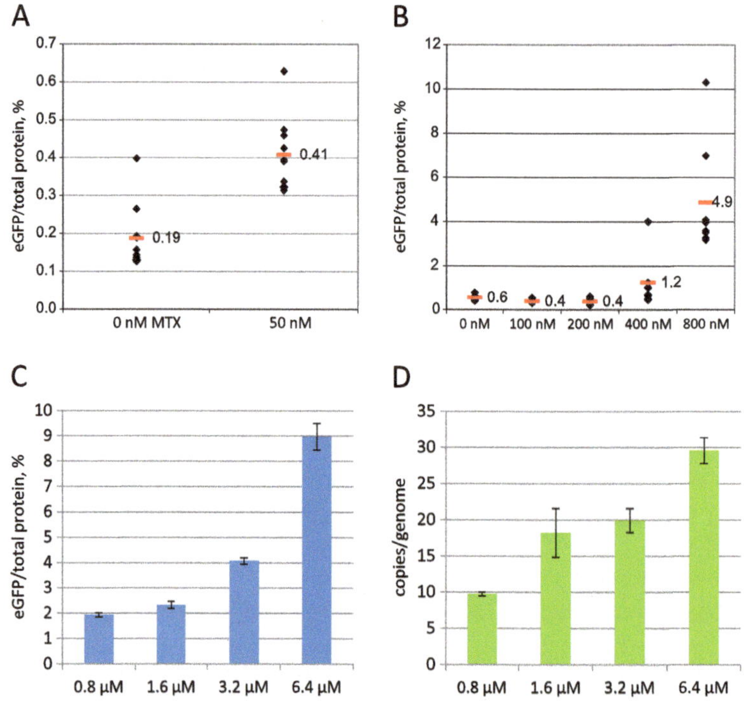

Figure 4 eGFP-expressing cell colonies obtained by elevated selection pressure and by targeted gene amplification. Red bars: mean eGFP levels for sets of cell colonies analysed. Concentrations represent final concentrations of MTX used. **A.** eGFP levels for 10 colonies obtained in the absence of MTX and in the presence of 50 nM MTX, colonies were obtained by the direct plating of transiently transfected cells. **B.** eGFP levels for 8 colonies obtained by growth in the presence of various MTX concentrations. Polyclonal stably transfected cell population was used for plating. **C.** Level of intracellular eGFP in polyclonal cell populations, obtained by primary selection in presence of 50 nM MTX and subsequent amplification in presence of various concentrations of MTX. Error bars indicate the standard deviation, n = 2. **D.** Number of copies of genome-integrated plasmids measured by Q-PCR for populations from panel C. Amplicons are located inside the eGFP ORF and one representative value experiment from three independent measurements is shown. Error bars represents standard deviations, n = 3.

200 nM MTX. The populations obtained were examined to determine the proportion of eGFP-expressing cells and eGFP levels in cell lysates (Figure 3). We found that for all three selection markers at all levels of drug selection pressure the resulting cell populations contained more than 75% of eGFP-positive cells. For the hygromycin and MTX resistance markers, less than 5% of the cells were eGFP-negative. The level of eGFP in the cell lysates was maximal for hygromycin selection, peaking as 8.9% of the total cellular protein with 0.5 mg/ml of hygromycin. In contrast, eGFP levels in the polyclonal cell populations obtained from transfection with p1.1eGFP or p1.1(EBVTR-)eGFP were much lower at 1.9% and 1.0%, respectively; however, eGFP expression levels for the p1.1 vector could potentially increase by eight-fold using the MTX-driven target gene amplification described above.

We also measured the intracellular eGFP distribution in polyclonal cell populations using FACS (Figure 5). Virtually no cells were eGFP-negative with DHFR and hygromycin selection markers, whereas with the neomycin resistance gene the level of eGFP-negative cells was inversely proportional to the concentration of G418 used. The mean eGFP level for the upper 10% of the eGFP-positive cells was not dependent on the antibiotic concentration for neomycin and zeocin selection, whereas with hygromycin selection the mean eGFP level was higher at higher antibiotic concentrations.

Analysis of the copy numbers of the genome-integrated plasmids using quantitative PCR revealed that the p1.2-Hyg-eGFP plasmid generated the maximum number of inserts, correlating with the highest expression level of eGFP. While the p1.2-Zeo-eGFP plasmid exhibited higher eGFP expression levels than p1.2-Neo-eGFP, it was present at half the copy number. In the case of plasmids containing the DHFR selection marker, the presence of the EBVTR element resulted in higher eGFP expression levels at lower numbers of genome inserts; this probably indicates that EBVTR drives integration events in areas of the genome that are transcriptionally active.

Conclusions

Creation of mammalian cell lines that express high levels of recombinant protein and maintain stable production levels over several months of cultivation is still a very time-consuming and labour-intensive process. Introduction of

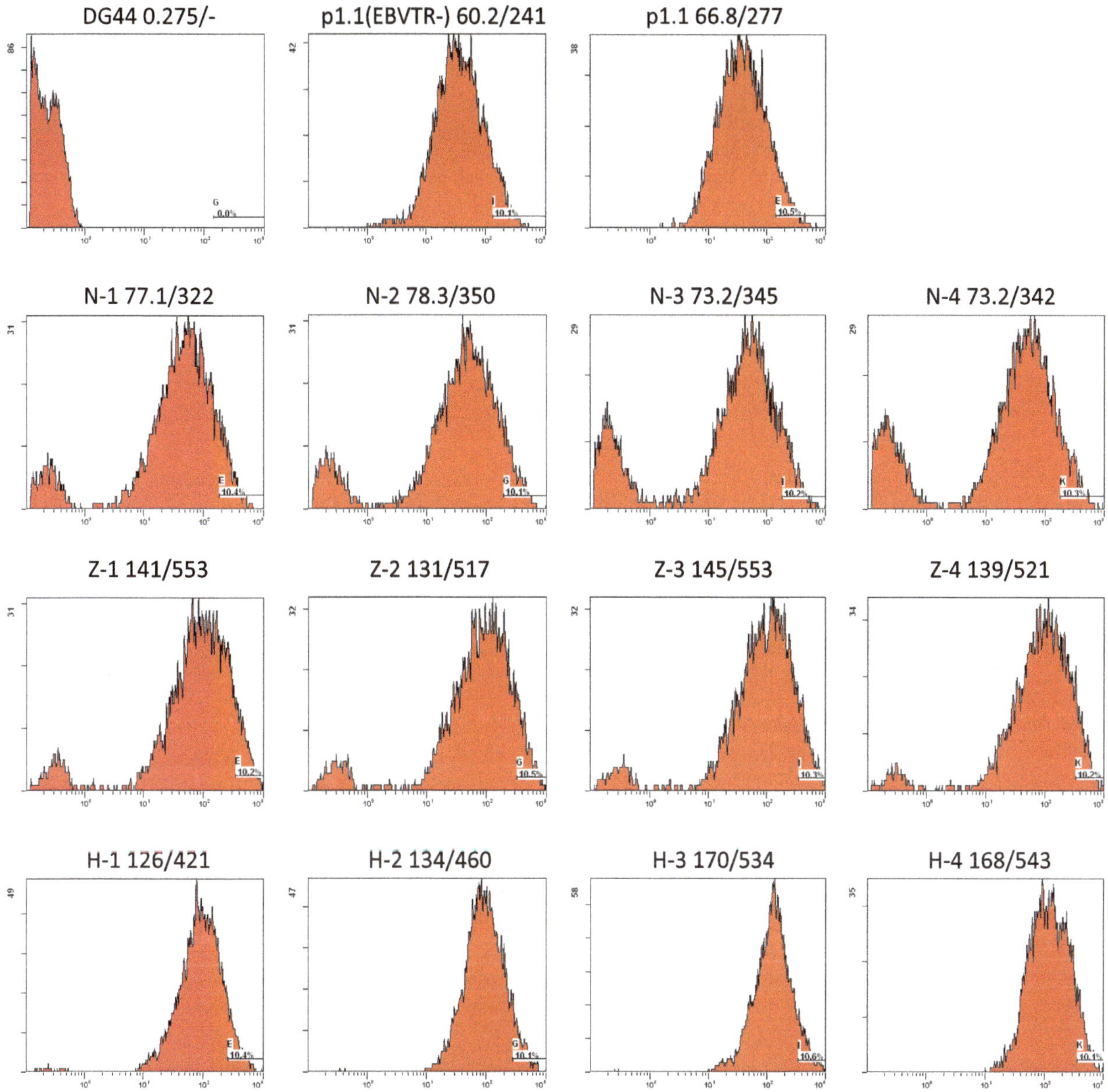

Figure 5 Distribution of the eGFP expression levels in cell populations as determined by FACS analysis. Codes for the corresponding cell populations are the same as in Figure 3. First number after the cell population code: mean level of eGFP in the sample; second number: mean level of eGFP in the upper 10% of the eGFP-positive cells.

EEF1A-based vectors superseding CMV-based types has enabled smaller numbers of cell clones to be screened and evaluated by increasing the mean level of target protein expression. We have modified existing EEF1A-based vectors by linking the DHFR selection marker and target gene in the bicistronic RNA, shortening the overall plasmid size, and adding an EBVTR element. The presence of an EBVTR element in the resulting p1.1 vector increased the stable transfection rate by a factor of 24, and increased the target protein expression level by eight-fold using a single round of MTX-driven target gene amplification. Two consecutive rounds of MTX-driven amplification, performed for suspension culture, resulted in the polyclonal cell population with the eGFP expression level comprising 9.0% of the total cytoplasmic protein. Compatible vectors bearing antibiotic resistance markers instead of the DHFR gene were created and found to be approximately equal to the DHFR-based vector for generation of highly productive cell populations. We found that the EEF1A-based vector, p1.2-Hygro, containing the hygromycin selection marker, allowed direct generation of a polyclonal cell population that was nearly devoid of eGFP-negative cells, while eGFP expression comprised up to 8.9% of the total cytoplasmic protein. This level of eGFP expression corresponds to only 30 copies of the target gene per single haploid genome, in contrast to CMV-based vectors that have thousands of copies per genome in highly productive lines [19]. The set of vectors developed herein allows generation of highly productive and stable cell clones with limited effort and such vectors may be employed to create cell lines for production of biosimilar pharmaceuticals. p1.1 or p1.2-based plasmids, stably transfected into polyclonal cell populations expressing large quantities of target proteins at a scale of $4–6*10^7$ cells, can be generated in less than one month by simple periodic passage of a culture from a shaking flask. This technique may be useful for obtaining milligram quantities of mutants of a protein of interest or for evaluation of several mAb clones. Cells from these polyclonal populations may be also used for direct development of industrially applicable clonal cell lines by limiting dilution.

Additional file

Additional file 1: Primer sequences used for cloning and sequencing the p 1.1 and p 1.2 expression vectors.

Abbreviations

CHO cells: Chinese hamster ovary (CHO) cells; MTX: Methotrexate; eGFP: Enhanced green fluorescent protein; EBVTR: Concatemer of the terminal repeats from the Epstein-Barr virus; PCR: Polymerase chain reaction; Kbp: Kilo base pairs (1000 base pairs); EEF1A: Chinese hamster Elongation factor 1 alpha; IRES: Internal ribosome entry site; DHFR: dihydrofolate reductase (EC 1.5.1.3); ORF: open reading frame; RFU: relative fluorescence units.

Competing interests

JAH, AGG and KGS declare that they have no competing interests. NAO, SVK and IIV are inventors of the patent RU2488633, which covers use of the p1.1 vector.

Authors' contributions

NAO developed the experimental strategy and plasmid design, performed cloning procedures, and drafted the manuscript. SVK performed the cell culture experiments and helped to draft the manuscript. IIV initiated the project, participated in its design and coordination and drafted the manuscript. NAO, JAH and IIV conducted the copy number determination experiments, SVK, JAH and IIV performed eGFP level determinations. AGG and KGS coordinated the project. All of the authors have read and approved the final manuscript.

Acknowledgments

The authors wish to thank Dr Sergei Khaidukhov for performing the FACS analysis and Dr Ivan Vorobiev (senior) for providing the guidelines for FACS data interpretation. We also thank Dr Ivan Zvyagin, Lev Usakin, Maria Kordyukova and Victoria Shender for taking part in the molecular cloning procedures. The work was supported by the following grants: RFBR 13-04-40277-H "Combinatorial chemistry and biology approaches to theranostics of multiple sclerosis"; RFBR 14-04-00647 "A Combinatorial biology approaches using yeast display for development of biocatalysts de novo"; Program of the Presidium of the Russian Academy of Sciences No24 "Detection of Viral Antigens as Possible Triggers of Autoimmune Neurodegeneration and Nanotechnologies and Nanomaterials"; Program of the Presidium of Russian Academy of Sciences "Fundamental science for medicine" - grant "Features of the degradation of antigens in neurodegenerative processes"; Scientific Schools 2046.2012.4 "Chemical Basis of Biocatalysis". Funding bodies didn't play any role in the design, collection, analysis, and interpretation of data; in the writing of the manuscript and in the decision to submit the manuscript for publication.

Author details

[1]Laboratory of Mammalian Cell Bioengineering, Centre "Bioengineering", Russian Academy of Sciences, 60-letija Oktyabrya 7, Moscow 117312, Russia. [2]Laboratory of Biocatalysis, Institute of Bioorganic Chemistry, Russian Academy of Sciences, Miklukho-Maklaya 16/10, Moscow 119971, Russia. [3]Kazan Federal University, Kazan, Republic of Tatarstan 420008, Russia.

References

1. Assaraf YG, Molina A, Schimke RT: **Sequential amplification of dihydrofolate reductase and multidrug resistance genes in Chinese hamster ovary cells selected for stepwise resistance to the lipid-soluble antifolate trimetrexate.** *J Biol Chem* 1989, **264**(31):18326–18334.
2. Running Deer J, Allison DS: **High-level expression of proteins in mammalian cells using transcription regulatory sequences from the Chinese hamster EF-1alpha gene.** *Biotechnol Prog* 2004, **20**(3):880–889.
3. Zimmermann J, Hammerschmidt W: **Structure and role of the terminal repeats of Epstein-Barr virus in processing and packaging of virion DNA.** *J Virol* 1995, **69**(5):3147–3155.
4. Cho MS, Tran VM: **A concatenated form of Epstein-Barr viral DNA in lymphoblastoid cell lines induced by transfection with BZLF1.** *Virology* 1993, **194**(2):838–842.
5. Cho MS, Chan SY: **Vectors having terminal repeat sequence of Epstein-Barr virus.** In *US Patent 6180108.* Washington, DC: U.S. Patent and Trademark Office; 2001.
6. Sun R, Spain TA, Lin SF, Miller G: **Autoantigenic proteins that bind recombinogenic sequences in Epstein-Barr virus and cellular DNA.** *Proc Natl Acad Sci U S A* 1994, **91**(18):8646–8650.
7. Matsuo T, Heller M, Petti L, O'Shiro E, Kieff E: **Persistence of the entire Epstein-Barr virus genome integrated into human lymphocyte DNA.** *Science* 1984, **226**(4680):1322–1325.
8. Leenman EE, Panzer-Grumayer RE, Fischer S, Leitch HA, Horsman DE, Lion T, Gadner H, Ambros PF, Lestou VS: **Rapid determination of Epstein-Barr virus latent or lytic infection in single human cells using in situ hybridization.** *Mod Pathol* 2004, **17**(12):1564–1572.
9. Hung SC, Kang MS, Kieff E: **Maintenance of Epstein-Barr virus (EBV) oriP-based episomes requires EBV-encoded nuclear antigen-1 chromosome-binding domains, which can be replaced by high-mobility group-I or histone H1.** *Proc Natl Acad Sci U S A* 2001, **98**(4):1865–1870.
10. Urlaub G, Chasin LA: **Isolation of Chinese hamster cell mutants deficient in dihydrofolate reductase activity.** *Proc Natl Acad Sci U S A* 1980, **77**(7):4216–4220.
11. Derouazi M, Martinet D, Besuchet Schmutz N, Flaction R, Wicht M, Bertschinger M, Hacker DL, Beckmann JS, Wurm FM: **Genetic characterization of CHO production host DG44 and derivative recombinant cell lines.** *Biochem Biophys Res Commun* 2006, **340**(4):1069–1077.
12. Weiner M, Gackstetter T, Costa G, Bauer C, Kretz K: **Site-directed Mutagenesis using PCR.** In *Molecular Biology: Current Innovations and Future Trends.* Edited by Griffin A, Griffin H. Wymondham, Norfolk, U.K: Horizon Scientific Press; 1995.
13. Orlova NA, Orlov AV, Vorobiev II: **A modular assembly cloning technique (aided by the BIOF software tool) for seamless and error-free assembly of long DNA fragments.** *BMC Res Notes* 2012, **5**:303.
14. Kozak M: **An analysis of 5'-noncoding sequences from 699 vertebrate messenger RNAs.** *Nucleic Acids Res* 1987, **15**(20):8125–8148.
15. Patterson GH, Knobel SM, Sharif WD, Kain SR, Piston DW: **Use of the green fluorescent protein and its mutants in quantitative fluorescence microscopy.** *Biophys J* 1997, **73**(5):2782–2790.
16. Ribeiro S, Mairhofer J, Madeira C, Diogo MM, Lobato da Silva C, Monteiro G, Grabherr R, Cabral JM: **Plasmid DNA size does affect nonviral gene delivery efficiency in stem cells.** *Cell Reprogram* 2012, **14**(2):130–137.
17. Wasley LC, Rehemtulla A, Bristol JA, Kaufman RJ: **PACE/furin can process the vitamin K-dependent pro-factor IX precursor within the secretory pathway.** *J Biol Chem* 1993, **268**(12):8458–8465.

18. Wajih N, Hutson SM, Owen J, Wallin R: **Increased production of functional recombinant human clotting factor IX by baby hamster kidney cells engineered to overexpress VKORC1, the vitamin K 2,3-epoxide-reducing enzyme of the vitamin K cycle.** *J Biol Chem* 2005, **280**(36):31603–31607.
19. Bebbington CR, Hentschel CC: **The use of vectors based on gene amplification for the expression of cloned genes in mammalian cells.** In *DNA Cloning.* Volume IIIth edition. Edited by Glover D. San Diego: Academic; 1987:163–188.

Isolation and characterization of a Sca-1$^+$/CD31$^-$ progenitor cell lineage derived from mouse heart tissue

Hao Wang[1†], Hao Chen[1†], Bei Feng[1,2], Xiang Wang[2,3], Xiaomin He[1], Renjie Hu[1], Meng Yin[1], Wei Wang[1], Wei Fu[1,2*] and Zhiwei Xu[1*]

Abstract

Background: Myocardial infarction remains the leading cause of mortality in developed countries despite recent advances in its prevention and treatment. Regenerative therapies based on resident cardiac progenitor cells (CPCs) are a promising alternative to conventional treatments. However, CPCs resident in the heart are quite rare. It is unclear how these CPCs can be isolated and cultured efficiently and what the effects of long-term culture in vitro are on their 'stemness' and differentiation potential, but this is critical knowledge for CPCs' clinical application.

Results: Here, we isolated stem cell antigen-1 positive cells from postnatal mouse heart by magnetic active cell sorting using an iron-labeled anti-mouse Sca-1 antibody, and cultured them long-term in vitro. We tested stemness marker expression and the proliferation ability of long-term cultured Sca-1$^+$ cells at early, middle and late passages. Furthermore, we determined the differentiation potential of these three passages into cardiac cell lineages (cardiomyocytes, smooth muscle and endothelial cells) after induction in vitro. The expression of myocardial, smooth muscle and endothelial cell-specific genes and surface markers were analyzed by RT-PCR and IF staining. We also investigated the oncogenicity of the three passages by subcutaneously injecting cells in nude mice. Overall, heart-derived Sca-1$^+$ cells showed CPC characteristics: long-term propagation ability in vitro, non-tumorigenic in vivo, persistent expression of stemness and cardiac-specific markers, and multipotent differentiation into cardiac cell lineages.

Conclusions: Our research may bring new insights to myocardium regeneration, for which even a small number of biopsy-derived CPCs could be enriched and propagated long term in vitro to obtain sufficient seed cells for cell injection or cardiac tissue engineering.

Keywords: Cardiac progenitor cell, Stem cell antigen-1, Differentiation, Multipotent, Self-renewal

Background

Despite recent progress in prevention and therapies, heart failure resulting from acute myocardial infarction or chronic myocardial ischemia remains the leading cause of mortality in developed countries [1]. The central cellular mechanism underlying myocardial dysfunction is an irreversible loss of viable cardiomyocytes and an inability of residual cardiomyocytes to compensate for this loss [2,3]. Heart transplantation remains the final therapeutic option for end-stage heart failure [4], but is limited by donor organ shortages and life-threatening complications, including organ rejection and side effects of pharmacological immune suppression. Regenerative therapies, including cell injection and myocardial tissue engineering, have been pursued as new possibilities to repair the damaged myocardium. Over the last decade, much research has focused on finding the ideal cell type to mediate myocardial repair. To date, cells derived from fetal or neonatal hearts have been applied to damaged myocardia to regenerate myocardial tissues in animal studies [5,6], but a similar approach in humans poses ethical issues and may be insufficient to

* Correspondence: fuweizhulu@163.com; zwxumd@gmail.com
†Equal contributors
[1]Department of Pediatric Cardiothoracic Surgery, Shanghai Children's Medical Center, School of Medicine, Shanghai Jiao Tong University, 1678 Dong Fang Road, Shanghai 200127, China
[2]Institute of Pediatric Translational Medicine, Shanghai Children's Medical Center, School of Medicine, Shanghai Jiao Tong University, 1678 Dong Fang Road, Shanghai 200127, China.
Full list of author information is available at the end of the article

provide the large number of cardiomyocytes needed for clinical use [7]. Developing methods to use stem cell-derived cardiomyocytes may be required.

The recent discovery of resident cardiac progenitor cells (CPCs) in the postnatal heart has marked a new era of cardiac regenerative medicine. Several laboratories have identified distinct populations of CPCs including side population (SP) cells [8], stem cell antigen-1-positive (Sca-1$^+$) cells [9], c-kit-positive (c-kit$^+$) cells [10], Islet-1-positive (Isl-1$^+$) cells [11,12], Wilms' tumor 1-positive (WT1$^+$) epicardial progenitor cells [13], and cardiosphere-derived cells (CDCs) [14]. However, the levels of resident CPCs in postnatal heart are limited and there is a clear negative correlation between CPC number and age [15]. Furthermore, in clinical practice, CPCs may be more likely to be isolated from catheter-based tiny biopsy specimens rather than large masses of cardiac tissue. Yet, large quantities of CPCs are needed for myocardium regeneration.

Goumans' team has developed a method for efficient isolation and expansion of human CPCs from cardiac surgical waste by magnetic cell sorting (MACS) and has developed a detailed protocol for very efficient in vitro differentiation of CPCs into cardiomyocytes (80–90% differentiation) [16]. However, it is unclear how many seed cells would be eventually obtained for regenerative therapies and whether their 'stemness' or differentiation potential would decline with long-term culture in vitro. In this report, we isolated Sca-1$^+$ cells from postnatal mouse heart by MACS using an iron-labeled anti-mouse Sca-1 antibody, and cultured the cells long-term in vitro. To monitor the stemness and differentiation potential of long-term cultured Sca-1$^+$ cells, we selected the 7th, 28th, and 53th passages to test early, middle and late phases of culture. We further identified these cells as Sca-1$^+$/CD31$^-$ CPC subpopulations, which are able to proliferate over the long-term in vitro without a detectable decline in stemness. In addition, they retain the potential to differentiate into multiple cardiac cell lineages.

Results

Isolation and in vitro culture of postnatal mouse heart-derived Sca-1$^+$ cells

MACS is an efficient system for cell isolation, especially for rare cells in tissues [17]. We used this system to isolate resident stem cells from postnatal mice hearts, based on the expression of the surface marker stem cell antigen Sca-1 (Figure 1A). To verify the efficiency of MACS using an iron-labeled anti-Sca-1 antibody, we performed FCM to analyze the enrichment of Sca-1$^+$ cells in samples before and after MACS. Pre-MACS, just after filtration through a 40-μm cell strainer, the whole cardiac cell suspension was composed of approximately 3.2% Sca-1$^+$ cells. After MACS, the Sca-1$^+$-enriched population contained approximately 85% Sca-1$^+$ cells, and the flow-through population contained only 0.1% Sca-1$^+$ cells (Figure 1B). Viewed with an inverted microscopy in ordinary light, cells in Sca-1$^+$-enriched populations were sparse, with quasi-circular or oval morphology and homogenous in size, while cells in flow-through populations were crowded, containing diverse cell types and sizes (Figure 1B). Thus, the efficiency and specificity of MACS were considered reasonable.

After seeding on gelatin-coated 12-well dishes, the long-term cell culture of Sca-1$^+$-enriched population was initiated. After about one week, small, flat, spindle-shaped cells appeared and grew gradually. Many of these cells were phase-bright under white light. On day 36, the primary cells (P0) had grown to approximately 90% confluence and were propagated to P1 at 1:2, that is, transferred from one well of the 12-well dish (4.5 cm^2) to one well in a 6-well dish (9.6 cm^2), also coated with gelatin. About 4 days after the first passage, cells of P1 were detached and further subcultured as P2 at 1:5; cells were transferred from one well of the 6-well dish (9.6 cm^2) to one 9-cm dish (49 cm^2) without gelatin coating. P2 cells were cultured in normal medium for 4 weeks, at which point they reached 90% confluence and were split at 1:2 to P3. P3 grew more rapidly and was split after about 1 week. From then on, the subcultured cells were passaged every 3–4 days at different ratios. The Sca-1$^+$ cells were passaged to over P90 and still grew well with phase-brightness (data not shown). There were no noticeable changes in gross morphology until at least P53 (representing a period of 245 days; Figure 1C).

Surface characteristics of Sca-1$^+$-enriched long-term cell culture

To identify the surface characteristics of heart-derived Sca-1$^+$-enriched cells and determine whether these would change during long-term in vitro culture, cells of P7, P28, and P53 were analyzed for stem cell antigen Sca-1 and other cell surface markers by FCM and IF staining. FCM analysis showed that the percentage of Sca-1$^+$ cells increased from 91.2% to 99.9% over P7 to P53. This indicates that not only do the Sca-1$^+$-enriched passages still mainly consist of Sca-1$^+$ cells, but the purity of Sca-1$^+$ cells increases with passaging. Few cells were detected that expressing c-kit, another stem cell marker; proportions varied from 0.3-0.6%. Flk-1, an endothelial progenitor cell marker, was also barely expressed. CD34, a hematopoietic progenitor cell marker [18], was initially expressed in 11.6% of P7 cells, which declined to 1.3% of cells at P53. CD31, a marker of endothelial cells [19,20] and CD45, a marker of the hematopoietic lineage [17] were also both rarely expressed (Figure 2).

To further characterize the Sca-1$^+$ cells, we double labeled them with antibodies against Sca-1 and CD31 (Figure 3), because Sca-1$^+$/CD31$^-$ CPCs show functional

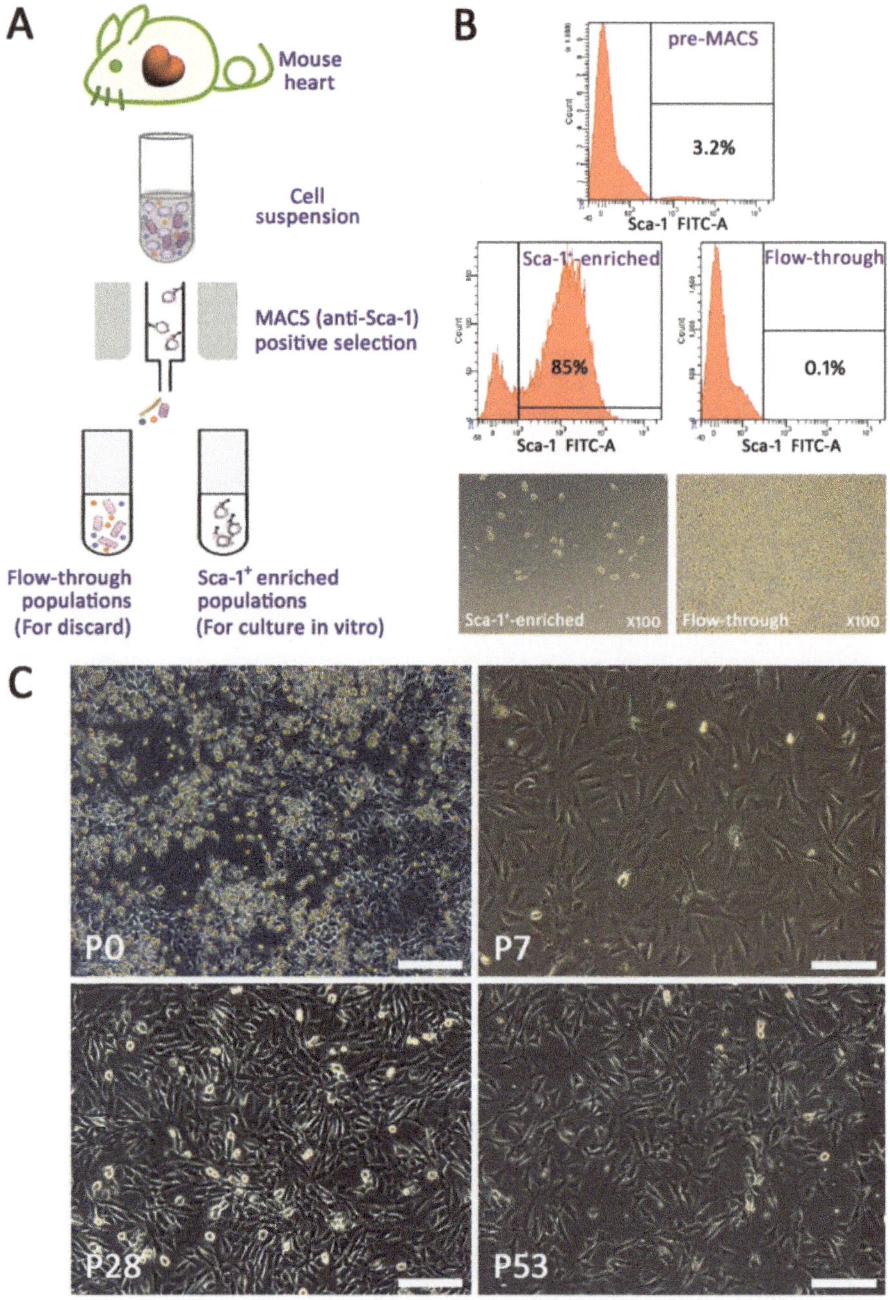

Figure 1 Isolation and enrichment of Sca-1$^+$ cells from the mouse heart and in vitro culture. A. Schematic representation of the isolation and enrichment of Sca-1$^+$ cell populations from the mouse heart. **B.** Efficiency and feasibility of MACS. The proportion of Sca-1$^+$ cells contained in pre-MACS population, Sca-1$^+$-enriched population, and flow-through population analyzed by flow cytometry. The lower two panels show the morphology of cells after MACS under ordinary light microscopy (×100). **C.** Cells of Sca-1$^+$-enriched populations cultured and passaged in vitro. Morphology of cells under inverted microscopy. Scale bar = 200 μm.

cardiomyogenic differentiation, and Sca-1$^+$/CD31$^+$ cells do not [21]. Based on FCM, there was strong expression of Sca-1 in cells at P7, P28, and P53, but few was Sca-1$^+$/CD31$^+$ cells (Figure 3A), which was confirmed by IF staining. IF staining showed that few cells expressed CD31 in Sca-1$^+$ enriched cell populations (Figure 3B). These data indicate that the mouse heart-derived Sca-1$^+$ cells isolated here represent a Sca-1$^+$/CD31$^-$ subpopulation.

Gene expression profile of Sca-1$^+$-enriched cells

To further characterize gene expression in the long-term cultured Sca-1$^+$-enriched cells, gene expression profiles

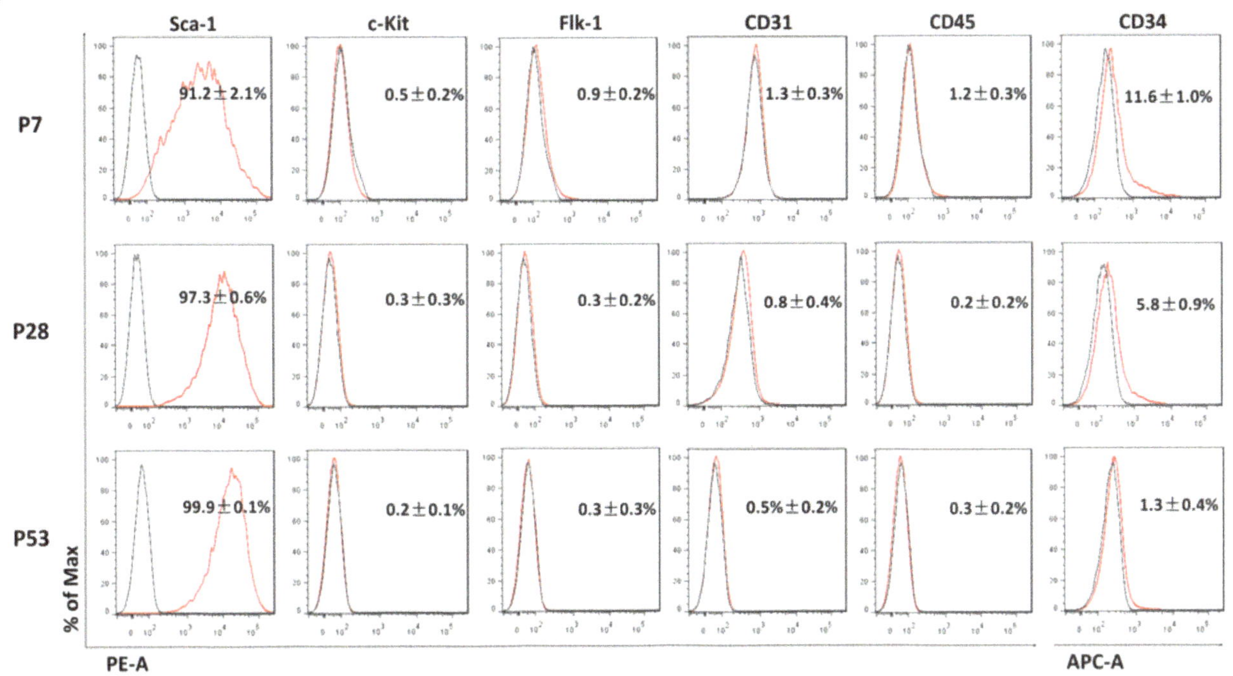

Figure 2 Surface marker expression profiles of three subcultures derived from Sca-1$^+$-enriched populations analyzed by flow cytometry. All populations were analyzed for the expression of the stem/progenitor cell markers Sca-1 and c-kit, endothelial progenitor cell marker flk-1, endothelial cell specific marker CD31, hematopoietic progenitor marker CD34, and hematopoietic lineage marker CD45. Black line, non-stained cells; red line, corresponding antibody. (n = 3).

of P7, P28, and P53 were generated by RT-PCR. Two pluripotent stem cell specific genes, Nanog and telomerase reverse transcriptase (TERT) [17], were both expressed; TERT is usually not detectable in cardiac fibroblasts [22]. Islet-1 (ISL-1), a marker of the secondary heart field [15], and TBX5, a marker of the primary heart field [17] were both detected, though the expression level of TBX5 was weaker than that of ISL-1. The cardiac-specific transcription factors GATA-4, Nkx-2.5, and MEF2C were all expressed in three different passages. In contrast, the mature cardiomyocyte structural gene cardiac α-myosin heavy chain (α-MHC) [17] was undetectable (Figure 4A). These results indicate that the long-term cultured cardiac-derived Sca-1$^+$-enriched cells show cardiac-specific features, but are not mature cardiomyocytes or cardiac fibroblasts.

In vitro proliferation capacity and tumorigenicity in vivo

The ability to proliferate is a significant feature of stem cells and progenitor cells. The proliferative abilities of P7, P28, and P53 cells were indirectly assessed by measuring the OD value of the supernatant incubated with CCK-8. Five sequential time points were chosen for testing: days 1 (24 hours after seeding), 3, 5, 7 and 9. At day 1 of each passage, the OD value was generally low and there was no statistical difference among these three passages. But from day 1 to day 3, it sharply increased. After day3, the OD value continued increasing slowly, peaked at day 7, then plateaued at this level for several days (Figure 4B). There was no statistical difference in proliferation between P7 and P28. However, the OD value at day 3, day 5, day 7, and day 9 of P53 was significantly higher than that of P7 and P28(*$p < 0.01$). And in all the comparisons between two adjacent time points in each passage, only day 1 and day 3 presented statistical difference ($^{\#}p < 0.01$). These data indicate that the proliferative ability has increased during long term culture, and the significant time point is day 3.

To exclude that Sca-1$^+$ cells were spontaneously transforming into tumorigenic cells, cells from P7, P28, and P53 were injected subcutaneously into SCID mice. Equivalent mouse R_1 ES cells were injected contralaterally as a positive control. No tumors developed at the site of implantation of P7, P28, or P53 cells over approximately 3 months. In contrast, sites injected with R_1 ES cells all developed neoplasms about 3 weeks after injection; 3 months later, these tumors were nearly 20 mm in diameter. The tumorigenic rates are shown in Additional file 1: Table S1.

Multipotent differentiation of mouse heart-derived Sca-1$^+$ cells into cardiac cell lineages

To estimate the differentiation potential of cells into the cardiomyocyte, smooth muscle, and endothelial lineages,

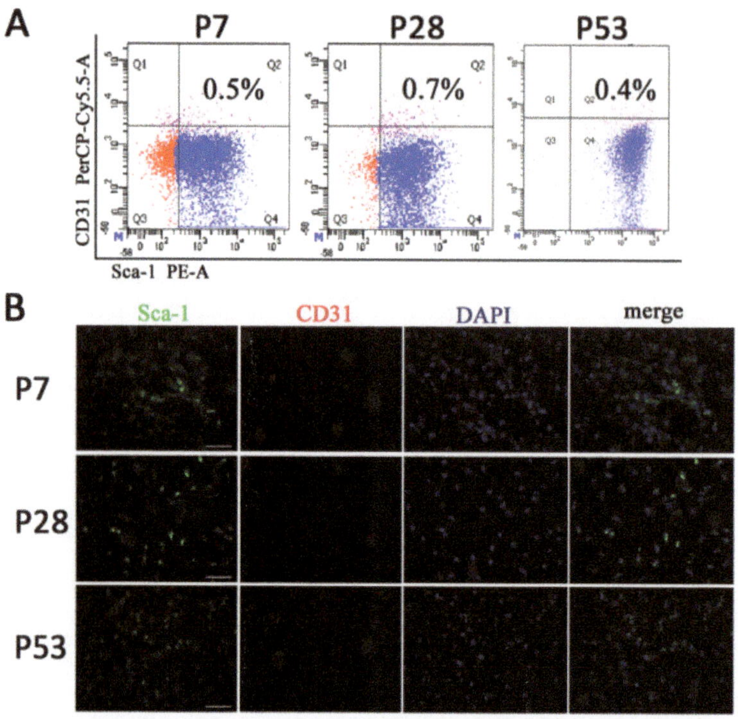

Figure 3 Analysis of CD31 expression in Sca-1$^+$ cells by FCM and identification of Sca-1$^+$-enriched cells by IF staining. A. Flow cytometry analyzed the expression of surface markers CD31 in Sca-1$^+$ cells. **B**. All cells were triple-stained for Sca-1 (green), CD31 (red), and DAPI (blue). Bar = 100 μm.

P7, P28, and P53 cells were cultured in specific differentiation medium, with simultaneously cultured in differentiation medium without growth factors serving as controls. When induction was completed, lineage-specific markers were analyzed by IF staining and RT-PCR.

After 2 weeks of induction to a cardiomyocyte fate, cMHC-positive and cTNT-positive cells were detected in all three passages, and no cMHC- or cTNT-positive cells were observed in control groups (Figure 5A,B). Though the differentiation potential in quantitative analysis of these three passages didn't present statistical difference among them, it was still low in each passage (Additional file 2: Figure S1A). This cardiomyogenic potential was confirmed by their gene expression profiles. Mature cardiomyocyte marker genes, including cardiac α-myosin heavy chain (α-MHC), cardiac β-myosin heavy chain (β-MHC), myosin light chain-2a (MLC-2a) and myosin light chain-2v (MLC-2v) were expressed in induced cells. In contrast, these genes were not expressed in control groups. After differentiation, the cardiac-specific transcription factor GATA4 was downregulated, while Nkx2.5 and MEF2C were upregulated (Figure 5C).

After 10 days of induction to a smooth muscle cell fate, SMA-positive and sMHC-positive cells were detected and the number of calponin-positive cells increased significantly compared to the control groups (Figure 6A-C). And the differentiation potential in quantitative analysis of these three passages didn't present statistical differences among them (Additional file 2: Figure S1B). This result was further verified by examining the gene expression of SMA and calponin. SMA was expressed in all induced passages and there was no expression in control groups, while the expression of calponin was upregulated after induction (Figure 6D).

After 2 weeks of induction to an endothelial fate, CD31-positive cells were detected in all passages, while none were detected in control groups (Figure 7A). And the differentiation efficiency of these three passages didn't present statistical differences among them (Additional file 2: Figure S1C). There was expression of the endothelial specific genes CD31, VE-cadherin, and vWF in all the induced populations, and none in the control groups. The expression level of the endothelial progenitor cell marker Flk-1 did not detectably change after differentiation (Figure 7B). Overall, these data indicate that mouse heart-derived Sca-1$^+$ cells are multipotent, retaining an ability to differentiate into different cardiac cell lineages, despite long-term propagation in vitro.

Discussion

In this article, we successfully isolate Sca-1$^+$ cells from postnatal mice hearts with the MACS system, using a microbead-labeled anti-Sca-1 antibody. The isolated Sca-1$^+$ cells express the stemness markers Nanog and TERT, the

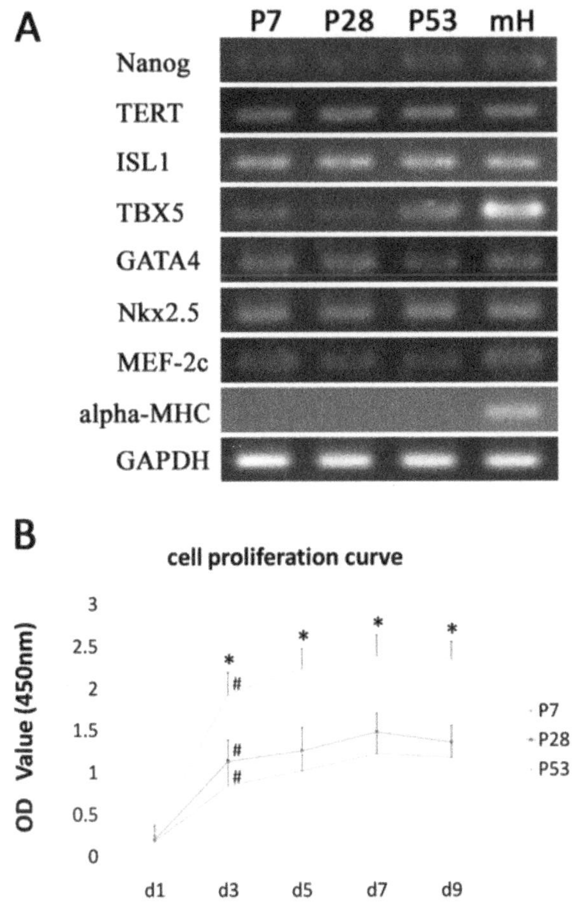

Figure 4 Gene expression profiles and proliferation abilities of Sca-1 high-expressing cell populations analyzed by RT-PCR and CCK-8. A. Expression of stem cell-related genes Nanog and TERT, cardiac precursor-related genes ISL-1 and TBX5, early cardiogenic genes GATA4, Nkx2.5, MEF2c, and mature cardiomyocyte specific gene α-MHC was analyzed by RT-PCR. RNA extracted from whole postnatal mouse heart (mH) was used as a positive control. (n = 3). **B**. Proliferation abilities were tested with CCK-8. Five successive time points were chosen: days 1 (24 hours after seeding), 3, 5, 7, and 9. Proliferation abilities of these cell populations were indirectly measured by the OD value at 450 nm. (n = 3). ($^{*}P < 0.01$ vs P7, P28; $^{\#}P < 0.01$ vs d1).

cardiac mesoderm markers ISL-1, TBX5, and the cardiac specific transcription factors GATA4, Nkx2.5, and MEF2C. They can be propagated in vitro for a long time without any significant changes in marker expression. Moreover, the isolated Sca-1$^+$ cells and their long-term descendants exhibit multipotent differentiation to cardiac cell lineages, including smooth muscle, endothelial cells and cardiomyocytes. So, the Sca-1$^+$ cells we isolated from mouse heart tissue should be identified as CPCs.

Isolation of a cardiac progenitor cell (CPC) population from the heart is challenging, particularly because of a lack of specific surface markers. Though Sca-1$^+$ cells, c-kit$^+$ cells, Isl-1$^+$ cells, WT1$^+$ cells, SP cells and CDCs have been extracted from postnatal cardiac tissue [8-14], it remains to be determined whether these individual CPC populations represent different intermediate stages of a single cell lineage or are independent cell lineages originating from different precursors. Some cell markers that could distinguish different lineages, such as Isl-1 and WT-1, are transcription factors, and therefore, transgenic animals would be necessary to genetically tag these proteins to use them as tools to isolate live cells [12,13]. Whether CDCs possess cardiomyogenic potential remains controversial [14,23,24]. Although c-kit$^+$ cardiac stem cells were the first stem cells identified in the rat heart [10], recent studies have suggested that Sca-1$^+$ progenitor cells are actually the predominant stem cell population there [9,25]: Sca-1$^+$ cells are 100–700 fold more frequent than c-kit$^+$ cells [10,26]. Although the biological hierarchy and correlation between cardiac SP cells and Scal$^+$ cells is still unclear, interestingly, a high percentage (>80%) of cardiac SP cells express the Scal antigen and only a very small percentage (approximately 3.6%) of cardiac Sca-1$^+$ cells is included in the SP fraction [9,21]. Furthermore, cardiac SP cells typically lack c-kit, in contrast to marrow SP cells [9,21]. For these reasons, Sca-1$^+$ cells in cardiac tissue may be the most populous CPCs or predominate over the long-term, and thus may be relatively easy to isolate from cardiac tissue. This is why we aimed to isolate Sca-1$^+$ CPCs from the mouse heart in our research. In 3- to 4-week-old mice, Sca-1$^+$ cells accounted for approximately 3.2% of total cardiac cells (Figure 1B), in line with the previously reported 3.0% [27] and 2.1% [9] of Sca-1$^+$ cells in 8-week-old and 6- to 12- week-old mice heart respectively; there is a clear negative correlation between CSC number and age [15].

Here, Sca-1$^+$ cardiac cells, including early, middle and late passages, showed limited expression of the common stemness marker c-Kit and the hematopoietic lineage markers CD34 and CD45, which characterize bone marrow-derived stem cells [21]. Moreover, they also showed limited expression of the endothelial lineage markers Flk-1 and CD31. However, during early stages of in vitro culture, such as P7, the expression of CD34 in enriched cells is somewhat higher than found in previous studies [17]. We ascribe this outcome to only one round of magnetic sorting. Since CD34 has been characterized as a marker of progenitor cells in several lineages, it may represent another subpopulation of CPC or yet another lineage. Based on FCM and IF staining (Figure 3), these cardiac progenitor cells belonged to a Sca-1$^+$/CD31$^-$ subpopulation, which has been reported to show functional cardiomyogenic differentiation [21]. Sca-1$^+$/CD31$^-$ CPCs may be promising starting material to produce cells for cardiac regeneration. However, little is known about these cells' stemness characteristics or differentiation potential after long-term culture in vitro. These features will determine whether it is possible to obtain sufficient seed cells

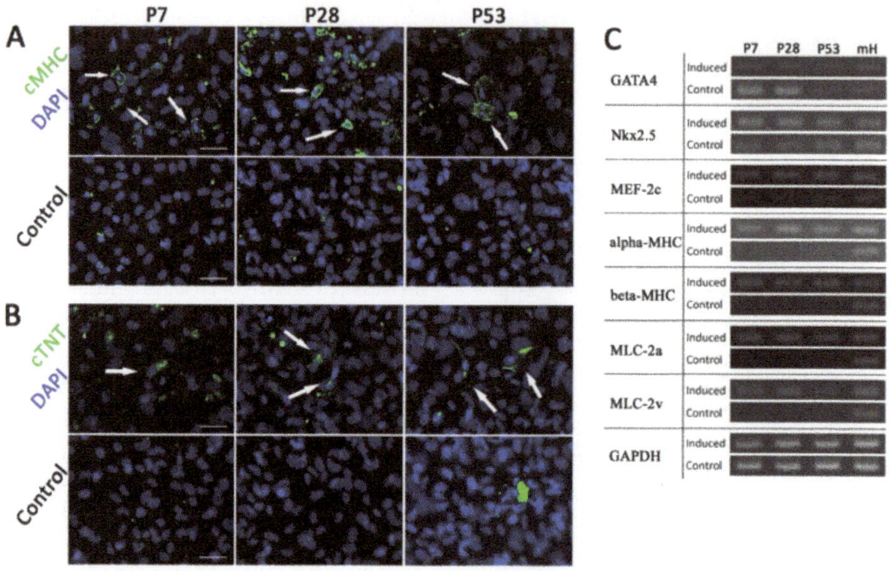

Figure 5 Differentiation potential of subcultured cells from Sca-1$^+$-enriched populations into cardiomyocyte-like cells in vitro. A-B. All cell populations were stained for the cardiomyocyte specific marker cardiac MHC (green), cardiac Troponin T (green) and DAPI (blue) after induction. White arrows indicate representative cells positive for each cardiomyocyte-specific antigen. Scale bar = 50 μm. **C.** Expression of cardiac-specific transcription factors GATA4 and Nkx2.5, early cardiogenic gene MEF2c, and cardiomyocyte-specific genes α-MHC, β-MHC, MLC-2a, and MLC-2v was analyzed by RT-PCR. RNA extracted from whole postnatal mouse heart (mH) was used as a positive control.

from these rare primary cells. Here, we monitor these features in cultured CPCs, propagating Sca-1$^+$/CD31$^-$ CPCs over 53 passages. If calculated from one primary isolated cell, with a splitting ratio of 1:3, there would be approximately 2×10^{14} cells at P28 and approximately 2×10^{25} cells at P53. These quantities would be sufficient to meet the requirements of cardiac tissue engineering.

Stemness characteristics include specific marker expression, proliferation ability, and differentiation potential. In gene expression profile analysis, Sca-1$^+$ cardiac cells were positive for the cardiac mesoderm precursor markers Isl-1 and TBX5, and cardiac-specific transcription factors Nkx2.5 and GATA4, but negative for the cardiomyocyte structure gene α-MHC (Figure 4A). Isl-1 is a marker of cardiac precursors in the developing as well as postnatal heart [15], and also is a marker of cells in the secondary heart field. Isl-1$^+$ cells are considered to be cardiovascular precursors with the ability to differentiate into the three main cardiac cell lineages [28]. In contrast, Tbx5 is expressed in the primary heart field [29]. Proliferation ability is another important stemness characteristic. Here, Sca-1$^+$ CPCs in early, middle and late passages proliferated rapidly during days 1–3 after seeding, with continuous growth that peaked at day 7. These results are consistent with previous reports [17]. There were significant differences in proliferation ability between P7, P28 and P53 during the period of rapid proliferation (Figure 4B). P53 showed the greatest proliferation ability of the three. Tateishi and colleagues have reported that knockdown of Sca-1 transcripts in CPCs, by inactivating Akt, leads to retarded ex vivo expansion and apoptosis. That is, Sca-1 signaling might be essential to promote CPC proliferation and survival [30]. In this study, the purity of Sca-1$^+$ cells was highest at P53, which may have contributed to their greater proliferation ability. To test whether cells became tumorigenic over long-term culture, Sca-1$^+$ cells were injected subcutaneously into nude mice: in over more than 3 months of observation, no neoplasms appeared.

We tested the differentiation potential of Sca-1$^+$/CD31$^-$ cells in early, middle, and late stages of culture. After differentiation, Sca-1$^+$ cells presented specific marker expression patterns of multiple cardiac cell lineages, such as sMHC, SMA and calponin for smooth muscle, CD31, vWF and VE-cadherin for endothelial cells, and cTNT, cMHC and MLC for cardiomyocytes (Figures 5, 6 and 7). There were no significant differences between early, middle and late passages in differentiation potential. This study nevertheless has some limitations. Although the cells did present cardiomyogenic differentiation potential, the frequency of cardiomyogenic differentiation was lower than expected. Though 5-azacytidine is a classic agent to induce bone marrow-derived cells to efficiently differentiate into cardiomyocytes, it appears less effective in the Sca-1$^+$ CPC population [17]. Previous research has shown that cardiomyogenic differentiation of Sca-1$^+$ CPCs induced by dexamethasone, dimethylsulfoxide, 5-azacytidine, oxytocine, TGF-β, Wnt1, Wnt5a, BMP2,

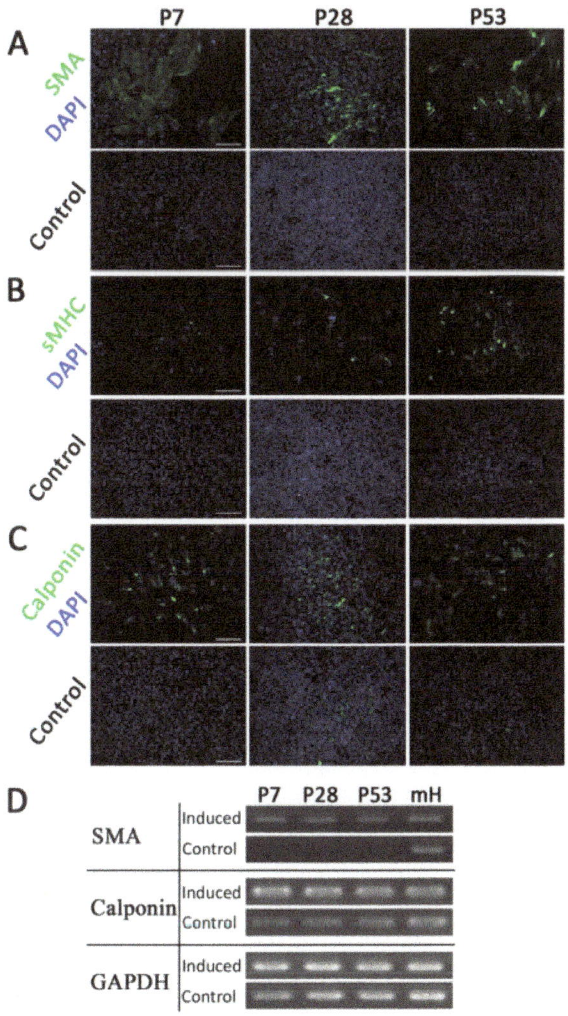

Figure 6 Differentiation potential of subcultured cells from Sca-1^{+}-enriched population into smooth muscle-like cells in vitro. **A-C**. All cell populations were stained for the smooth muscle specific marker SMA (green), smooth muscle MHC (green), calponin (green) and DAPI (blue) after induction. Scale bar = 200 μm. **D**. Expression of smooth muscle specific genes α-SMA and calponin was analyzed by RT-PCR. RNA extracted from whole postnatal mouse heart (mH) was used as a positive control.

and FGF4, individually or in combinations, failed to generate mature, beating cardiomyocytes. Only stimulation with BMP2 combined with FGF-4 has been shown to induce the expression of mature cardiac markers [17]. Therefore, we also utilized BMP2 combined with FGF-4 to induce Sca-1+ CPCs for cardiomyogenic differentiation in our research. Regardless of a failure to generate beating cardiomyocytes, the ratio of cardiomyogenic marker expression was still low (Additional file 2: Figure S1A). Further studies are required to identify efficient inducing factors. The function of differentiated cells from those Sca-1^{+}/CD31^{-} progenitor cells should also be investigated in the future.

Conclusions

We have isolated postnatal mouse heart-derived Sca-1^{+}/CD31^{-} CPCs by MACS, which can be propagated in vitro over the long-term while retaining stemness characteristics and multiple differentiation potentials. This research may present a possibility that in future clinical studies on cardiac regeneration we can provide a sufficient source of seed cells from little human heart tissue.

Methods

Experimental animals and ethics statement

Wild-type ICR mice (3–4 weeks old) and nude mice (3–4 weeks old) were used for cell isolation and subcutaneous cell injection, respectively. All mice were purchased from the Shanghai Laboratory Animal Center (Shanghai Institutes for Biological Sciences, Chinese Academy of Sciences). The Animal Care Committee of Shanghai Children's Medical Center approved all procedures and protocols. The investigation conformed to the Guide for the Care and Use of Laboratory Animals of the Shanghai Children's Medical Center.

Cell isolation

Whole hearts were extracted from either male or female ICR mice, split, and washed several times with ice-cold PBS to remove the residual blood. The hearts were minced and digested with 0.1% collagenase type A (Roche) at 37°C for 60 minutes, then filtered through a 40 μm cell strainer (BD Falcon). Half of the filtered cardiac cell suspension was used for flow cytometry (FCM) analysis; the other half was centrifuged at 300 g for 5 min at 4°C to remove the blood cells and debris in the supernatant, then resuspended in cold M-buffer saline (containing 1% BSA; Miltenyi Biotec). After incubation for 5 min on ice (to remove settled cardiomyocytes), the supernatant was collected in a new tube and re-centrifuged. The sample was resuspended and an aliquot taken to determine the cell number. The sample was then centrifuged at 300 g for 10 min at 4°C. The supernatant was aspirated completely, and the cell pellet was resuspended in 90 μl cold M-buffer, followed by incubation with 10 μl FITC-conjugated monoclonal anti-mouse Sca-1 antibody (Miltenyi Biotec) at 4°C for 10 min. The cells were washed 3 times with cold buffer, re-centrifuged, resuspended in 80 μl buffer, then incubated with 20 μl microbeads-conjugated monoclonal anti-FITC antibody(Miltenyi Biotec) for 15 min at 4°C. Cells were then passed through a magnetic cell sorting (MACS) column in a Miltenyi magnetic field to isolate a Sca-1^{+}-enriched cell suspension. All manipulations were conducted with aseptic technique and MACS was performed in the dark. Samples of the Sca-1^{+}-enriched suspension and the flow-through suspension were subsequently used for flow cytometry. The residual Sca-1^{+}-enriched cell suspension was used for further culture. The cell isolation

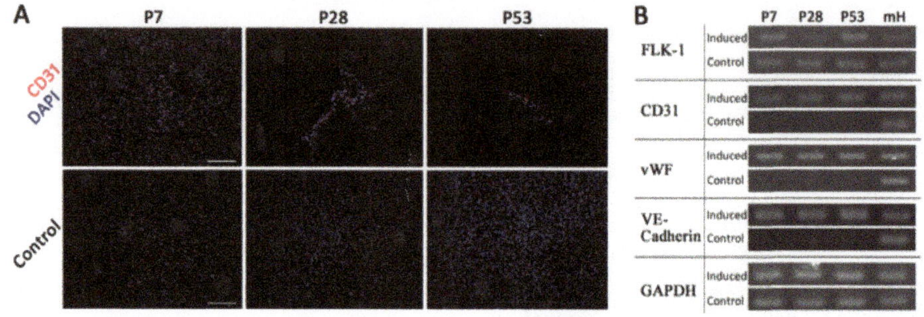

Figure 7 Differentiation potential of subcultured cells from Sca-1$^+$-enriched populations into endothelial-like cells in vitro. A. All cell populations were stained for the endothelial cell marker CD31 (red) and DAPI (blue) after induction. Scale bar = 200 μm. **B**. Expression of the endothelial progenitor gene flk-1 and the endothelial cell specific genes CD31, vWF, and VE-Cadherin was analyzed by RT-PCR. RNA extracted from whole postnatal mouse heart (mH) was used as a positive control.

experiment was repeated three independent times, and each time 3–5 mice were used.

Cell culture

The Sca-1$^+$-enriched cells were seeded into 0.1% gelatin-coated 12-well culture plates and cultured in growth medium [16] composed of EBM-2 medium (EGM-2 Single Quots; Lonza) and M199 medium (Gibco) at the ratio of 1:3, plus 10% fetal bovine serum (FBS; Gibco), 1% penicillin-streptomycin (P/S; Gibco), 1% MEM nonessential amino acids (Gibco), and 10 ng/ml basic fibroblast growth factor (bFGF; Roche). Cells were cultured at 37°C in humid air with 5% CO_2. Cells were given at least 3 days to attach before refreshing the medium and were passaged at 80-90% confluence. Cells were washed with PBS, detached with 0.25% trypsin-EDTA (Gibco), then split at ratios from 1:2 to 1:5. When the cells covered a surface larger than 10 cm^2, they were converted to cultures in normal medium that contained low glucose Dulbecco's modified Eagle's medium (DMEM; Gibco), 10% FBS, and 1% P/S, without bFGF.

Flow cytometry (FCM) analysis

FCM analysis was performed on the whole cardiac cell suspension filtered through a 40-μm strainer (pre-MACS), the Sca-1$^+$-enriched cell suspension (post-MACS) and the flow-through cell suspension (post-MACS) to evaluate the effectiveness of MACS and the purity of the Sca-1$^+$-enriched population. The pre-MACS suspension was further divided into two equal parts, one of which was used for a blank control. Other cell preparations were resuspended in 100 or 200 μl buffer and incubated with FITC-conjugated anti-Sca-1 antibody (Miltenyi Biotec) at 4°C for 30 minutes.

To determine whether the expression of surface markers in Sca-1$^+$-enriched cells changed during long-term culture in vitro, cells from three passages (P7, P28 and P53) were detached using 0.25% trypsin/EDTA solution. Cells from each passage were counted and divided into several samples ($1\text{-}5 \times 10^5$ cells/sample). For immunofluorescence labeling, each cell sample was resuspended in 100 μl buffer and incubated for 30 minutes at 4°C with PE-labeled anti-Sca-1 (eBioscience), anti-c-kit (CD117; eBioscience), anti-Flk-1 (eBioscience), anti-CD31 (eBioscience), anti-CD45 (eBioscience) antibodies, or APC-labeled anti-CD34 (eBioscience) antibody with dilutions as suggested by the manufacturer. The cells were subsequently washed 3 times with buffer, resuspended in 100 μl cold buffer, then analyzed in a FACSCanto IIflow cytometer (BD Biosciences). Non-stained cells from each passage were used as controls. And this experiment was performed three independent replicates.

Differentiation assays

Three cell populations (P7, P28, and P53) were cultured in specific differentiation medium for induction into different cardiac cell lineages. Cells were initially seeded at a density of 1×10^4 cells/cm^2 in normal medium, then converted into specific differentiation medium when they approached 80-90% confluence. For smooth muscle cell induction, cells were cultured in normal medium supplemented with 10 ng/ml transforming growth factor-β1 (TGF-β1; Peprotech) for 10 days. For endothelial cell induction, cells were cultured in EBM-2 medium (Lonza) plus the EGM-2 Single Quots kit (Lonza) but without hydrocortisone, plus 20 ng/ml vascular endothelial growth factor-165 ($VEGF_{165}$, Peprotech), and were grown on fibronectin-coated dishes for 14 days. For cardiomyocyte induction, cells were cultured with DMEM/F12 (Gibco) supplemented with 2% FBS, 2% B27, 100 ng/ml bone morphogenetic protein-2 (BMP-2; Peprotech), and 100 ng/ml fibroblast growth factor-4 (FGF-4; Peprotech) on gelatin-coated dishes for 14 days. In control group, cells were cultured in the same medium for differentiation without growth factors. All the media were changed every 3 days.

Cell proliferation analysis

Cells (P7, P28, P53) were seeded at a density of 1 × 10^4 cells/cm^2 on 24-well culture plates and cultured in normal medium at 37°C in humid air with 5% CO_2. Cell counting kit-8 (CCK-8; Dojindo) antigen was added to the cell supernatant at the ratio of 1:10 for a 2 h incubation on days 1 (24 hours after initial seeding), 3, 5, 7, and 9. Subsequently, 100 µl of supernatant from each well was carefully transferred to a 96-well plate to measure the optical density(OD) value at a wavelength of 450 nm. It was measured at the five time points with 6 technical replicates of each passage. The living cell number can be indirectly presented by the OD value as they are positively correlated. And this experiment was also performed three independent replicates.

Analysis of mRNA expression by reverse transcription polymerase chain reaction (RT-PCR)

Total RNA was extracted from P7, P28, P53, differentiated cells, and whole mouse heart tissue using TriZol reagent (Invitrogen), according to the manufacturer's instructions. Integrity of the RNA samples was verified with a NanoDrop 2000 (Thermo Scientific). Total purified RNA samples were treated with RNase-free DNase (Qiagen) to eliminate DNA contamination and were stored at −80°C until use. RT-PCR was carried out in two steps, and reverse transcription was performed using PrimeScript RTase (TaKaRa). An equal amount of each sample of cDNA was amplified with 2× Taq Master Mix (OFFO BioPharm, Shanghai). The primers used for polymerase chain reaction (PCR) are shown in Additional file 3: Table S2. The reaction was performed with 38 cycles, with denaturation at 95°C for 30s, annealing from 54-70°C for 30s, and extension at 72°C for 30s. After the last cycle, PCR products were run on a 1-2% agarose TAE gel (OFFO BioPharm, Shanghai). Expression levels were evaluated by the intensity of the band stained with ethidium bromide. Quantified values were normalized against the housekeeping gene, GAPDH. The RT-PCR tests were replicated three times.

Immunofluorescence (IF) staining

Cells from P7, P28, and P53 were analyzed with IF staining for the surface markers Sca-1 and CD31. Induced and control group cells were labeled with IF for cardiac cell lineage-specific markers. Cells were washed with PBS and fixed with 4% paraformaldehyde for 15 min at room temperature (RT), permeabilized if noted with 0.3% Triton X-100 (Sigma) for 20 min at RT, washed with PBS, blocked with 10% donkey or rabbit serum (Multisciences Biotech, China) in PBS for 30 min at 37°C, and then incubated for 2 hours at RT with primary antibodies against Sca-1 (Millipore), CD31 (eBioscience), cardiac troponin T (cTNT; Santa Cruz), cardiac-myosin heavy chain (cMHC; Abcam), smooth muscle actin (SMA; Epitomics), smooth muscle-myosin heavy chain (sMHC; Abcam), or calponin-1 (Abcam). After rinsing with PBS, cells were subsequently incubated with Alexa Fluor 488-conjugated donkey or rabbit-originated secondary antibodies (Life Technologies, Invitrogen) or PE conjugated streptavidin (eBioscience) for 30 min at RT. Nuclei were counter-stained with DAPI (1:1000; Enzo) in PBS for 1 min at RT. Immunostaining was observed and photoed by an inverted fluorescence microscope (Leica DMI3000B, Germany) and analyzed by Image-Pro Plus 7.0 software (Media Cybernetics, America).

Tumorigenic assays

2 × 10^6 cells from P7, P28, and P53 were injected subcutaneously into one dorsal side of severe combined immune deficiency (SCID) mice. As a positive control, an equal quantity of mouse R_1 embryonic stem (ES) cells were inoculated subcutaneously into the opposite dorsal side of the same mice. Details of the cell type, cell concentration, injection volume and injection points are shown in Additional file 1: Table S1.

Statistical analysis

All data are expressed as mean ± SD. Significance between two comparisons was evaluated by independent-samples *T* test. Significance between multiple comparisons was evaluated by one-way ANOVA. Bonferroni post-hoc tests were used to identify differences. Statistical values were calculated using the SPSS 17.0 software. A value of $P < 0.05$ was considered statistically significant.

Additional files

Additional file 1: Table S1. Tumorigenic Assay.

Additional file 2: Figure S1. Quantitative analysis of differentiation potential of subcultured cells from Sca-1$^+$-enriched populations into cardiac cell lineages in vitro. A, cMHC or cTNT positive cells were calculated after induction to cardiomyocyte-like cells. (n = 10). B, SMA, sMHC or calponin positive cells were calculated after induction to smooth muscle-like cells. (n = 10). C, CD31 positive cells were calculated after induction to endothelial-like cells. (n = 10). The positive rate was presented as ratio of positive cell number to total cell number (*p < 0.01 vs control).

Additional file 3: Table S2. Primers used for reverse transcription PCR.

Competing interest

The authors declare that they have no competing interest.

Authors' contributions

Conceived and designed the experiments: HW HC WF ZX. Performed the experiments: HW HC BF XW XH RH MY. Analyzed the data: HW HC WW WF. Drafted the manuscript: HW HC WF ZX. All authors read and approved the final manuscript.

Acknowledgements

This study was supported by National Natural Science Fund of China (81370117,81170123,31200735,81271726,80170151), Shanghai Natural Science Fund for Youth Scholars(12ZR1446500),Science and Technology Development Fund of Shanghai Pudong(PKJ2012-Y48), the Biomedical

Engineering fund of Shanghai Jiao Tong University (YG2012MS36, YG2012MS35), the College Young Teachers' Training and Funding Project of Shanghai(ZZjdyx12117,ZZjdyx12124, ZZjdyx12120) and the College Young Teachers' Training and Funding Project of Shanghai Jiao Tong University School of Medicine.

Author details

[1]Department of Pediatric Cardiothoracic Surgery, Shanghai Children's Medical Center, School of Medicine, Shanghai Jiao Tong University, 1678 Dong Fang Road, Shanghai 200127, China. [2]Institute of Pediatric Translational Medicine, Shanghai Children's Medical Center, School of Medicine, Shanghai Jiao Tong University, 1678 Dong Fang Road, Shanghai 200127, China. [3]Department of Hematology/Oncology, Shanghai Children's Medical Center, School of Medicine, Shanghai Jiao Tong University, 1678 Dong Fang Road, Shanghai 200127, China.

References

1. Lloyd-Jones D, Adams RJ, Brown TM, Carnethon M, Dai S, De Simone G, Ferguson TB, Ford E, Furie K, Gillespie C, Go A, Greenlund K, Haase N, Hailpern S, Ho PM, Howard V, Kissela B, Kittner S, Lackland D, Lisabeth L, Marelli A, McDermott MM, Meigs J, Mozaffarian D, Mussolino M, Nichol G, Roger VL, Rosamond W, Sacco R, Sorlie P, *et al*: **Executive summary: heart disease and stroke statistics–2010 update: a report from the American Heart Association.** *Circulation* 2010, **121**(7):948–954.
2. Braunwald E, Pfeffer MA: **Ventricular enlargement and remodeling following acute myocardial infarction: mechanisms and management.** *Am J Cardiol* 1991, **68**(14):1D–6D.
3. Wencker D, Chandra M, Nguyen K, Miao W, Garantziotis S, Factor SM, Shirani J, Armstrong RC, Kitsis RN: **A mechanistic role for cardiac myocyte apoptosis in heart failure.** *J Clin Invest* 2003, **111**(10):1497–1504.
4. Koerner MM, Durand JB, Lafuente JA, Noon GP, Torre-Amione G: **Cardiac transplantation: the final therapeutic option for the treatment of heart failure.** *Curr Opin Cardiol* 2000, **15**(3):178–182.
5. Soonpaa MH, Koh GY, Klug MG, Field LJ: **Formation of nascent intercalated disks between grafted fetal cardiomyocytes and host myocardium.** *Science* 1994, **264**(5155):98–101.
6. Sekine H, Shimizu T, Hobo K, Sekiya S, Yang J, Yamato M, Kurosawa H, Kobayashi E, Okano T: **Endothelial cell coculture within tissue-engineered cardiomyocyte sheets enhances neovascularization and improves cardiac function of ischemic hearts.** *Circulation* 2008, **118**(14 Suppl):S145–S152.
7. Matsuura K, Haraguchi Y, Shimizu T, Okano T: **Cell sheet transplantation for heart tissue repair.** *J Control Release* 2013, **169**(3):336–340.
8. Oyama T, Nagai T, Wada H, Naito AT, Matsuura K, Iwanaga K, Takahashi T, Goto M, Mikami Y, Yasuda N, Akazawa H, Uezumi A, Takeda S, Komuro I: **Cardiac side population cells have a potential to migrate and differentiate into cardiomyocytes in vitro and in vivo.** *J Cell Biol* 2007, **176**(3):329–341.
9. Oh H, Bradfute SB, Gallardo TD, Nakamura T, Gaussin V, Mishina Y, Pocius J, Michael LH, Behringer RR, Garry DJ, Entman ML, Schneider MD: **Cardiac progenitor cells from adult myocardium: homing, differentiation, and fusion after infarction.** *Proc Natl Acad Sci U S A* 2003, **100**(21):12313–12318.
10. Beltrami AP, Barlucchi L, Torella D, Baker M, Limana F, Chimenti S, Kasahara H, Rota M, Musso E, Urbanek K, Leri A, Kajstura J, Nadal-Ginard B, Anversa P: **Adult cardiac stem cells are multipotent and support myocardial regeneration.** *Cell* 2003, **114**(6):763–776.
11. Bu L, Jiang X, Martin-Puig S, Caron L, Zhu S, Shao Y, Roberts DJ, Huang PL, Domian IJ, Chien KR: **Human ISL1 heart progenitors generate diverse multipotent cardiovascular cell lineages.** *Nature* 2009, **460**(7251):113–117.
12. Domian IJ, Chiravuri M, van der Meer P, Feinberg AW, Shi X, Shao Y, Wu SM, Parker KK, Chien KR: **Generation of functional ventricular heart muscle from mouse ventricular progenitor cells.** *Science* 2009, **326**(5951):426–429.
13. Smart N, Bollini S, Dube KN, Vieira JM, Zhou B, Davidson S, Yellon D, Riegler J, Price AN, Lythgoe MF, Pu WT, Riley PR: **De novo cardiomyocytes from within the activated adult heart after injury.** *Nature* 2011, **474**(7353):640–644.
14. Messina E, De Angelis L, Frati G, Morrone S, Chimenti S, Fiordaliso F, Salio M, Battaglia M, Latronico MV, Coletta M, Vivarelli E, Frati L, Cossu G, Giacomello A: **Isolation and expansion of adult cardiac stem cells from human and murine heart.** *Circ Res* 2004, **95**(9):911–921.
15. Laugwitz KL, Moretti A, Lam J, Gruber P, Chen Y, Woodard S, Lin LZ, Cai CL, Lu MM, Reth M, Platoshyn O, Yuan JX, Evans S, Chien KR: **Postnatal isl1+ cardioblasts enter fully differentiated cardiomyocyte lineages.** *Nature* 2005, **433**(7026):647–653.
16. Smits AM, van Vliet P, Metz CH, Korfage T, Sluijter JP, Doevendans PA, Goumans MJ: **Human cardiomyocyte progenitor cells differentiate into functional mature cardiomyocytes: an in vitro model for studying human cardiac physiology and pathophysiology.** *Nat Protoc* 2009, **4**(2):232–243.
17. Takamiya M, Haider KH, Ashraf M: **Identification and characterization of a novel multipotent sub-population of Sca-1(+) cardiac progenitor cells for myocardial regeneration.** *PLoS One* 2011, **6**(9):e25265.
18. McKinney-Freeman SL, Naveiras O, Yates F, Loewer S, Philitas M, Curran M, Park PJ, Daley GQ: **Surface antigen phenotypes of hematopoietic stem cells from embryos and murine embryonic stem cells.** *Blood* 2009, **114**(2):268–278.
19. Wu SM, Fujiwara Y, Cibulsky SM, Clapham DE, Lien CL, Schultheiss TM, Orkin SH: **Developmental origin of a bipotential myocardial and smooth muscle cell precursor in the mammalian heart.** *Cell* 2006, **127**(6):1137–1150.
20. Lai L, Alaverdi N, Maltais L, Morse HC 3rd: **Mouse cell surface antigens: nomenclature and immunophenotyping.** *J Immunol* 1998, **160**(8):3861–3868.
21. Pfister O, Mouquet F, Jain M, Summer R, Helmes M, Fine A, Colucci WS, Liao R: **CD31- but Not CD31+ cardiac side population cells exhibit functional cardiomyogenic differentiation.** *Circ Res* 2005, **97**(1):52–61.
22. Leri A, Barlucchi L, Limana F, Deptala A, Darzynkiewicz Z, Hintze TH, Kajstura J, Nadal-Ginard B, Anversa P: **Telomerase expression and activity are coupled with myocyte proliferation and preservation of telomeric length in the failing heart.** *Proc Natl Acad Sci U S A* 2001, **98**(15):8626–8631.
23. Andersen DC, Andersen P, Schneider M, Jensen HB, Sheikh SP: **Murine "cardiospheres" are not a source of stem cells with cardiomyogenic potential.** *Stem Cells* 2009, **27**(7):1571–1581.
24. Smith RR, Barile L, Cho HC, Leppo MK, Hare JM, Messina E, Giacomello A, Abraham MR, Marban E: **Regenerative potential of cardiosphere-derived cells expanded from percutaneous endomyocardial biopsy specimens.** *Circulation* 2007, **115**(7):896–908.
25. Matsuura K, Nagai T, Nishigaki N, Oyama T, Nishi J, Wada H, Sano M, Toko H, Akazawa H, Sato T, Nakaya H, Kasanuki H, Komuro I: **Adult cardiac Sca-1-positive cells differentiate into beating cardiomyocytes.** *J Biol Chem* 2004, **279**(12):11384–11391.
26. Leri A, Kajstura J, Anversa P: **Cardiac stem cells and mechanisms of myocardial regeneration.** *Physiol Rev* 2005, **85**(4):1373–1416.
27. Nagai T, Matsuura K, Komuro I: **Cardiac side population cells and Sca-1-positive cells.** *Methods Mol Biol* 2013, **1036**:63–74.
28. Moretti A, Caron L, Nakano A, Lam JT, Bernshausen A, Chen Y, Qyang Y, Bu L, Sasaki M, Martin-Puig S, Sun Y, Evans SM, Laugwitz KL, Chien KR: **Multipotent embryonic isl1+ progenitor cells lead to cardiac, smooth muscle, and endothelial cell diversification.** *Cell* 2006, **127**(6):1151–1165.
29. Cai CL, Liang X, Shi Y, Chu PH, Pfaff SL, Chen J, Evans S: **Isl1 identifies a cardiac progenitor population that proliferates prior to differentiation and contributes a majority of cells to the heart.** *Dev Cell* 2003, **5**(6):877–889.
30. Tateishi K, Ashihara E, Takehara N, Nomura T, Honsho S, Nakagami T, Morikawa S, Takahashi T, Ueyama T, Matsubara H, Oh H: **Clonally amplified cardiac stem cells are regulated by Sca-1 signaling for efficient cardiovascular regeneration.** *J Cell Sci* 2007, **120**(Pt 10):1791–1800.

Establishment of a reborn MMV-microarray technology: realization of microbiome analysis and other hitherto inaccessible technologies

Harshita Sharma[1], Yasunori Kinoshita[1], Seiichi Fujiu[1,2], Shota Nomura[1], Mizuho Sawada[2], Shamim Ahmed[1,6], Masaki Shibuya[3], Kosaku Shirai[3], Syota Takamatsu[4], Tsuyoshi Watanabe[4], Hitoshi Yamazaki[5], Ryohei Kamiyama[1], Tetsuya Kobayashi[1], Hidenao Arai[1], Miho Suzuki[1], Naoto Nemoto[1], Ki Ando[1,7], Hidekazu Uchida[1], Koichiro Kitamura[2], Osamu Takei[3] and Koichi Nishigaki[1*]

Abstract

Background: With the accelerating development of bioscience, the problem of research cost has become important. We previously devised and developed a novel concept microarray with manageable volumes (MMV) using a soft gel. It demonstrated the great potential of the MMV technology with the examples of 1024-parallel-cell culture and PCR experiments. However, its full potential failed to be expressed, owing to the nature of the material used for the MMV chip.

Results: In the present study, by developing plastic-based MMVs and associated technologies, we introduced novel technologies such as C2D2P (in which the cells in each well are converted from DNA to protein in 1024-parallel), NGS-non-dependent microbiome analysis, and other powerful applications.

Conclusions: The reborn MMV-microarray technology has proven to be highly efficient and cost-effective (with approximately 100-fold cost reduction) and enables us to realize hitherto unattainable technologies.

Keywords: Microarray, Multi-parallel reactions, Multi-conditioner, Lysozyme crystallization, Microbiome analysis

Background

For drug discovery, high-throughput screening is ultimately required. This is similar to those in various other scientific fields in which optimal conditions and combinations must be verified. The rising cost of reagents has motivated the development of microminiaturization technologies such as microarrays and microfluidics. For example, microplates have evolved from 96–384 wells to up to 3456 wells, lowering assay cost owing to the reduction of the amount of reagents and samples [1,2]. However, even the 3456-well plate requires μL volumes for handling, due to adsorption and evaporation losses.

In recent years, many digital PCR systems such as the Fluidigm and OpenArray systems have provided a leap toward the miniaturization of assay vessels by permitting gene expression analysis at the nL scale [3,4]. Moreover, picoliter-droplet digital PCR system is capable of analyzing millions of picoliter-sized droplets for molecular genotyping [5]. However, these systems are highly specialized for specific purposes, requiring specific liquid handlers and digital PCR instruments [6]. In addition, multistep reactions are, in principle, difficult to perform on these platforms, confining their applicability to molecular genotyping and a few other assays.

Microfluidics and bead-based technologies have been developed and are currently still expanding their possibilities [7-10]. These technologies have the potential to deal with minute amounts of sample and, particularly in microfluidics, support multistep reactions and detection. However, there are some difficulties in microfluidics in supporting the high density of parallel operations owing to the difficulty in supplying a uniform force to drive liquids [11], avoiding losses from surface adsorption, and

* Correspondence: koichi@fms.saitama-u.ac.jp
[1]Department of Functional Materials Science, Graduate School of Science and Engineering, Saitama University, 255 Shimo-okubo, Saitama 338-8570, Japan
Full list of author information is available at the end of the article

other operations [12,13]. In bead technology, parallel bead handling requires highly specialized machines that cannot be readily used for general purposes [14], although they are useful for specific purposes such as random multiple sequencing [15]. In summary, current innovative methods have their own merits but require further development to reach their goal such as high cost performance.

Here, we describe a reborn MMV-microarray technology, i.e., microarray with manageable volumes, operated without pipettes. Pipette-free operation offers several benefits, for instance, rapid operation, handling of nanoliter volumes, avoiding adsorption and evaporation losses during operations, and dispensing with elaborate equipment. This technology was originally developed using polyacrylamide gel [16,17] and had the main drawbacks of fragility and difficulty of accurate handling, although it has already enabled us to culture bacteria and perform PCR reactions [17]. The method reported here could overcome these defects, thereby strengthening the MMV technology.

Microarray technology has permitted a high density of parallelism and the development of various methods [18,19]. However, this technology has the constraint that it essentially comprises surface reactions and is difficult to use for successive and independent multistep reactions. The MMV technology solves this problem and has a potential to support novel applications. In the present study, this potential to enhance the basic MMV technology is described.

Results

Sub-μL liquid handling technique

In this study, we developed novel necessary tools and principal technologies for pipette-free sub-μL volume handling in MMV, which were not available for MMV made from gel (Figure 1a). The main principle of MMV is "aperture-to-aperture transfer." Parallel transfer of thousands of solutions was performed by centrifugation of a face-to-face stack of two MMVs with each well tightly contacting the other via an adhesive spacer (Figure 1b). Initial charging with sample was performed as shown in Figure 1c. Similarly, various operation techniques have been developed, including those described in our previous report [17], and are collectively presented in Table 1 (for detail, see Additional file 1: Table S1, Additional file 2: Figure S1, Additional file 3: Figure S2, Additional file 4: Figure S3, Additional file 5: Figure S4, Additional file 6: Figure S5, Additional file 7: Figure S6, Additional file 8: Figure S7, Additional file 9: Figure S8, Additional file 10: Figure S9).

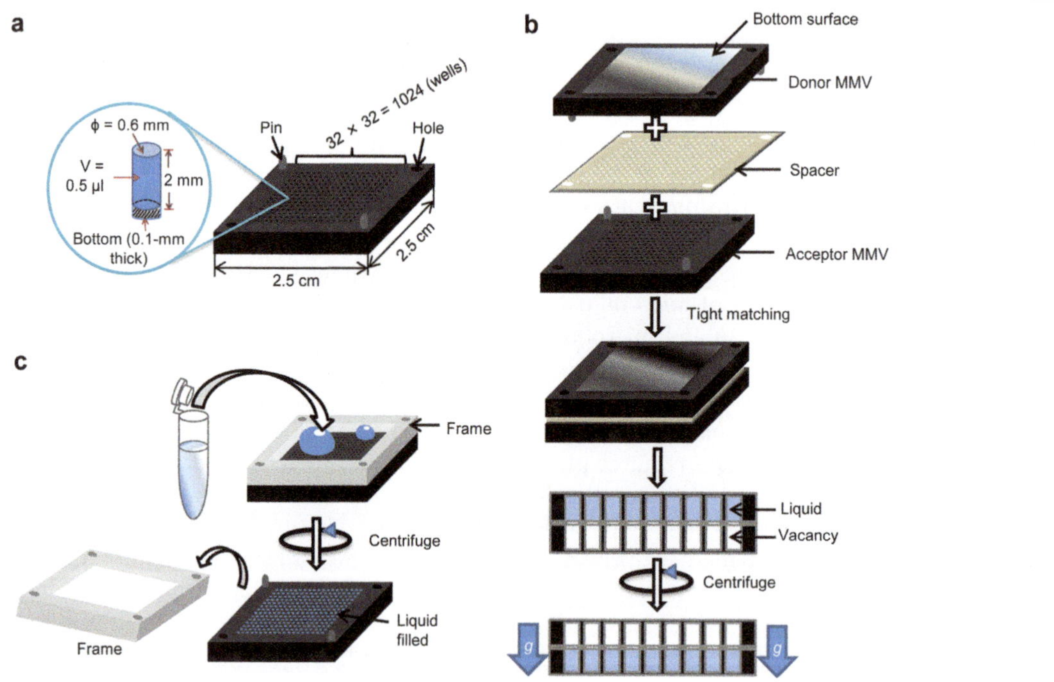

Figure 1 MMV chip construction and basic operation. (a) Dimensions of an MMV chip. The MMV chip has a dimension of 2.5 cm × 2.5 cm × 0.2 cm and is fabricated with 1024 wells of 0.6-mm diameter and 2-mm depth, giving a volume of 0.5 μL of solution. **(b)** Usual transfer (Z-mode) of solution. Well-to-well transfer of solution is performed using a silicone/urethane spacer that enables a leak-proof flow of solution from the donor to the acceptor MMV. **(c)** Schematic representation of initial input of solution into wells (I-mode). A urethane/silicone frame is used to confine the solution to the top surface of the MMV.

Table 1 Basic transfer operations needed for MMV

Operation (Pipette-dependence mode[a])				Purpose	Details
PF	1	Input (I-mode)		Liquid charging into MMV wells	Figure 1c
	2	Output (O-mode)	a. Transfer	Complete ejection of liquid in wells	Figure 1b and Additional file 6: Figure S5
			b. Washing		
	3	Square transfer (S-mode)		Fractional volume transfer	Additional file 2: Figure S1
	4	Z-direction transfer (Z-mode)		Normal transfer of liquid into wells	Figure 1b and Additional file 5: Figure S4
	5	X-direction transfer (X-mode)		Division of liquid in all wells	Additional file 7: Figure S6
	6	Magnetic beads transfer (M-mode)	a. Beads recovery	Beads recovery	Additional file 8: Figure S7
			b. Division of liquid	Division of liquid in all wells	Additional file 9: Figure S8
	7	Filtering (F-mode)		Filter-selective transfer	Figure 2a
PD	8	Pipette-dependent transfer (P-mode)	a. Manual	Manual transfer of liquid or beads	Additional file 10: Figure S9
			b. Robotic	Robotic transfer of liquid or beads	Additional file 3: Figure S2

[a]Pipette-free (PF) or pipette-dependent (PD).

For example, selective transfer of solutions (F-mode) to specific wells was achieved by exploiting specifically patterned filters (Figure 2a). In the MMV square transfer (S-mode) method, different volumes of solutions were transferred using the MMV-matching thin filter device (Additional file 2: Figure S1). Although in general, "well-to-well transfer" had been adopted in the MMV technology and could accomplish most necessary operations, we also developed an automated liquid handling (P-mode) system to achieve versatility and manageability, expanding the ability of the MMV system (Additional file 3: Figure S2).

Generation of replicas of cell-to-cell (C2C) and DNA-to-DNA (D2D)

To demonstrate the efficient multistep parallel transfer of solutions, MMVs containing DNA and cells in desired patterns were fabricated and used to test replica generation (Additional file 4: Figure S3a and b). The MMV chip containing *E. coli* cells in a pattern (cherry blossom) was subjected to successive five rounds of forward [from the donor MMV (original) to the acceptor MMV (replica)] and backward transfer [transfer back of solution from the acceptor MMV (replica) to the donor MMV (original)] with replacement of each acceptor MMV chip. The vacant acceptor chips were then filled with Davis media and incubated at 37°C for 21 h. As shown in Figure 3a, every replica chip reproduces the original cherry blossom pattern beginning with the vacant state, demonstrating the effectiveness of the replica generation technique. As the original itself can be readily amplified, we can produce replicas perpetually. Close observation of these patterns indicated that 1% or fewer of the wells stochastically failed to amplify cells within this incubation time. In an auxiliary experiment, a very small amount of aliquot (approximately 0.1% of the total solution) was shown to be left in each vacant well (Additional file 4: Figure S3c and d) and served as a seed for the next bacterial culture. Therefore, although the current system is subject to stochastic errors at a level of 0.1%, it can be used with careful monitoring of each well's state (presence of an aliquot) under a microscope. Otherwise, it can be safely used by making multiple copies (3–4) of each sample. This situation can be greatly improved with the use of a much finer, though more expensive casting mold to generate a next-generation MMV chip.

Similar repetitive forward and backward transfer experiments were performed with DNA solution, and vacant MMV chips were filled with a reaction buffer and then subjected to PCR. As shown in Figure 3b, the 10th replica was completely replicated, showing that a DNA replica array was also successfully generated. In an independent experiment, cross-contamination between wells was shown to be eliminated by the introduction of a packing spacer (of the same shape as the MMV chip and with a thickness of 0.8 mm) (Figure 1b and see Additional file 5: Figure S4 for actual unit transfer). To our knowledge, this is the first success in perpetual DNA microarray replica formation (DNA microarray replication effective only once for all was reported in a previous study) [20].

Generation of 1024 conditions (multiple-conditioner)

To exploit the array nature of MMV, we generated 1024 different conditions by applying the "2^N method" [17] (Additional file 11: Figure S10 and Additional file 12). To investigate the crystallization conditions of lysozyme, ten types of template MMVs were used to generate 2^{10} (=1024) different conditions (Figure 2a). The conditions ranged from pH 2.9 to 9.6 and from 0 to 1.5 M NaCl (Additional file 13: Figure S11a). Typical crystal images that were obtained are shown in Figure 2b and Additional file 13: Figure S11, and they comprised type I (0.1–0.2-mm large tetragonal), type II (tetragonal larger than 0.2 mm),

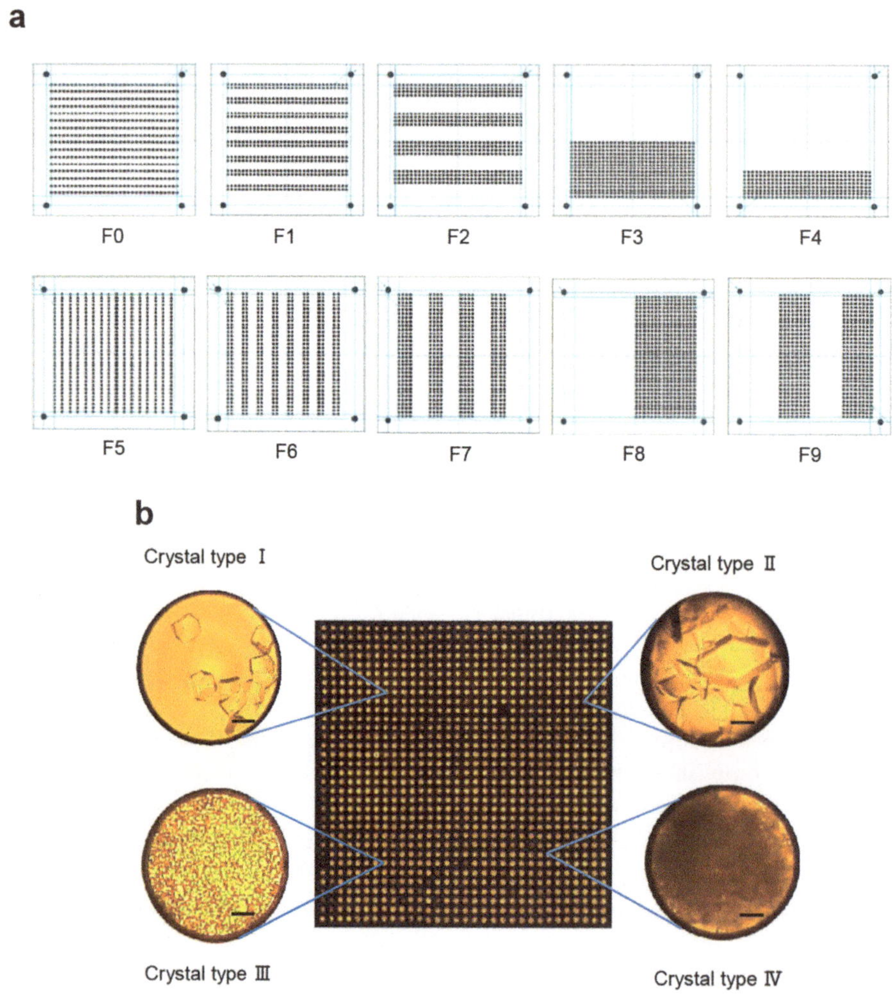

Figure 2 Generation of 1024 different conditions. (a) A set of filters used to generate multiple conditions in an MMV chip. In each of these filters, if the N-bit is 1 ("go") for the $(i + 1)$-th least significant bit in the binary number, then the corresponding position of filter F_i is a hole (allowing the liquid to pass through). On selective addition of 10 species of materials to an acceptor MMV using these filters, 1024 different conditions (corresponding to 1024 different binary numbers) were generated (for detail, see Additional file 11: Figure S10). **(b)** Lysozyme crystallization in MMV. The 1024 different conditions were generated for determining the optimum crystallization conditions. A combination of pH differences (2.9–9.6) and NaCl concentration differences (0–1.5 M) was examined (see Methods). In the inset, magnified images are shown for some MMV wells. The scale bar is 100 μm.

type III (0.03 mm or smaller microcrystalline), and type IV (needle-like) crystals. This result showed the most optimal conditions (pH 8.6 and 0.6 M NaCl) for generating the largest crystal (0.65 mm; Additional file 13: Figure S11a and b) of lysozyme as well as the conditions under which other types of crystals can be obtained. This crystal size of lysozyme seems to be relatively large, although one of 1.6 mm has recently been reported [21]. The sample amount required for a single condition was in the mg range for that experiment, but only 10 μg (100-fold smaller) for ours.

Microbiome analysis without dependence on NGS

Identifying all the members of microbiome found in an ecosystem is important for genuine understanding. Accordingly, next-generation sequencing (NGS), which is a powerful, but expensive technique has been used for microbiome analysis [22-24]. The amount of information obtained using this method is often too great for the mere identification of organisms.

Another method could identify the species of organisms with cost-effective performance, i.e., genome profiling (GP) [25]. In this study, we investigated the possibility of MMV-dependent GP analysis of microbiomes. Its entire procedure is shown in Figure 4a. Several crucial features are included in this method; they are as follows: *i)* limiting the dilution of organisms constituting a microbiome entity into a single cell, *ii)* performing single-cell PCR by random PCR, *iii)* performing GP analysis using micro-temperature gradient gel electrophoresis (μTGGE), and *iv)* obtaining the sequence information of each cloned

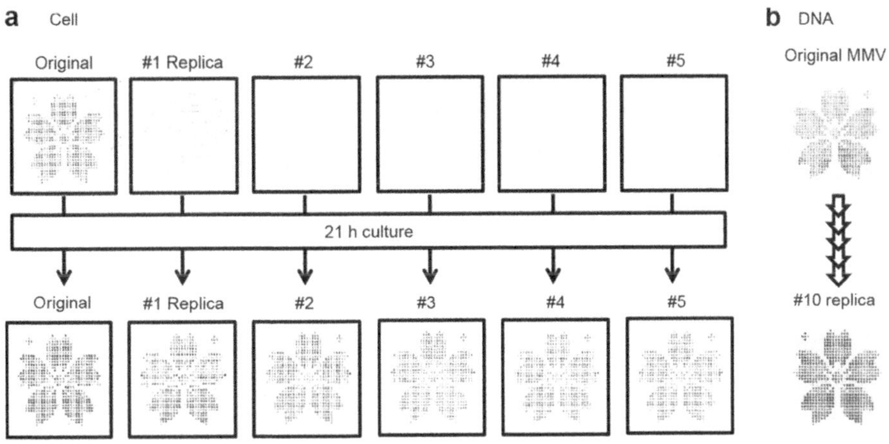

Figure 3 Generation of the MMV replica. (a) Cell replica. The liquid in the original MMV containing *E. coli* cells (harboring the GFP plasmid) in which the wells collectively resemble a cherry blossom pattern was transferred and returned, leaving a minute droplet in each replica MMV (upper rows #1 ~ 5). All these MMVs were added with culture media and incubated at 37°C for 21 h to regenerate cells (lower row). **(b)** DNA replica. The replica image is that of the 10th replication (PCR amplified), whereas the original MMV is shown just after the 10 rounds of replica generation operation. Both are stained with SYBR green I.

organism. The first feature can be achieved with the aid of multiple MMV microvessels to reduce the quantity of reagents required. In this study, dispersion of some of the flocculated bacteria was found to be a problem. However, this problem was substantially addressed, as shown in Methods (enzymatic treatment and sonication, although approximately 1% of the number of flocs remained after this treatment). The second feature addresses the difficulty of single-cell PCR [26] and the lack of a common PCR primer for all microbes. Fortunately, random PCR is in effect, which is a universal PCR that can be applied to any organism using the same primer [27]. Thus, we could easily overcome this problem. Much experience with single-cell PCR has been accumulated both worldwide [28-30] and in our laboratory [26].

Thus, it is no longer difficult to perform single-cell PCR, if precautions are taken to avoid adsorption loss and DNA contamination. The third feature (GP analysis using μTGGE) is supported by prior experience [25,26,31-34]. The final feature can be realized by another benefit of the GP method, i.e., the existence of easily assignable (from its band pattern in the electrophoresis) and PCR-recoverable DNA, namely, a commonly conserved genetic fragment (ccgf) [31] that can be amplified from DNA extracted from a gel band using the same primer adopted for random PCR.

With solutions to each technical challenge, oral microbes were used as a test case and subjected to the entire NGS-non-dependent microbiome analysis (NNMA) shown in Figure 4a.

Figure 4b clearly demonstrates that some MMV wells generated distinct genome profiles (original data in Additional file 14: Figure S12) by the GP method, suggesting a successful microbial cell distribution in different MMV wells and extraction and amplification of DNA in each well. The DNAs recovered from the μTGGE gel as ccgfs were further analyzed by conventional cloning and sequencing analysis, and the resulting DNA sequences were subjected to BLAST analysis (Figure 4c). In this analysis, *Leuconostoc citreum, Haemophilus parainfluenzae, Rothia dentocariosa, R. mucilaginosa, Aspergillus nidulans, Cronobacter sakazakii,* and *Psychrobacter sp.* were tentatively identified based on sequence similarity with organisms in the NCBI database (see Additional file 15: Table S2). Because all these microbes can be assigned as stable or transient oral inhabitants, this finding strongly supported the inference that NNMA analysis was performed without obstacles.

The genome profiles and species identification dots (*spiddos*) that were obtained in this analysis (Figure 4b) were subjected to clustering analysis (Figure 4d), in which clones were assigned to possible species based on their own ccgf DNA sequence (Figure 4c and Additional file 15: Table S2).

In this experiment, not all but only some clones (47 of them) appearing in an MMV plate were processed to test the feasibility of the entire NNMA process because this number was sufficient to confirm experimental success. The same microbiome analysis was performed three times, with all tests performed using possible clones and their genome profiles (*spiddos*) (Additional file 16: Table S3), thereby supporting the effectiveness of the NNMA method.

Discussion

In this paper, we have established the basic technology for the MMV handling and three typical applications of the MMV method: replica generation, generation of

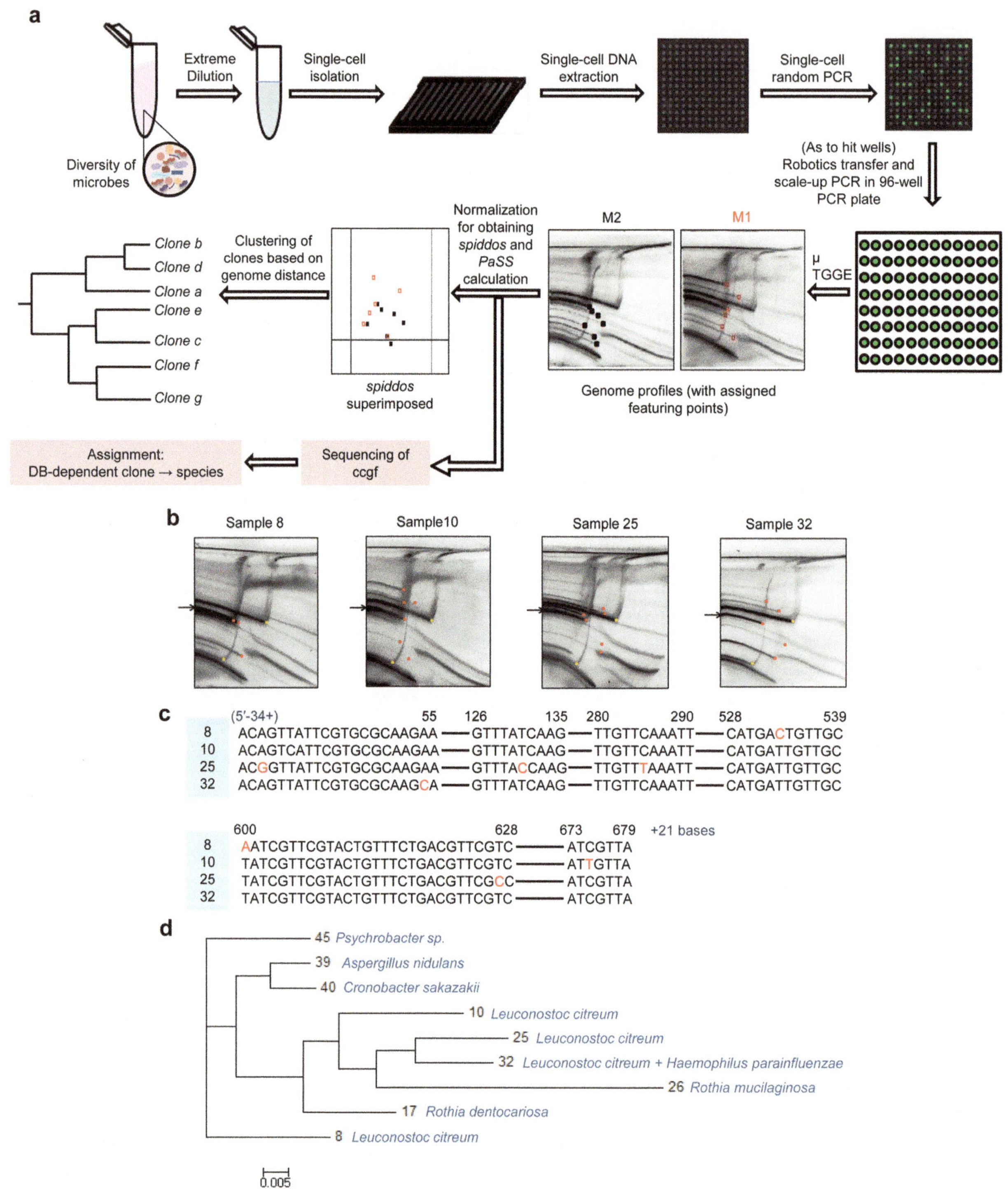

Figure 4 (See legend on next page.)

(See figure on previous page.)
Figure 4 Schematic representation of NGS-non-dependent microbiome analysis (NNMA). (a) A sample containing a complex mixture of microbes was serially diluted to a concentration corresponding to one or fewer cell/well as expectation value (in Poisson distribution). DNA extraction and random PCR were performed on the same MMV chip. The PCR product was scaled up using PCR in a 96-well plate and analyzed using µTGGE. In the genome profile obtained, feature points were assigned and processed using computer-aided normalization, generating species identification dots (*spiddos*). Based on *spiddos*, a genome distance, d_G [defined as 1 – *PaSS* (pattern similarity score)], was obtained for each pair of microbes (Additional file 12), and a clustering tree was then generated using d_G. **(b)** Genome profiles of four samples with feature points assigned. Red dots represent feature points (pre-*spiddos*) and yellow dots, internal reference points. Arrow shows possible commonly conserved genetic fragment (ccgf). **(c)** Partial sequences of ccgfs. Point mutations are shown with red letters. Completely matching regions are shown with lines. **(d)** Clustering tree for nine samples. Here, tentative microbial species are assigned from the sequence obtained for ccgf. The tree was constructed using Phylip 3.69 and MEGA 5.1 software.

1024 conditions, and microbiome analysis. The original solution (containing DNA or cells) can be repeatedly used for replica generation. According to the requirement, replicas containing defective wells can be repaired by the operation of a nanoliter-dispensing robot (see Additional file 3: Figures S2 and Additional file 17: Figure S13). In effect, this technology enables us to produce DNA, cell, and even protein arrays easily and endlessly. The quality of the replica thus generated is sufficiently reliable if MMV is loaded with samples in multiples of 10 (because the expected defect rate is <1 in 10, implying that more than 9 of 10 wells are intact). In addition, the use of a higher-quality MMV chip mold (a matter of cost) is empirically known to improve the defect rate greatly. Moreover, monitoring the content of each individual well (parameters such as volume, temperature, and conductivity) is within our scope and achieved by semiconductor technology.

Another important point is that the success of cell-to-cell (C2C) and DNA-to-DNA (D2D) replications implies that by combining cell-to-DNA (C2D) (Figure 4) and DNA-to-peptide/protein (D2P) [17] (Additional file 18: Figure S14), we can achieve C2D2P, thus beginning with cells and ending with a protein (peptide) microarray, an outcome that was never attainable before.

We generated 1024 conditions for lysozyme crystallization. This application of MMV is unexpectedly powerful and widely applicable because it circumvents lengthy and time-consuming preparation processes and spares reagents. The crystallization study described here was performed only to demonstrate the utility and capacity of the 1024-well MMV technology (because such research has already been performed [17,21,35-37]), and to show that a high volume of information can be readily obtained with high performance. In this experiment, we obtained crystals that were larger than 200 µm in a much smaller volume (0.5 µL) than that used in previous studies [21,35,36], thus sparing samples and reagents by about 100-fold in comparison with conventional approaches, although a more competitive (~150-fold) reduction technique has been reported that employs 81 microwells and a specially fabricated apparatus [37]. Additionally, we evaluated the reproducibility of the optimal crystallization conditions at the 50 µL scale, resulting in crystals that were around 0.65 mm long (Additional file 13: Figure S11b). If necessary, a more statistically reliable and realistic approach can be applied to examine each condition in quadruplex or octuplex. Even in an octuplex test, more than 100 conditions can be readily examined in a single MMV chip test, with only 4 µL (0.5 µL × 8) consumed per condition.

The conditions that can be examined are not only those for crystallization of proteins, but also those for the combinatorial effect of biological factors (hormones, cytokines, and others) on induction of cell differentiation such as iPS (induced pluripotent stem cell) induction. In the field of medicinal chemistry, more effective drug combinations (synergy effect) may be easily found owing to the readiness in generating multiple conditions with ease and low cost. Because chemical and biological experiments originally require surveys over a wide experimental range and multiple assays, the problems of cost, labor, and time-consuming processes, which have inhibited desirable experiments, can be revolutionarily solved by the MMV approach.

Furthermore, the NNMA system is a potential method for identifying the component species of microbiomes. To date, we have had no effective method to analyze a huge microbiome population mainly because we lacked an appropriate container to accommodate a large number of microbes separately and efficiently, i.e., accommodating a single organism per well and enabling a series of DNA extraction processes, PCR amplification, and DNA sequence analysis in multiples with an affordable cost in reasonable time. This situation was, for the first time, overcome by the advent of NGS, a technology worthy of the intensive attention of microbiologists. However, NGS is not always effective; it is very expensive, and the raw data generated is generally too difficult to process except by bioinformatics specialists. A significant amount of data accordingly remains unprocessed. In contrast, the NNMA system enables microbiologists to design and complete their study on their own.

The following potential advantage of NNMA over NGS further supports this approach: NNMA can deal with an organism as a whole, whereas NGS yields a mixture of genome fragments for analysis. Thus, the latter renders it impossible to further investigate the organism, as it is theoretically difficult to detect extreme minorities such as one or fewer in 1000, whereas NNMA offers the possibility of detecting, in principle, a minority of 1 in 10000 if twenty or more 1024-well MMV plates are used in the dilution with the expectation of 0.5 cell/well. This is the essential difference between the NGS and NNMA approaches.

A few points remain to be improved for the NNMA analysis, including an increase in the output of μTGGE with sample scale reduction. The realization of scale reduction will eliminate the scale-up step required in the current NNMA process (Figure 4a). Another goal is to develop a means of complete deflocculation (although this is a classical and difficult problem [38,39]), because in the present experiment, flocs were still present (Additional file 19: Figure S15), consistent with our observation that DNAs from a single well of MMV are sometimes assigned to different species (Additional file 15: Table S2). This observation is noteworthy because the current NNMA system, although it remains to be improved in techniques outside the MMV technology proper, has already succeeded in extracting DNAs from microbes, from either single or flocculated cell(s), and amplifying, analyzing, and assigning them to species using MMV-based technology—a feat never achieved by conventional approaches such as microplate-based technologies.

In this study, all necessary basic skills and tools for the MMV operations have been described. Moreover, highly usable application methods have been proposed.

In addition, there are other potentially powerful applications. Among them are the function-based selection of possible drug molecules termed as panning on microarray MMV (POMM) (Additional file 18: Figure S14) and All-In-One and All-At-Once ($A^I/_AO$) (Table 2). The former (POMM) is a multistep reaction experiment, comprising PCR, replica formation, enzymatic reactions (transcription/translation), and fluorescence detection supported by multiple operations of well-to-well transfers. The latter is simple and wide ranging, such that the diversities of its usage are open to researchers, particularly when pre-charged MMVs with various samples such as buffers, DNAs, enzymes, bead-bound antibodies, and cells are available from commercial sources in the future. If necessary, all one has to do is to purchase a pre-charged MMV with diverse conditions and overlay it on an experimenter's MMV (containing one's own samples). Similarly, an analyst can identify the optimal medium for culturing a particular cell (iPS or cancer cell) using an MMV of a combinatorial set of factors (generated by the 2^N method).

The MMV technology has the merits of simple operation, rapidity, economy, reduced error rate, and flexibility to both classical microscope-based technologies (phase-contrast, fluorescence, and others) and emerging semi-conductor-based technologies (Additional file 20: Figure S16).

Conclusions

The plastic-based MMV chip enabled us to develop an easy-to-handle and quantitative system in handling sub-μL volumes. It also facilitated the development of a set of necessary unit technologies for MMV operation such as input, output, division, filtering of solutions, and others. This development has led to genuinely novel methods

Table 2 MMV applications developed to date

Application	Content	Level achieved[b]	Comment
Multiple (1024) conditions generation (Multiple-conditioner)	Generation of 1024 different conditions in MMV for lysozyme crystallization.	+++	Applicable to iPS primary induction factor screening
Semi-infinite replica Formation (D2D, C2C, and D2D2P)	Replication of DNA and cells in MMV perpetually. Besides, DNA processed to protein (D2D2P) replica.	+++	Proteins replica can be perpetually generated by two steps of D2D and D2P, i.e., D2D2P.
NGS-non-dependent microbiome analysis (NNMA)	Single-cell isolation, DNA extraction, single-cell random PCR of microbiome samples in MMV, and processing of PCR products by the Genome Profiling (GP) method.	+++	Serves as single cell isolation and analysis tool
Multistep function-based screening (POMM)	Operation of all steps involved in DNA amplification, *in vitro* transcription and translation; identification of functional peptides in MMV.	++	Screening tool with samples addressed
All-In-One/All-At-Once assay ($A^I/_AO$)	Screening of apoptosis-inducing peptides against cancer cells.	+	Applicable to monoclonal antibody screening

[b]+++, successfully achieved; ++, whole process developed with tentative confirmation; +, system developed with successful preliminary experiment.

such as 1024-parallel Cell-to-DNA-to-Protein/Peptide (C2D2P) and NNMA.

In addition, MMV can reduce experimental time and cost drastically (Additional file 21: Table S4).

Methods

The most basic nature of the MMV technology can be represented by the facts that the liquid in an MMV well does not fall by the gravity and is less volatile due to the narrow aperture area, easily thermo-conductive due to the small volume, and readily mixed by convection.

MMV chip

Polycarbonate (PC) MMV chips were developed in a joint project, involving most of the authors, which was sponsored by JST, and were finally obtained from Enplas Corporation. The MMV chip has an overall dimension of 2.5×2.5 cm^2, with diameter, depth, and volume of 0.6 mm, 2 mm, and 0.5 μL, respectively (Figure 1a). PC MMV is classified into the two following types according to the chip's bottom: *i)* PC molded and *ii)* transparent sheet pasted. For microscopic observation, the second type is used. Silicone rubber or PDMS was also used for fabricating flexible well-size MMV chips [17].

Charging and discharging of liquid into MMV (I-mode)

Initial charging of sample into wells can be achieved by centrifugation force as shown in Figure 1c; with excess solution covering the top surface of MMV, it was centrifuged at 1400 *g* for 1 min to fill all the wells with solution and evacuate the remaining solution from MMV wells (Figure 1c).

As I-mode is a kind of dump transfer, a specific one-by-one transfer is carried out by P-mode as written in Table 1. In order to prevent evaporation, MMV was sealed with silicone tape (Rescue Tape), and to remove liquid from MMV wells, centrifugation was applied to the MMV placed upside down (at 1400 *g* for 2 min) (Additional file 6: Figure S5). One MMV chip can be repeatedly reused by several rounds of washing with ultrapure water and/or 70% ethanol by centrifugation and ultrasonic vibration. Liquid can be transferred from one well of an MMV to another well of an opposing MMV by centrifugation. A silicone-PET-adhesive spacer (0.2-mm thick with 0.6-mm wide holes) (Finetech) was fixed on the donor MMV and then the acceptor MMV was firmly laid against it (Figure 1b). The liquid moves from the donor to the acceptor MMV through the thin channel in the spacer without evaporation or adsorption losses.

Further, to add a droplet solution, S-mode operation (Additional file 2: Figure S1) was applied. This type of spacer was repeatedly used, and, if necessary, the thickness and/or hole patterns of the spacer were changed. Layers from bottom to top, namely MMV (used as support), a bottom-side-adhesive film attached to the MMV, and a spacer of desired thickness and hole pattern, were constructed and matched to charge the wells with the solution in the square method.

Selective addition to specific wells (F-mode)

For this purpose, different types of filter spacers were produced (Finetech) and used (examples in Figure 2a). These spacers are adhesive and contain different surface patterns of holes and without holes, enabling selective transfer of samples to the acceptor MMV. As an alternative solution, template MMVs having a specific pattern of open and closed wells can be used for the same purpose. Such templates could be formed using PDMS and a laser patterning machine [17].

MMV coating

To prevent adsorption of biomolecules on MMV surfaces, MMVs were mainly surface coated with a solution of 0.1% (w/v) bovine serum albumin (BSA), obtained from Sigma-Aldrich, by centrifugation (at 1400 *g* for 1 min). These were used for cell culturing, PCR, and other operations. The aperture of this MMV was sealed with silicone tape and incubated overnight at 4°C. After incubation, the solution was removed by centrifugation (O-mode), and the MMV was dried at room temperature for 1 h.

Replica formation

A typical example of replica formation is shown in Additional file 4: Figure S3, where a cherry blossom pattern was formed on MMV using two colors (Elements 1 and 2). Element 1 was charged in E1-MMV using a #1 filter with a thickness of 0.3 mm and a hole diameter of 0.5 mm for expression of the cherry blossom pattern. E2-MMV contained Element 2 in the complementary pattern of Element 1. Here, Element 1 was a DNA (Aβ 42-binding peptide gene)-containing PCR mixture, whereas Element 2 contained no DNA. The PCR mixture comprised 1× FB I buffer (Takara), 250 μM dNTP mixture, 0.04 μM CA primer (5′-CAACACACCACCCACCC AAC-3′), and 0.025 U/μL SpeedSTAR HS DNA polymerase (Takara). Both E1- and E2-MMVs were subjected to PCR and then combined to generate the original MMV. This original MMV was used for the replica formation by a stamping method wherein a spacer with a thickness of 0.8 mm and hole diameter of 0.6 mm (Finetech) was interposed between the original (donor) and the acceptor MMVs. The centrifugal transfer (1300 *g* for 1 min) from the donor to the acceptor MMV was reversed just after the first transfer, resulting in the generation of an MMV in which a very small aliquot remained. This aliquot can serve as a seed for PCR amplification (DNA), thus reproducing a replica.

A replica of cells is generated by a similar operation as described above. For this demonstration, GFP-plasmid-harboring *Escherichia coli* cells were used and cultured in LB broth [0.5% (w/v) yeast extract, 1% (w/v) tryptone, and 1% (w/v) NaCl with 0.1 µg/µL ampicillin and 1 mM IPTG]. Here, Element 1 corresponds to GFP-harboring *E. coli* and Element 2 to LB medium without cells.

MMV setup in a conventional thermocycler

In a conventional thermocycler (Bio-Rad C1000 Touch), for better heat conduction, a copper plate was inserted between the PCR block and the MMV chip. In addition, MMV was sealed with silicone tape, occluded with aluminum foil, and installed in the thermocycler. Silicone rubber was used to closely pack the MMV to protect against pressure.

Imaging of MMV reaction products

The staining of PCR products with SYBR Green I (Lonza) was performed using a shallow well MMV (0.1 µL/well) according to the manufacturer's instructions. Before the transfer (1300 *g* for 1 min) of the dye from the shallow MMV to the PCR-product-containing MMV, the contents of the latter were reduced in advance by dry air-driven evaporation (lid was left open on a clean bench for few minutes) to make space for the solution to be transferred. Further, MMV with stained DNA was monitored using a fluoroimager (Molecular Imager FX, Bio-Rad). MMV with *E. coli* cells was visualized by FITC staining and monitored using the same fluoroimager.

Lysozyme crystallization

We applied the 2^N method to test 1024 different conditions for lysozyme crystallization (Additional file 11: Figure S10 and Additional file 12). Lyophilized powder of lysozyme (Sigma-Aldrich) was dissolved in 5 mM MES buffer (pH 6.6) [final concentration of 28% (w/v), final volume of 1 mL]. This solution was carefully vortexed to eliminate any visible flocs of lysozyme and filtered through a membrane filter (0.2-µm pore size), and the supernatant was collected for use in the crystallization experiment. Stepwise different pH values were generated with filter spacers of F0, F1, F2, F3, and F4 (Figure 2a; S-mode operation) charged with approximately 40 nL of 1 M sodium citrate buffer (pH 2.9), 1 M sodium acetate buffer (pH 3.9), 1 M MES-NaOH (pH 6.6), 1 M Tris–HCl (pH 8.5), and 1 M glycine buffer (pH 9.6), respectively. For the condition of a different ionic strength, filter spacers of F5, F6, F7, F8, and F9 were charged with approximately 40 nL of 1 M, 2 M, 3 M, 4 M, and 5 M NaCl solution, respectively. To equalize the volume of solution added to each well, distilled water was added to the wells not charged with any of F0–F9 using the templates ($\bar{F0}-\bar{F9}$) complementary to F0–F9 (the well position of F_i is not a well in $\bar{F}i$ and inversely, a nonwell position of F_i is assigned as a well in $\bar{F}i$; $i = 0, 1, 2$). Finally, lysozyme solution (280 mg/mL) was added (approximately 40 nL) to the wells. The crystallization was performed by leaving the MMV in the middle of a closed container (10 cm × 10 cm × 10 cm) with its interior soaked with 500 mL of 2 M ammonium sulfate at 20°C for one day to one week. Each well was monitored with an inverted microscope (IM; Olympus, Tokyo, Japan).

NGS-non-dependent microbiome analysis

Sample collection

A saliva sample was collected from one of the authors (with their consent), by themselves 5 h after brushing and with no intake of food in this period. This sample-collection was performed with the permission of the Saitama University Ethical Committee on Human-related Research. For sample collection, two small, sterile sponge pieces were kept bitten in the mouth for 5 min and then dipped in 1 mL of Dulbecco's phosphate buffered saline [DPBS (–)] (pH 7.4) (Wako Pure Chemical Industries, Ltd.). The saliva was squeezed out of the sponge pieces and was dissolved in PBS and subjected to centrifugation for 5 min at 1470 *g*. The supernatant was discarded; the pellet was dissolved in 200 µL of DPBS (–) buffer and stored in 13% glycerol at –80°C.

Deflocculation and cell counting

To dissolve the flocs of cells, 1 µL of the oral sample was treated with 9 µl of DPBS (–) buffer and 200 pg/µL proteinase K. This solution was sonicated for 40 s, incubated at 55°C for 1 h and at 95°C for 5 min, and then again sonicated for 40 s. This deflocculation-treated solution was subjected to staining with 10000× diluted SYBR gold solution, and cells were counted using a Neubauer improved cell counting chamber.

Single-cell DNA extraction

The oral microbes were suspended in bacterial cell lysis reaction buffer containing 10 mM Tris–HCl (pH 8.0), 0.025% BSA, 0.10% Tween 20 (Wako Pure Chemical Industries, Ltd.), and 0.75% polyethylene glycol 8000 (PEG8000, Sigma-Aldrich) to obtain a concentration of 300 cells/700 µL and to lyse the cells. Moreover, achromopeptidase [0.004% (w/v), Wako Pure Chemical Industries, Ltd.] solution was added to this extraction mixture. The resulting solution was immediately charged into a shallow well-type MMV (0.14 µL/well capacity) and then transferred to a full volume-type MMV (0.5 µL/well capacity) by Z-mode transfer and incubated at 37°C for 1 h and then at 95°C for 10 min.

Single-cell random PCR

PCR mixture containing 200 μM dNTPs (N = G, A, T, or C), 0.7 μM primer pfM 19 (5′-CAGGGCGCGTAC-3′), 10 mM Tris–HCl (pH 8.3), 50 mM KCl, 1.5 mM $MgCl_2$, 0.03 U/μL Taq DNA polymerase (Takara), and an additional solution of 0.025% BSA, 0.75% PEG 8000, and 0.1% Tween 20 were loaded into the MMV treated with bacterial cell lysis reaction buffer. A dodecamer primer, pfM 19, was used for random PCR. The MMV chip was sealed tightly with silicone tape and aluminum foil to prevent evaporation of the reaction mixture during PCR.

To minimize the amplification of undesirable DNAs such as primer dimers, the PCR program comprising the two following series was chosen: *i*) 15 cycles of denaturation at 94°C for 30 s, annealing at 26°C for 60 s, and extension at 72°C for 60 s and *ii*) 20 cycles of denaturation at 94°C for 30 s, annealing at 60°C for 60 s, and extension at 72°C for 60 s. PCR was performed using an MMV PCR machine (Lifetech) or a conventional thermocycler (Bio-Rad C1000 Touch).

After PCR, the MMV was stained with SYBR green I and monitored using a fluoroimager. MMV wells showing high fluorescence (dark pixels) were selected for scale-up PCR and transferred to a 96-well PCR plate. For scale-up PCR, a PCR mixture that composed of 200 μM dNTPs (N = G, A, T, or C), 0.7 μM primer pfM 19, 10 mM Tris–HCl (pH 8.3), 50 mM KCl, 1.5 mM $MgCl_2$, and 0.02 U/μL Taq DNA polymerase (Takara) was used. Samples were amplified by 15 cycles of denaturation at 94°C for 30 s, annealing at 26°C for 60 s, and extension at 72°C for 60 s.

PCR product analysis

Random PCR products were analyzed using μTGGE, which can extract sequence-specific information of double-stranded DNA without sequencing (Figure 4b and Additional file 12). Each profile was assigned genome-specific feature points called *spiddos* [40], and *PaSS* (pattern similarity score) was calculated (Additional file 12). The genome distance d_G, defined as the value of 1 – *PaSS*, was subjected to clustering analysis to construct a phylogenetic tree using the neighbor-joining method with Phylip 3.69 and MEGA 5.1 software.

Additional files

Additional file 1: Table S1. Unit technologies and accessories of the MMV system.

Additional file 2: Figure S1. Transfer of solution in a square (thin filter) to MMV wells (S-mode operation). Here, the holes of a thin filter act as squares. The filling process is explained in the Figure 1c legend.

Additional file 3: Figure S2. Robotic transfer of solution to an MMV well (P-mode operation). A robot developed for P-mode transfer in MMV operations. Both MMV-to-MMV and MMV-to-microplate or other transfers can be performed by this machine. This robot (manufactured by Lifetech) comprises robotic dispenser arms (1), three platforms [one for microplate (2) and two for MMVs (5, 6)], a tip or syringe stand (3), a container to discard used tips (4), and a tip-position sensor (7). The whole system is controlled by a computer (not shown). The tip-position sensor is required to adjust and control the fine (10 μm or less) 3D positions of tips.

Additional file 4: Figure S3. Generation of MMV replicas. **(a)** An example of preparing an original MMV made of two types of wells (Elements 1 and 2) in a specific pattern. Element 1-containing MMV (E1-MMV) was prepared using a specific pattern filter, and similarly, Element 2-containing MMV (E2-MMV) was prepared with a pattern complementary to that of E1-MMV. These two MMVs were combined by transferring the contents of E1-MMV to E2-MMV. **(b)** The contents of the original MMV were transferred to the vacant acceptor MMV by centrifugation and then reversed as shown, leaving a small amount of solution (seed) in each well of replica MMV. These seeds may be DNA or cells depending on the type of replica formation. **(c)** Picture of wells containing tiny droplets. **(d)** Actual microscopic image of droplet containing MMV wells.

Additional file 5: Figure S4. Actual checker-pattern experiment for verification of MMV solution transfer operation (Z-mode). Methylene blue dye solution is charged into a checker-pattern packing spacer (filled and empty wells alternatively) on a small area of the donor MMV and transferred to the acceptor MMV by Z-mode transfer, thus transferring solution only to corresponding wells without cross-contamination.

Additional file 6: Figure S5. Ejection of MMV contents (O-mode operation: washing). **(a)** View of a polycarbonate MMV. **(b)** Schematic representation of solution discharge from an MMV chip. An MMV chip covered with a sheet of filter and tissue paper is centrifuged to eject the contained solutions and can be repeatedly washed by the same process.

Additional file 7: Figure S6. Division of solutions in MMV (X-mode operation). A filter patterned with two small holes per well was attached to an empty acceptor MMV and was tightly bound to a filled (donor) MMV. These sets of MMVs were subjected to centrifugation directed parallel to the MMV surface. On centrifugation, the liquids were divided into two portions in the facing donor and acceptor MMV wells.

Additional file 8: Figure S7. Magnetic bead recovery from solutions (M-mode operation). Magnetic beads bound to the desired molecules can be recovered by the following steps. Mixing of beads-containing solution can be performed by sealing the solution in MMV with silicone tape and the alternative attractive force of magnets. The recovered magnetic beads bind the desired molecule 'A' on their surface via the 'anti-A' molecule directly bound to the bead.

Additional file 9: Figure S8. Magnetizable bead-assisted division of liquid (M-mode operation). Using packing with holes smaller than the bead diameter, the solution in the 70% volume of MMV can be retained during centrifugation [step b) to c)].

Additional file 10: Figure S9. Pipette-dependent transfer of solution in an MMV well (P-mode operation). (a) Image of a pipette, tip, and MMV chip. (b) Close-up view of manual pipette operation along MMV wells.

Additional file 11: Figure S10. Generation of diverse conditions. N-times of 2 state (2^N) method. Each well of an MMV can be uniquely assigned by a binary number composed of N-bits (here, N = 4). Here, "0" or "1" at each position of a binary number corresponds to the absence or presence, respectively, of a specific component input using the corresponding plate. Thus, "0000" means the absence and "1111" the presence of all four components. Similarly, "0101" implies the presence of only the second and fourth components.

Additional file 12: 2^N method, genome profiling, μTGGE, sequencing analysis, and peptide aptamer selection: panning on microarray MMV (POMM) methods are described in detail.

Additional file 13: Figure S11. Phase-diagram-like presentation of lysozyme crystals. **(a)** Different conditions composed of pH (2.9–9.6) and ionic strength (NaCl: 0–1.5 M) generated different types of lysozyme crystals. Each shape of crystal in MMV wells is depicted in different colors. Four main types of crystals (large and small tetragonal crystals and microcrystalline and needle-like crystals) were observed. **(b)** Microscopic image of a crystal (0.65 mm in length) obtained by the lysozyme crystallization reproducibility experiment (at 50 μL scale) performed under one of the conditions generated in the MMV chip, i.e., 0.6 M NaCl and pH 8.6. The scale bar is 100 μm.

Additional file 14: Figure S12. Original genome profiles obtained from random PCR-successful microbial samples appearing in the MMV-PCR experiment. Samples that have been successfully processed up to clustering analysis (Figure 4d in text) are shown. Sample feature points (pre-*spiddos*) and internal reference points are indicated by red and yellow dots, respectively.

Additional file 15: Table S2. BLASTN-aided tentative assignment of species for each DNA band in the NNMA experiment.

Additional file 16: Table S3. Basic data obtained for three trials of the NNMA experiments.

Additional file 17: Figure S13. Well-to-well transfer by nanoliter dispenser robot. Transfer of 0.1 μL volume of solution from upper donor wells (A, B, and C) to lower acceptor wells (A', B', and C') by robotic transfer in the same MMV. In this experiment, around 5% solution (0.005 μL) is left after transfer as shown in the crescent shape (upper row) due to the difficulty of withdrawing all the solution.

Additional file 18: Figure S14. Micro-high-throughput screening (μ-HTS) of Aβ-binding peptides by panning on microarray MMV (POMM). A DNA library of candidate sequences was transferred to the MMV. Each candidate sequence contained the T7 promoter region and was tagged with 3× FLAG. MMV PCR was performed, and MMV was replicated. The replica MMV was stored (at −20°C) for future replication and screening experiments. For subsequent steps, the original MMV was used. DNA sequences were *in vitro* transcribed and translated (IVT) (see Additional file 12). Selectively Aβ 42-binding peptides were extracted using streptavidin magnetic beads and biotin-conjugated Aβ 42. MMV was thoroughly washed to eliminate everything except binding peptides. FITC-labeled anti-FLAG antibody solution was added, MMV was washed, and peptides were released by proteinase K treatment. MMV was visualized in the detection unit (CCD camera or TRF unit or laser scanner with FITC filter). In addition, MMV wells stained with FITC indicated the presence of Aβ-42 binding peptides, resulting in fluorescence, whereas in the absence of Aβ-binding peptide, FITC-labeled anti-FLAG antibody did not bind and was washed out in the initial steps, thus resulting in no fluorescence.

Additional file 19: Figure S15. Deflocculation of an oral microbiome sample. **(a)** Microscopic view of SYBR gold-stained negative (PBS buffer only), positive (oral microbiome sample without treatment), heat (positive sample with heat treatment only), and proteinase K (positive sample with proteinase K treatment only) samples. **(b)** Phase-contrast microscopic images of an oral microbiome sample treated with proteinase K only and with both proteinase K and sonication (for 1 min). The scale bar is 50 μm for both **(a)** and **(b)**. The ratio of single to flocculated cells was approximately 100:1.

Additional file 20: Figure S16. A trial semiconductor-based apparatus for the evaluation of solution conductivity in MMV wells. The conductivity of the solution in a particular well can be selectively monitored by laser-light illumination through light-transparent semiconductor ITO and photoconductive polymer membrane. The figure shows the redox reaction occurring on the surface of the photoconductive polymer membrane.

Additional file 21: Table S4. Tentative cost comparison of the reagents required in MMV and a 96-well microplate (for 1000 reactions). **(a)** Cost comparison for a multistep function-based screening experiment. **(b)** Cost comparison for the experiment of apoptosis detection in HeLa cells. The prices were taken from those in 2012 in Japan for both tables **(a)** and **(b)**.

Abbreviations

MMV: Microarray with manageable volumes; GP: Genome profiling; μTGGE: Micro-temperature gradient gel electrophoresis; *Spiddos*: Species identification dots; *PaSS*: Pattern similarity score; NGS: Next generation sequencing; NNMA: NGS-non-dependent microbiome analysis; ccgf: Commonly conserved genetic fragment; C2C: Cell-to-cell; D2D: DNA-to-DNA; C2D: Cell-to-DNA; D2P: DNA-to-peptide/protein; C2D2P: Cell-to-DNA-to-peptide/protein; iPS: Induced pluripotent stem cell.

Competing interests

Nine of the authors are affiliated with four commercial organizations: SF, MiSa and KK with Janusys Corporation, MaS, KS and OT with Lifetech Co., Ltd., ST and TW with Enplas Corporation, and HY with Finetech Corporation. The corresponding author (KN) was financially supported in 2010 by Lifetech Co. Two of the authors (KK and NN) hold shares of Janusys Corporation. The following patents related to this paper are applied by some of the authors: Patent Application Publication No. 2013–195370 (YK, SA, ST, HY, OT, KN). Patent Application No. 2013–177214 (YK, SF, HY, KK, OT, KN). The authors declare that they have no other competing interests.

Authors' contributions

HS performed the NNMA experiment and wrote/co-edited the manuscript. YK and SF developed the MMV basic handling technologies and performed the lysozyme crystallization and replica generation experiments, respectively. SN, MiSa, and SA initiated and developed the NNMA, MMV-cell-based assay, and MMV-based peptide screening experiments, respectively. MaS, KS, and OT chiefly contributed to manufacturing the MMV robot. ST, TW, and HY devised the MMV chips and accessories. RK and TK performed the confirmation experiments. HA discussed and performed essential calculations. MiSu and NN discussed and directed this study. KA and HU established the MMV-fluorescence detection. KK and OT (again) contributed to jointly organizing, discussing, and directing this joint research. KN conceived, organized, and directed the whole research and edited and revised the manuscript. All authors read and approved the final manuscript.

Acknowledgments

This research was supported by grants from the Japan Science and Technology Agency, Development of Systems and Technology for Advanced Measurement and Analysis and partly by a grant for the City Area Program (Saitama Metropolitan Area) from the Ministry of Education, Culture, Sports, Science, and Technology.

Author details

[1]Department of Functional Materials Science, Graduate School of Science and Engineering, Saitama University, 255 Shimo-okubo, Saitama 338-8570, Japan. [2]Janusys Corporation, Saitama Industrial Technology Center, 3-12-18 Kamiaoki, Kawaguchi, Saitama 334-0844, Japan. [3]Lifetech Co., Ltd., 4074 Miyadera, Iruma City, Saitama 358-0014, Japan. [4]Enplas Corporation, 2-30-1 Namiki, Kawaguchi City, Saitama 332-0034, Japan. [5]Finetech Corporation, 1-7-1, Asagaya-minami Suginami-ku, Tokyo, Japan. [6]Present address: Department of Biochemistry and Molecular Biology, Shahjalal University of Science and Technology, Sylhet, Bangladesh. [7]Present address: Department of Electrical and Electronic Engineering, Tokyo Denki University, 5 Senjyu-Asahi-cho, Adachi-ku, Tokyo 120-8551, Japan.

References

1. Sundberg SA: **High-throughput and ultra-high-throughput screening: solution- and cell-based approaches.** *Curr Opin Biotechnol* 2000, **11**:47–53.
2. Kornienko O, Lacson R, Kunapuli P, Schneeweis J, Hoffman I, Smith T, Alberts M, Inglese J, Strulovici B: **Miniaturization of whole live cell-based GPCR assays using microdispensing and detection systems.** *J Biomol Screen* 2004, **9**:186–195.
3. Jang JS, Simon VA, Feddersen RM, Rakhshan F, Schultz DA, Zschunke MA, Lingle WL, Kolbert CP, Jen J: **Quantitative miRNA expression analysis using Fluidigm microfluidics dynamic arrays.** *BMC Genomics* 2011, **12**:144.
4. van Doorn R, Klerks MM, van Gent-Pelzer MPE, Speksnijder AGCL, Kowalchuk GA, Schoen CD: **Accurate quantification of microorganisms in PCR-inhibiting environmental DNA extracts by a novel internal amplification control approach using Biotrove OpenArrays.** *Appl Environ Microbiol* 2009, **75**:7253–7260.
5. Didelot A, Kotsopoulos SK, Lupo A, Pekin D, Li X, Atochin I, Srinivasan P, Zhong Q, Olson J, Link DR, Laurent-Puig P, Blons H, Hutchison JB, Taly V: **Multiplex picoliter-droplet digital PCR for quantitative assessment of DNA integrity in clinical samples.** *Clin Chem* 2013, **59**:815–823.

6. Baker M: **Digital PCR hits its stride.** *Nat Methods* 2012, **9**:541–544.
7. Vyawahare S, Griffiths AD, Merten CA: **Miniaturization and parallelization of biological and chemical assays in microfluidic devices.** *Chem Biol* 2010, **17**:1052–1065.
8. Srisa-Art M, de Mello AJ, Edel JB: **High-throughput DNA droplet assays using picoliter reactor volumes.** *Anal Chem* 2007, **79**:6682–6689.
9. Gan R, Yamanaka Y, Kojima T, Nakano H: **Microbeads display of proteins using emulsion PCR and cell-free protein synthesis.** *Biotechnol Prog* 2008, **24**:1107–1114.
10. Boulanger J, Muresan L, Tiemann-Boege I: **Massively parallel haplotyping on microscopic beads for the high-throughput phase analysis of single molecules.** *PLoS One* 2012, **7**:e36064.
11. Kim DS, Lee KC, Kwon TH, Lee SS: **Micro-channel filling flow considering surface tension effect.** *J Micromech Microeng* 2002, **12**:236.
12. Zhang YH, Ozdemir P: **Microfluidic DNA amplification - a review.** *Anal Chim Acta* 2009, **63**:115–125.
13. Delamarche E, Juncker D, Schmid H: **Microfluidics for processing surfaces and miniaturizing biological assays.** *Adv Mater* 2005, **17**:2911–2933.
14. Lilliehorn T, Nilsson M, Simu U, Johansson S, Almqvist M, Nilsson J, Laurell T: **Dynamic arraying of microbeads for bioassays in microfluidic channels.** *Sensor Actuator B Chem* 2005, **106**:851–858.
15. Margulies M, Egholm M, Altman WE, Attiya S, Bader JS, Bemben LA, Berka J, Braverman MS, Chen YJ, Chen Z, Dewell SB, Du L, Fierro JM, Gomes XV, Godwin BC, He W, Helgesen S, Ho CH, Irzyk GP, Jando SC, Alenquer MLI, Jarvie TP, Jirage KB, Kim JB, Knight JR, Lanza JR, Leamon JH, Lefkowitz SM, Lei M, Li J, *et al*: **Genome sequencing in microfabricated high-density picolitre reactors.** *Nature* 2005, **437**:376–380.
16. Salimullah M, Mori M, Nishigaki K: **High-throughput three-dimensional gel electrophoresis for versatile utilities: a stacked slice-gel system for separation and reactions (4SR).** *Genomics Proteomics Bioinformatics* 2006, **4**:26–33.
17. Kinoshita Y, Tayama T, Kitamura K, Salimullah M, Uchida H, Suzuki M, Husimi Y, Nishigaki K: **Novel concept microarray enabling PCR and multistep reactions through pipette-free aperture-to-aperture parallel transfer** *BMC Biotechnol* 2010, **10**:71.
18. Dill K, McShea A: **Recent advances in microarrays.** *Drug Discov Today Technol* 2005, **2**:261–266.
19. Zhou SM, Cheng L, Guo SJ, Zhu H, Tao SC: **Functional protein microarray: an ideal platform for investigating protein binding property.** *Front Biol* 2012, **7**:336–349.
20. Kim J, Crooks RM: **Replication of DNA microarrays prepared by in situ oligonucleotide polymerization and mechanical transfer.** *Anal Chem* 2007, **79**:7267–7274.
21. Hekmat D, Hebel D, Joswig S, Schmidt M, Weuster-Botz D: **Advanced protein crystallization using water-soluble ionic liquids as crystallization additives.** *Biotechnol Lett* 2007, **29**:1703–1711.
22. Yatsunenko T, Rey FE, Manary MJ, Trehan I, Dominguez-Bello MG, Contreras M, Magris M, Hidalgo G, Baldassano RN, Anokhin AP, Heath AC, Warner B, Reeder J, Kuczynski J, Caporaso JG, Lozupone CA, Lauber C, Clemente JC, Knights D, Knight R, Gordon JI: **Human gut microbiome viewed across age and geography.** *Nature* 2012, **486**:222–227.
23. Liu B, Faller LL, Klitgord N, Mazumdar V, Ghodsi M, Sommer DD, Gibbons TR, Treangen TJ, Chang YC, Li S, Stine OC, Hasturk H, Kasif S, Segre D, Pop M, Amar S: **Deep sequencing of the oral microbiome reveals signatures of periodontal disease.** *PLoS ONE* 2012, **7**:e37919.
24. Wang J, Qi J, Zhao H, He S, Zhang Y, Wei S, Zhao F: **Metagenomic sequencing reveals microbiota and its functional potential associated with periodontal disease.** *Sci Rep* 2013, **3**:1843.
25. Nishigaki K, Naimuddin M, Hamano K: **Genome profiling: a realistic solution for genotype-based identification of species.** *J Biochem* 2000, **128**:107–112.
26. Kouduka M, Matsuoka A, Nishigaki K: **Acquisition of genome information from single-celled unculturable organisms (radiolaria) by exploiting genome profiling (GP).** *BMC Genomics* 2006, **7**:135.
27. Sakuma Y, Nishigaki K: **Computer prediction of general PCR products based on dynamical solution structures of DNA.** *J Biochem* 1994, **116**:736–741.
28. Kurimoto K, Yabuta Y, Ohinata Y, Ono Y, Uno KD, Yamada RG, Ueda HR, Saitou M: **An improved single-cell cDNA amplification method for efficient high-density oligonucleotide microarray analysis.** *Nucleic Acids Res* 2006, **34**:e42.
29. Frumkin D, Wasserstrom A, Itzkovitz S, Harmelin A, Rechavi G, Shapiro E: **Amplification of multiple genomic loci from single cells isolated by laser micro-dissection of tissues.** *BMC Biotechnol* 2008, **8**:17.
30. White AK, Heyries KA, Doolin C, VanInsberghe M, Hansen CL: **High-throughput microfluidic single-cell digital polymerase chain reaction.** *Anal Chem* 2013, **85**:7182–7190.
31. Naimuddin M, Kurazono T, Nishigaki K: **Commonly conserved genetic fragments revealed by genome profiling can serve as tracers of evolution.** *Nucleic Acids Res* 2002, **30**:e42.
32. Kouduka M, Sato D, Komori M, Kikuchi M, Miyamoto K, Kosaku A, Naimuddin M, Matsuoka A, Nishigaki K: **A solution for universal classification of species based on genomic DNA.** *Int J Plant Genomics* 2006, **2007**:Article ID 27894.
33. Ahmed S, Komori M, Tsuji-Ueno S, Suzuki M, Kosaku A, Miyamoto K, Nishigaki K: **Genome profiling (GP) method based classification of insects: congruence with that of classical phenotype-based one.** *PLoS One* 2011, **6**:e23963.
34. Hamano K, Tsuji-Ueno S, Tanaka R, Suzuki M, Nishimura K, Nishigaki K: **Genome profiling (GP) as an effective tool for monitoring culture collections: a case study with Trichosporon.** *J Microbiol Methods* 2012, **89**:119–128.
35. Judge RA, Jacobs RS, Frazier T, Snell EH, Pusey ML: **The effect of temperature and solution pH on the nucleation of tetragonal lysozyme crystals.** *Biophys J* 1999, **77**:1585–1593.
36. Hebel D, Urdingen M, Hekmat D, Weuster-Botz D: **Development and scale up of high-yield crystallization processes of lysozyme and lipase using additives.** *Cryst Growth Des* 2013, **13**:2499–2506.
37. Li F, Robinson H, Yeung ES: **Automated high-throughput nanoliter-scale protein crystallization screening.** *Anal Bioanal Chem* 2005, **383**:1034–1041.
38. Rickard AH, Gilbert P, High NJ, Kolenbrander PE, Handley PS: **Bacterial coaggregation: an integral process in the development of multi-species biofilms.** *Trends Microbiol* 2003, **11**:94–100.
39. Nagaoka S, Hojo K, Murata S, Mori T, Ohshima T, Maeda N: **Interactions between salivary *Bifidobacterium adolescentis* and other oral bacteria: in vitro coaggregation and coadhesion assays.** *FEMS Microbiol Lett* 2008, **281**:183–189.
40. Naimuddin M, Kurazono T, Zhang Y, Watanabe T, Yamaguchi M, Nishigaki K: **Species-identification dots: a potent tool for developing genome microbiology.** *Gene* 2000, **261**:243–250.

Novel human recombinant antibodies against Mycobacterium tuberculosis antigen 85B

Manon Fuchs[1], Susanne Kämpfer[2], Saskia Helmsing[1], Ralf Spallek[2], Wulf Oehlmann[2], Wiebke Prilop[2], Ronald Frank[3,4], Stefan Dübel[1], Mahavir Singh[2] and Michael Hust[1*]

Abstract

Background: Tuberculosis is the leading cause of death due to bacterial infections worldwide, mainly caused by *Mycobacterium tuberculosis*. The antigen 85 complex comprises a set of major secreted proteins of *M. tuberculosis*, which are potential biomarkers for diagnostic.

Results: In this work, the first human single chain fragment variable (scFv) antibodies specific for the tuberculosis biomarker 85 B were selected by phage display from naïve antibody gene libraries (HAL7/8). Produced as scFv-Fc in mammalian cells, these antibodies were further characterized and analysed for specificity and applicability in different tuberculosis antigen detection assays. Sandwich detection of recombinant 85 B was successful in enzyme linked immunosorbent assay (ELISA), lateral flow immunoassay and immunoblot. Whereas detection of *M. tuberculosis* cell extracts and culture filtrates was only possible in direct ELISA and immunoblot assays. It was found that the conformation of 85 B, depending on sample treatment, influenced antigen detection.

Conclusions: Recombinant antibodies, selected by phage display, may be applicable for 85 B detection in various assays. These antibodies are candidates for the development of future point of care tuberculosis diagnostic kits. Using 85 B as a biomarker, the antigen conformation influenced by sample treatment is important.

Background

Approximately 8.7 million people worldwide fell ill with tuberculosis (TB) in 2011 and 1.4 million deaths were reported [1]. Worldwide TB ranks as the second major cause of death from an infectious disease. One third of the world population is estimated to be infected with *M. tuberculosis* (Mtb), yet they remain asymptomatic. This is defined as latent TB infection (LTBI) [2]. Only 66% of the TB-cases worldwide are correctly diagnosed [1]. The gold standard in TB diagnosis remains the preparation of liquid cultures in selective media from sputum or tissue/body fluid specimens [3]. This is followed by further Mtb specific tests or drug susceptibility testing (i.e. nucleic acid amplification tests such as the Gene Xpert MTB/RIF [4]).

Diagnosis of TB in most low- and middle-income countries continues to rely on sputum smear microscopy for acid-fast bacilli (Ziel-Neelsen stain) [1]. This technique detects only 40 – 60% of pulmonary TB cases and is not able to differentiate between Mtb and other ubiquitous mycobacteria [5,6]. This is less sensitive in children, HIV co-infected patients and in patients with extrapulmonary TB [7,8]. The diagnosis of TB in developing countries is limited by equipment and infrastructure [1]. Therefore, in these countries, a simple diagnostic tool without the need for sophisticated instruments is needed. Accuracy, simplicity, affordability and technical robustness are important factors for a point of care (POC) TB-test. The main advantage of this is a rapid diagnosis, which allows initiation of treatment while the patient is still accessible [9]. Several anti-tuberculosis antibody detection systems are available. However the WHO recommended against the use of these assays because of their lack of sensitivity and specificity [10]. Direct detection of Mtb antigens in human specimens would allow specific diagnosis of active TB to be made, independent of the host's immune response. Furthermore, the use of specific antibodies to Mtb antigens in a lateral flow immuno assay (LFIA) would potentially provide a rapid POC test in a cost effective, easy-to-use format. Potential target antigens for POC TB-detection in human samples should be selected by the following

* Correspondence: m.hust@tu-bs.de
[1]Technische Universität Braunschweig, Institut für Biochemie und Biotechnologie, Spielmannstr.7, 38106 Braunschweig, Germany
Full list of author information is available at the end of the article

criteria: substantial expression by bacteria *in vivo*, presence in the extracellular environment or on the mycobacterial cell wall, and resistance to degradation by host enzymes [11]. The 85 complex is a major secretion product of Mtb [12,13] which comprises three variant 85 proteins (A, B and C) that are encoded by three different genes (fbpA, fbpB and fbpC2, [14]). Antigens 85 A, 85 B and 85 C possess mycoyltransferase activity and are involved in the formation of trehalose monomycolate (TMM) and trehalose dimycolate (TDM), which are components of the mycobacterial cell wall [15,16]. Besides the 85 complex there is another 85 protein, known as 85 D or FbpC1. 85 D shows structural similarity to the 85 complex proteins, yet does not possess mycoyltransferase activity [15] and its function remains unknown. Disruption of fbpA results in the inability of Mtb to replicate within macrophages, indicating a key role in virulence [17]. Furthermore, it was shown that 85 complex proteins interact with gelatine-binding sites of human fibronectin, enhancing complement-mediated phagocytosis by macrophages [18-20]. Antibodies specific for only one of the 85 proteins would allow for further investigation of their individual roles in the pathogenesis of TB. However only antibodies cross reacting with all other 85 complex proteins have been reported so far [21-23].

The presence of the 85 complex was demonstrated in human serum [24], urine [20], cerebrospinal fluid [25] and sputum [26] of TB infected individuals. This makes it a prominent marker for TB. Although other mycobacteria express similar 85 complex proteins, these antigens could be used as TB markers in combination with other antigens to enhance specificity.

For the future development of diagnostic assays, recombinant antibodies generated by phage display are an alternative to polyclonal and monoclonal antibodies [27-32]. Naïve antibody libraries allow generation of antibodies to epitopes which are not recognized by any immune system. In addition, recombinant human antibodies could be used in therapeutic applications. In this work, single chain Fv (scFv) were isolated from a human naïve antibody gene library using phage display [33] and produced as scFv-Fc (yumab) [34]. These antibodies were further characterized and analysed for their suitability in an antigen detection assay targeting Mtb 85 B.

Results

Selection of human antibodies against 85 B

For antibody selection, panning was performed on immobilized 85 B followed by a screening ELISA using soluble scFv (see Additional file 1: Figure S1). Finally, five unique scFv were confirmed by DNA sequencing. The corresponding human germline sequences according to VBASE2 [35] are displayed in Table 1. All α-85 B antibodies contain a variable gene segment of the heavy chain (HV) of subfamily V_H3. Four binders have a lambda light chain ($V_L\lambda$) and one has a kappa light chain ($V_L\kappa$). All $V_L\lambda$ clones have a joining segment of the heavy chain (HJ) of subfamily J4, a joining segment of the light chain (LJ) of subfamily J3, and a variable gene segment of the heavy chain (HV) of subfamily V_H3 gene 30.

Cloning and production of scFv-Fc (Yumabs)

The scFv were subcloned into pCSE2.5-hIgG1-Fc-XP [34], produced in 50 mL scale and purified from the culture supernatant via Protein A. The purified scFv-Fc (Yumabs) were analysed by SDS-PAGE, silver staining, α-human IgG(Fc) immunoblot and reducing gel analysis via Tape Station. No degradation was detected (data not shown). The purity of the obtained antibody preparations was determined to 93.4 – 96.9%.

Validation of antigen binding

The antigen binding of the Yumabs was analysed by titration ELISA (Figure 1A). The antigen binding was confirmed for all Yumabs. Antigen detection limits of the α-85 B scFv-Fc were determined by antigen titration ELISA (Figure 1B). About 5 ng/mL were detected by MFU50-C10, ~ 10 ng/mL by MFU50-A10 and MFU50-E2, and ~ 30 ng/mL by MFU50-D4 and MFU50-D7.

Epitope mapping

To determine whether the antibodies bind to linear or conformational epitopes, the α-85 B scFv-Fc were analysed by immunoblot (data not shown). All antibodies

Table 1 Comparison of heavy and light chain gene segments of α-85 B antibodies

	Heavy chain			Light chain	
Clone	HV	D	HJ	LV	LJ
MFU50-A10	IGHV3-30*04	IGHD6-19*01	IGHJ4*02	IGLV8-61*01	IGLJ3*02
MFU50-C10	IGHV3	IGHD2-15*01	IGHJ6*02	IGKV1-12*01	IGKJ2*01
MFU50-D4	IGHV3-30*18	IGHD3-3*02	IGHJ4*02	IGLV2-14*01	IGLJ3*01
MFU50-D7	IGHV3-30*04	IGHD4-17*01	IGHJ4*02	IGLV3-21*02	IGLJ3*02
MFU50-E2	IGHV3-30*03	IGHD6-13*01	IGHJ4*02	IGLV7-43*01	IGLJ3*02

Abbreviations: HV variable (V) gene segment of the heavy chain, *D* diversity gene segment, *HJ* joining (J) gene segment of the heavy chain, *LV* variable gene segment of the light chain, *LJ* joining gene segment of the light chain.

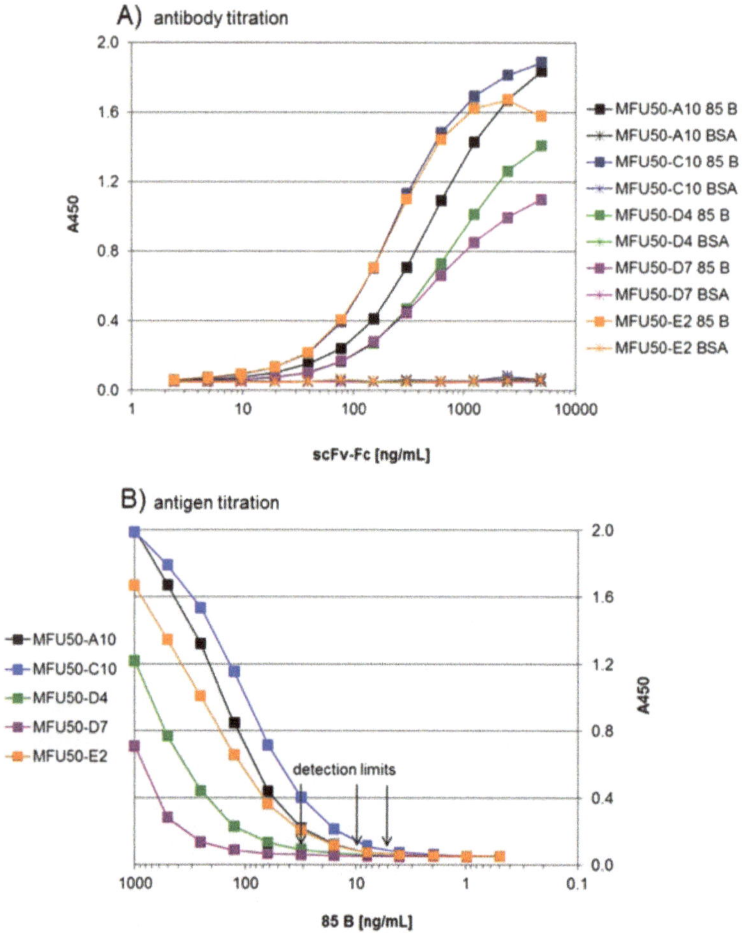

Figure 1 Analysis of antigen binding. A) Antibody titration and **B)** antigen titration in ELISA with α-85 B scFv-Fc. A) Dilution series of scFv-Fc were used for detection of directly coated antigen (85 B) or BSA (negative control). Detection of bound scFv-Fc with goat α-human IgG(Fc)-HRP and development with TMB. B) Different dilutions of antigen (85 B) were directly coated to the wells. Antigen detection with a scFv-Fc concentration of half maximal saturation. Detection of bound scFv-Fc with goat α-human IgG(Fc)-HRP, development with TMB. Negative control BSA A450 = 0.05.

recognized linear epitopes (data not shown). Epitope mapping of the α-85 B scFv-Fc was carried out on PepSpot membranes (Figure 2A). Through the overlap of the peptide sequences the epitopes were determined (Figure 2B). The crystal structure of antigen 85 B was determined by Anderson et al. [36] and the epitopes were visualized on the 3D structure of the antigen (Figure 2C). The distance between epitope regions "AFSRPGPLV(EYL)" and "SPAVYL" was computed to ~ 2.5 nm, between "SPAVYL" and "SSDPAWERN(DPT)" to ~ 4 nm and between "SSDPAWERN(DPT)" and "AFSRPGPLV(EYL)" to ~ 5 nm. Considering the width of an antibody (~4 nm, [37]), sandwich detection was likely targeting epitopes on different sites of the antigen.

Cross reactions

Sequence comparison of the epitopes of the α-85 B antibodies with corresponding sequences of the other 85 complex antigens provided information about possible cross reactions (Table 2). The complete epitope "AFSRPGPLV (EYL)" is present in antigen 85 A and 85 C, but not in 85 D. Homologous regions of the epitopes "SPAVYL" and "SSDPAWERN(DPT)" can be found in 85 A, C, and D. Experimental study of 85 complex cross reactivity was carried out by indirect ELISA (Figure 3). MFU50-A10 reacted with 85 A and 85 D. MFU50-D4 displayed a weak cross reactivity with 85 A as predicted. MFU50-E2 bound antigen 85 B weakly and showed no cross reactions. MFU50-D7 reacted weakly with 85 B. No reaction with the other 85 complex antigens was detected. MFU50-C10 was also not cross reactive. A summary of the expected cross reactions in comparison to determined cross reactions is outlined in Table 2.

Detection of 85 B in Mtb cell extracts and culture filtrates

The antibody binding to Mtb/BCG cell extracts and culture filtrates was analysed by indirect ELISA (Figure 4). In this assay only MFU50-A10 reacted with Mtb culture

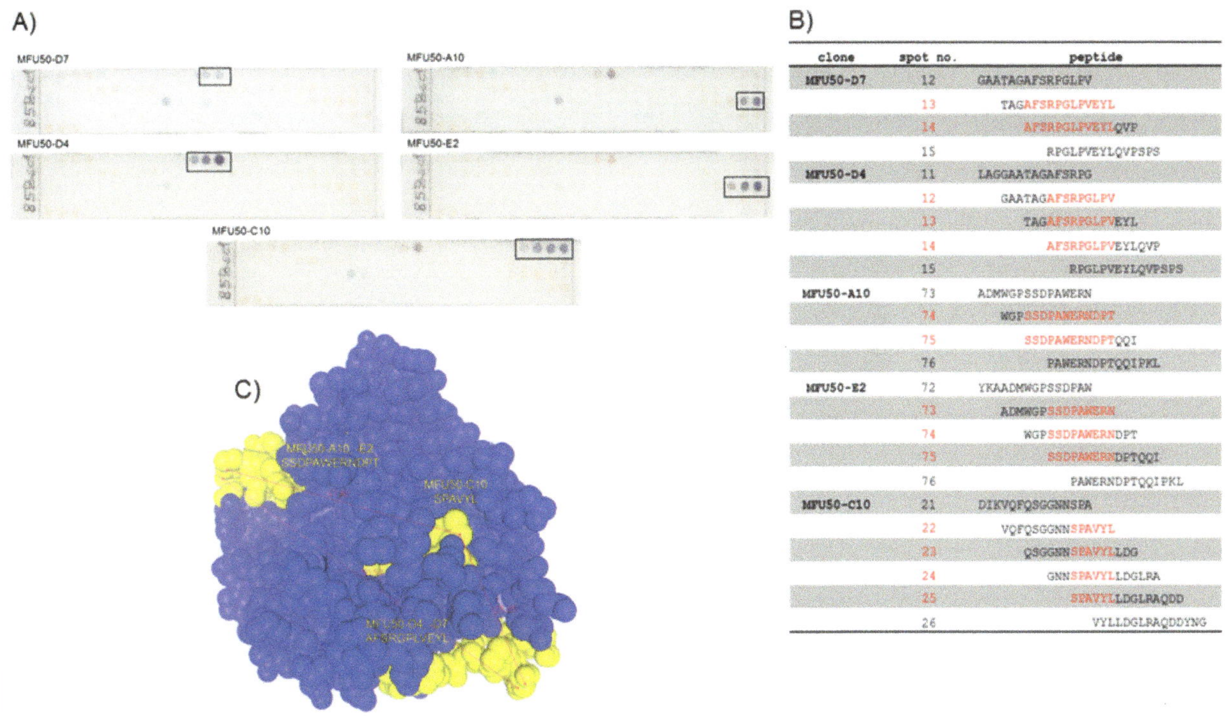

Figure 2 Analysis of the epitope. A) Epitope mapping of α-85 B scFv-Fc with PepSpot membrane, **B)** Amino acid sequences of PepSpots and **C)** 3 D structure of 85 B and α-85 B antibody epitopes. A) Membrane was incubated with respectively 5 µg scFv-Fc, detection of bound scFv-Fc with goat α-human IgG(Fc)-HRP, development with TMB. Immunostained spots are circled. No cross reaction of the secondary antibody with the PepSpots was determined (data not shown). B) Amino acid sequences of peptides on 85 B PepSpot membrane. The immunostained spots and identified epitopes are highlighted. C) 3D structure of antigen 85 B, epitopes of α-85 B antibodies. Pdb1f0n (Resolution 1.9 Å, [36]) was modified with 3D molecule viewer (Invitrogen). Only protein chain A is shown, atoms are displayed as space filling balls, epitopes are marked in yellow, distance measurement in red.

Table 2 Cross reactions of α-85 B antibodies with other 85 complex antigens

sequence comparison (corresponding sequences to α-85 B epitopes)					
Clone	85 A	85 B	85 C*	85 D	Expected cross reaction
MFU50-A10	KEDPAWQRNDPL	SSDPAWERNDPT	SSDPAWKRNDPM	SDPA	85 A, 85 C, 85 D
MFU50-C10	SPALYL	SPAVYL	PHAVYL	PHAVYL	85 A, 85 C, 85 D
MFU50-D4	AFSRPGLPV	AFSRPGLPV	AFSRPGLPV	-	85 A, 85 C
MFU50-D7	AFSRPGLPVEYL	AFSRPGLPVEYL	AFSRPGLPVEYL	-	85 A, 85 C
MFU50-E2	KEDPAWQRN	SSDPAWERN	SSDPAWKRN	SDPA	85 A, 85 C, 85 D
Reaction with 85 complex antigens in ELISA					
Clone	85 A	85 B	85 C*	85 D	Determined cross reaction
MFU50-A10	Strong	Strong	-	Strong	85 A, 85 D
MFU50-C10	None	Strong	-	None	None
MFU50-D4	Weak	Strong	-	None	Weak 85 A
MFU50-D7	None	Weak	-	None	None
MFU50-E2	None	Weak	-	None	None

*Purified antigen 85 C was not available for immunological assays, nevertheless it is listed for completeness of the illustration.

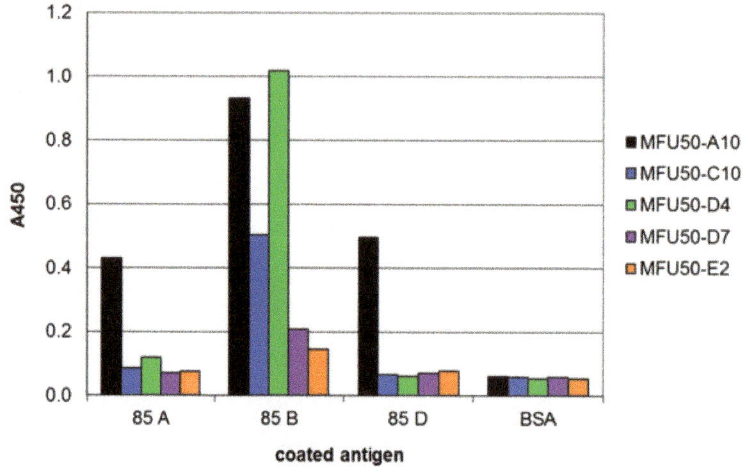

Figure 3 Cross reactions of α-85 B scFv-Fc with other 85 complex antigens in ELISA. 100 ng of the antigens were directly coated to the wells. 85 B detection with α-85 B scFv-Fc (using concentrations at half maximal saturation), detection of bound scFv-Fc with goat α-human IgG (Fc)-HRP, development with TMB.

filtrate and Mtb/BCG cell extracts. The other antibodies bound only to recombinant 85 B. A weak cross reaction of MFU50-A10 with 7H9 medium was detected.

Development of a sandwich ELISA

Due to the availability of five different antibodies with three different epitopes on the target antigen a sandwich α-85 B assay was performed. Therefore MFU50-A10 (epitope "SSDPAWERNDPT"), MFU50-C10 (epitope "SPAVYL") and MFU50-D4 (epitope "AFSRPGLPV") scFv-Fc were conjugated to HRP. Sandwich ELISA detection of the antigen was carried out with MFU50-A10-HRP, MFU50-C10-HRP or MFU50-D4-HRP as detection antibodies. Capturing was conducted with corresponding antibodies recognizing different epitopes (than the detection antibody) on the target (Figure 5A, B, C). Blocking reagent only was used as a negative control ($A_{450} = 0.02 - 0.05$). Sandwich detection of the recombinant antigen was successful with all combinations. The antigen detection limit for capturing with MFU50-A10, MFU50-C10, MFU50-D4 or MFU50-D7 was determined to ~ 10 ng/mL, independent from the detection antibody. Capturing with MFU50-E2 resulted in a detection limit of ~ 25 ng/mL. The most suitable combination was obtained with MFU50-A10 as the capture antibody and MFU50-C10-HRP as detection antibody-conjugate. Mtb culture filtrate (Mtb cultivated in 7H9 + ADC + Tween for 3 months at 37°C) was analysed by α-85 B sandwich titration ELISA (Figure 6A, red curve). This resulted in weak binding barely distinguishable from the medium control (Figure 6A, pink curve). Mtb culture

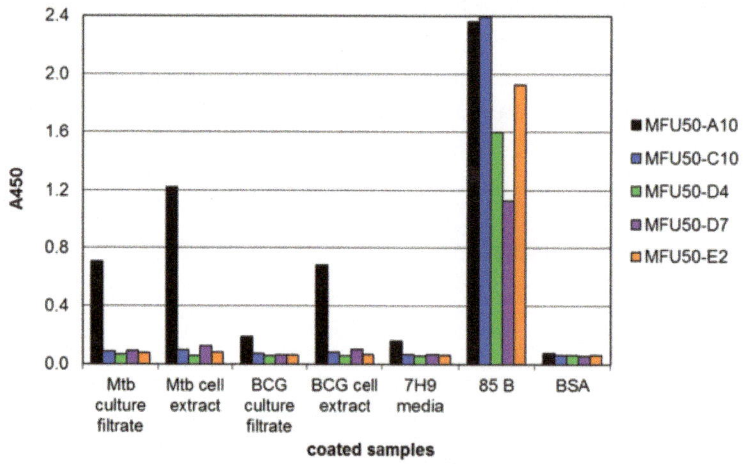

Figure 4 Reaction of α-85 B scFv-Fc with Mtb/BCG cell extracts and culture filtrates determined by indirect ELISA. 100 ng/100 μL of the different samples were directly coated to the well, reaction was determined with a scFv-Fc concentration of 2.5 μg/mL. Detection of bound scFv-Fc with goat α-human IgG(Fc)-HRP. Development with TMB.

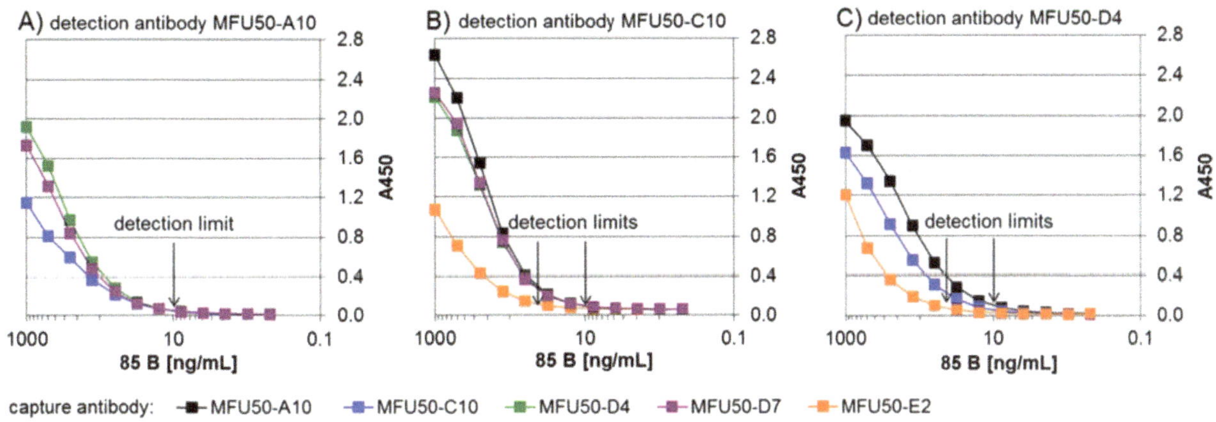

Figure 5 A-85 B sandwich antigen titration ELISA. 100 ng of capture scFv-Fc were directly coated to the wells. After blocking different dilutions of antigen were applied and incubated. Detection of bound antigen with: **A)** MFU50-A10-scFv-Fc-HRP at ~0.05 μg/mL and development with TMB. Negative control (BSA) $A_{450} = 0.02$, **B)** MFU50-C10-scFv-Fc-HRP at ~0.05 μg/mL and development with TMB. Negative control (BSA) $A_{450} = 0.05$, **C)** MFU50-D4-scFv-Fc-HRP at ~0.05 μg/mL and development with TMB. Negative control (BSA) $A_{450} = 0.02$.

filtrate was further analysed by direct α-85 B ELISA (Figure 6A, black curve). In this assay there was no increase of the signal by increasing of sample volume. Additionally, Mtb cell extract was analysed by α-85 B sandwich titration ELISA (Figure 6B, red curve). This showed no binding distinguishable from the negative control except for the highest concentration (Figure 6B, grey curve). However, in a direct α-85 B ELISA specific binding to Mtb cell extract was observed (Figure 6B, black curve).

Effect of sample treatment on 85 B detection

The failed detection of 85 B in Mtb cell extract in sandwich ELISA compared to the successful detection in direct ELISA was further analysed. It was hypothesized that the sample treatment, especially autoclaving, was the cause for differing antigen detection. Cell extracts were prepared by autoclaving at 121°C for 20 min. The influence of sample pretreatment by heat on 85 B detection was investigated by direct ELISA (Figure 7). When 85 B was boiled, antigen binding was reduced. Further examination of the susceptibility of 85 B to heat was carried out by analytical SEC. The theoretical molecular mass of an 85 B monomer was computed with the ExPASy prot param tool [38] to 34.6 kDa. The untreated sample of 85 B displayed a small monomer peak at 33.8 kDa and a dominant multimer peak with a molecular mass out of the measurement range but definitely greater than 669 kDa (Figure 8A). In comparison the boiled 85 B

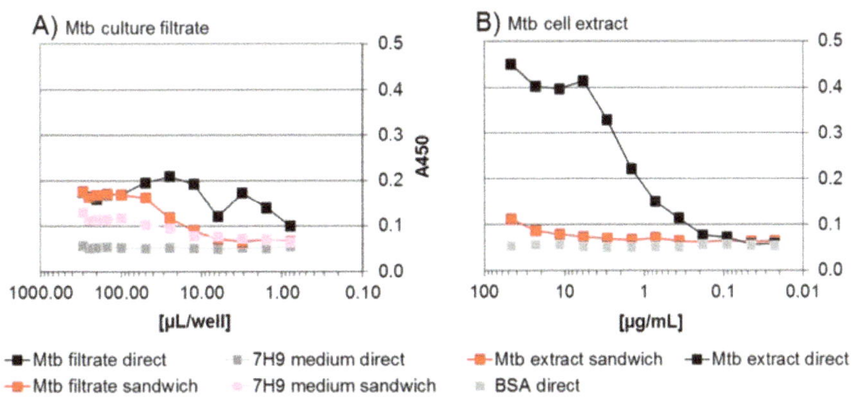

Figure 6 Mtb culture filtrate and Mtb cell extract titration in α-85 B direct and sandwich ELISA. A) various volumes of Mtb culture filtrate or **B)** different dilutions of Mtb cell extract were directly coated to the wells. After blocking antigen was detected using MFU50-C10-scFv-Fc-HRP at ~ 0.05 μg/mL and development with TMB. Sandwich ELISA: 100 ng of capture scFv-Fc MFU50-A10 were directly coated to the wells, after blocking A) various volumes of Mtb culture filtrate or B) different dilutions of Mtb cell extract were applied and incubated. Detection of bound antigen with MFU50-C10-scFv-Fc-HRP at ~0.05 μg/mL and development with TMB. Negative controls: BSA, M. smegmatis cell extract, M. vaccae cell extract all A450 = 0.05.

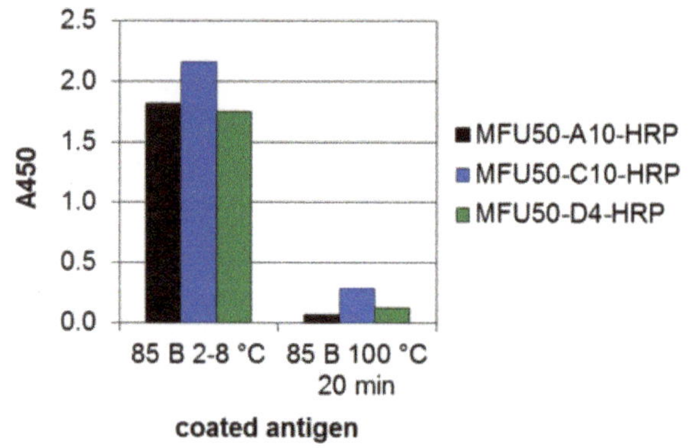

Figure 7 Detection of boiled 85 B by α-85 B direct ELISA. 100 ng of antigens in PBS (boiled for 20 min at 100°C or directly taken from storage at 2 – 8°C) were coated to the wells. After blocking bound antigen was detected using MFU50-A10/-C10/-D4-scFv-Fc-HRP at ~0.05 μg/mL and development with TMB. Negative control BSA A450 = 0.02.

sample offered a dominant peak at the size of a monomer (34.6 kDa), multiple peaks with lower molecular mass and a few non-dominant multimer peaks (Figure 8B). Interestingly, aggregation was greater in the untreated sample than in the boiled sample. It was concluded that the aggregation improved sandwich detection, and if multimers were detected by sandwich ELISA, the use of the same antibody for capturing and detection should be possible. Experimental study by sandwich ELISA (Figure 9) verified this assumption. In the untreated as well as in the boiled sample, sandwich detection with only one antibody was possible (Figure 9, arrows).

Development of a lateral flow immuno assay

To develop a Lateral Flow Immuno Assay (LFIA) all α-85 B scFv-Fc were conjugated to 40 nm colloidal gold. Sandwich 85 B detection was performed with all available antibodies for capturing and all antibody-gold conjugates for detection. The most suitable combination, MFU50-A10 as capture antibody and MFU50-D4-gold as detection antibody, was further analysed. A procedure that allowed sensitive antigen detection and low background combined with good feasibility was developed (data not shown). The 85 B detection limit was determined by sandwich LFIA to ≤ 5 ng/mL (Figure 10). Mtb cell extract or culture filtrate (concentrated and unconcentrated) were analysed by α-85 B sandwich LFIA. Unfortunately, 85 B was not detectable in any sample in this assay.

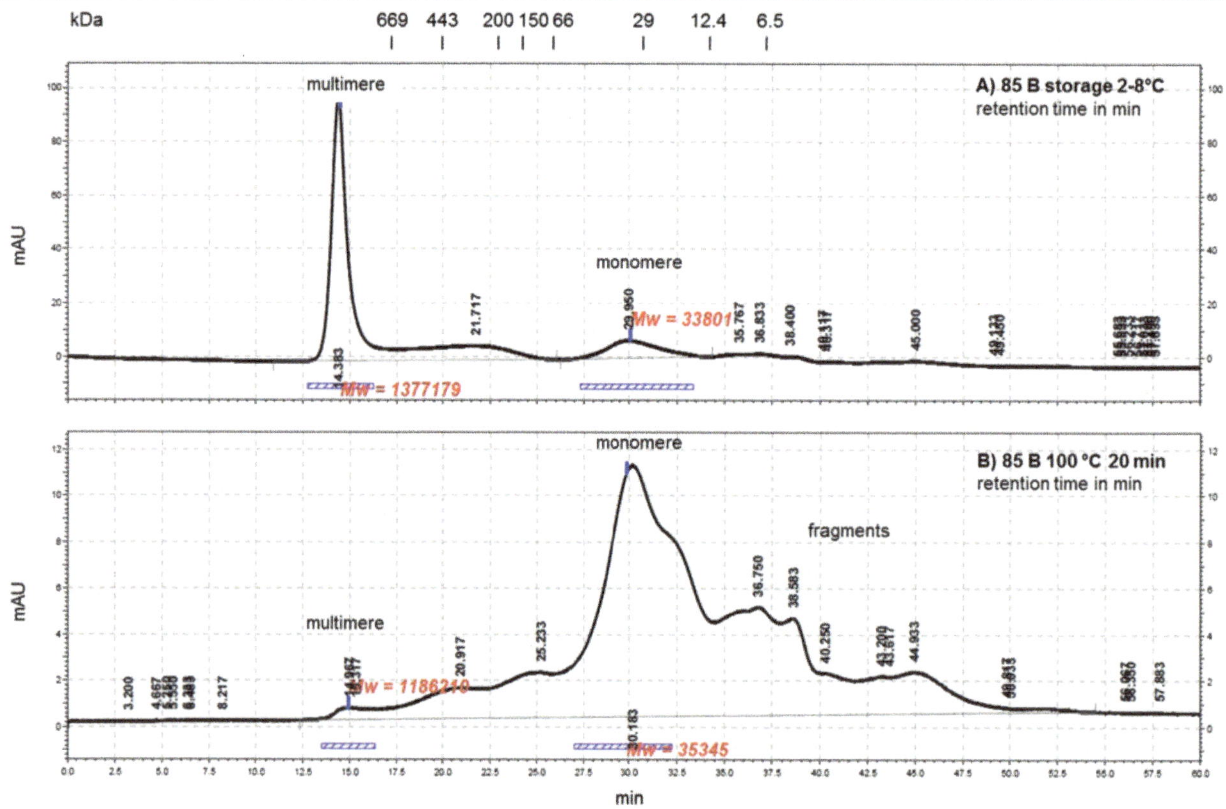

Figure 8 Analytical SEC of antigen 85 B (batch 09-2/1). A) Sample stored at 2 – 8°C, **B)** sample boiled for 20 min at 100°C. 500 μL of sample were applied to Superdex 200 10/300 GL column. Running buffer was PBS pH 7.4. Standards: 1. 669 kDa 17 min, 2. 443 kDa 20 min, 3. 200 kDa 23 min, 4. 150 kDa 24 min, 5. 66 kDa 26 min, 6. 29 kDa 31 min, 7. 12.4 kDa 34 min, 8. 6.5 kDa 37 min.

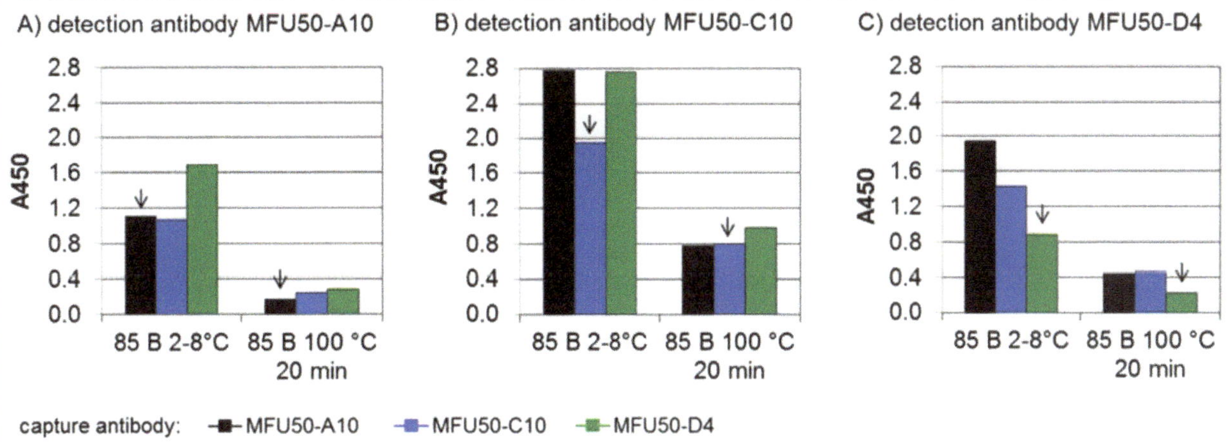

Figure 9 Detection of 85 B multimers in α-85 B sandwich ELISA. 100 ng of capture scFv-Fc were coated to the wells. After blocking 100 ng of 85 B in PBS (untreated or boiled for 20 min at 100°C) were applied to the wells and incubated. Bound antigen was detected using **A)** MFU50-A10 **B)** MFU50-C10 and **C)** MFU50-D4-scFv-Fc-HRP at ~0.05 μg/mL and development with TMB. Negative control BSA A_{450} = 0.05. Multimer-detection is indicated by arrows.

Development of an immunoblot assay for the detection of 85B

Sandwich detection of antigen 85 B in Mtb culture filtrates and cell extracts was complex in previously described sandwich ELISA and LFIA. As an alternative a direct α-85 B immunoblot assay was developed to avoid the challenge of sample pretreatment in sandwich detection. Furthermore, Mtb was cultured in Sauton's minimal medium and culture filtrates were concentrated to increase the antigen quantity. Weekly samples were taken, concentrated 20 to 30 fold and analysed by reducing gel analysis via Tape Station and α-85 B immunoblot. No 85 B was detected in 7 and 16 days old cultures by either means (data not shown). After 22 days a protein band at ~30 kDa (according to Tape Station analysis, Figure 11B) was recognized by MFU50-A10 in immunoblot (Figure 11A). Additionally, 85 B expression was found in 61 and 70 days old cultures (data not shown). The Tape Station and α-85 B immunoblot results for 22 and 76 days old cultures are given in Figure 11.

Discussion

Tuberculosis is the leading cause of death due to bacterial infections worldwide [1]. State of the art TB detection

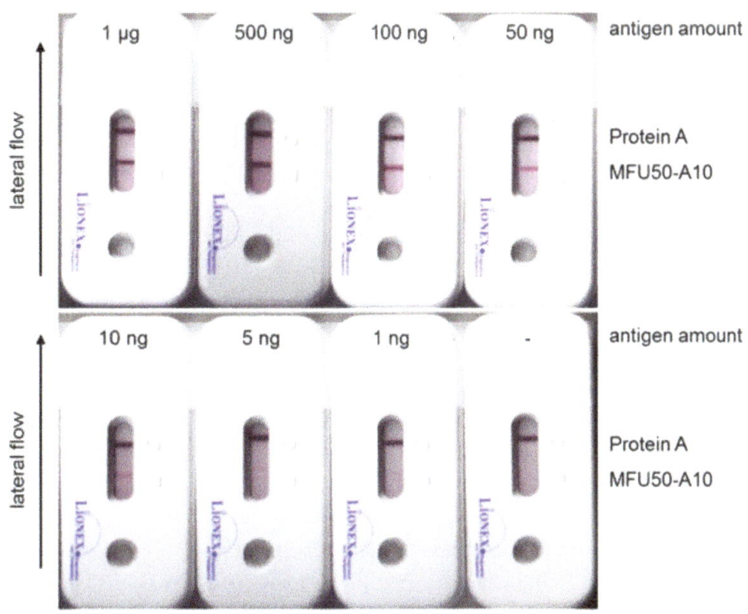

Figure 10 A-85 B sandwich LFIA, determination of detection limit. 100 μL sample (7H9 medium spiked with 85 B) + 50 μL 2.2 mM borax pH 8.8 containing 8% BSA were applied on sample pad, read out after 20 min.

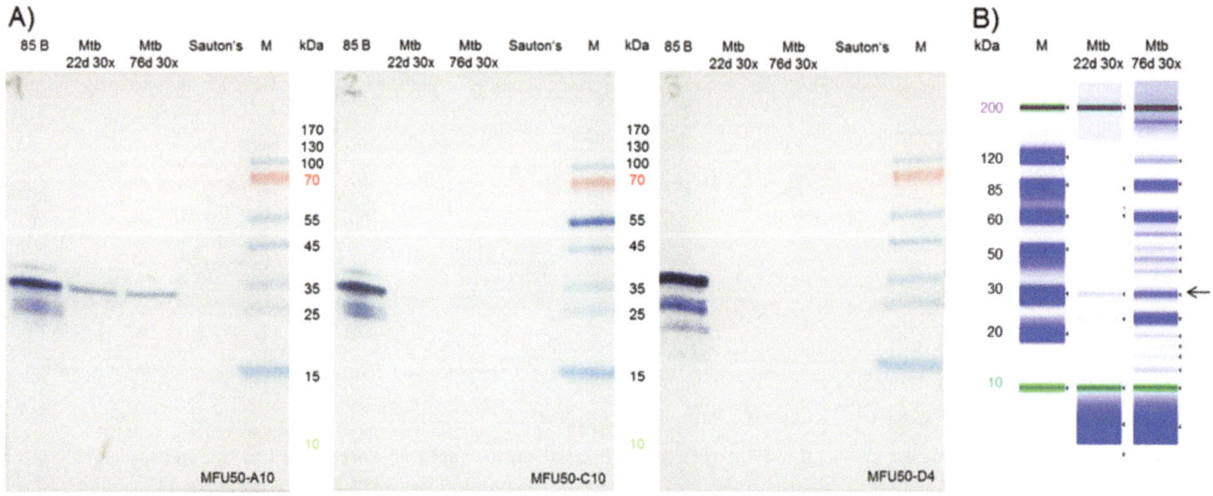

Figure 11 Detection of 85B by immunoblot. A) A-85 B immunoblot and **B)** Tape Station analysis of concentrated Mtb culture filtrates (22 d and 76 d at 37°C). A) 100 ng of recombinant 85 B, 25 μL of concentrated Mtb culture filtrates (after 22 or 76 days at 37°C in Sauton's medium) and 8 μL of marker (M) were separated on reducing 12% SDS-PAGE and electro blotted to a PVDF membrane. After blocking, blots were incubated with scFv-Fc-HRP, development with TMB. B) Samples were directly analysed by reducing gel analysis via Tape Station.

is mostly time consuming or needs expensive infrastructure [39]. In developing countries, a rapid, cheap and easy to use POC TB test is needed [9]. The WHO does not recommend the detection of antibodies against Mtb [10]. The only available assay for direct Mtb antigen detection is that against mycobacterial Lipoarabinomannan (LAM). However this test displays poor sensitivity apart from in patients with advanced HIV/AIDS [40].

In this work, we isolated and characterized specific antibodies against Mtb antigen 85 B. These antibodies were proved suitable for 85 B sandwich detection in ELISA and LFIA. In addition, an Mtb immunoblot assay was developed.

Five unique antibodies were selected from the human naïve libraries HAL7/8 [33]. The examined antibodies displayed homogeneity in the subfamily of the variable gene segment of the heavy chain (all V_H3). This correlates with the overrepresentation of V_H1 and V_H3 in the HAL7/8 libraries [41] and *in vivo* [42]. Four antibodies have a lambda light chain, one a kappa light chain. According to Løset et al. [43] lambda scFv are expressed in higher yields in *E. coli* compared to kappa scFv, which may lead to an advantage in phage display [33,41]. The antibody MFU50-C10 has a germline combination of V_H3 and $V_\kappa1$, which is common *in vivo* and in naïve libraries [33,42,44]. The lambda germline sequences present in the other antibodies are from the subfamilies 2, 3, 7 and 8. Subfamilies Vλ1,2 and 3 are dominant *in vivo*, whereupon Vλ7 and 8 are rare *in vivo* [42,44]. In the HAL7 library Vλ7 and 8 are more frequent than *in vivo* [33].

Ferrara et al. [22] selected 48 antibodies directed against the Mtb 85 complex in a combination of yeast and phage display. None of these antibodies were specific for an individual 85 antigen but, cross reacted with all other 85 complex proteins. Landowski et al. [45] isolated an α-85 B chicken IgY which cross reacted with 85 complex proteins. Drowart et al. [21] generated seven monoclonal α-85 complex antibodies, which all cross reacted with other mycobacterial species however none were specific for antigen 85 B. This study shows the isolation of the first human recombinant antibodies (MFU50-C10, MFU50-D7 and MFU50-E2) specific for antigen 85 B. Furthermore four α-85 A antibodies (MFU12-D8, MFU53-A3, MFU53-F3 and MFU53-G2) and three α-85 D antibodies (MFU51-A6, MFU51-B10 and MFU53-C2) were generated (unpublished data). MFU12 -D8 was found to be specific for antigen 85 A and MFU51-B10 was specific for 85 D. These antibodies allow for the discrimination between individual components of the 85 complex. Antibody phage display using human naive antibody gene libraries allows for the selection against non-immunogenic proteins and epitopes. We propose these epitopes, which would allow discrimination between the different 85 proteins, are not immunogenic. We propose this because of the fact that no antibodies, specific for one component of the 85 complex, were isolated by hybridoma technology [21,23].

Easy applicability to *in vitro* assays was showed in this study by conjugation to colloidal gold or HRP. Due to the recombinant nature of the explored antibodies, they can easily be altered to different formats, fused to different Fc-parts [34] or fused to markers such as green fluorescent protein (GFP, [46]) for *in vivo* experiments. Thereby the role of particular 85 proteins in Mtb cell wall biosynthesis and evasion of the host's immune response may be investigated.

All antibodies generated in this study recognized continuous sequences of the antigen, and the corresponding epitopes were determined by screening overlapping peptides immobilized on a cellulose membrane. MFU50-D4 recognized the epitope "AFSRPGLPVEYL" and MFU50-D7 recognized the epitope "AFSRPGLPV". This epitope region seems to be a potent T cell antigen since synthetic peptides including this sequence were found to induce cytokine release or immune cell proliferation in peripheral blood of individuals with varying TB status [47-50]. However only weak human B cell responses against these peptides were reported [49,51]. Interestingly, no antibodies against this epitope were generated by immunization. In this work, human antibodies were successfully generated by screening phage display libraries, fortifying the advantage of phage display technology over conventional immunization methods. Protein sequence comparison of the 85 complex proteins revealed the presence of the epitope in antigen 85 A and 85 C, suggesting cross reactivity. In ELISA and immunoblot analysis MFU50-D7 showed no 85 A binding, however MFU50-D4 showed slight cross reactivity with 85 A. During this study 85 C was not available for examination. Furthermore, a protein blast search (BLASTP, [52]) disclosed the existence of the entire epitope in several different mycobacterial strains (i.e. *M. bovis, M. ulcerans, M. marinum, M. smegmatis, M. vaccae, etc.*), suggesting cross reactivity.

The α-85 B antibodies MFU50-A10 and MFU50-E2 recognize the epitopes "SSDPAWERNDPT" and "SSDPAWERN" respectively. Landowski et al. [45] generated an oligoclonal chicken IgY antibody against the peptide "SSDPAWERNDPT" (epitope ID 60953, IEDB) and demonstrated 55% sensitivity and 85% specificity for detection of circulating 85 B in human blood by an immunoblot approach. Shen et al. [51] reported the synthetic peptide "GPSSDPAWERNDPTQ QIPKL" (epitope ID 21797, IEDB) was recognized by sera (IgG) of TB + individuals. Similar peptides containing the epitope "SSDPAWE RNDPT" were found to induce cytokine release or T cell proliferation in various assays with samples of TB + individuals [49,50,53-55]. The entire amino acid sequence "SSDPAWERNDPT" is present only in 85 B of *M. tuberculosis* and *M. bovis* BCG. Homologous sequences in Mtb 85 A or 85 C and in 85 complex proteins from other mycobacterial species (identified by BLASTP) differed primarily by substitution of one amino acid. Cross reactivity of MFU50-A10 with Mtb 85 A and *M. bovis* BCG cell extract and culture filtrate was detected by indirect ELISA and immunoblot. Landowski et al. [45] found a "SSDPAWERNDPT" specific chicken IgY antibody to be cross reactive with Mtb 85 A and 85 C as well as 85 complex proteins of *M. avium*. The reaction of MFU50-A10 with 85 A, B and D indicates a possible smaller epitope than determined by epitope mapping. The amino acid sequence "DPA" is present in antigen 85 A, B and C, but the adjacent amino acids differ. In 85 B the "DPA" is flanked by serine on the N-terminal site and tryptophan on the C-terminal site ("SDPAWE"). In 85 A glutamic acid is N-terminal and tryptophan C-terminal ("EDPAWQ"). In 85 D serine is N-terminal and alanine C-terminal ("SDPAAM"). It seems possible that the epitope of MFU50-A10 is "SDPAW" or "DPAWE", which would mean that two amino acids are different in 85 A or 85 D respectively. This would explain the weaker recognition of 85 A and 85 D. Furthermore, the comparison of the 3D structures of antigen 85 A and 85 B (Figure 12) revealed that the glutamic acid in 85 B forms a pin structure in front of the "DPAW" region. This formation is missing in 85 A, instead of glutamic acid there is glutamine. Glutamic acid can mediate strong electrostatic attractions and hydrogen bridges through the loaded carboxylgroup, whereupon glutamine is uncharged and can only mediate hydrogen bridges through the amino- and the ketogroup [56]. Thus, it is reasoned that the glutamic acid pin structure is needed structurally for correct docking, and physico-chemically for full strength binding to the antigen.

MFU50-C10 recognizes the epitope "SPAVYL", which is close to the suggested active site of antigen 85 B [36], offering a possible inhibitory effect by steric hindrance [57]. Synthetic peptides including this epitope were found to induce cytokine release and T cell proliferation in peripheral blood mononuclear cells of TB + humans [49,50,54,58] and antibodies against these peptides were detected in sera of TB + individuals [51,59]. Sequence comparison of Mtb 85 complex proteins revealed no homology in this area. According to BLASTP equivalent sequences are present in several other mycobacterial species (i.e. *M. bovis, M. marinum, M. leprae, M. vaccae, M. ulcerans, M. avium*, etc.). No cross reactivity with 85 A and 85 D was detected by indirect ELISA and immunoblot, suggesting specificity for antigen 85 B. Surprisingly, no cross reactions with *M. bovis* BCG cell extract and culture filtrate were detected in an indirect ELISA.

Sandwich ELISA detection of recombinant 85 B with the α-85 B Yumabs was successful in all performed assays. The most suitable combination was capturing with MFU50-A10 and detection with MFU50-C10-HRP, reaching a detection limit of ~10 ng/mL. Sandwich LFIA detection of recombinant 85 B was successful capturing with MFU50-A10 and detecting with MFU50-D4-gold. A detection limit of 5 ng/strip (0.03 ng/mL) was evaluated. There are no published reports on single antigen 85 B sandwich detection, only whole 85 complex sandwich assays [20,22-24,39,45,60].

Sandwich detection of recombinant 85 B was enhanced by multimeric antigen conformation, and reduced by

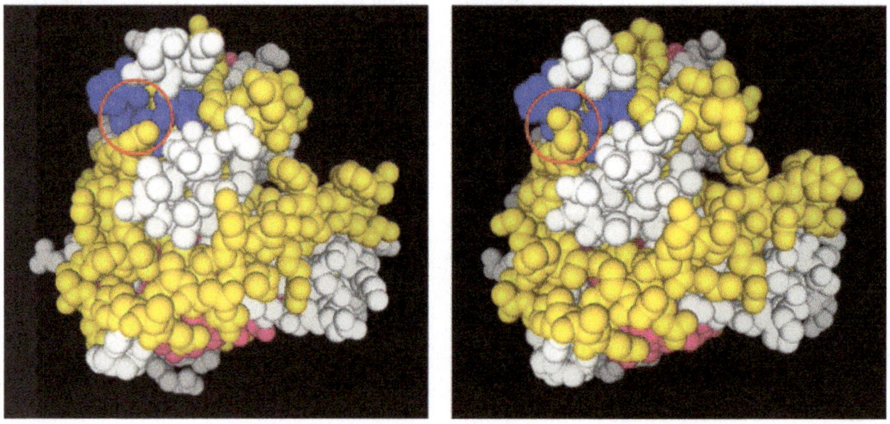

Figure 12 3 D structures of antigens A) 85 A and B) 85 B around the amino acid sequence "DPAW". A) Pdb1sfr (Resolution 2.7 Å, [75]) and B) Pdb1f0n (Resolution 1.9 Å, [36]) were modified with 3D molecule viewer (Invitrogen). Respectively only protein chain A is shown, atoms are displayed as space filling balls, "DPAW" is marked in blue, glutamic acid in 85 B is encircled in comparison to glutamine in 85 A.

monomeric antigen conformation. The small size of the antigen (~5 nm diameter, [36]) may be responsible for this finding. Considering the length of a human scFv of ~ 4.5 nm [37] and the Fc-mediated homodimerisation of scFv-Fc fusions [34,61], sterical inhibition by the capture antibody is possible. The commercially available immunochromatographic assays for MPT64 antigen detection in Mtb cultures are sandwich assays with monoclonal antibodies [62]. The Capilia TB Test Kit [63] uses only one monoclonal antibody for a sandwich assay, indicating multimer detection. Recombinant antigen MPT64 showed multimerization by disulphide bonds in two unrelated studies [64,65] and no difference in immunogenicity compared to native antigen [64,66]. These findings imply multimerization of native MPT64 and that these aggregates are detected in sandwich assays. Detection of 85 B in culture filtrates was only possible with MFU50-A10 in direct ELISA and immunoblot. MFU50-A10 was cross reactive with 85 A and 85 D in ELISA and immunoblotting. It has been reported to cross react with 85 C by Landowski et al. [45].

Conclusions

This work identifies the first recombinant human scFv antibodies specific for antigen 85 B selected by phage display from naïve antibody gene libraries (HAL7/8). These antibodies allow the discrimination between 85 complex proteins, and showed suitability for the establishment of different assays for Mtb 85 B detection. Antigen conformation influenced by sample treatment was shown to be important. The presented antibodies are candidates for the future development of a POC TB diagnostic kit. However, for this purpose an affinity maturation of the antibodies would be necessary to improve the sensitivity.

Material and methods

All chemicals used were p.a. purity grade and purchased from Sigma Aldrich, Merck or Carl Roth (all located in Germany) or as indicated otherwise.

Antigen purification

Genes fbpA, B, D coding for proteins of antigen 85 complex of *M. tuberculosis* were amplified by PCR and inserted into pET expression vectors (Novagen, Germany) by site directed cloning (Table 3).

Exchanges L78Q and S196T within the derived amino acid sequence of 85B, and exchange F54L within 85C were detected by DNA sequencing.

Recombinant 85A was produced in *E. coli* strain BL21 (DE3) [pLysS] (Novagen, Germany) by growth in LB [67] containing 400 µg/mL ampicillin and 34 µg/mL chloramphenicol. Expression was induced by addition of 1 mM IPTG at mid-log-phase. Cells were harvested 4 hours after induction. 85 B and 85 C were produced using BL21(DE3) (Novagen, Germany) as host strains grown in LB containing 50 µg/mL kanamycin. After induction with 1 mM IPTG at mid-log-phase, cells were incubated overnight and then harvested.

Table 3 Cloned genes of M. tuberculosis antigen 85 complex and properties of the recombinant gene products

Gene	Vector	Gene product		
		Antigen	Signal sequence	His-tag
fbpA (Rv3804c)	pET26b	85 A	No	N-terminal
fbpB (Rv1886c)	pET22b	85 B	Yes	no
fbpD (Rv3803c)	pET28c	85 D	Yes	N-terminal

85 B purification

Cells were suspended in 10 mM tris/HCl, pH 8.0 containing 1% Triton X-100 and 10 mM EDTA (10 mL per g wet weight). After cell disruption by sonification and centrifugation the pellet was washed once using the same buffer. The pellet was denatured in 20 mM tris/HCl, pH 8.0, 1% Triton X-100 buffer containing 8 M urea and 2 mM DTT.

After centrifugation the supernatant containing the denatured 85 B antigen was bound on a Q-Sepharose High Performance column (GE Healthcare) and eluted by a linear gradient from 0 to 500 mM NaCl. Fractions containing 85 B antigen were pooled and underwent a buffer exchange step on a Sephadex G25 fine column (GE Healthcare) into the buffer described above without Triton X-100. The protein containing pool was subjected to another chromatography step on a Q-Sepharose High Performance column (GE Healthcare). Again protein elution was performed by a linear gradient from 0 – 500 mM NaCl. Pure antigen 85 B containing fractions were pooled and underwent a final refolding and buffer exchange step on a Sephadex G25 fine column (GE Healthcare) into 10 mM NH_4HCO_3, 100 µM DTT, pH 7,9.

The final protein preparation is a mixture of protein containing the signal sequence and lacking the signal sequence. Approximately 90% of the protein shows the signal sequence and approximately 10% lacks the signal sequence.

85 A purification

Cells were suspended in 20 mM tris/HCl, 100 mM NaCl, pH 8.0 (10 mL per gram wet weight). After cell disruption by sonification and centrifugation, imidazole was added to the supernatant to a final concentration of 5 mM. This solution was applied on a Ni-NTA Superflow column (GE Healthcare) and eluted by a linear gradient from 0 to 500 mM imidazole. Fractions containing 85 A antigen were pooled and underwent a buffer exchange step on a Sephadex G25 fine column (GE Healthcare) into 20 mM tris/HCl, pH 8.0. Chromatography using a Q-Sepharose High Performance column (GE Healthcare) was performed on the pooled protein sample. Protein elution was performed using a linear gradient from 0 – 500 mM NaCl. Pure antigen 85 A containing fractions were pooled and underwent a final buffer exchange step on a Sephadex G25 fine column (GE Healthcare) into 10 mM NH_4HCO_3, pH 7.9.

85 D purification

Cells were suspended in 20 mM tris/HCl, 100 mM NaCl, pH 8.0 containing 1% Triton X-100 (10 mL per gram wet weight). After cell disruption by sonification and centrifugation the pellet was washed once using 20 mM tris/HCl, 100 mM NaCl, pH 8.0 containing 1% Triton X-100. Afterwards the pellet was denatured in 20 mM tris/HCl, 100 mM NaCl, pH 8.0 containing 8 M urea. After centrifugation the supernatant containing the denatured 85 D antigen underwent a refolding step on a Sephadex G25 fine column (GE Healthcare) into 20 mM tris/HCl, 100 mM NaCl, pH 8.0. The protein peak of this refolding step was collected and then bound on a Q-Sepharose High Performance column (GE Healthcare). Protein elution was performed by a linear gradient from 0 to 300 mM NaCl. Fractions containing pure 85 D antigen were pooled and underwent a buffer exchange step on a Sephadex G25 fine column (GE Healthcare) into 10 mM NH_4HCO_3, pH 7,9. For final endotoxin removal the protein solution was passed four times over a 1 mL Profos endotrap red column (Hyglos GmbH).

Selection of human antibodies using phage display

ScFv were isolated in vitro from the human naïve phage libraries HAL7/8 by panning on decreasing amounts (10, 3 and 1 µg) of immobilized recombinant antigen 85 B according to [68].

Production of soluble scFv antibodies for screening ELISA

The identification of monoclonal binders was performed as described in [69] with the following modifications: 96-well polypropylene (PP) micro titer plates (Greiner, Germany) containing 150 µL 2xYT-G/A (2xYT [67] supplemented with 100 mM glucose and 100 µg/mL ampicillin) were inoculated with colonies from the titration plate of the third panning round. Micro titer plates (MTP) were incubated overnight at 37°C and 850 rpm in a MTP shaker (PST-60HL-4, Lab4you, Austria). A volume of 180 µL 2xYT-G/A per well in PP-MTP was inoculated with 10 µL of the overnight culture and grown at 37°C and 850 rpm for two hours. Bacteria were harvested by centrifugation for 10 min at 3,220 × g and 180 µL supernatant were removed. The pellets were resuspended in 180 µL 2xYT supplemented with 50 mM sucrose and 100 µg/mL ampicillin + 50 µM IPTG and incubated at 30°C and 850 rpm overnight. Bacteria were pelleted by centrifugation for 15 min at 3,220 × g and 4°C. The scFv-containing supernatant was transferred to a new PP-MTP and stored at 4°C prior to analysis.

Screening ELISA

96 wells of Microlon MTP (Greiner, Germany) were coated with 100 ng of recombinant 85 B in 100 µL PBS buffer pH 7.4 or BSA in 100 µL PBS buffer pH 7.4 (Carl Roth, Germany) as a negative control [67] overnight at 4°C. After coating, the wells were blocked with 300 µL PBST-B (PBS supplemented with 0.1% (w/v) Tween-20 and 1% (w/v) BSA) for 2 h at RT. This was followed by three washing steps with PBST0.05 (PBS supplemented with 0.05% (w/v) Tween-20). For identification of binders, supernatants containing monoclonal scFv were incubated

in the antigen coated plates for 1.5 h at RT followed by three PBST washing cycles. Bound scFv were detected using mouse α-c-Myc-tag 9E10 (culture supernatant, 100 μL, 1:1,000 in PBST-B; 1.5 h at RT) followed by goat α-mouse IgG (Fc)-HRP (A0168, Sigma-Aldrich, Germany) (100 μL, 1:30,000 in PBST-B; 45 min at RT). After three washing steps with PBST0.05 the reactions were visualized with 100 μL 3,3',5,5'-tetramethylbenzidine (TMB, Seramun, Germany) as a substrate. The staining reaction was terminated by addition of 100 μL 0.2 M H_2SO_4. Absorbance at 450 nm (620 nm reference) was measured using MRX ELISA reader (Dynatec, Germany).

Antibody titration ELISA

For each antibody, 24 wells of Greiner Microlon MTP were coated with 100 ng antigen in PBS at 4°C overnight. BSA was coated as a negative control. After coating, the wells were blocked with PBST-B for 2 h at RT, followed by three washing steps with PBST0.05. Twelve dilutions of antibody (differing, depending on the antibody) in PBST-B were applied in duplicates on the antigen and BSA controls and incubated for 1.5 h at RT. Bound scFv-Fc were detected using goat α-human IgG (Fc)-HRP (A0170, Sigma-Aldrich, Germany) (100 μL, 1:130,000 in PBST-B; 45 min at RT). The assay was further processed as described above.

Antigen titration ELISA

Antigen in PBS at twelve dilutions (1 μg/mL to 0.5 ng/mL) were coated in duplicates to wells of Greiner Microlon MTP at 4°C overnight. BSA coated wells were used as a negative control. After coating, the wells were blocked with PBST-B for 2 h at RT, followed by three washing steps with PBST. Antigen detection was carried out with a concentration of antibody at half maximal saturation (determined by antibody titration ELISA) in PBST-B for 1.5 h at RT followed by three PBST washing cycles. Bound scFv-Fc were detected using goat α-human IgG (Fc)-HRP (100 μL, 1:130,000 in PBST-B; 45 min at RT). The assay was further processed as described above.

Direct ELISA

Analyte (antigen or antibody) was coated to the surface of Greiner Microlon 96 Well MTP at various concentrations in PBS buffer overnight at 4°C. Detection of antigen was carried out with an scFv-Fc antibody conjugated to HRP (by EZ-Link Plus Activated Peroxidase Kit, Pierce, Germany, according to the manufacturer's instructions) diluted in PBST-B for 1 h at RT. Detection of antibody was carried out with Goat a-human IgG (Fc)-HRP diluted (100 μL, 1:130,000 in PBST-B) 45 min at RT. The assay was further processed as described above.

Sandwich antigen titration

100 ng of capture antibodies were coated to the surface of 96 wells of Greiner Microlon MTP in PBS buffer overnight at 4°C. After coating, the wells were blocked with PBST-B for 2 h at RT, followed by three washing steps with PBST. Twelve dilutions of antigen in PBST-B were applied in duplicates onto the antibody coated wells and incubated for 1.5 h at RT, followed by three washing steps with PBST. Detection of bound antigen was performed with HRP conjugated scFv-Fc (exact dilutions described in results) in PBST-B for 1.5 h at RT, followed by three washing steps with PBST. The assay was further processed as described above.

Epitope Mapping

The protein sequence of 85 B (Rv1886c, UniProt ID P0C5B9, sequence source: http://genolist.pasteur.fr/TubercuList/) was divided into overlapping peptide fragments, each consisting of 15 amino acids, with an offset of three amino acids. This array of peptides was synthesized by the SPOT technique [70,71] on an amino-pegylated cellulose membrane (AIMS Scientific Products GmbH, Wolfenbüttel, Germany) as described previously [72]. Peptides are N-terminally acetylated and remain covalently attached to the membrane via their carboxy-terminus. The membrane bound peptide array was probed with the α-85 B antibodies for binding according to established procedures [30,72].

Immunoblot

Proteins were separated on 12% SDS-PAGE [73,74] and semi-dry blotted onto polyvinylidene fluoride (PVDF) membranes (Carl Roth, Germany) according to the manufacturer's instructions. Blocking was performed with MPBST (PBST0.1 supplemented with 2% (w/v) dry milk) for minimum 30 min at RT. Subsequent incubation with protein-specific antibody was carried out in MPBST for a minimum of 60 min at RT. Three washing steps with PBST0.1 for 5 min were performed. The PVDF membrane was incubated with a secondary antibody coupled to HRP in MPBST for at least 45 min at RT. The blot was washed another three times with PBST0.1 and then developed with TMB peroxidase membrane substrate (Seramun, Germany) until an adequate signal was obtained. Development was stopped by three short washing steps with deionized water.

LFIA assembly + execution

5 – 40 μL cm^{-1} of colloidal gold-antibody conjugates were dispensed onto 8 mm glass fibre pads (Millipore, Germany) using the xyz-dispenser (Biodot, USA) and dried at 37°C. 1 μL cm^{-1} of different concentrations of capture antibody solution (sandwich assay) or antigen (direct assay) in PBS were dispensed in a line onto Unisart CN 95 nitrocellulose

membranes (20 mm, Sartorius, Germany) that were already assembled onto 300 mm backing cards (Jieyi Biotechnology, China), and dried at 37°C. The glass fibre conjugate pads were laminated onto the backing cards, overlapping the nitrocellulose membrane at the connection point for ~ 0.5 – 1 mm. Then cellulose fibre pads (Millipore, Germany) were laminated onto the backing cards as sample and wicking pad, overlapping the nitrocellulose membrane at the connection point for ~ 0.5 – 1 mm. The assembled cards were cut into strips of 0.4 cm width with the CM4000 guillotine-cutter (Biodot) and the test strips were placed into Lateral Flow Strip Test (LFST) cassettes (Jieyi Biotechnology, China).

Several experiments were carried out to determine a procedure that would allow sensitive antigen detection and a low background combined with good feasibility (data not shown). The best option for a manageable LFIA was to add a diluent to the sample before applying it to the sample well. Therefore 1/3 sample volume (50 μL) of conjugation buffer (2 mM borax pH 8.8) enriched with 8% (w/v) BSA were mixed with 100 μL sample and applied to the LFIA. After 15 – 20 min results were optically evaluated.

Yumab production + purification

All scFv isolated from antibody-phage display were subcloned into pCSE2.5-hIgG1-Fc-XP and produced as scFv-Fc (yumab) in HEK293-6E cells (National Research Council (NRC), Biotechnological Research Institute (BRI), Montreal, Canada). HEK293-6E cells were cultured in chemical defined F17 medium (Invitrogen, Life Technologies, Darmstadt, Germany) supplemented with 1 g/L pluronic F68 (Applichem, Darmstadt, Germany), 4 mM L-glutamine (PAA) and 25 mg/L G418 (PAA) as previously described [34]. DNA was transiently transfected into 25 mL HEK293-6E cells in 125 mL Erlenmeyer shake flasks. After 48 hours of cultivation at 110 rounds per minute (rpm) in a Minitron orbital shaker (Infors, Bottmingen, Switzerland) at 37°C and 5% CO_2 atmosphere, one volume culture medium and a final concentration of 0.5% (w/v) of tryptone N1 (TN1, Organotechnie S.A.S., La Courneuve, France) were added. ScFv-Fc were purified via UNOsphere SUPrA column (Biorad, Munich, Germany) using a Profinia automate (Biorad, Hercules, California, USA), according to the manufacturer's instructions.

Analytical SEC

For analytical size exclusion chromatography (SEC) a Knauer PLATINblue HPLC Plus system (Knauer, Germany) with a Superdex 200 10/300 GL column (GE Healthcare, USA) was used according to the manufacturer's instructions. The column was equilibrated with the Gel Filtration HMW/LMW Calibration Kits (GE Healthcare, USA) according to the manufacturer's instructions.

Tape Station

Protein solutions were analysed with the Screen Tape P200 Protein Standard Kit (Agilent, USA) under reducing conditions on a 2200 Tape Station system (Agilent, USA) according to the manufacturer's instructions. The "P200 molecular weight standard" (Agilent) was used as a molecular marker.

Additional file

Additional file 1: Figure S1. Screening ELISA for 85 B binding scFv in HAL7/8. Culture supernatants containing soluble scFv of 92 single clones (3. panning round) were screened for their ability to bind antigen (85 B) and BSA (negative control). On positions H9 and H12 an anti-lysozyme antibody was used on lysozyme as control for scFv production and ELISA detection system. Detection of bound scFv with mouse α-c-Myc-tag 9E10 IgG followed by goat α-mouse IgG(Fc)-HRP, development with TMB.

Competing interests

The authors have no competing financial or non-financial interests. This study was financially supported by the German Federal Ministry of Commerce, Grant No. KF 2584301. All aspects of cooperation and the generated results by the partner organisations have been mentioned in the cooperation agreement of the grant as required by the German Federal Ministry of Commerce. We did not receive any other financial or personal contribution by any other third party. Currently, there are no plans to file a patent. Following the rules of the German Federal Ministry of Commerce, LIONEX reserves its rights to develop any product based on the results or materials of this manuscript but agrees to provide the materials to research organisations only, exclusively for research use.

Authors' contributions

MF performed most of the experiments, participated in the design of the study, analysed data and drafted the manuscript. SK, SH, WP, RS, WO and RF performed some of the experiments, analysed data and provided material. SD participated in the design of the study and analysed data. MS and MH conceived the project and wrote the grant application, participated in the design of the study, analysed data and drafted the manuscript. All authors read and approved the final manuscript.

Acknowledgements

We kindly acknowledge the technical support of Susanne Daenicke and Franziska Resch. Further, we acknowledge the financial support of ZIM BMBF (contract No. KF 2584301). We also thank Elizabeth O'Brien for the careful correction of this manuscript.

Author details

[1]Technische Universität Braunschweig, Institut für Biochemie und Biotechnologie, Spielmannstr.7, 38106 Braunschweig, Germany. [2]Lionex GmbH, Salzdahlumer Str. 196, 38126 Braunschweig, Germany. [3]Helmholtz Zentrum für Infektionsforschung, Abteilung Chemische Biologie, Inhoffenstr. 7, 38124 Braunschweig, Germany. [4]Leibniz-Institut für Molekulare Pharmakologie, Robert-Roessle-Str. 10, 13125 Berlin, Germany.

References

1. WHO: *Global Tuberculosis Report 2012*. Geneva, Switzerland: 2012.
2. Dye C, Scheele S, Dolin P, Pathania V, Raviglione MC: **Consensus statement. Global burden of tuberculosis: estimated incidence, prevalence, and mortality by country. WHO Global Surveillance and Monitoring Project.** *JAMA* 1999, **282:**677–686.
3. Kumar V, Robbins SL: *Robbins basic pathology*. 8th edition. Philadelphia, PA: Saunders/Elsevier; 2007.
4. Helb D, Jones M, Story E, Boehme C, Wallace E, Ho K, Kop J, Owens MR, Rodgers R, Banada P, Safi H, Blakemore R, Lan NTN, Jones-Lopez EC, Levi M,

Burday M, Ayakaka I, Mugerwa RD, McMillan B, Winn-Deen E, Christel L, Dailey P, Perkins MD, Persing DH, Alland D: **Rapid Detection of Mycobacterium tuberculosis and Rifampin Resistance by Use of On-Demand, Near-Patient Technology.** *J Clin Microbiol* 2010, **48:**229–237.

5. Rai DR, Kshetry NT, Bhargava D, Pokhrel BM: **Comparison of Ziehl-Neelsen staining microscopy and immunochromatographic tuberculosis test for diagnosis of pulmonary tuberculosis.** *J Inst Med* 2006, **28:**15–18.
6. van Deun A, Portaels F: **Limitations and requirements for quality control of sputum smear microscopy for acid-fast bacilli.** *Int J Tuberc Lung Dis* 1998, **2:**756–765.
7. Steingart KR, Ng V, Henry M, Hopewell PC, Ramsay A, Cunningham J, Urbanczik R, Perkins MD, Aziz MA, Pai M: **Sputum processing methods to improve the sensitivity of smear microscopy for tuberculosis: a systematic review.** *Lancet Infect Dis* 2006, **6:**664–674.
8. Perkins MD, Cunningham J: **Facing the crisis: improving the diagnosis of tuberculosis in the HIV era.** *J Infect Dis* 2007, **196**(Suppl 1):S15–S27.
9. Dheda K, Ruhwald M, Theron G, Peter J, Yam WC: **Point-of-care diagnosis of tuberculosis: Past, present and future.** *Respirology* 2013, **18:**217–232.
10. WHO: *An International Roadmap for Tuberculosis Research.* Geneva, Switzerland: WHO; 2011.
11. Flores LL, Steingart KR, Dendukuri N, Schiller I, Minion J, Pai M, Ramsay A, Henry M, Laal S: **Systematic Review and Meta-Analysis of Antigen Detection Tests for the Diagnosis of Tuberculosis.** *Clin Vaccine Immunol* 2011, **18:**1616–1627.
12. Wiker HG, Harboe M: **The Antigen 85 Complex: a Major Secretion Product of Mycobacterium tuberculosis.** *Microbiol Rev* 1992, **56:**648–661.
13. Sonnenberg MG, Belisle JT: **Definition of Mycobacterium tuberculosis culture filtrate proteins by two-dimensional polyacrylamide gel electrophoresis, N-terminal amino acid sequencing, and electrospray mass spectrometry.** *Infect Immun* 1997, **65:**4515–4524.
14. Wiker HG, Sletten K, Nagai S, Harboe M: **Evidence for three separate genes encoding the proteins of the mycobacterial antigen 85 complex.** *Infect Immun* 1990, **58:**272–274.
15. Kremer L, Maughan W, Wilson R, Dover L, Besra GS: **The M. tuberculosis antigen 85 complex and mycolyltransferase activity.** *Lett Appl Microbiol* 2002, **34:**233–237.
16. Belisle JT, Vissa VD, Sievert T, Takayama K, Brennan PJ, Besra GS: **Role of the major antigen of Mycobacterium tuberculosis in cell wall biogenesis.** *Science* 1997, **276:**1420–1422.
17. Armitige LY, Jagannath C, Wanger AR, Norris SJ: **Disruption of the genes encoding antigen 85A and antigen 85B of Mycobacterium tuberculosis H37Rv: effect on growth in culture and in macrophages.** *Infect Immun* 2000, **68:**767–778.
18. Abou-Zeid C, Ratliff T, Wiker H, Harboe M: **Characterization of Fibronectin-Binding Antigens Released by Mycobacterium tuberculosis and Mycobacterium bovis BCG.** *Infect Immun* 1988, **56:**3046–3051.
19. Abou-Zeid C, Garbe T, Lathigra R, Wiker HG, Harboe M, Rook G, Young B: **Genetic and Immunological Analysis of Mycobacterium tuberculosis Fibronectin-Binding Proteins.** *Infect Immun* 1991, **59:**2712–2718.
20. Bentley-Hibbert SI, Quan X, Newman T, Huygen K, Godfrey HP: **Pathophysiology of antigen 85 in patients with active tuberculosis: antigen 85 circulates as complexes with fibronectin and immunoglobulin G.** *Infect Immun* 1999, **67:**581–588.
21. Drowart A, de Bruyn J, Huygen K, Damiani G, Godfrey HP, Stelandre M, Yernault JC, Van Vooren JP: **Isoelectrophoretic characterization of protein antigens present in mycobacterial culture filtrates and recognized by monoclonal antibodies directed against the Mycobacterium bovis BCG antigen 85 complex.** *Scand J Immunol* 1992, **36:**697–702.
22. Ferrara F, Naranjo LA, Kumar S, Gaiotto T, Mukundan H, Swanson B, Bradbury AR, Andrew RM: **Using phage and yeast display to select hundreds of monoclonal antibodies: application to antigen 85, a tuberculosis biomarker.** *PLoS One* 2012, **7:**e49535.
23. Mukundan H, Kumar S, Price DN, Ray SM, Lee Y, Min S, Eum S, Kubicek-Sutherland J, Resnick JM, Grace WK, Anderson AS, Hwang SH, Cho SN, Via LE, Barry C3, Sakamuri R, Swanson BI: **Rapid detection of Mycobacterium tuberculosis biomarkers in a sandwich immunoassay format using a waveguide-based optical biosensor.** *Tuberculosis (Edinb)* 2012, **92:**407–416.
24. Kashyap RS, Rajan AN, Ramteke SS, Agrawal VS, Kelkar SS, Purohit HJ, Taori GM, Daginawala HF: **Diagnosis of tuberculosis in an Indian population by an indirect ELISA protocol based on detection of Antigen 85 complex: a prospective cohort study.** *BMC Infect Dis* 2007, **7:**74.
25. Kashyap RS, Dobos KM, Belisle JT, Purohit HJ, Chandak NH, Taori GM, Daginawala HF: **Demonstration of components of antigen 85 complex in cerebrospinal fluid of tuberculous meningitis patients.** *Clin Diagn Lab Immunol* 2005, **12:**752–758.
26. Wallis RS, Perkins M, Phillips M, Joloba M, Demchuk B, Namale A, Johnson JL, Williams D, Wolski K, Teixeira L, Dietze R, Mugerwa RD, Eisenach K, Ellner JJ: **Induction of the antigen 85 complex of Mycobacterium tuberculosis in sputum: a determinant of outcome in pulmonary tuberculosis treatment.** *J Infect Dis* 1998, **178:**1115–1121.
27. Kirsch MI, Hülseweh B, Nacke C, Rülker T, Schirrmann T, Marschall HJ, Hust M, Dübel S: **Development of human antibody fragments using antibody phage display for the detection and diagnosis of Venezuelan equine encephalitis virus (VEEV).** *BMC Biotechnol* 2008, **8:**66.
28. Meyer T, Stratmann-Selke J, Meens J, Schirrmann T, Gerlach GF, Frank R, Dübel S, Strutzberg-Minder K, Hust M: **Isolation of scFv fragments specific to OmpD of Salmonella Typhimurium.** *Vet Microbiol* 2011, **147:**162–169.
29. Meyer T, Schirrmann T, Frenzel A, Miethe S, Stratmann-Selke J, Gerlach GF, Strutzberg-Minder K, Dübel S, Hust M: **Identification of immunogenic proteins and generation of antibodies against Salmonella Typhimurium using phage display.** *BMC Biotechnol* 2012, **12:**29.
30. Schütte M, Thullier P, Pelat T, Wezler X, Rosenstock P, Hinz D, Kirsch MI, Hasenberg M, Frank R, Schirrmann T, Gunzer M, Hust M, Dübel S: **Identification of a Putative Crf Splice Variant and Generation of Recombinant Antibodies for the Specific Detection of Aspergillus fumigatus.** *PLoS One* 2009, **4:**e6625.
31. Chan C, Zhao B, Cazenave-Gassiot A, Pang S, Bendt A, Wenk M, Macary P, Hanson B: **Novel phage display-derived mycolic acid-specific antibodies with potential for tuberculosis diagnosis.** *J Lipid Res* 2013, **54:**2924–2932.
32. Lillo AM, Ayriss JE, Shou Y, Graves SW, Bradbury AR: **Development of phage-based single chain Fv antibody reagents for detection of Yersinia pestis.** *PLoS One* 2011, **6:**e27756.
33. Hust M, Meyer T, Voedisch B, Rülker T, Thie H, El-Ghezal A, Kirsch MI, Schütte M, Helmsing S, Meier D, Schirrmann T, Dübel S: **A human scFv antibody generation pipeline for proteome research.** *J Biotechnol* 2011, **152:**159–170.
34. Jäger V, Büssow K, Wagner A, Weber S, Hust M, Frenzel A, Schirrmann T: **High level transient production of recombinant antibodies and antibody fusion proteins in HEK293 cells.** *BMC Biotechnol* 2013, **13:**52.
35. Retter I, Althaus H, Münch R, Müller W: **VBASE2, an integrative V gene database.** *Nucleic Acids Res* 2004, **33:**D671.
36. Anderson DH, Harth G, Horwitz MA, Eisenberg D: **An interfacial mechanism and a class of inhibitors inferred from two crystal structures of the Mycobacterium tuberculosis 30 kda major secretory protein (antigen 85B), a mycolyl transferase.** *J Mol Biol* 2001, **307:**671–681.
37. Boehm MK, Woof JM, Kerr MA, Perkins SJ: **The Fab and Fc fragments of IgA1 exhibit a different arrangement from that in IgG: a study by X-ray and neutron solution scattering and homology modelling.** *J Mol Biol* 1999, **286:**1421–1447.
38. Artimo P, Jonnalagedda M, Arnold K, Baratin D, Csardi G, de Castro E, Duvaud S, Flegel V, Fortier A, Gasteiger E, Grosdidier A, Hernandez C, Loannidis V, Kuznetsov D, Liechti R, Moretti S, Mostaguir K, Redaschi N, Rossier G, Xenarios I, Stockinger H: **ExPASy: SIB bioinformatics resource portal.** *Nucleic Acids Res* 2012, **40:**W597.
39. Bekmurzayeva A, Sypabekova M, Kanayeva D: **Tuberculosis diagnosis using immunodominant, secreted antigens of Mycobacterium tuberculosis.** *Tuberculosis* 2013, **93:**381–388.
40. Lawn SD: **Point-of-care detection of lipoarabinomannan (LAM) in urine for diagnosis of HIV-associated tuberculosis: a state of the art review.** *BMC Infect Dis* 2012, **12:**103.
41. Frenzel A, Fröde D, Meyer T, Schirrmann T, Hust M: **Generating Recombinant Antibodies for Research, Diagnostics and Therapy Using Phage Display.** *Curr Biotechnol* 2012, **1:**33–41.
42. Knappik A, Ge L, Honegger A, Pack P, Fischer M, Wellnhofer G, Hoess A, Wolle J, Plückthun A, Virnekas B: **Fully synthetic human combinatorial antibody libraries (HuCAL) based on modular consensus frameworks and CDRs randomized with trinucleotides.** *J Mol Biol* 2000, **296:**57–86.
43. Loset GA, Lobersli I, Kavlie A, Stacy JE, Borgen T, Kausmally L, Hvattum E, Simonsen B, Hovda MB, Brekke OH: **Construction, evaluation and refinement of a large human antibody phage library based on the IgD and IgM variable gene repertoire.** *J Immunol Methods* 2005, **299:**47–62.
44. Schofield DJ, Pope AR, Clementel V, Buckell J, Chapple SD, Clarke KF, Conquer JS, Crofts AM, Crowther SR, Dyson MR, Flack G, Griffin GJ, Hooks Y,

Howat WJ, Kolb-Kokocinski A, Kunze S, Martin CD, Maslen GL, Mitchell JN, O'Sullivan M, Perera RL, Roake W, Shadbolt SP, Vincent KJ, Warford A, Wilson WE, Xie J, Young JL, McCafferty J: **Application of phage display to high throughput antibody generation and characterization.** *Genome Biol* 2007, **8:**R254.

45. Landowski CP, Godfrey HP, Bentley-Hibbert SI, Liu X, Huang Z, Sepulveda R, Huygen K, Gennaro ML, Moy FH, Lesley SA, Haak-Frendscho M: **Combinatorial Use of Antibodies to Secreted Mycobacterial Proteins in a Host Immune System-Independent Test for Tuberculosis.** *J Clin Microbiol* 2001, **39:**2418–2424.
46. Kremer L, Baulard A, Estaquier J, Poulain-Godefroy O, Locht C: **Green fluorescent protein as a new expression marker in mycobacteria.** *Mol Microbiol* 1995, **17:**913–922.
47. Huygen K, Lozes E, Gilles B, Drowart A, Palfliet K, Jurion F, Roland I, Art M, Dufaux M, Nyabenda J: **Mapping of TH1 helper T-cell epitopes on major secreted mycobacterial antigen 85A in mice infected with live Mycobacterium bovis BCG.** *Infect Immun* 1994, **62:**363–370.
48. Launois P, DeLeys R, Niang MN, Drowart A, Andrien M, Dierckx P, Cartel JL, Sarthou JL, Van Vooren JP, Huygen K: **T-cell-epitope mapping of the major secreted mycobacterial antigen Ag85A in tuberculosis and leprosy.** *Infect Immun* 1994, **62:**3679–3687.
49. Roche PW, Peake PW, Billman-Jacobe H, Doran T, Britton WJ: **T-cell determinants and antibody binding sites on the major mycobacterial secretory protein MPB59 of Mycobacterium bovis.** *Infect Immun* 1994, **62:**5319–5326.
50. Commandeur S, Lin MY, van Meijgaarden KE, Friggen AH, Franken KL, Drijfhout JW, Korsvold GE, Oftung F, Geluk A, Ottenhoff TH: **Double- and monofunctional CD4(+) and CD8(+) T-cell responses to Mycobacterium tuberculosis DosR antigens and peptides in long-term latently infected individuals.** *Eur J Immunol* 2011, **41:**2925–2936.
51. Shen G, Behera D, Bhalla M, Nadas A, Laal S: **Peptide-based antibody detection for tuberculosis diagnosis.** *Clin Vaccine Immunol* 2009, **16:**49–54.
52. Altschul SF, Gish W, Miller W, Myers EW, Lipman DJ: **Basic local alignment search tool.** *J Mol Biol* 1990, **215:**403–410.
53. Lightbody KA, Girvin RM, Pollock DA, Mackie DP, Neill SD, Pollock JM: **Recognition of a common mycobacterial T-cell epitope in MPB59 of Mycobacterium bovis.** *Immunology* 1998, **93:**314–322.
54. Valle MT, Megiovanni AM, Merlo A, Li Pira G, Bottone L, Angelini G, Bracci L, Lozzi L, Huygen K, Manca F: **Epitope focus, clonal composition and Th1 phenotype of the human CD4 response to the secretory mycobacterial antigen Ag85.** *Clin Exp Immunol* 2001, **123:**226–232.
55. D'Souza S, Rosseels V, Romano M, Tanghe A, Denis O, Jurion F, Castiglione N, Vanonckelen A, Palfliet K, Huygen K: **Mapping of murine Th1 helper T-Cell epitopes of mycolyl transferases Ag85A, Ag85B, and Ag85C from Mycobacterium tuberculosis.** *Infect Immun* 2003, **71:**483–493.
56. Murphy K, Travers P, Walport M, Janeway C: *Janeway's immunobiology.* 7th edition. New York: Garland Science; 2008.
57. Arnon R: **Enzyme inhibition by antibodies.** *Acta Endocrinol Suppl (Copenh)* 1975, **194:**133–153.
58. Mustafa AS, Abal AT, Shaban F, El-Shamy AM, Amoudy HA: **HLA-DR binding prediction and experimental evaluation of T-cell epitopes of mycolyl transferase 85B (Ag85B), a major secreted antigen of Mycobacterium tuberculosis.** *Med Princ Pract* 2005, **14:**140–146.
59. Gaseitsiwe S, Valentini D, Mahdavifar S, Magalhaes I, Hoft DF, Zerweck J, Schutkowski M, Andersson J, Reilly M, Maeurer MJ: **Pattern recognition in pulmonary tuberculosis defined by high content peptide microarray chip analysis representing 61 proteins from M. tuberculosis.** *PLoS One* 2008, **3:**e3840.
60. Nguyen YH, Ma X, Qin L: **Rapid identification and drug susceptibility screening of ESAT-6 secreting Mycobacteria by a NanoELIwell assay.** *Sci Rep* 2012, **2:**635.
61. Powers DB, Amersdorfer P, Poul M, Nielsen UB, Shalaby MR, Adams GP, Weiner LM, Marks JD: **Expression of single-chain Fv-Fc fusions in Pichia pastoris.** *J Immunol Methods* 2001, **251:**123–135.
62. Yin X, Zheng L, Lin L, Hu Y, Zheng F, Hu Y, Wang Q: **Commercial MPT64-based tests for rapid identification of Mycobacterium tuberculosis complex: A meta-analysis.** *J Infect* 2013, **67:**369–377.
63. Ngamlert K, Sinthuwattanawibool C, McCarthy KD, Sohn H, Starks A, Kanjanamongkolsiri P, Anek-vorapong R, Tasaneeyapan T, Monkongdee P, Diem L, Varma JK: **Diagnostic performance and costs of Capilia TB for Mycobacterium tuberculosis complex identification from broth-based culture in Bangkok, Thailand.** *Trop Med Int Health* 2009, **14:**748–753.
64. Geisbrecht BV, Nikonenko B, Samala R, Nakamura R, Nacy CA, Sacksteder KA: **Design and optimization of a recombinant system for large-scale production of the MPT64 antigen from Mycobacterium tuberculosis.** *Protein Expr Purif* 2006, **46:**64–72.
65. Chu TJ, Yuann JP: **Expression, purification, and characterization of protective MPT64 antigen protein and identification of its multimers isolated from nontoxic Mycobacterium tuberculosis H37Ra.** *Biotechnol Appl Biochem* 2011, **58:**185–189.
66. Oettinger T, Holm A, Haslov K: **Characterization of the delayed type hypersensitivity-inducing epitope of MPT64 from Mycobacterium tuberculosis.** *Scand J Immunol* 1997, **45:**499–503.
67. Sambrook J, Russell D: *Molecular Cloning: A Laboratory Manual.* 3rd edition. New York: Cold Spring Harbor Laboratory Press; 2001.
68. Schirrmann T, Hust M: **Construction of human antibody gene libraries and selection of antibodies by phage display.** *Methods Mol Biol* 2010, **651:**177–209.
69. Hust M, Steinwand M, Al-Halabi L, Helmsing S, Schirrmann T, Dübel S: **Improved microtitre plate production of single chain Fv fragments in Escherichia coli.** *N Biotechnol* 2009, **25:**424–428.
70. Frank R: **Spot-synthesis: an easy technique for the positionally addressable, parallel chemical synthesis on a membrane support.** *Tetrahedron* 1992, **48:**9217–9232.
71. Frank R: **The SPOT-synthesis technique. Synthetic peptide arrays on membrane supports–principles and applications.** *J Immunol Methods* 2002, **267:**13–26.
72. Frank R, Dübel S: **Analysis of protein interactions with immobilized peptide arrays synthesized on membrane supports.** In *Protein-Protein Interactions, A Molecular Cloning Manual.* 2nd edition. Edited by Golemis E, Adams PD. Cold Spring Harbor, N.Y: Cold Spring Harbor Laboratory Press; 2005.
73. Laemmli UK: **Cleavage of structural proteins during the assembly of the head of bacteriophage T4.** *Nature* 1970, **227:**680–685.
74. Schägger H, von Jagow G: **Tricine-sodium dodecyl sulfate-polyacrylamide gel electrophoresis for the separation of proteins in the range from 1 to 100 kDa.** *Anal Biochem* 1987, **166:**368–379.
75. Ronning DR, Vissa V, Besra GS, Belisle J, Sacchettini JC: **Mycobacterium tuberculosis Antigen 85A and 85C Structures Confirm Binding Orientation and Conserved Substrate Specificity.** *J Biol Chem* 2004, **279:**36771–36777.

16

Identification of an antibody fragment specific for androgen-dependent prostate cancer cells

Ryan M Williams[1,2], Cyrus J Hajiran[3], Sara Nayeem[1] and Letha J Sooter[1*]

Abstract

Background: Prostate cancer is the most-diagnosed non-skin cancer among males in the US, and the second leading cause of cancer-related death. Current methods of treatment and diagnosis are not specific for the disease. This work identified an antibody fragment that binds selectively to a molecule on the surface of androgen-dependent prostate cancer cells but not benign prostatic cells.

Results: Antibody fragment identification was achieved using a library screening and enrichment strategy. A library of 10^9 yeast-displayed human non-immune antibody fragments was enriched for those that bind to androgen-dependent prostate cancer cells, but not to benign prostatic cells or purified prostate-specific membrane antigen (PSMA). Seven rounds of panning and fluorescence-activated cell sorting (FACS) screening yielded one antibody fragment identified from the enriched library. This molecule, termed HiR7.8, has a low-nanomolar equilibrium dissociation constant (K_d) and high specificity for androgen-dependent prostate cancer cells.

Conclusions: Antibody fragment screening from a yeast-displayed library has yielded one molecule with high affinity and specificity. With further pre-clinical development, it is hoped that the antibody fragment identified using this screening strategy will be useful in the specific detection of prostate cancer and in targeted delivery of therapeutic agents for increased efficacy and reduced side effects.

Keywords: Prostate cancer, Antibody fragment, scFv, Library screening, Yeast

Background

The overall breadth of prostate cancer and problems associated with it render it necessary to develop novel therapeutics and diagnostics for the disease. Prostate cancer is the most-diagnosed non-skin cancer in the United States, with an estimated 233,000 new diagnoses in 2014 alone [1]. It is the second leading cause of cancer-related deaths among males in the U.S., with an estimated 29,480 mortalities in 2014 [1]. The scope of this disease portends the necessity in developing improved clinical tools for its treatment.

It is necessary to develop specifically targeted therapeutics and diagnostics to further aid in treatment of prostate cancer. Current practices for prostate cancer are partially effective; however they are not specific for the disease, causing many unwanted side effects and over-diagnoses. Therapeutic side effects can be serious and leave the possibility for recurrence in a more aggressive, androgen-independent form [2-4]. For many non-localized cancers, chemotherapies are used which can be effective, however specific delivery would be more effective and cause fewer side effects [5-8]. In addition to non-specific therapeutics, prostate cancer diagnosis using prostate specific antigen (PSA) blood levels is no longer recommended for use by the United States Preventive Services Task Force (USPSTF) [9-13]. This is because serum PSA levels are raised not only due to cancerous prostates, but due to benign prostatic conditions such as prostatic intraepithelial neoplasia (PIN), benign prostate hyperplasia (BPH), and prostatitis [14]. Studies have shown the specificity of the PSA test in prostate cancer diagnosis to be just 24%, meaning there is a 76% overdiagnosis rate, and that it will prevent just one prostate cancer-related death in greater than 1000 men [11,15]. Many other biomarkers identified that may have some diagnostic potential, however are not specific for prostate cancer [16-18]. Therefore, it is necessary to develop molecular targeting mechanisms for clinical use against prostate cancer.

* Correspondence: lsooter@hsc.wvu.edu
[1]Department of Basic Pharmaceutical Sciences, West Virginia University, 1 Medical Center Drive, PO Box 9530, Morgantown, WV 26506, USA
Full list of author information is available at the end of the article

An increasingly popular method to develop disease-specific targeting molecules is through antibody fragment library screening. Previous library screening methodologies have been adapted for use in identifying molecules that bind to whole-cell targets [19,20]. Additionally, screening methods exist for utilizing antibody fragment libraries displayed on bacteriophage, bacteria, and yeast [21-23]. Antibody fragments that bind to cell surfaces have largely been selected from phage-displayed libraries, with examples including those that bind to ovarian, breast, and hepatocellular carcinoma cells [24-26]. It is therefore possible to identify antibody fragments which bind to the surface of prostate cancer cells.

The work described here enriches an antibody fragment library for molecules which bind to androgen-dependent prostate cancer cells. Additionally, multiple stringent negative enrichments using targets to which the antibody fragment should not bind were selected against. This work has identified an antibody fragment which binds to androgen-dependent prostate cancer cells, and not to various benign prostate cells. It utilized a human non-immune single-chain Fragment variable (scFv) library displayed on the surface of *Sacchoromyces cerevisiae*. It is hoped that the obtained scFv will be useful for both specific treatment and diagnosis of prostate cancer and serves as proof-of-principle for future screening of disease-specific antibody fragments.

Results

Antibody fragment screening

In order to obtain androgen-dependent prostate cancer cell-specific antibody fragments, seven rounds of screening and enrichment were completed with a yeast-displayed scFv library [23] (Figure 1, Table 1). Sequencing of a random, representative sample of the scFv-encoding plasmid in the initial naïve library, as well as after Round 2 and Round 3 of panning, was performed. Out of the 30–35 sequences identified, there were no duplicate sequences in any library or between the libraries. This suggests that no particular sequence was selected for in these rounds.

After three rounds, FACS-based screening was performed. In Round 4 (+), there were two separate populations of yeast binding to target cells which were kept separate through the remaining rounds of screening and termed 'Hi' and 'Lo' (Additional file 1: Figure S1). After Round 4, the Lo population was comprised of 29% HiR7.8 (Figure 2). Interestingly, HiR7.6 and HiR6.8 were not identified in the Lo population. HiR7.8 comprised 100% of the population in Rounds 6 and 7. This suggests that HiR7.8 out-competed all other molecules in the Lo population. It should be noted that HiR7.6 and HiR7.8 are both truncated sequences containing only the heavy chain, while HiR6.8 is a full-length scFv (Table 2).

In the Round 4 Hi population, however, HiR7.6 comprised 52% of the population, while HiR6.8 was 26%. Interestingly, HiR7.8 did not appear in the Round 4 Hi population. In the Round 6 population, 77% of the library was HiR7.6 while 3% was HiR6.8. HiR7.8 now comprised 20% of the library. In Round 7, HiR6.8 was no longer detected, while HiR7.6 was now just 16% of the library. HiR7.8 comprised 84% of the library, suggesting it strongly outcompeted HiR7.6. It is also of note that HiR7.6 and HiR7.8 did not appear in any previous library sequencing, however HiR6.8 did appear once in the Round 3 library. This enrichment and convergence upon HiR7.8 validates the sequence enrichment capabilities of this dual-method screening strategy.

Affinity binding assays of scFvs

Once HiR7.8 and HiR6.8 was identified for further characterization, each was cloned into a secretion vector, expressed, purified, and verified by polyacrylamide gel electrophoresis (PAGE). These were used in fluorescent saturation binding assays with LNCaP cells as described in the Methods, plotted, and fit with nonlinear regression analysis (Figure 3). HiR6.8 exhibited an equilibrium

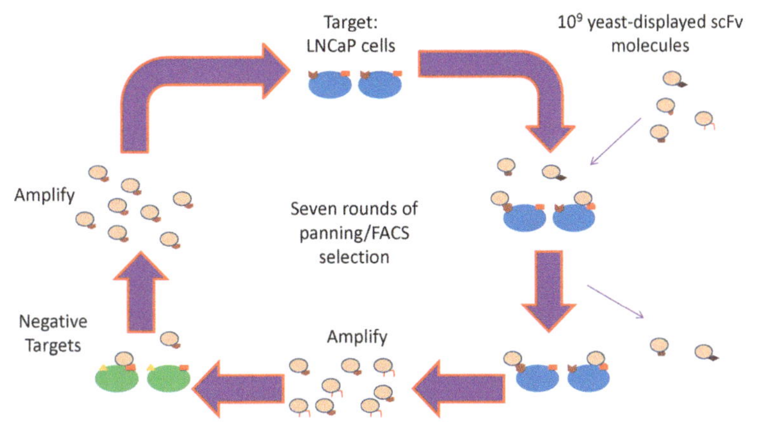

Figure 1 Library screening and enrichment strategy utilized to identify a prostate cancer cell-specific antibody fragment.

Table 1 In vitro selection scheme for selecting an androgen-dependent prostate cancer cell-specific scFv

Selection round	Target	Yeast number	Human cell number	Incubation time (hours)	Incubation volume (mL)	Separation method
1 (+)	LNCaP	1×10^{10}	$\sim 8 \times 10^6$	3	15	Panning
1 (−)	HGPIN	1×10^{10}	$\sim 7 \times 10^6$	0.5	15	Panning
	BPH-1		$\sim 9 \times 10^6$	0.5		
	BHPrE1		$\sim 7.5 \times 10^6$	0.5		
2 (+)	LNCaP	1×10^{10}	$\sim 9 \times 10^6$	2	15	Panning
2 (−)	HGPIN	1×10^{10}	$\sim 8.5 \times 10^6$	1	15	Panning
	BPH-1		$\sim 9.5 \times 10^6$	1		
	BHPrE1		$\sim 8 \times 10^6$	1		
3 (+)	LNCaP	1×10^{10}	$\sim 9 \times 10^6$	1	15	Panning
3 (−)	HGPIN	1×10^{10}	$\sim 7 \times 10^6$	2	15	Panning
	BPH-1		$\sim 9 \times 10^6$	2		
	BHPrE1		$\sim 8 \times 10^6$	2		
4 (+)	LNCaP	1×10^7	1×10^6	0.5	2.5	FACS
4 (−) a	HGPIN	1×10^7	1×10^6	0.5	1.5	FACS
4 (−) b	BPH-1	1×10^7	1×10^6	0.5	2	FACS
4 (−) c	BHPrE1	1×10^7	1×10^6	0.5	1.5	FACS
4 (−) d	PSMA	1×10^7	1 nmol protein	0.5	1	FACS
5 (+)	LNCaP	1×10^7	1×10^6	0.5	5	FACS
6 (+)	LNCaP	1×10^7	1×10^5	0.5	2.75	FACS
6 (−) a	HGPIN	1×10^7	1×10^6	0.5	2.5	FACS
6 (−) b	BPH-1	1×10^7	1×10^6	0.5	2.5	FACS
6 (−) c	BHPrE1	1×10^7	1×10^6	0.5	2.5	FACS
6 (−) d	PSMA	1×10^7	1 nmol protein	0.5	2	FACS
7 (+)	LNCaP	1×10^7	1×10^3	0.5	2.5	FACS

dissociation constant (K_d) of 33.2 +/− 22.2 nM. A K_d of 27.3 +/− 15.9 nM was obtained for HiR7.8.

Binding of each scFv for androgen-dependent prostate cancer cells is further evidenced by fluorescent micrographic analysis. Binding of each scFv is clear when pre-incubated with a secondary fluorescent antibody and then with target cells and imaged for fluorescent staining (Figure 4C and D). Fluorescence is much brighter than target cells incubated with fluorescent secondary-only control (Figure 4B) and comparable to the positive control (Figure 4A). Interestingly, both antibodies have a horseshoe-like staining pattern for each cell. Therefore, fluorescence may be visualized in the cytosol surrounding the nucleus.

Specificity binding assays of scFvs

Purified scFvs were each assayed for their ability to bind to cell lines other than LNCaP. These included the

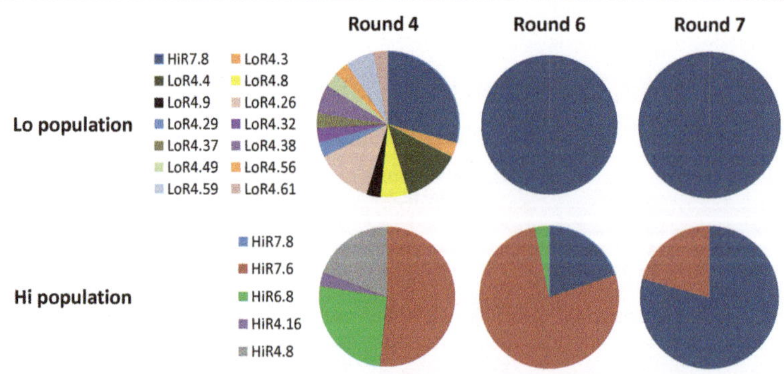

Figure 2 Sequence enrichment in the enriched scFv library in both the Lo and Hi populations beginning in Round 4.

Table 2 Amino acid sequences of HiR6.8, HiR7.6, and HiR7.8

Sequence name	Sequence	Predicted molecular weight
HiR6.8	DVPDYALQASGGGGSGGGGSGGGGS*ASQVQLVQSGAEVKKPGASVKVSCKASGYTFTDYYMHWVRQAPGQGLEWMGWINPNSG* *GTNYAQKFQGRVTMTRDTSISTAYMELSSLRSDDTAIYYCARDADSGSMSAIYWYFNLWGRGTLVTVSSGILGSGGGGSGGGGSGGGGS* **DIQLTQSPSSLSASVGDRVTITCRASQSISRFLNWYQQKPGKAPKLLIYGGSSLQSGVPSRFSGGGSGTDFTLTISSLQPE** **DFATYYCQQSYSKFWTFGQGTKVEIKSGIL**EQKLISEEDL	30.6 kDa
HiR7.6	DVPDYALQASGGGGSGGGGSGGGGS*ASQVQLQESGPGLVKPSGTLSLTCAVSGGSISSSNWWSWVRQPPGKELEWIGEIHHSGS* *TNYNPSLKSRVTISVDKSKNQFSLKMRSVTAADTAVYYCARVEEWPYDALDMWGQGTMVTVSSEF*	15.6 kDa
HiR7.8	DVPDYALQASGGGGSGGGGSGGGGS*ASQVQLQESGPGLVKPSQTLSLTCTVSGDSIYSSGHYWSWVRQHPGKGLEWIGYIYAS* *GRTYYNPSLESRVTMSVDTSKNQSSLKLTSVTAADTAVYYCARDDSRTWYKAFDTWGQGTMVTVSSEF*	16.0 kDa

Italicized amino acids represent the heavy chain. Bold amino acids represent the light chain. HiR7.6 and HiR7.8 do not contain a light chain. Non-italicized or bolded amino acids represent scFv structural elements.

negative targets in the screening strategy, HGPIN, BPH-1, and BHPrE1 (Figure 5A & B). They also included androgen-independent prostate cancer cell lines DU-145 and PC-3 as well as normal prostate epithelium cell lines RWPE-1 and NHPrE1 (Figure 5C & D).

From these experiments, it is clear that HiR6.8 binds to all three negative cell lines as well as the other four lines assayed (Figure 5A & B). Binding to many of these lines was as good or better than binding to LNCaP cells. This is a likely explanation as to why HiR6.8 was eliminated from both populations, in spite of its low dissociation constant. HiR7.8, however, showed excellent selectivity for LNCaP cells (Figure 5C & D). Binding to HGPIN, BPH-1, and BHPrE1 cells all were much less than to the target cancer cells. Subtraction of background secondary antibody MFI from scFv binding resulted in negative numbers due to binding of the secondary antibody to the scFv and not the cell lines. Additionally, binding to the four cell lines not used in screening and enrichment was significantly less relative to binding to target cells. In sum, this suggests that HiR7.8 is highly specific to androgen-dependent prostate cancer cells and therefore will be useful in both the therapeutic targeting and disease detection of prostate cancer.

Discussion

The library screening and enrichment strategy used in this work focused on multiple, stringent negative enrichment targets to identify a scFv specific for androgen-dependent prostate cancer cells. Three rounds of panning were performed to remove scFvs with very low target affinity or that bound to molecules highly expressed on other cell surfaces [27]. There was a large amount of diversity present in the library however, so a more stringent separation method, FACS, was employed [28]. Initial use of FACS would have been unsuccessful as prior work has shown the efficiency of FACS is poor at low

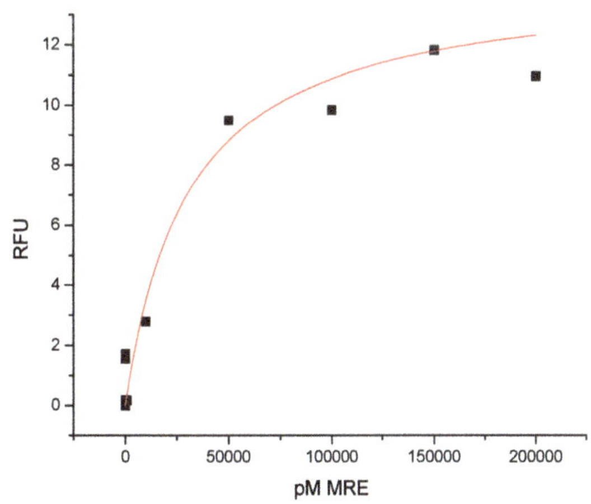

HiR6.8 (K_d +/- Std Error)	HiR7.8 (K_d +/- Std Error)
33.2 +/- 22.2 nM	27.3 +/- 15.9 nM

Figure 3 Equilibrium dissociation constants for HiR7.8 and HiR6.8. The graph depicts a representative saturation binding curve for HiR7.8 fit with nonlinear regression analysis. The table reports the dissociation constant (K_d) of both scFvs in nM averaged from three assays with standard errors.

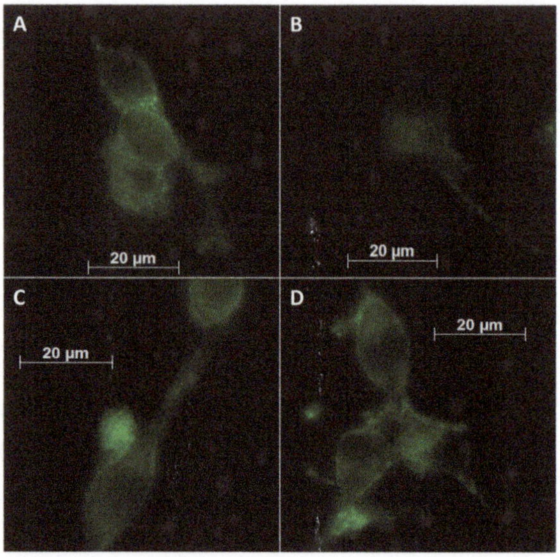

Figure 4 Representative fluorescent micrographs of LNCaP cells bound by identified scFvs. Images are at 630X magnification. **(A)** Image of cells tagged with fluorescent anti-PSMA antibody. **(B)** Image of cells incubated with anti-HA secondary fluorescent antibody. **(C)** Image of cells incubated with HiR6.8 and secondary antibody. **(D)** Image of cells incubated with HiR7.8 and secondary antibody.

concentrations of the binding population [29,30]. Thus, removal of non-binding molecules increased the concentration of binding scFv in the population, allowing successful FACS isolation due to its single-cell and quantitative nature, as has previously been described [30,31]. Additionally, the initial library size was much too great to efficiently analyze single cells in a reasonable time span. Thus, panning followed by FACS was utilized, and the strong enrichment of the identified random sample of the library validates this strategy.

From the obvious enrichment through seven rounds of screening (Figure 2), HiR7.8 was chosen as the most likely prostate cancer cell-specific antibody fragment and underwent further characterization. However, as a representative of scFvs present in earlier rounds of screening and to determine why it was outcompeted, HiR6.8 was also chosen for further characterization. It is also important to note that HiR6.8 is a full-length scFv, comprising both a heavy and light chain, with a predicted molecular weight of approximately 30.6 kDa for the scFv itself. HiR7.8 and HiR7.6, however, only contain a heavy chain, as they are truncated just before the chain linker. They have a predicted molecular weight of approximately 16.0 and 15.6 kDa, respectively, and have 81%

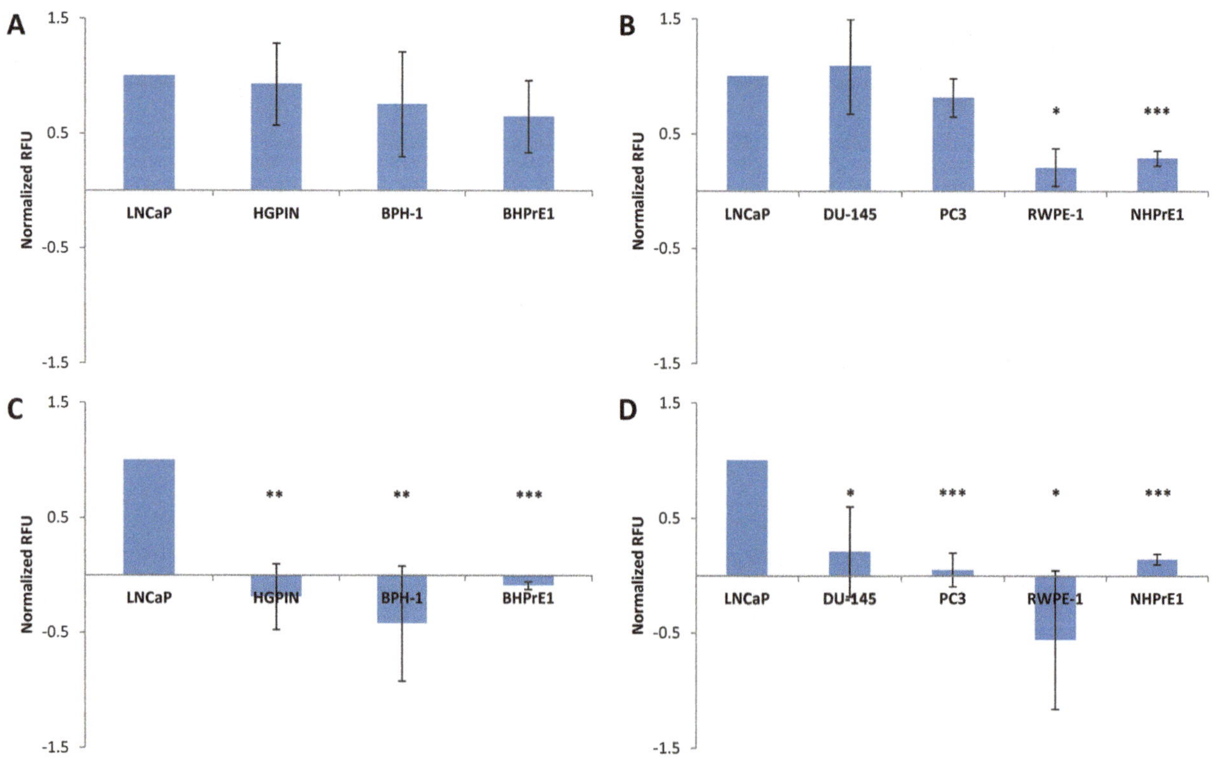

Figure 5 Cross-binding assays of HiR6.8 and HiR7.8. (A) HiR6.8 binding to negative target cell lines used throughout screening and **(B)** cell lines not used as negative targets in screening. **(C)** HiR7.8 binding to negative target cell lines used throughout screening and **(D)** cell lines not used as negative targets in screening. Average values are graphed and normalized to target cells with errors bars representing standard deviations for three assays. * = $p < 0.05$, ** = $p < 0.01$, *** = $p < 0.001$.

sequence identity and 86% sequence similarity. This may be relevant *in vivo*, as it has been shown that smaller scFvs have greater tumor penetration ability than larger antibodies [32]. It is possible that truncated scFvs will suffer from decreased stability. Future work will assay stability in envisioned applications and, if necessary, perform mutagenesis to select stable variants [33] or insert it into the scaffold of a stable full antibody [34].

The dissociation constants for each of these scFvs are in the low-nanomolar range, which is similar to previous reports of scFvs isolated from similar screening strategies [35-37]. More importantly, it is in a similar range to other antibody fragments that have strong pharmacokinetic properties [38,39]. It is possible that further mutagenesis and screening may select for scFvs with higher affinity for the cell surface target [37,40].

It may be possible that fluorescent staining images represent cellular uptake of the scFv-secondary antibody complex due to the cytoplasmic pattern. This has previously shown to be possible and even likely with scFvs alone or in complex with other molecules [41-43]. If these scFvs are taken up by androgen-dependent prostate cancer cells, this may have an important role in the clinical application of these scFvs. The potential ability to use these scFvs as therapeutic targeting agents in the form of antibody-drug conjugates (ADCs) is enhanced by scFv internalization [44]. Potential uptake of the scFv by androgen-dependent prostate cancer cells will be explored in future studies.

Biologically, the selectivity of HiR7.8 for androgen-dependent prostate cancer cells is important. HiR6.8 clearly binds to other cell lines tested, and this is likely the cause of its depletion from the library as the stringency of FACS enrichment increased. This depletion seems to validate the powerful selectivity of the FACS screening strategy. There is an absence of binding to all benign cell lines studied, whether they represent normal prostatic cells, BPH, or PIN. This suggests that the cell surface molecule HiR7.8 is binding to is not displayed on the surface of benign prostatic cells. This is supported by previous studies that have identified many genes differentially expressed between the benign conditions represented here and prostate cancer cells [45-47]. Additionally, there is no binding to the androgen-independent prostate cancer cell lines DU-145 or PC-3. This suggests that the cell surface antigen is displayed in the earlier, androgen-dependent stage of prostate cancer. This is possible considering previous studies have found a wide array of genes that are differentially expressed from androgen dependent to independent disease progression, many of which are downregulated [48-50]. In fact, downregulation of expression levels from LNCaP to both DU-145 and PC-3 cells has been found before [51-54] Future work will determine binding to other androgen-dependent prostate cancer cells utilizing antigen capture techniques coupled with mass spectrometry. Determining binding will be aided by identifying the exact cell surface molecule to which HiR7.8 is binding, as has been done before with similar cell surface antigen-binding molecules [55,56]. This type of work has even identified novel cell surface molecules previously unknown to be involved in disease progression [57]. It is therefore possible that this work not only created a novel targeting agent, but also identified a novel target.

The high affinity and specificity of HiR7.8 for androgen-dependent prostate cancer cells suggest it has potential in future therapeutic and diagnostic applications in prostate cancer. Though not determined here, future work will determine its binding to tissue representing androgen-dependent prostate cancer and other prostatic diseases which will determine the translational potential of the molecule. It is likely, however, that HiR7.8 will bind to androgen-dependent prostate cancer tissue from patients based on binding specificity and affinity similar to previously-identified molecules [58]. The selective scFv will likely be useful for drug delivery as has been shown before [59-62]. It has previously been noted that scFvs have much less immunogenicity than full antibodies in part due to their size and lack of an Fc component [63-65]. Additionally, it will likely have use in specific diagnosis of prostate cancer whether *in* or *ex vivo* [66-69]. Finally, identification of the cell surface markers to which these scFvs bind may possibly identify novel proteins, new functions, or expression patterns that aid in treatment and diagnosis of prostate cancer and will be done in future work. This cell surface molecule and its expression pattern in androgen-dependent prostate cancers will determine the translation of the identified scFv beyond the cells used here. While it is likely to be useful in a defined set of prostate cancers, the antibody fragment screening strategy described here also serves as a proof-of-principle for selection of other disease-specific antibody fragments.

Conclusions

This work has selected a scFv antibody fragment, HiR7.8, with high affinity for androgen-dependent prostate cancer cells. Furthermore, the selected molecule has a high selectivity for the target cells and not for benign prostatic cells or for androgen-independent prostate cancer cells. These characteristics will allow the selected scFv to be useful for both therapeutic and diagnostic applications in the clinical treatment of prostate cancer with further pre-clinical development. With targeted therapeutics and more specific diagnostics, it is possible that increased efficacy and reduced side effects of treatments as well as earlier disease detection will be realized by the scFv obtained here.

Methods

A library screening and enrichment process was used in order to obtain an androgen-dependent prostate cancer cell-specific antibody fragment (Figure 1). A library of non-immune human single-chain Fragment variable (scFv) antibody fragments, displayed on the surface of *Saccharomyces cerevisiae,* was a generous gift from Dr. Dane Wittrup (Massachusetts Institute of Technology; Cambridge, MA) [23]. Seven rounds of screening were completed, enriching for those scFvs which bound to androgen-dependent prostate cancer cells and subtracting those that bound to benign prostate cell lines as well as the protein PSMA.

Cell culture and materials

In order to obtain a prostate cancer cell-specific scFv, prostatic cell lines were used. For general maintenance, each line was passaged every 5–7 days in a T75 cell culture dish with media changed every 2–3 days. The cells were grown in a 37°C incubator with 5% carbon dioxide and humidity. The LNCaP cell line was used as a model of androgen-dependent prostate cancer and was the target of positive enrichment. It was obtained from the American Type Culture Collection (ATCC) (Manassas, VA) and cultured in RPMI 1640 with L-Glutamine and 25 mM HEPES (Cellgro; Manassas, VA) and 10% Fetal Bovine Serum (FBS) (Fisher Scientific; Pittsburgh, PA) and 1X antibiotic/antimycotic mixture (ab/am) (Cellgro) [70]. The High Grade Prostatic Intraepithelial Neoplasia (HGPIN) cell line was a generous gift from Dr. Mark Stearns (Drexel University; Philadelphia, PA) and was cultured in Defined KSFM (Gibco; Grand Island, NY) with 5% FBS and 1X ab/am [71]. The Benign Prostate Hyperplasia (BPH-1) cell line was a generous gift from Dr. Simon Hayward (Vanderbilt University; Nashville, TN) and was cultured in RPMI-1640 with L-Glutamine and 25 mM HEPES and 10% FBS and 1X ab/am [72]. The intermediate prostate stem cell line BHPrE1 was also a generous gift from Dr. Simon Hayward and cultured in DMEM/F12 (Cellgro) supplemented with 5% FBS, 1X ab/am, 1% insulin/transferrin/selenium (Gibco), 0.4% bovine pituitary extract (Sigma; St. Louis, MO), 5 ng/mL epidermal growth factor (Gemini Bio-Products; West Sacramento, CA), and 1X ab/am [73]. The androgen-independent DU-145 prostate cancer cell line was obtained from ATCC and cultured in EMEM (Cellgro) with 10% FBS and 1X ab/am [74]. The androgen-independent prostate cancer cell line PC-3 was also obtained from ATCC and cultured in F12K media (Cellgro) with 10% FBS and 1X ab/am [75]. The normal prostatic epithelium cell line RWPE-1 was obtained from ATCC and cultured in Defined KSFM (Gibco) plus 1X ab/am [76]. The early prostate stem cell line NHPrE1 was a generous gift from Dr. Simon Hayward (Vanderbilt University) and cultured in the same media as BHPrE1 [73].

scFv library and growth

A human non-immune scFv library with 10^9 diversity displayed on the surface of *Sacchoromyces cerevisiae* was utilized [23,28]. The yeast library was chosen due to its amenability to FACS screening and the ability of yeast to display post-translationally modified proteins due to their eukaryotic nature. The library was amplified and expression induced as previously described [23,28]. Before each screening incubation, expression was verified by tagging with a monoclonal anti-HA tag antibody conjugated to either DyLight 488 (Columbia Biosciences; Columbia, MD) or AlexaFluor 488 (Invitrogen; Grand Island, NY). The samples were run on either a Cell Lab Quanta SC (Beckman Coulter; Brea, CA) or a FACSCalibur (BD Biosciences; San Jose, CA) flow cytometer equipped with a 488 nm argon laser and 525 nm emission filter.

Library screening

Seven rounds of screening were performed in order to obtain a scFv specific for androgen-dependent prostate cancer cells (Table 1). The first three rounds of screening were performed by panning and the last four by fluorescence-activated cell sorting (FACS).

For Round 1(+) screening, androgen-dependent LNCaP prostate cancer cells were grown to 80-90% confluency and the media was removed. The cells were gently washed with calcium- and magnesium-free phosphate-buffered saline (PBS). The cells were then incubated with 10^{10} yeast from the naïve library in 15 mL yeast screening buffer (YSB) containing PBS, 0.5% bovine serum albumin (BSA) and 1% FBS. The library was placed into the flask containing prostate cells and placed on a 37°C shaker at 25 RPM for three hours. After incubation, yeast not bound to the cells were removed, and the LNCaP cells were gently washed three times with 15 mL YSB and confluence of remaining attached cells was visually confirmed. 100 mL yeast amplification media was added to the flask to allow for amplification of yeast bound to the prostatic cells. This was grown overnight, and this enriched library was prepared for a round of negative enrichment. For Round 1(–) screening, scFv-expressing yeast were suspended in YSB and incubated with rinsed HGPIN cells at 80-90% confluency for 30 minutes at 37°C with shaking at 25 RPM. The supernatant containing yeast-displayed scFvs not bound to HGPIN cells was removed and added to rinsed BPH cells under the same conditions, removed again and added to rinsed normal prostatic epithelium BHPrE1 cells under the same conditions. The serially-incubated supernatant was centrifuged to obtain yeast, which were amplified. Two more rounds of panning were performed in this manner, with decreasing incubation times for positive rounds and increasing incubation times for negative rounds.

The ensuing four rounds of screening were performed using FACS-based separation (Additional file 1: Figure S1). For Round 4(+), LNCaP cells grown to 80-90% confluency were fluorescently dyed with CFSE (Invitrogen) according to manufacturer's instructions. They were then dissociated from the flask with Cellstripper reagent (Cellgro) to prevent cell surface protein digestion associated with trypsinization. Cells were then suspended in YSB and counted with a Scepter equipped with 60 μm sensors (Millipore; Billerica, MA). In Round 4, all yeast were fluorescently dyed with Syto61 (Invitrogen) according to manufacturer's instructions. A total of 10^7 yeast were suspended in YSB and mixed with 10^6 LNCaP in 2.5 mL. They were mixed by inversion for 30 minutes at 37°C and placed on ice before FACS sorting. The sample was then sorted with a FACSAria (BD Biosciences), with excitation at 488 nm from a sapphire solid state laser and 633 nm from a HeNe laser and 525 nm and 650 nm emission filters. Events identified as bound yeast and prostate cells were collected. In Round 4(+), there were two separate populations of bound cells which were collected and amplified separately. These were named the 'Hi' and 'Lo' populations and kept separate through the following rounds of screening and subjected to the identical conditions.

In Round 4(−)a, for each of the Hi and Lo populations, yeast were prepared for screening in the same manner and incubated with 10^6 HGPIN cells dyed the same with inversion for 30 minutes at 37°C in YSB. The sample was subjected to FACS and events that indicated yeast that were not bound to HGPIN cells were collected and amplified. Those were then incubated with 10^6 dyed BPH cells for Round 4(−)b and subjected to FACS with unbound yeast collected and amplified. The enriched library was then prepared for Round 4(−)c and subjected to FACS after incubation with 10^6 dyed BHPrE1 cells. Unbound yeast were collected and amplified, then prepared for screening and incubated with 1 nmol recombinant full-length PSMA protein, which is expressed in non-prostatic normal and tumor tissue, as well as normal prostate cells [77] (Abnova #H00002346-P01; Walnut, CA). This protein is 107 kDa and 719 amino acids with a GST tag and was pre-incubated with AlexaFluor 488-conjugated anti-GST tag antibody (Invitrogen). The yeast which showed single fluorescence signals corresponding to Syto61 were kept and amplified. Positive enrichment rounds continued through Round 7, with yeast fluorescent tagging being done with anti-HA tag antibody conjugated to AlexaFluor 647 (Invitrogen) with decreasing target cell concentrations. This tagging method was used for a negative enrichment performed in Round 6.

Sequencing of scFv library

A representative sample of scFv genes from the naïve library (Round 0) was sequenced to determine diversity of the library. Additionally, this was completed for the enriched post-Round 2(−) and post-Round 3(−) libraries. It was also performed for each of the Hi and Lo populations following Round 4(−), Round 6(−) and Round 7(+). To do this, yeast were plated onto amplification media agar, and individual colonies were chosen for polymerase chain reaction (PCR) amplification.

Yeast colonies were picked and placed into double distilled water and boiled, which served as template for the PCR reaction. The reaction ingredients were as follows: 400 nM forward (5′-GTACGAGCTAAAAGTACAGTG-3′) and reverse (5′-TAGATACCCATACGACGTTC-3′) pPNL6 primers (Eurofins MWG Operon), 250 μM deoxynucleotide triphosphates, 5% dimethyl sulfoxide, 1X Phusion Reaction Buffer (New England Biolabs, Ipswich, MA), 2 units Phusion High-Fidelity DNA polymerase (New England Biolabs), and double distilled water to 100 μL. Reactions conditions were: initial denaturation at 98°C for 30 seconds; 35 cycles of 98°C for 10 seconds, 50°C for 30 seconds, and 72°C for 30 seconds; and final extension at 72°C for 10 minutes. Results were analyzed using agarose gel electrophoresis and those PCRs that contained bands corresponding to the scFv gene were purified using a PCR purification kit (IBI Scientific; Peosta, IA) and sent for DNA sequencing (Eurofins MWG Operon; Huntsville, AL) using both the forward and reverse pPNL6 primers. In total, 30–35 sequences were obtained for each enriched library noted above, including for individual Hi and Lo populations separately. Analysis was then performed by translating the DNA sequence to protein using the ExPASy translate tool (Swiss Institute of Bioinformatics; http://web.expasy.org/translate/). Hemaglutttanin, c-Myc, and linker protein tag sequences, landmarks of the scFv expression scaffold, were identified to ensure sequence quality [23]. Sequences were compared for similarities and duplicates within an enriched library were identified to determine diversity of a random sample of the library.

Secretion and purification of selected scFvs

From the Rounds 6(−) and 7(+) sequences, one scFv, HiR7.8, was chosen for further study due to its abundance in the enriched libraries. HiR6.8 was also chosen to be representative of scFvs present earlier in the screening process and to determine why it was outcompeted from the population by HiR7.8. These sequences were subcloned into the pPNL9 secretion vector in YVH10 yeast using gap repair essentially as previously described [28]. For truncated scFv sequences, a modified reverse PCR primer (5′- GGGTTAGGGATAGGCTTACCGAACTCT GAAGAGACGGTGACC-3′) was used. After growth of the yeast containing the scFv sequence within the secretion vector, the supernatant was recovered and scFv purified with Ni-NTA resin following the manufacturer's protocol (Thermo Scientific; West Palm Beach, FL). Purified scFv was analyzed by polyacrylamide gel electrophoresis as previously reported, and the concentration of the scFv was determined using spectroscopy with a NanoDrop (Thermo Scientific). The scFv was then diluted to a working concentration of 10 μM in PBS.

Affinity binding assays with scFvs

In order to determine the binding affinity of the secreted scFvs with the target androgen-independent prostate cancer cells, saturation binding assays were performed essentially as previously described [36,40]. LNCaP cells were grown to 80-90% confluency and removed from the flask with CellStripper reagent. They were counted and 2×10^5 cells were placed into a 500 μL total volume of YSB. For saturation binding assays, the appropriate amount of scFv was incubated on ice for 30 minutes with 2 μL of the monoclonal anti-HA antibody conjugated to DyLight 488 (Columbia Biosciences). Concentrations of 0, 1, 10, 100, 1000, 10000, 50000, 100000, 150000, 200000 pM of the scFv were used. The pre-incubated scFv and secondary antibody were mixed with LNCaP cells and placed on a rotisserie at 37°C for 30 minutes. After this period, the cells were washed and run on a FACSCalibur (BD Biosciences) as previously described. The mean fluorescence intensity (MFI) for each sample was recorded. The MFI for the 0 pM incubation, which had only the secondary fluorescent antibody, was subtracted from the MFI of each incubation. These data were graphed using Origin 8 software (OriginLab Corporation; Northampton, MA). The data were fit with a nonlinear regression model of single-event binding given by the equation, $Y = ((B_{max}*X)/(K_d + X)) + NS*X$, where B_{max} is maximum binding, K_d is the dissociation constant, and NS is nonspecific binding [78]. Each assay was performed in triplicate, with the K_d values averaged and the standard errors of the Origin-obtained K_d values were combined.

Fluorescent imaging of scFv binding

In order to visualize binding of each scFv, fluorescent images were taken. Target androgen-independent prostate cancer cells were grown in a 6-well culture dish on number 1.5 microscopic cover glasses to ~90% confluency. For each secreted scFv, a 100 nM concentration was pre-incubated with anti-HA antibody conjugated to DyLight 488 as described for binding assays. Cells incubated with anti-PSMA AlexaFluor488 antibody served as a positive control (BioLegend; San Diego, CA) and cells incubated with only anti-HA DyLight 488 antibody served as negative control. Cells were incubated at 37°C for 30 minutes with shaking at 25 RPM. The supernatant was removed and the cells were washed, formalin-fixed, and mounted onto microscope slides. Cells were imaged on a Zeiss AxioImager Z2 Fluorescent Microscope (Zeiss; Thornwood, NY) using an Argon laser at 488 nm for excitation with a band pass emission filter at 505–530 nm.

Specificity binding assays of scFvs

In order to determine the specificity of scFv binding, cross-binding assays were performed with the cells used as negative targets. LNCaP, HGPIN, BPH-1, and BHPrE1 cells were counted and 2×10^5 cells were placed into 500 μL YSB. Concentrations of 0 or 100 nM scFv were incubated with the secondary anti-HA AlexaFluor 488 antibody for 30 minutes on ice. Cell were prepared as described above and then run on a FACSCalibur and the MFI recorded. The MFI of the 0 pM incubation, which had only the secondary fluorescent antibody, served as background fluorescent labeling and was subtracted from the 100 nM incubation. Data were normalized to target cell binding, which was set to 100%, and graphed to show comparative binding between the cell lines.

In order to determine binding to other prostatic cell lines not used throughout screening, cross-binding assays were performed. The target androngen-dependent prostate cancer cells, the androgen-independent prostate cancer cell lines DU-145 and PC-3, and the normal prostatic cell lines RWPE-1 and NHPrE1 were used for this experiment. The cells were prepared as noted above and run on a FACSCalibur. Each set of cross-binding assays were performed in triplicate and significance in the differences of means were obtained using a student's *T*-test.

Additional file

Additional file 1: Figure S1. Representative FACS plots from the selection. For all, the X-axis is CFSE fluorescence, which was used to stain the target cell line and the Y-axis is Syto 61 or Alexa647 (anti-HA antibody) which was used to stain yeast expressing the library. Q1 shows stained yeast alone, Q2 shows events representing yeast bound to the target cell line, Q3 shows stained cells alone, and Q4 shows unstained cells and debris. **A)** FACS plot from Round 4(+) selection with the target LNCaP cell line. The origination of the "Hi" and "Lo" populations are shown and so named due to amount of yeast staining present in the events. **B)** FACS plots from Round 5(+) selection with the target LNCaP cell line. Left shows sorting of the yeast binding to the LNCaP cell line in the Lo population and right shows sorting of the yeast binding to the LNCaP cell line in the Hi population. **C)** FACS plots from Round 6(-)c selection with the non-target BHPrE1 cell line. Left shows sorting of yeast that did not bind to BHPrE1 cells in the Lo population and right shows sorting of yeast that did not bind to the BHPrE1 cells in the Hi population.

Competing interests

The authors declare no commercial or financial conflict of interests.

Authors' contributions

RMW conceived of the manuscript and designed all experiments, performed the experiments, analyzed the data, and drafted the manuscript. CJH and SN performed experiments and analyzed the data. LJS conceived of the manuscript and designed all experiments, analyzed the data, and drafted the manuscript. All authors read and approved the manuscript.

Acknowledgements

We acknowledge support from the National Science Foundation (Cooperative Agreements 1003907 and 0554328). RMW was supported in part by a pre-doctoral fellowship from the American Foundation for Pharmaceutical Education. A portion of the flow cytometry experiments were performed in the West Virginia University Flow Cytometry Core Facility, which is supported in part by NIH grants RR032138 and RR020866. The United States Army Research Laboratory generously allowed use of their Beckman flow cytometer for a portion of the experiments. Imaging experiments and image analysis were performed in the West Virginia University Imaging Facility, which is supported in part by the Mary Babb Randolph Cancer Center and NIH grant P20 RR016440.

Author details

[1]Department of Basic Pharmaceutical Sciences, West Virginia University, 1 Medical Center Drive, PO Box 9530, Morgantown, WV 26506, USA. [2]Current address: Memorial Sloan Kettering Cancer Center, Molecular Pharmacology & Chemistry Program, 1275 York Ave., New York, NY 10065, USA. [3]Department of Biology, West Virginia University, 53 Campus Drive, PO Box 6057, Morgantown, WV 26506, USA.

References

1. Siegel R, Ma J, Zou Z, Jemal A: **Cancer statistics, 2014.** *CA Cancer J Clin* 2014, **64**(1):9–29.
2. Chou R, Dana T, Bougatsos C, Fu R, Blazina I, Gleitsmann K, Rugge JB: **Treatments for localized prostate cancer: systematic review to update the 2002 US preventive services task force recommendation.** *Evid Synthesis* 2011, **91**:12-05161-EF-1.
3. Yao SL, Lu-Yao G: **Population-based study of relationships between hospital volume of prostatectomies, patient outcomes, and length of hospital stay.** *J Natl Cancer Inst* 1999, **91**(22):1950–1956.
4. Feldman BJ, Feldman D: **The development of androgen-independent prostate cancer.** *Nat Rev Cancer* 2001, **1**(1):34–45.
5. Yagoda A, Petrylak D: **Cytotoxic chemotherapy for advanced hormone-resistant prostate cancer.** *Cancer* 2006, **71**(S3):1098–1109.
6. Extra JM, Rousseau F, Bruno R, Clavel M, Le Bail N, Marty M: **Phase I and pharmacokinetic study of Taxotere (RP 56976; NSC 628503) given as a short intravenous infusion.** *Cancer Res* 1993, **53**(5):1037–1042.
7. Pienta KJ: **Preclinical mechanisms of action of docetaxel and docetaxel combinations in prostate cancer.** *Semin Oncol* 2001, **28**(S15):3–7. Elsevier.
8. Pienta KJ, Smith DC: **Advances in prostate cancer chemotherapy: a New Era Begins.** *CA Cancer J Clin* 2005, **55**(5):300–318.
9. Lin K, Croswell JM, Koenig H, Lam C, Maltz A: **Prostate-specific antigen-based screening for prostate cancer: an evidence update for the US preventive services task force.** *Evid Synthesis* 2011, **90**:12-05160-EF-1.
10. Chou R, Croswell JM, Dana T, Bougatsos C, Blazina I, Fu R, Gleitsmann K, Koenig HC, Lam C, Maltz A: **Screening for prostate cancer: a review of the evidence for the US Preventive Services Task Force.** *Ann Intern Med* 2011, **155**(11):762.
11. Moyer VA: **Screening for prostate cancer: US Preventive Services Task Force recommendation statement.** *Ann Intern Med* 2012, **157**(2):120–134.
12. Barry MJ: **Screening for prostate cancer—the controversy that refuses to die.** *N Engl J Med* 2009, **360**(13):1351–1354.
13. Pollack CE, Noronha G, Green GE, Bhavsar NA, Carter HB: **Primary care Providers' response to the US preventive services task force draft recommendations on screening for prostate cancer.** *Arch Intern Med* 2012, **172**(8):668–670.
14. Woolf SH: **Screening for prostate cancer with prostate-specific antigen—an examination of the evidence.** *N Engl J Med* 1995, **333**(21):1401–1405.
15. Thompson IM, Chi C, Ankerst DP, Goodman PJ, Tangen CM, Lippman SM, Lucia MS, Parnes HL, Coltman CA Jr: **Effect of finasteride on the sensitivity of PSA for detecting prostate cancer.** *J Natl Cancer Inst* 2006, **98**(16):1128–1133.
16. Williams R, Naz R: **Novel biomarkers and therapeutic targets for prostate cancer.** *Front Biosci (Schol Ed)* 2010, **2**:677–684.
17. Makarov DV, Loeb S, Getzenberg RH, Partin AW: **Biomarkers for Prostate Cancer.** In *Annual Review of Medicine*, Volume 60. Palo Alto: Annual Reviews; 2009:139–151.
18. Prensner JR, Rubin MA, Wei JT, Chinnaiyan AM: Beyond PSA: **The next generation of prostate cancer biomarkers.** *Sci Transl Med* 2012, **4**(127):127rv3.
19. Morris KN, Jensen KB, Julin CM, Weil M, Gold L: **High affinity ligands from in vitro selection: complex targets.** *Proc Natl Acad Sci U S A* 1998, **95**(6):2902–2907.
20. Guo KT, Ziemer G, Paul A, Wendel HP: **CELL-SELEX: Novel perspectives of aptamer-based therapeutics.** *Int J Mol Sci* 2008, **9**(4):668–678.
21. Huse WD, Sastry L, Iverson SA, Kang AS, Alting-Mees M, Burton DR, Benkovic SJ, Lerner RA: **Generation of a large combinatorial library of the immunoglobulin repertoire in phage lambda.** *Science* 1989, **246**(4935):1275–1281.
22. Gunneriusson E, Samuelson P, Uhlen M, Nygren PA, Stahl S: **Surface display of a functional single-chain Fv antibody on staphylococci.** *J Bacteriol* 1996, **178**(5):1341–1346.
23. Boder ET, Wittrup KD: **Yeast surface display for screening combinatorial polypeptide libraries.** *Nat Biotechnol* 1997, **15**(6):553–557.
24. Figini M, Obici L, Mezzanzanica D, Griffiths A, Colnaghi MI, Winter G, Canevari S: **Panning phage antibody libraries on cells: isolation of human Fab fragments against ovarian carcinoma using guided selection.** *Cancer Res* 1998, **58**(5):991–996.
25. Jakobsen CG, Rasmussen N, Laenkholm AV, Ditzel HJ: **Phage display derived human monoclonal antibodies isolated by binding to the surface of live primary breast cancer cells recognize GRP78.** *Cancer Res* 2007, **67**(19):9507–9517.
26. Yu B, Ni M, Li WH, Lei P, Xing W, Xiao DW, Huang Y, Tang ZJ, Zhu HF, Shen GX: **Human scFv antibody fragments specific for hepatocellular carcinoma selected from a phage display library.** *World J Gastroenterol* 2005, **11**(26):3985–3989.
27. Wang XX, Shusta EV: **The use of scFv-displaying yeast in mammalian cell surface selections.** *J Immunol Methods* 2005, **304**(1–2):30–42.
28. Feldhaus MJ, Siegel RW, Opresko LK, Coleman JR, Feldhaus JM, Yeung YA, Cochran JR, Heinzelman P, Colby D, Swers J, Graff C, Wiley HS, Wittrup KD: **Flow-cytometric isolation of human antibodies from a nonimmune Saccharomyces cerevisiae surface display library.** *Nat Biotechnol* 2003, **21**(2):163–170.
29. Jung ST, Jeong KJ, Iverson BL, Georgiou G: **Binding and enrichment of Escherichia coli spheroplasts expressing inner membrane tethered scFv antibodies on surface immobilized antigens.** *Biotechnol Bioeng* 2007, **98**(1):39–47.
30. Qiu J-K, Jung S-T, Georgiou G, Hang H-Y: **Enrichment of Escherichia coli spheroplasts displaying scFv antibodies specific for antigens expressed on the human cell surface.** *Appl Microbiol Biotechnol* 2010, **88**(6):1385–1391.
31. Mazor Y, Van Blarcom T, Carroll S, Georgiou G: **Selection of full-length IgGs by tandem display on filamentous phage particles and Escherichia coli fluorescence activated cell sorting screening.** *FEBS J* 2010, **277**(10):2291–2303.
32. Colcher D, Pavlinkova G, Beresford G, Booth B, Choudhury A, Batra S: **Pharmacokinetics and biodistribution of genetically-engineered antibodies.** *Q J Nucl Med* 1998, **42**(4):225–241.
33. Brockmann E-C, Cooper M, Strömsten N, Vehniäinen M, Saviranta P: **Selecting for antibody scFv fragments with improved stability using phage display with denaturation under reducing conditions.** *J Immunol Methods* 2005, **296**(1):159–170.
34. Asano R, Watanabe Y, Kawaguchi H, Fukazawa H, Nakanishi T, Umetsu M, Hayashi H, Katayose Y, Unno M, Kudo T: **Highly effective recombinant format of a humanized IgG-like bispecific antibody for cancer immunotherapy with retargeting of lymphocytes to tumor cells.** *J Biol Chem* 2007, **282**(38):27659–27665.
35. Böldicke T, Tesar M, Griesel C, Rohde M, Gröne HJ, Waltenberger J, Kollet O, Lapidot T, Yayon A, Weich H: **Anti-VEGFR-2 scFvs for Cell Isolation. Single-Chain Antibodies Recognizing the Human Vascular Endothelial Growth Factor Receptor-2 (VEGFR-2/flk-1) on the Surface of Primary Endothelial Cells and Preselected CD34+ Cells from Cord Blood.** *Stem Cells* 2001, **19**(1):24–36.
36. Benedict CA, MacKrell AJ, Anderson WF: **Determination of the binding affinity of an anti-CD34 single-chain antibody using a novel, flow cytometry based assay.** *J Immunol Methods* 1997, **201**(2):223–231.
37. Schier R, Bye J, Apell G, McCall A, Adams GP, Malmqvist M, Weiner LM, Marks JD: **Isolation of High-affinity Monomeric Human Anti-c-*erb*B-2 Single chain Fv Using Affinity-driven Selection.** *J Mol Biol* 1996, **255**(1):28–43.
38. Adams GP, Schier R, Marshall K, Wolf EJ, McCall AM, Marks JD, Weiner LM: **Increased affinity leads to improved selective tumor delivery of single-chain Fv antibodies.** *Cancer Res* 1998, **58**(3):485–490.

39. Jackson H, Bacon L, Pedley R, Derbyshire E, Field A, Osbourn J, Allen D: **Antigen specificity and tumour targeting efficiency of a human carcinoembryonic antigen-specific scFv and affinity-matured derivatives.** *Br J Cancer* 1998, **78**(2):181.
40. Boder ET, Midelfort KS, Wittrup KD: **Directed evolution of antibody fragments with monovalent femtomolar antigen-binding affinity.** *Proc Natl Acad Sci* 2000, **97**(20):10701–10705.
41. He J, Wang Y, Feng J, Zhu X, Lan X, Iyer AK, Zhang N, Seo Y, VanBrocklin HF, Liu B: **Targeting prostate cancer cells in vivo using a rapidly internalizing novel human single-chain antibody fragment.** *J Nucl Med* 2010, **51**(3):427–432.
42. Gao C, Mao S, Ronca F, Zhuang S, Quaranta V, Wirsching P, Janda KD: **De novo identification of tumor-specific internalizing human antibody–receptor pairs by phage-display methods.** *J Immunol Methods* 2003, **274**(1):185–197.
43. Nielsen UB, Kirpotin DB, Pickering EM, Drummond DC, Marks JD: **A novel assay for monitoring internalization of nanocarrier coupled antibodies.** *BMC Immunol* 2006, **7**(1):24.
44. Schrama D, Reisfeld RA, Becker JC: **Antibody targeted drugs as cancer therapeutics.** *Nat Rev Drug Discov* 2006, **5**(2):147–159.
45. Zeng L, Rowland RG, Lele SM, Kyprianou N: **Apoptosis incidence and protein expression of p53, TGF-beta receptor II, p27Kip1, and Smad4 in benign, premalignant, and malignant human prostate.** *Hum Pathol* 2004, **35**(3):290–297.
46. Chen W, Pang B, Yang B, Zhou J, Sun Y: **Differential proteome analysis of conditioned medium of BPH-1 and LNCaP cells.** *Chin Med J Beijing* 2011, **124**(22):3806–3809.
47. Chakrabarti R, Robles LD, Gibson J, Muroski M: **Profiling of differential expression of messenger RNA in normal, benign, and metastatic prostate cell lines.** *Cancer Genet Cytogenet* 2002, **139**(2):115–125.
48. Amler LC, Agus DB, LeDuc C, Sapinoso ML, Fox WD, Kern S, Lee D, Wang V, Leysens M, Higgins B, Martin J, Gerald W, Dracopoli N, Cordon-Cardo C, Scher HI, Hampton GM: **Dysregulated expression of androgen-responsive and nonresponsive genes in the androgen-independent prostate cancer xenograft model CWR22-R1.** *Cancer Res* 2000, **60**(21):6134–6141.
49. Karan D, Kelly DL, Rizzino A, Lin MF, Batra SK: **Expression profile of differentially-regulated genes during progression of androgen-independent growth in human prostate cancer cells.** *Carcinogenesis* 2002, **23**(6):967–976.
50. Chen Q, Watson JT, Marengo SR, Decker KS, Coleman I, Nelson PS, Sikes RA: **Gene expression in the LNCaP human prostate cancer progression model: progression associated expression in vitro corresponds to expression changes associated with prostate cancer progression in vivo.** *Cancer Lett* 2006, **244**(2):274–288.
51. Yang M, Loda M, Sytkowski AJ: **Identification of genes expressed differentially by LNCaP or PC-3 prostate cancer cell lines.** *Cancer Res* 1998, **58**(16):3732–3735.
52. Liu Z, Marquez M, Nilsson S, Holmberg AR: **Comparison of protein expression in two prostate cancer cell-lines, LNCaP and DU145, after treatment with somatostatin.** *Oncol Rep* 2009, **22**(6):1451.
53. Aalinkeel R, Nair MPN, Sufrin G, Mahajan SD, Chadha KC, Chawda RP, Schwartz SA: **Gene expression of angiogenic factors correlates with metastatic potential of prostate cancer cells.** *Cancer Res* 2004, **64**(15):5311–5321.
54. Okamura K, Koike H, Matsui H, Suzuki K: **Gene Expression Profiles of Prostate Cancer Cell Lines, LNCaP, PC-3 and DU-145, Assessed by cDNA Microarray.** *Kitakanto Med J* 2008, **58**(4):363–369.
55. Rérole AL, Gobbo J, De Thonel A, Schmitt E, de Barros JPP, Hammann A, Lanneau D, Fourmaux E, Deminov O, Micheau O: **Peptides and aptamers targeting HSP70: a novel approach for anticancer chemotherapy.** *Cancer Res* 2011, **71**(2):484–495.
56. Robert R, Jacobin-Valat MJ, Daret D, Miraux S, Nurden AT, Franconi JM, Clofent-Sanchez G: **Identification of human scFvs targeting atherosclerotic lesions.** *J Biol Chem* 2006, **281**(52):40135–40143.
57. Berezovski MV, Lechmann M, Musheev MU, Mak TW, Krylov SN: **Aptamer-facilitated biomarker discovery (AptaBiD).** *J Am Chem Soc* 2008, **130**(28):9137–9143.
58. Ni X, Zhang Y, Ribas J, Chowdhury WH, Castanares M, Zhang Z, Laiho M, DeWeese TL, Lupold SE: **Prostate-targeted radiosensitization via aptamer-shRNA chimeras in human tumor xenografts.** *J Clin Invest* 2011, **121**(6):2383–2390.
59. Senter PD: **Potent antibody drug conjugates for cancer therapy.** *Curr Opin Chem Biol* 2009, **13**(3):235–244.
60. Regino C, Wong K, Milenic D, Holmes E, Garmestani K, Choyke P, Brechbiel M: **Preclinical evaluation of a monoclonal antibody (3C6) specific for prostate-specific membrane antigen.** *Curr Radiopharm* 2009, **2**(1):9–17.
61. Sievers EL, Linenberger M: **Mylotarg: antibody-targeted chemotherapy comes of age.** *Curr Opin Oncol* 2001, **13**(6):522–527.
62. Stebbing J, Copson E, O'Reilly S: **Herceptin (trastuzamab) in advanced breast cancer.** *Cancer Treat Rev* 2000, **26**(4):287–290.
63. Kanter G, Yang J, Voloshin A, Levy S, Swartz JR, Levy R: **Cell-free production of scFv fusion proteins: an efficient approach for personalized lymphoma vaccines.** *Blood* 2007, **109**(8):3393–3399.
64. Begent R, Chester K: **Single-chain Fv antibodies for targeting cancer therapy.** *Biochem Soc Trans* 1997, **25**(2):715–716.
65. Clark M: **Antibody humanization: a case of the 'Emperor's new clothes'?** *Immunol Today* 2000, **21**(8):397–402.
66. Elsässer-Beile U, Reischl G, Wiehr S, Bühler P, Wolf P, Alt K, Shively J, Judenhofer MS, Machulla HJ, Pichler BJ: **PET imaging of prostate cancer xenografts with a highly specific antibody against the prostate-specific membrane antigen.** *J Nucl Med* 2009, **50**(4):606–611.
67. Veiseh O, Gunn JW, Zhang M: **Design and fabrication of magnetic nanoparticles for targeted drug delivery and imaging.** *Adv Drug Deliv Rev* 2010, **62**(3):284–304.
68. Johnston WW, Szpak CA, Lottich SC, Thor A, Schlom J: **Use of a monoclonal antibody (B72. 3) as a novel immunohistochemical adjunct for the diagnosis of carcinomas in fine needle aspiration biopsy specimens.** *Hum Pathol* 1986, **17**(5):501–513.
69. Nagrath S, Sequist LV, Maheswaran S, Bell DW, Irimia D, Ulkus L, Smith MR, Kwak EL, Digumarthy S, Muzikansky A: **Isolation of rare circulating tumour cells in cancer patients by microchip technology.** *Nature* 2007, **450**(7173):1235–1239.
70. Horoszewicz JS, Leong SS, Kawinski E, Karr JP, Rosenthal H, Chu TM, Mirand EA, Murphy GP: **LNCaP model of human prostatic carcinoma.** *Cancer Res* 1983, **43**(4):1809–1818.
71. Wang M, Liu A, Garcia FU, Rhim JS, Stearns ME: **Growth of HPV-18 immortalized human prostatic intraepithelial neoplasia cell lines. Influence of IL-10, follistatin, activin-A, and DHT.** *Int J Oncol* 1999, **14**(6):1185–1195.
72. Hayward S, Dahiya R, Cunha G, Bartek J, Deshpande N, Narayan P: **Establishment and characterization of an immortalized but non-transformed human prostate epithelial cell line: BPH-1.** *In Vitro CellDev Biol Anim* 1995, **31**(1):14–24.
73. Jiang M, Strand DW, Fernandez S, He Y, Yi Y, Birbach A, Qiu Q, Schmid J, Tang DG, Hayward SW: **Functional remodeling of benign human prostatic tissues in vivo by spontaneously immortalized progenitor and intermediate cells.** *Stem Cells* 2009, **28**(2):344–356.
74. Stone K, Mickey D, Wunderli H, Mickey G, Paulson D: **Isolation of a human prostate carcinoma cell line (DU 145).** *Int J Cancer* 1978, **21**(3):274–281.
75. Kaighn M, Narayan KS, Ohnuki Y, Lechner J, Jones L: **Establishment and characterization of a human prostatic carcinoma cell line (PC-3).** *Invest Urol* 1979, **17**(1):16–23.
76. Bello D, Webber M, Kleinman H, Wartinger D, Rhim J: **Androgen responsive adult human prostatic epithelial cell lines immortalized by human papillomavirus 18.** *Carcinog* 1997, **18**(6):1215–1223.
77. Silver DA, Pellicer I, Fair WR, Heston W, Cordon-Cardo C: **Prostate-specific membrane antigen expression in normal and malignant human tissues.** *Clin Cancer Res* 1997, **3**(1):81–85.
78. Langan TJ, Nyakubaya VT, Casto LD, Dolan TD, Archer-Hartmann SA, Yedlapalli SL, Sooter LJ, Holland LA: **Assessment of aptamer-steroid binding using stacking-enhanced capillary electrophoresis.** *Electrophoresis* 2012, **33**(5):866–869.

AAV2-mediated follistatin overexpression induces ovine primary myoblasts proliferation

Mahmood Nazari[1], Fatemeh Salabi[2], Li Zhang[1*], Fuping Zhao[1], Caihong Wei[1] and Lixin Du[1*]

Abstract

Background: Follistatin (FST) has been shown to bind to some TGF-β family members and can function as a potent myostatin (MSTN) antagonist. Recent studies have revealed that over-expression of FST by adeno-associated viruses increases muscle growth in mice, humans and nonhuman primates. In the present study, to determine the effect of FST on ovine primary myoblast (OPM) proliferation, FST was over-expressed using an adeno-associated virus serotype 2 (AAV 2) vector.

Results: Western blot results showed that AAV induced the expression of FST protein in transduced OPM cells. Real-time quantitative PCR results indicated that over-expression of FST resulted in a dramatic increase in Akt I and CDK2 expression and a decrease in p21 expression. Moreover, cell cycle analysis confirmed that FST down-regulated p21, a CDK inhibitor, and increased the level of CDK2 expression in OPM cells. Hence, follistatin positively regulated the G1 to S progression. Our results showed that FST induced proliferation through a down-regulation of p21, as only the p21 expression level was down-regulated as a result of FST over-expression in myoblasts, whereas no change was observed in the level of p57 expression.

Conclusions: These results expanded our understanding of the regulation mechanism of FST in ovine primary myoblasts. Our results provide the first evidence that the AAV viral system can be used for gene transfer in ovine myoblast cells. Moreover, the results showed that an AAV vector can successfully induce the expression of FST in OPM cells in vitro. These findings demonstrated that FST over-expression induces proliferation through a down-regulation of the p21 gene under proliferating conditions.

Keywords: Follistatin, Over-expression, Adeno-associated virus, Myoblasts, Proliferation

Background

Currently, somatic cell nuclear transfer or pronuclear injection is used to generate transgenic domestic animals. These methods are inefficient, and in some species, they are associated with a high risk of developmental abnormalities in the resulting offspring [1]. Recently, virus-mediated transfer has been employed to generate transgenic animal. Viral vectors provide an alternative, efficient mechanism of delivery. However, there is some controversy regarding the efficiency of transmission of exogenous DNA by viral vectors, such as retroviruses, adenoviruses and lentiviruses [2]. In contrast to retroviruses, recombinant adenovirus-based vectors are able to infect both dividing and non-dividing cells [3]. Adeno-associated viruses are small, non-pathogenic, dependent parvoviruses that can integrate in a site-specific manner into chromosomes [4,5]. The rAAV genome can integrate into the host chromosome, facilitating long-term transduction [6]. Recombinant AAV preparations are stable and can be produced at high titers of more than 10^{12} particles per ml [7]. Adeno-associated viral (AAV) vectors transduce a variety of somatic tissues, including skeletal muscle, without eliciting an immune response in mice [8]. Moreover, recent reports have indicated that adeno-associated virusesare capable of integrating the FST gene into the host chromosome and facilitating long-term transduction [9,10]. Follistatin (FST) has been demonstrated to be a potent antagonist of other members of the TGF-β family, including myostatin [11]. In addition, FST has been shown to be both safe and effective in mice and in nonhuman primates [10]. Studies have also shown that FST is capable of controlling muscle mass through

* Correspondence: zhangli@263.net; lxdu@263.net
[1]National Center for Molecular Genetics and Breeding of Animal, Institute of Animal Sciences, Chinese Academy of Agricultural Sciences, Beijing 100193, People's Republic of China
Full list of author information is available at the end of the article

pathways independent of the myostatin signaling cascade [12]. Myostatin (MSTN) negatively regulates myoblast differentiation through activation of the Smad, Akt, p38MAPK and p21 pathways [13-15]. Antagonists of MSTN have shown considerable promise for enhancing muscle mass and strength. MSTN inhibits proliferation and differentiation of myoblasts, limiting the growth rate and muscle mass in mammals [12,16]. Recent studies have highlighted the potential benefit of inhibiting MSTN, which results in a double muscle phenotype in MSTN-deficient cattle [17] and MSTN-knockout mice [18]. In particular, because sheep are an economically important animal, breeding double muscle sheep is of high economic value. However, AAV-mediated FST gene transfer has not been reported in sheep, whereas there are several reports of FST gene transfer in sheep by other vectors, such as lentiviral vectors [19]. The objective of the current study was to use a recombinant adeno-associated virus serotype 2 (rAAV2) carrying follistatin to explore the effects of FST on ovine primary myoblast (OPM) proliferation in vitro. In the present study, we tested the hypothesis that an AAV virus carrying the FST gene is capable of inducing myoblast proliferation in vitro. Additionally, because we recognized that FST is expressed in OPM cells, we verified whether over-expression of FST influences the proliferation of myoblasts irrespective of myostatin. Finally, RT-qPCR analysis was carried out to detect the transcript levels of FST, p21, CDK2, AkT I and ActRII B, which are genes that are thought to play roles that are redundant with MSTN for regulating muscle mass. Our results demonstrated that the FST gene is capable of inducing the proliferation of OPM in vitro. Moreover, the findings indicated that in OPM, over-expression of FST induced proliferation both directly or indirectly through down-regulation of the p21 gene under proliferating conditions.

Results

Expression analysis of ovine follistatin and myostatin

To determine whether FST and MSTN mRNA are expressed in OPM, total RNA and protein were isolated from OPM cells and from ovine fetal skeletal muscle to be used as a control and analyzed using RT-PCR. Agarose gel electrophoresis confirmed the expression of FST (Figure 1A) and MSTN mRNA (Figure 1B). The ovine FST and MSTN appeared as a single band on 1.5% (w/v) agarose gels. Electrophoresis of the PCR products showed the 1144 and 151 bp fragments for FST and MSTN, respectively. Next, western blot analyses were performed on protein extracts to detect FST and MSTN protein weight and their properties. Total protein was extracted from OPM cells that were cultured in growth media in 12-well plates for 72 h. The ovine FST protein appeared as a single band with a molecular weight of 66 kDa in SDS–PAGE (Figure 1C). Moreover, anti-MSTN antibodies specifically recognized the two following bands: precursor and processed forms of MSTN (Figure 1D). These two bands corresponded to the unprocessed full-length protein (52-kDa) and the mature processed MSTN (26-kDa). However, this 52-kDa precursor myostatin was not detectable in muscle extracts. These results indicated that OPM cells are capable of synthesizing FST and MSTN mRNA and protein in vitro.

To confirm the RT- qPCR results, FST and MSTN protein expression were also measured by western blot analysis for OPM cultured in GM over an 8-day time-course proliferation. Western blot results indicated that MSTN and FST protein expression was up-regulated during proliferation. FST protein expression was decreased at day 8, whereas MSTN expression remained constant at the end of culture period (Figure 1E).

Cloning of ovine follistatin

The full-length sequence of the gene encoding FST in OPMs, containing a 1035 bp segment encoding the 344 amino acid, was cloned into a pGEM-T vector and sequenced (GenBank Accession No. KF833357). From the follistatin CDS sequence, the amino acid sequence was deduced using software from NCBI. Multiple alignments of the amino acid sequences were produced via DNA-MAN 6.0, as shown in Figure 2A. The deduced amino acid sequence of the ovine FST is highly homologous to amino acid sequences of follistatins from ovine (*Ovisaries*, NP_001244022.1, 100%), goat (*caprahircus*, ADJ53355.1, 99%), bovine (*Bos Taurus*, NP_786995.2, 99%) and human (*Homo sapiens*, NP_037541.1, 92%) ovaries.

Follistatin protein is synthesized in transduced myoblast cells

Construction of the expression vector is shown in Figure 2B. FST cDNA was subcloned into pAAV-IRES-GFP, resulting in the recombinant expression plasmid pAAV-CFS-FLAG (Figure 2B), which was verified via DNA sequencing. Our results confirmed the expression of ovine FST in the transduced OPM cells. As shown in Figure 2C, anti-flag monoclonal antibodies specifically recognized a 66-kDa band for ovine FST in the transduced OPMs. The result here for recombinant FST driven by CMV promoter in transduced cells was similar with observed band in Figure 1C for endogenous FST.

Changes in mRNA concentration during myoblast proliferation

To characterize the growth rate of the OPM cells, the cells were seeded at 2×10^4 cells/well in 6-well plates and harvested and counted for eight days. The result indicated that the cells entered the logarithmic phase of

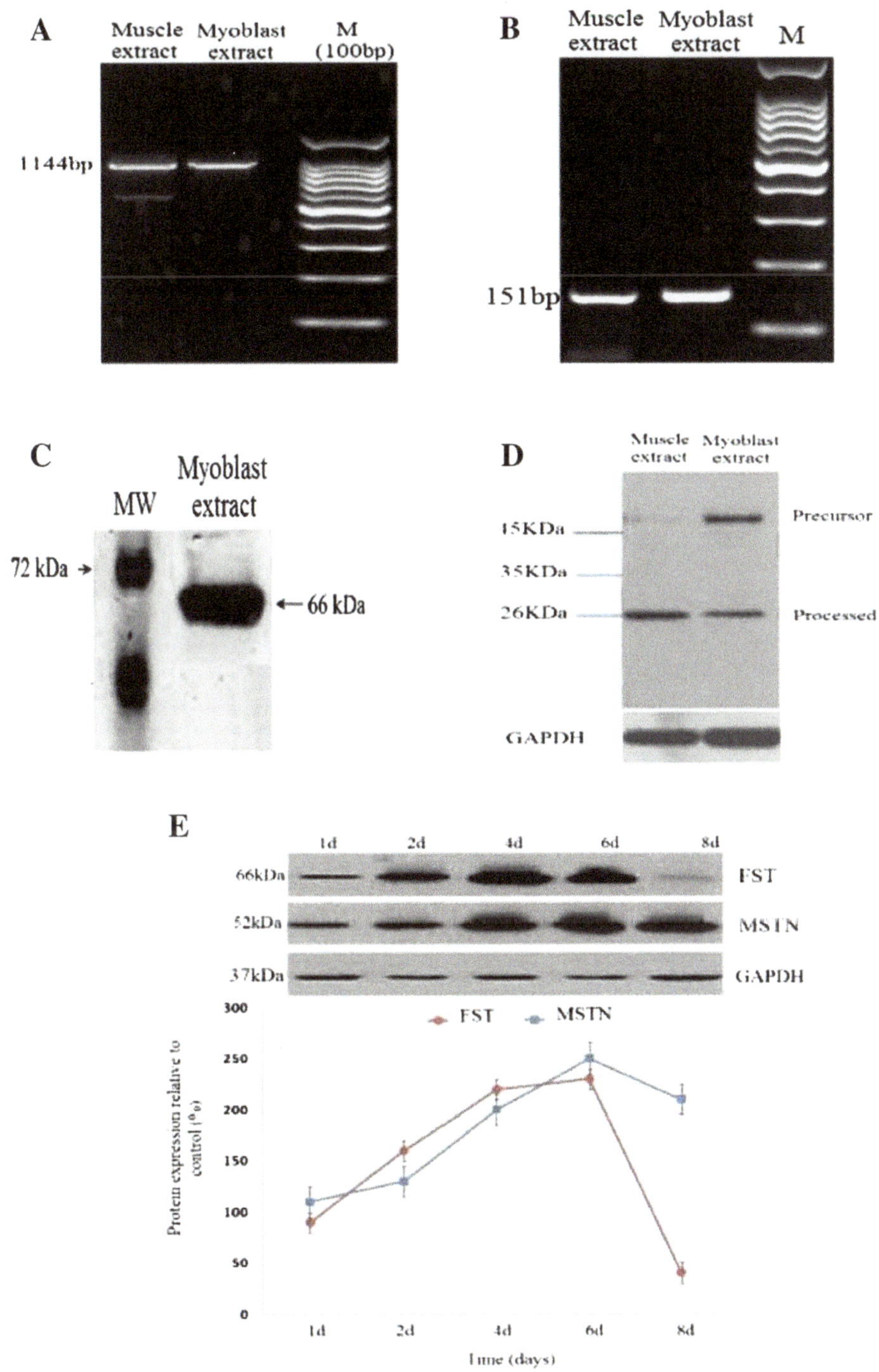

Figure 1 Detection of follistatin and myostatin mRNA and protein in ovine primary myoblast cells. **A**, determination of follistatin mRNA and myostatin. **B**, mRNA expression by RT-PCR in primary muscle myoblasts and ovine fetal skeletal muscle, as a control. M is 100 bp DNA ladder marker. **C**, detection of ovine follistatin protein in primary myoblasts via western blot analysis. 15 μg of total protein was resolved by 12% SDS-PAGE. Polyclonal anti-follistatin (Santa Cruz) was the primary antibody used in the western blot. **D**, detection of myostatin protein in ovine primary myoblast cells and ovine fetal skeletal muscle extracts by western blot analysis. Twenty micrograms of total protein from ovine fetal muscle and 60-day-old ovine primary myoblasts was resolved by 10% SDS-PAGE, and myostatin protein was detected with rabbit anti-myostatin antibodies. Precursor and processed forms of myostatin are indicated. Molecular masses of the western positive bands are indicated. **E**, Densitometry analysis of protein expression for western blots (myostatin and follistatin) was normalized to GAPDH expression in ovine primary myoblast cells over a time-course of 8 days.

A

Ovis (myoblast)	MARPRHQPGGLCLLLLLLCQFMEDRSAQAGNCWPRQAKNGRCQVLYKTELSKEECCSTGR	60
Ovis (ovary)	MARPRHQPGGLCLLLLLLCQFMEDRSAQAGNCWPRQAKNGRCQVLYKTELSKEECCSTGR	60
Bos Taurus	MARPRHQPGGLCLLLLLLCQFMEDRSAQAGNCWLRQAKNGRCQVLYKTELSKEECCSTGR	60
Capra hircus	MARPRHQPGGLCLLLLLLCQFMEDRSAQAGNCWLRQAKNGRCQVLYKTELSKEECCSTGR	60
Homo sapiens	MVRARHQPGGLCLLLLLLCQFMEDRSAQAGNCWLRQAKNGRCQVLYKTELSKEECCSTGR	60
Ovis (myoblast)	LSTSWTEEDVNDNTLFKWMIFNGGAPNCIPCKETCENVDCGPGKKCRMNKKNKPRCVCAP	120
Ovis (ovary)	LSTSWTEEDVNDNTLFKWMIFNGGAPNCIPCKETCENVDCGPGKKCRMNKKNKPRCVCAP	120
Bos Taurus	LSTSWTEEDVNDNTLFKWMIFNGGAPNCIPCKETCENVDCGPGKKCRMNKKNKPRCVCAP	120
Capra hircus	LSTSWTEEDVNDNTLFKWMIFNGGAPNCIPCKEACENVDCGPGKKCRMNKKNKPRCVCAP	120
Homo sapiens	LSTSWTEEDVNDNTLFKWMIFNGGAPNCIPCKETCENVDCGPGKKCRMNKKNKPRCVCAP	120
Ovis (myoblast)	DCSNITWKGPVCGLDGKTYRNECALLKARCKEQPELQVQYQGKCKKTCRDVFCPGSSTCV	180
Ovis (ovary)	DCSNITWKGPVCGLDGKTYRNECALLKARCKEQPELQVQYQGKCKKTCRDVFCPGSSTCV	180
Bos Taurus	DCSNITWKGPVCGLDGKTYRNECALLKARCKEQPELEVQYQGKCKKTCRDVFCPGSSTCV	180
Capra hircus	DCSNITWKGPVCGLDGKTYRNECALLKARCKEQPELQVQYQGKCKKTCRDVFCPGSSTCV	180
Homo sapiens	DCSNITWKGPVCGLDGKTYRNECALLKARCKEQPELEVQYQGRCKKTCRDVFCPGSSTCV	180
Ovis (myoblast)	VDQTNNAYCVTCNRICPEPTSSEQYLCGNDGVTYPSACHLRKATCLLGRSIGLAYEGKCI	240
Ovis (ovary)	VDQTNNAYCVTCNRICPEPTSSEQYLCGNDGVTYPSACHLRKATCLLGRSIGLAYEGKCI	240
Bos Taurus	VDQTNNAYCVTCNRICPEPTSSEQYLCGNDGVTYPSACHLRKATCLLGRSIGLAYEGKCI	240
Capra hircus	VDQTNNAYCVTCNRICPEPTSSEQYLCGNDGVTYPSACHLRKATCLLGRSIGLAYEGKCI	240
Homo sapiens	VDQTNNAYCVTCNRICPEPASSEQYLCGNDGVTYSSACHLRKATCLLGRSIGLAYEGKCI	240
Ovis (myoblast)	KAKSCEDIQCTGGKKCLWDFKIGRGRCSLCGELCPESKSEEPVCASDNATYASECAMKEA	300
Ovis (ovary)	KAKSCEDIQCTGGKKCLWDFKIGRGRCSLCGELCPESKSEEPVCASDNATYASECAMKEA	300
Bos Taurus	KAKSCDDIQCTGGKKCLWDFKVGRGRCSLCGELCPESKSEEPVCASDNATYASECAMKEA	300
Capra hircus	KAKSCEDIQCTGGKKCLWDFKVGRGRCSLCGELCPESKSEEPVCASDNATYASECAMKEA	300
Homo sapiens	KAKSCEDIQCTGGKKCLWDFKVGRGRCSLCDELCPDSKSDEPVCASDNATYASECAMKEA	300
Ovis (myoblast)	ACSSGVLLEVKHSGSCNSISEDTEDEEEDEDQDYSFPISSILEW	344
Ovis (ovary)	ACSSGVLLEVKHSGSCNSISEDTEDEEEDEDQDYSFPISSILEW	344
Bos Taurus	ACSSGVLLEVKHSGSCNSISEDTEDEEEDEDQDYSFPISSILEW	344
Capra hircus	ACSSGVLLEVKHSGSCNSISEDTEDEEEDEDQDYSFPISSILEW	344
Homo sapiens	ACSSGVLLEVKHSGSCNSISEDTEEEEEDEDQDYSFPISSILEW	344

B **C**

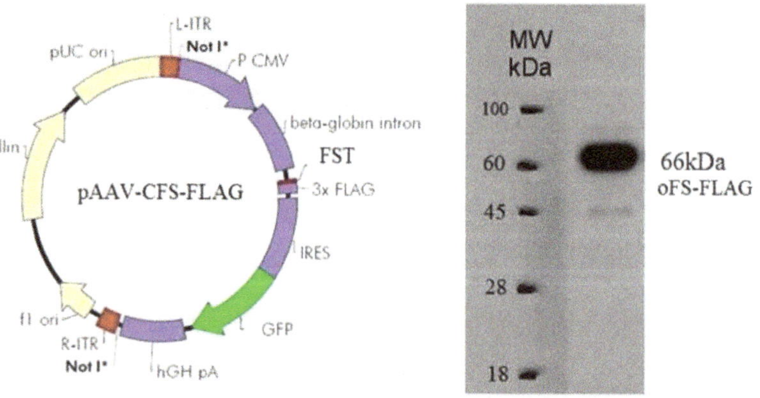

Figure 2 Box shade diagram showing similarity of deduced amino acid sequences encoding ovine follistatin in primary myoblast cells. **A**, Multiple sequence alignment of proteins was carried out using the DNAMAN 6.0 method based on amino acid sequences of follistatins from Ovisaries (NP_001244022.1), Bos Taurus (NP_786995.2), Capra hircus (ADJ53355.1) and Homo sapiens (BC004107.2). **B**, Gene map of recombinant plasmid pAAV-CFS-FLAG: ITR: inverted terminal repeat; CMV promoter: Cytomegalovirus promoter; FST: coding region for follistatin; IRES: internal ribosome entry site; GFP: green fluorescent protein; hGHpA: human growth hormone (hGH) polyA signal; **C**, Detection of FST-FLAG proteins in transduced and normal ovine primary myoblasts by western blot analysis. 15 mg of total protein was resolved by 12% SDS-PAGE. Monoclonal anti-Flag (Sigma) were the primary antibody used in the western blot.

growth after 2 days and grew extremely slowly after 6 days. Some fused cells were observed after 6 days. Eight days after incubation, many of the cells were fused and had a long tubular shape (data not shown). Next, to gain insight into the role of follistatin during OPM proliferation and to assess the expression of p21, p57, Act RIIB, Akt I, CDK2, MSTN and FST genes during OPM proliferation, total RNA was isolated from the OPM cells and harvested at days 2, 4, 6 and 8. In addition, the cells were harvested at day 1 to be used as a control. Then, the mRNA concentration was quantified by RT-qPCR. We noted that there were no significant changes in Act RIIB mRNA expression (Figure 3F). As shown in Figure 3, FST, MSTN, Akt I and CDK2 mRNA concentrations were up-regulated during OPM proliferation. The amounts of these genes gradually increased at day 2 ($P < 0.01$) after plating and then sharply

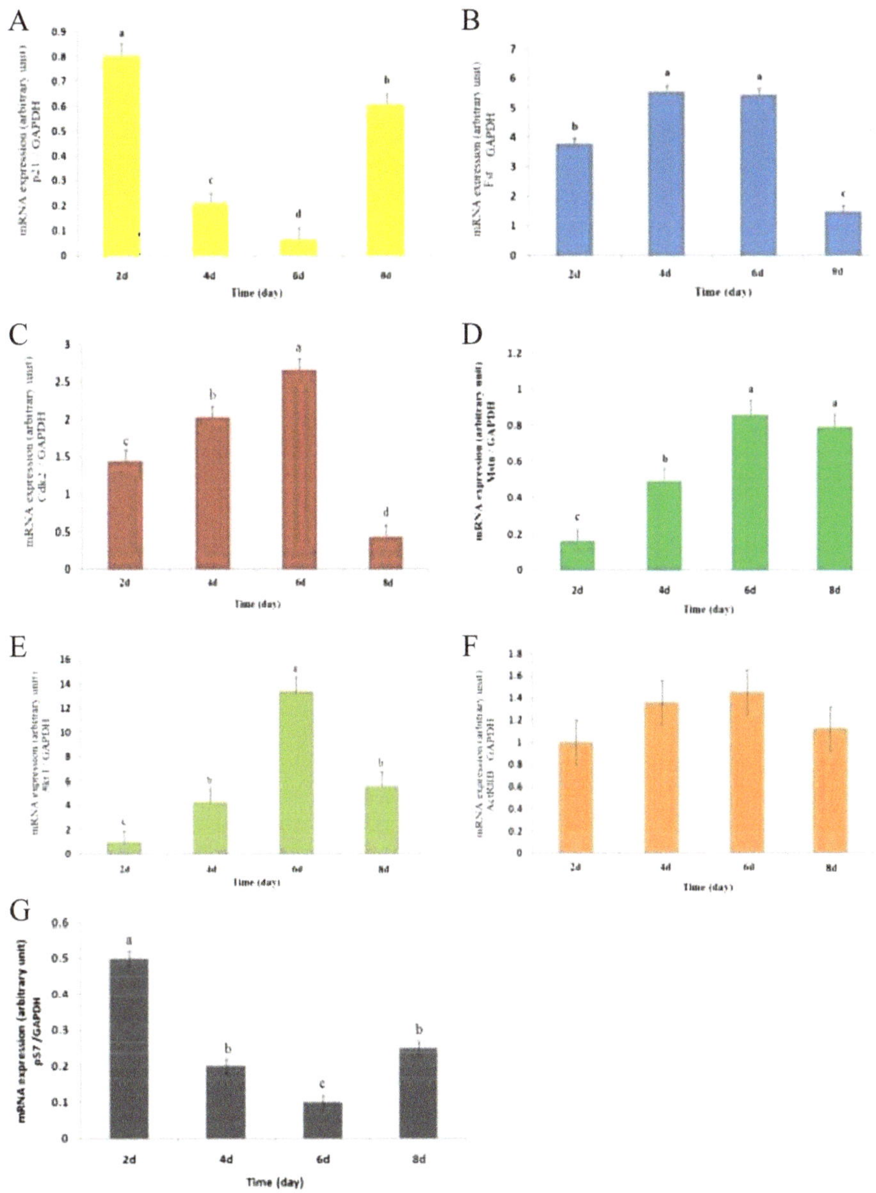

Figure 3 Real time quantitative PCR results of p21 (A), FST (B), CDK2 (C), MSTN (D), Akt I (E), ActRIIB (F)and p57 (G) mRNA expression during proliferation of ovine primary myoblast. Cells cultured in 6-well plates for eight days and harvested at days 2, 4, 6 and 8. The mRNA concentrations were quantified by real-time quantitative PCR. The amount of mRNAs was normalized to the amount GAPDH mRNA. Data were analyzed by one-way ANOVA ($P < 0.05$, $n = 8$). Statistical differences are indicated: a > b > c > d.

increased up to day 6. FST, Akt I and CDK2 mRNA levels declined at day 8 ($P < 0.01$) after plating, whereas the level of MSTN mRNA remained constant until the end of the culture period. As shown in Figure 3, the maximum expression levels of FST and CDK2 were observed at day 6. Both genes began to show increased expression after 2 days, reaching a maximum expression at day 6, and exhibited low expression at day 8. Moreover, the results showed that the p21 and p57 mRNA concentrations were down-regulated during myoblast proliferation (Figure 3A). Interestingly, the minimum expression levels of p21 and p57 were observed at day 6 (Figure 3A). However, the levels of p21 and p57 mRNA were both increased at day 8, which paralleled the increase in MSTN mRNA concentration and the reduction in FST mRNA concentration at day 8. Taken together, these findings strengthened the hypothesis that FST may be involved in the proliferation of myoblasts.

Effects of ovine FST over-expression on gene expression

In this experiment, total RNA was extracted from the transduced OPM to detect p21, p57, Myo D, Akt I, CDK2 and Myf5 gene expression using RT-qPCR under proliferating conditions. We noted that AAV-CFs-FLAG exhibited a 12.5-fold reduction in p21 mRNA expression ($P < 0.01$). Moreover, the expression of CDK2 exhibited a 6.1-fold increase ($P < 0.01$) compared to the control. In addition, Akt I mRNA expression exhibited a dramatic 41-fold increase ($P < 0.01$).These data also revealed that the expression of Myf5 mRNA exhibited a 2.5-fold decrease ($P < 0.01$). In addition, there were no significant changes in Myo D and p57 mRNA concentrations compared to the control (Figure 4).

Follistatin over-expression increases proliferation of ovine primary myoblasts

To determine the effect of ovine FST protein on OPM cell proliferation, we cultured OPM cells for 48, 72 and 96 h in growth media infected with or without the AAV-CFs-FLAG virus. The cells were fixed and stained with Giemsa (Figure 5A, a–f). OPM proliferation was monitored for alterations using a WST-8 assay with Cell-Counting Kit-8 (CCK-8) reagents. As shown in Figure 5B, the optical density (at 450 nm) was significantly increased in the transduced OPMs cultured in GM media after 48, 72 and 96 h compared with the control, indicating that FST significantly increased proliferation ($P< 0.01$). However, the FST protein induced the growth of myoblasts in a time-dependent manner with half-maximal induction occurring at the 72 h time point (Figure 5B). In addition, FST did not cause an increase in myotube formation and overt differentiation (Figures 5A, a–f).These data indicated that when OPM cells were incubated in GM with the AAV-CFs-FLAG virus for 96 h, there was a steady increase in cell number compared to the control (Figure 5A, a–f). The results strongly suggested that AAV-CFs-FLAG-expressing FST induced a significant effect on cell myoblast proliferation, suggesting that it is involved in modulating myoblast proliferation.

FST protein interferes at the G1 phase of the cell cycle

The cell cycle was analyzed using a fluorescence-activated cell sorter (FACS) after staining with propidium iodide (Figure 6). The cells were transduced in growth medium for 72 and 96 h. The percentage of cells in the G1, S, and G2/M phases was significantly different under the various conditions ($P< 0.05$). The detailed values of the cell cycle analysis are summarized in Table 1. As shown in Table 1, the percentage of cells in G1 was significantly decreased and the percentage in S phase increased with AAV-CFs-FLAG transduction. The FACS analysis also showed a dramatic increase in the number of cells in S phase (9.23 ± 1.60% to 27.10 ± 1.81%), which was accompanied by a decrease in the percentage of cells in G1 (85.45 ± 2.12% to 63.43 ± 1.95%) phase of the cell cycle (Figure 6; Table 1). In addition, there were no significant changes in AAV-GFP during the G1 and S phases compared to non-transduced cells (Figure 6B). Furthermore, AAV-CFs-FLAG transduction reduced the G1 phase in a time-dependent manner, with the maximum reduction at 96 h post-infection (Figure 6D). These data suggest that follistatin reduced the G1 phase in growth medium. The results confirmed that the increase in myoblast

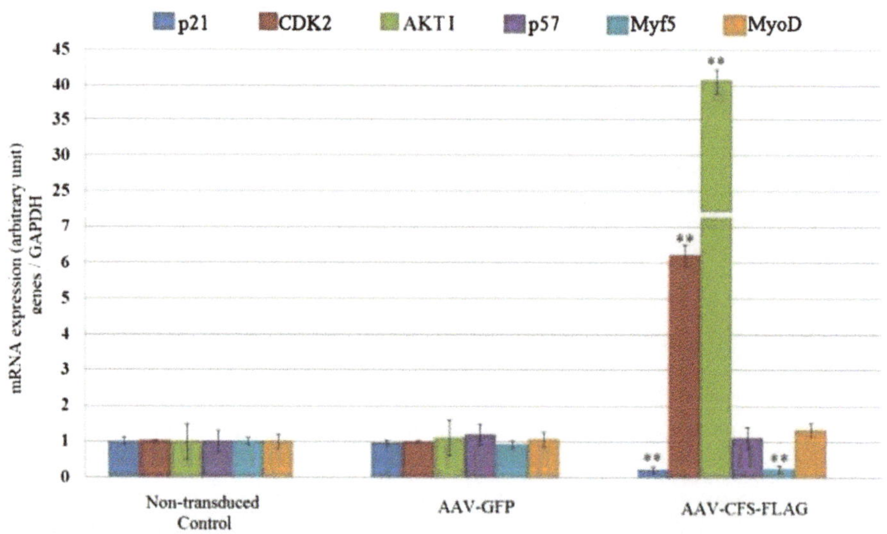

Figure 4 Real time quantitative PCR results of p21, MyoD, Akt I, CDK2, p57 and Myf5 genes expression in transduced primary myoblasts cultured under proliferating conditions. Total RNA was isolated after 96 h from the OPM transduced with AAV-CFS-FLAG and the non-transduced OPMs (control). Gene expression was determined using real-time qPCR and normalized to GAPDH mRNA. The non-transduced myoblasts and AAV-GFP were selected as the control and negative control, respectively. **, $P < 0.01$ with non-transduced control.

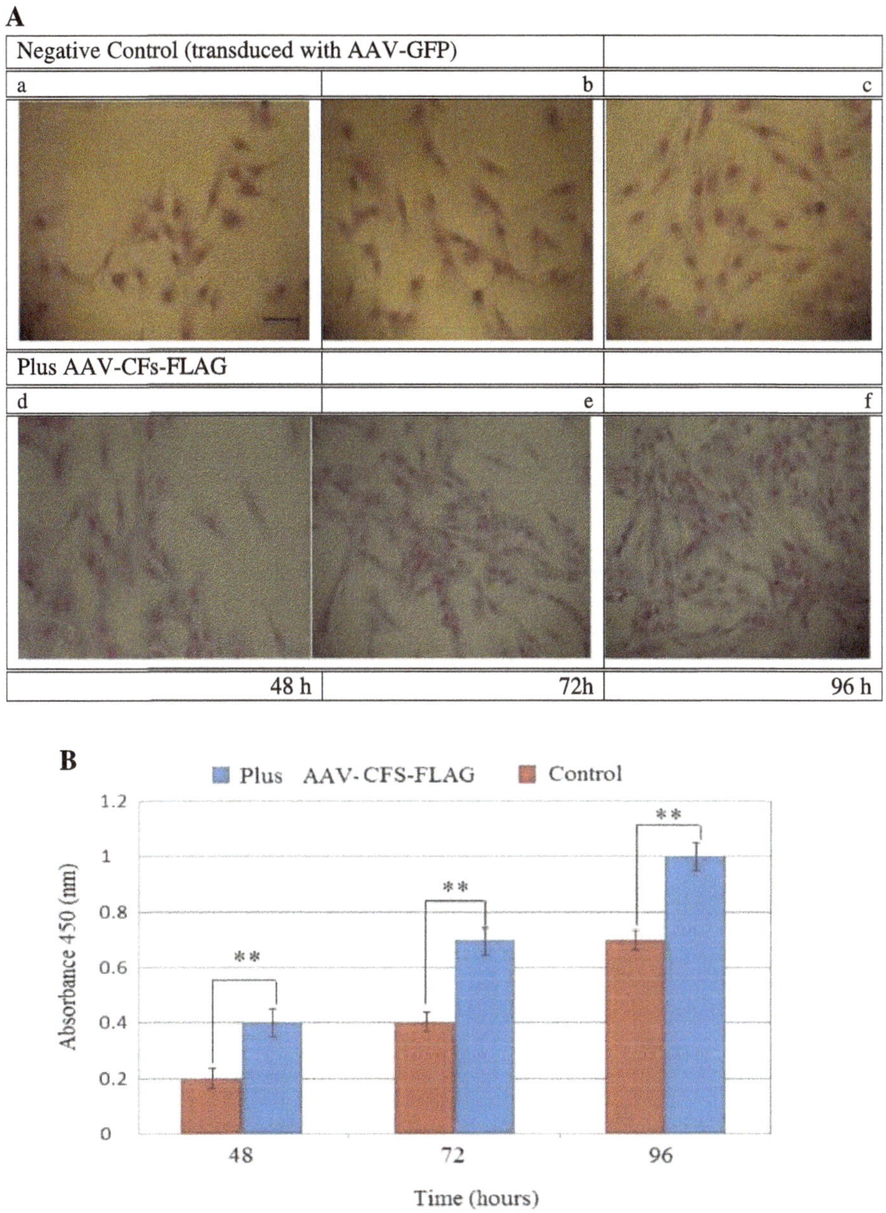

Figure 5 Induction of ovine primary myoblast proliferation by AAV-CFs-FLAG virus expressing FST protein in vitro. A, phase contrast pictures show the morphology of ovine primary myoblast cells after follistatin over-expression, as compared to negative control cells. After Giemsa staining, pictures were taken under the same magnification over a time-course of 4 days. Myoblasts were grown for 48, 72 and 96 h in growth media without or with AAV-CFs-FLAG virus transduction. Cells were fixed and stained with Giemsa. Bar =100 μM. **B**, absorbance (at 450 nm) was read at 48, 72 and 96 hours after plating in growth media without or with AAV-CFs-FLAG virus. Data were analyzed by one-way ANOVA (**P <0.01, n = 8).

proliferation was due to a decrease in the number of cells in G1 phase (Figure 6; Table 1).

Discussion

Recent studies have shown that human myoblast proliferation is induced by an over-expression of FST in cell culture. In addition, FST forms a complex with MSTN, preventing its interaction with its receptor and thus blocking the MSTN signal [20]. In the present study, ovine FST was cloned into the AAV2 vector, and then AAV2 virus particles of high titer were obtained. The OPM cells were infected by recombinant AAV vectors. After total RNA and protein isolation, real-time qPCR and western blotting were performed for the OPMs cultured in the GM. Furthermore, a cell proliferation assay was performed using Cell-Counting Kit-8 (CCK-8) reagents. The cells were stained with propidium iodide and were analyzed using a flow cytometer for cell cycle

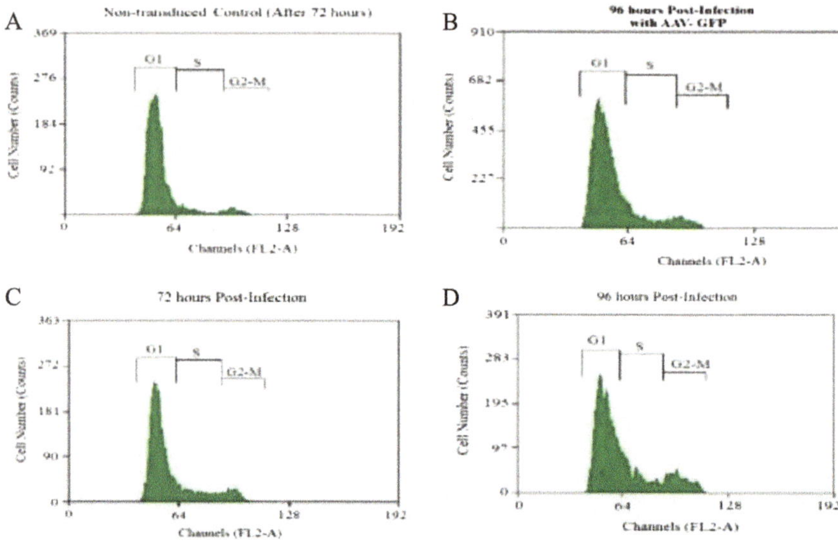

Figure 6 Effect of follistatin over-expression on ovine primary myoblast cell cycle progression. A, cell cycle distribution was measured by flow cytometry using propidium iodide stain and the percent of the cell cycle phase is shown in a bar graph form with the G1 (85.45%) and S (9.23%) phases. **B**, cell cycle distribution was measured by flow cytometry and the percent of the cell cycle phase is shown in a bar graph form with the G1 (81.65%) and S (11.49%) phases. **C**, cell cycle distribution was measured by flow cytometry using propidium iodide stain and the percent of the cell cycle phase is shown with the G1 (73.93%), and S (16.24%). **D**, cell cycle distribution was measured by flow cytometry and the percent of the cell cycle phase is shown with the G1 (63.43%), and S (27.10%).

progression. The results in Figure 1 show that the amino acid sequence deduced from the cloned full-length DNA sequence of FST contained an N-terminal domain (residues 1 ~ 63), domain 1 (64 ~ 136), domain 2 (137 ~ 211), domain 3 (212 ~ 288), and a C-terminal region (288 ~ 315). Therefore, the sequence indeed encoded the longest form of FST [21]. The western blot analyses showed that the ovine FST protein appeared as a single band with a molecular weight of 66-kDa in SDS–PAGE (Figure 2B), consistent with previous studies [19]. In this study, we observed the precursor and the processed MSTN at 52 and 26-kDa proteins, respectively (Figure 2F). These observations were consistent with previous reports [16,22-24]. These results indicated that the MSTN protein was indeed synthesized and proteolytically processed in the OPMs.

To understand the regulation mechanism of MSTN and FST protein and their interaction, real-time qPCR analysis was performed on the OPM cells undergoing proliferation. Moreover, to confirm the RT- qPCR results, FST and MSTN protein expression were also measured by western blot analysis. FST and MSTN mRNA concentrations were up-regulated at mRNA (Figure 3B and D) and protein level (Figure 1E), whereas p21 and p57 mRNA concentrations were down-regulated during OPM proliferation (Figure 3A and G). Despite the increased expression of MSTN mRNA, p21 and p57 mRNA expression was decreased in this condition. It has been thought that increased expression of MSTN leads to an up-regulation of p21 and p57 mRNA expression [23]. The lack of consistency demonstrated that the FST protein, as a binding protein for MSTN, may prevent MSTN from binding to receptors on the myoblast cells and thus block the MSTN signal [20]. However, the RT-qPCR results showed that FST over-expression only down-regulated p21 mRNA expression and had no effect on p57 mRNA expression. These results confirmed that FST induces proliferation through a down-regulation of p21 because only p21 levels

Table 1 The percentage of cells in G1, G2/M and S-phases under different conditions (mean ± S.E.M)

Conditions	G1 (%)	G2 /M (%)	S (%)
Non-transduced control (after 96 h)	85.45 ± 2.12^{a}	5.32 ± 0.87^{b}	9.23 ± 1.60^{c}
96 hours Post-Infection with AAV-GFP	81.65 ± 2.41^{a}	6.86 ± 1.12^{ab}	11.49 ± 1.34^{c}
72 hours Post-Infection with AAV-CFs-FLAG	73.93 ± 1.85^{b}	9.83 ± 0.56^{a}	16.24 ± 1.45^{b}
96 hours Post-Infection with AAV-CFs-FLAG	63.43 ± 1.95^{c}	9.47 ± 0.93^{a}	27.10 ± 1.81^{a}

Cell cycle analysis was performed at 72 and 96 hours after those cells was transduced with AAV-CFs-FLAG and the percentage of cells in G1, G2/M and S-phase were determined. The percentage of cells in G1 phase was greatly significant 96 hours AAV post-infection. Data were analyzed via one-way ANOVA ($P < 0.05$). Statistical differences are indicated: a > b > c.

were down-regulated as a result of FST over-expression in myoblasts (Figure 4), whereas no change was observed in the level of p57 expression [23].

Because FST can lead to increased muscle mass [9], it is possible that FST functions by increasing the number of myoblasts during development. To examine this possibility, we incubated myoblast cultures with an AAV vector carrying the FST gene. The results indicate that FST indeed induces the proliferation of exponentially growing myoblasts (Figure 5A). Moreover, because the surviving myoblasts at the end of the assay reflect the net result of cell proliferation, we performed a flow cytometry analysis to assess the cell cycle progression in the transduced myoblasts. The flow cytometry analysis showed that when myoblasts were transduced with the AAV-CFS-FLAG virus, a decreased number of myoblast cells was arrested inG1phase, and therefore, more myoblast cells made the transition to S (DNA synthesis) phase (Figure 6; Table 1). Consistent with this finding, others have suggested that MSTN can inhibit myoblast proliferation [23,25,26]. Investigations into the mechanisms revealed that one of the targets of MSTN signaling is the cell cycle. In cell culture studies, MSTN inhibited the progression of myoblasts at the G1 phase of the cell cycle. This cell cycle arrest was shown to be mediated by p21, which was the only CDK inhibitor to be up-regulated by MSTN. In addition, myostatin treatment also specifically reduced the levels of CDK2, the combination of which resulted in markedly reduced activity of CDK2. This, in turn, led to the hypophosphorylation of retinoblastoma protein and thus inhibited the progression of cells into the S phase of the cell cycle [23]. Here, in the OPMs infected with AAV-CFs-FLAG, there was indeed a decrease in p21 expression at the transcription level and an increase in CDK2 expression (Figure 4) under proliferating conditions, which led to a decrease in the arrest of myoblasts in the G1-phase of the cell cycle. FST likely forms a complex with MSTN, preventing its interaction with its receptor and thus blocking the MSTN signal [20]. Thus, the observed increase in the number of myoblast cells is the result of deregulated (increased) myoblast proliferation. These observations are consistent with the model proposed by Thomas et al. [23] regarding the role of MSTN during muscle growth [23]. We propose the following model for the role of FST in myoblast proliferation regulation. In response to MSTN signaling, consistent with the up-regulation of p21, CDK2 expression is down-regulated, which causes the hypophosphorylation of the Rb protein and G1 arrest. Therefore, these results confirmed the role of FST in myoblast proliferation regulation.

Conclusions

Our results provide the first evidence that an AAV viral system can be used for gene transfer in sheep. Moreover, the results showed that an AAV vector can successfully induce expression of FST in OPM cells in vitro. In addition, our findings demonstrated that FST induces proliferation through a down-regulation of the p21 gene under proliferating conditions in ovine primary myoblasts. In the future, it would be interesting to determine whether an AAV virus can be used to produce transgenic sheep for meat production.

Methods

Nucleic acid manipulation

DNA was purified and manipulated, as described previously [27].

Chemicals, strains, plasmids and media

T4 deoxyribonucleic acid (DNA) ligase and restriction endonucleases were purchased from New England Biolabs (Beverly, MA) and Fermentase (Life Sciences, Thermo Fisher), respectively. Escherichia coli DH5 alpha (Invitrogen, CA, USA) was used as a host for the plasmid cloning experiments. Bacteria were grown in Luria-Bertani (LB) medium containing 5 g/L yeast extract, 10 g/L tryptone and 10 g/L NaCl at pH = 7.0 and 37°C. The pAAV-IRES-GFP plasmids, consisting of a CMV promoter and a FLAG tag, the pAAV-GFP control, and the pAAV-RC and pHelper vectors were purchased from Agilent Technologies Company (La Jolla, CA). The pGM-T plasmid vector for the direct insertion of a blunt PCR product into a plasmid vector was obtained from Invitrogen Company (CA, USA). Plasmids were amplified in DH5 alpha bacteria, purified (Gene JET Plasmid Maxiprep Kit; Fermentas, Life Sciences, Thermo Fisher) and dissolved in nuclease-free water. Dulbecco's Modified Eagle's Medium (DMEM) was supplemented with 100 U/ml penicillin and 100 μg/ml streptomycin. Fetal bovine serum (Hyclone, Logan, UT) and horse serum (Hyclone, Logan, UT) were also added to DMEM to prepare specific media.

Cloning of the full-length follistatin sequence

Ovine full-length FST cDNA was isolated from a primary myoblast cDNA library, and its sequence was determined. To detect the full-length FST cDNA sequence, the total RNA was extracted from OPMs using a GeneJET RNA Purification Kit (Fermentas, Life Sciences, Thermo Fisher). Then, the cDNA was synthesized using a Maxima First Strand cDNA Synthesis Kit (Fermentas, Life Sciences, Thermo Fisher) according to the manufacturer's protocol. The DNA encoding FST was amplified via PCR from cDNA using the forward primer 5′-GCACCG CTTGCCAGAGCCAC -3′ and the reverse primer 5′-GGTGGAGGGTTTACCACTCTAGAATGGA -3′, which were designed based on the availability of bovine sequences (Gen Bank Accession No. NM_175801). The PCR reaction was carried out in a final volume of 20 μl containing the DNA template and 10 µl of commercially available PCR master mix (PyroStart™ Fast PCR Master Mix (2×), Fermentas, Life Sciences, Thermo Fisher). After brief centrifugation, the PCR amplification was carried out in an ABI 9700 PCR system (Applied Biosystems, Foster City, CA, USA). The conditions for the PCR amplification were as follows: initial denaturation was performed at 94°C for 4 min, followed by 35 cycles at 95°C for 30 s, 60°C annealing for 30 s, and 72°C extension for 1 min, and a final extension at 72°C for 10 min. The amplified DNA was ligated into a pGM-T vector and transformed into DH5a (E. coli). The DNA sequence was determined after digestion by a restriction enzyme, electrophoresis, and sequencing. The amino acid sequence of the cloned DNA sequence was determined using software from NCBI. Multiple alignments of the amino acid sequence were produced using DNAMAN 6.0.

Construction of the pAAV-CFS-FLAG plasmid vector

A complete open reading frame (ORF) of 1035 base pairs of the ovine FST gene was amplified via PCR from full-length cDNA using the forward primer 5′-AA**GAATTC**CCTCAGGATGGCCCGTCCTA-3′ (P1) containing an EcoRI site (bold) and the reverse primer 5′-G**CTCGAG**GGTTTTCCACTCTAGAATGGA -3′ (P2) containing an XhoI site (bold). The primers were designed based on the availability of ovine sequences (Gene Bank Accession No. KF833357). The PCR reactions and conditions were carried out as described above. The PCR products and the pAAV-IRES-GFP vector (Agilent Technologies Company, La Jolla, CA) were digested with EcoRI and XhoI (Fermentas, Life Sciences, Thermo Fisher), recovered through agarose gel electrophoresis, and then ligated by T4 DNA ligase (New England Biolabs, Beverly, MA). New recombinant plasmids, which consisted of a CMV promoter and a FLAG tag, carrying the ovine FST gene were termed pAAV-CFS-FLAG. The ligated plasmid vector was transformed into E. coli DH5 alpha. A single colony of the transformed bacteria was selected and transferred to 3 ml of Luria-Bertani (LB) medium containing 100 µg/mL ampicillin and incubated overnight under vigorous shaking (200 rpm) at 37°C. The culture was used to extract the recombined plasmid. The presence of the FST protein was confirmed via digestion with a restriction enzyme, electrophoresis, and sequencing.

Packaging AAV vectors

AAV particles were produced by co-transfection of the recombinant pAAV-CFS-FLAG vector (20 µg) with two helper plasmids (20 µg each of pAAV-RC and pHelper) into HEK 293 cells seeded in a 225-cm^2 flask. Briefly, 1 h before transfection, half of the medium was replaced with fresh medium, and 20 µg of vector and of each helper plasmid was added to 4 ml of 300 mM $CaCl_2$. This solution was added to 4 ml of 2× HBS and mixed immediately by gently inverting three times. The mixture was immediately pipetted into HEK 293 cells in 40 ml of DMEM medium with 10% FCS. The plate was incubated at 37°C for 6 h. Three days after transfection, 1 ml of 0.5 M EDTA was add and incubated for 3 min at room temperature. The supernatant was collected at 24, 48, and 72 h via centrifugation at 1000 × g for 10 min at 4°C. Next, the supernatant was removed and the cells of the pellet were resuspended in 2 ml of TBS. The cells were freeze-thawed that had been suspended in TBS three times by placing them alternately in a dry ice/ethanol bath until the suspension was completely frozen and in a water bath of 37°C until it was completely thawed. Then, the tissue debris was removed via centrifugation at 20,000 g for 10 min, and the supernatant was collected. The collected supernatant was filtered through a 0.45-µm low protein binding filter (Millipore, Billerica, MA, USA). Finally, a ViraBind™ Adeno-Associated Virus Purification Mega Kit (Cell Biolabs, INC, San Diego, CA, USA) was used according to manufacturer's protocol. The titer of the virus concentrate was determined by transducing 293 cells with serial dilutions. The suspended virus particles were diluted using serial dilutions of 10^{-1}, 10^{-2}, 10^{-3}, and 10^{-4}. After culturing for 48 h, the titers were determined by counting the number of GFP-positive colonies using a fluorescence microscope. The titers were expressed as transducing units/mL (TU/mL). The virus preparation used for transgenesis had a final titer of greater than 10^{12} transducing units/mL.

Ovine primary myoblast culture

Primary myoblast cultures were obtained from 60-day-old sheep fetuses, as previously described [28]. Briefly, hind limb skeletal muscle mass was excised, cut into small pieces, placed in DMEM containing 20% FBS and 10% Me2SO, and frozen in liquid nitrogen for subsequent myoblast cultures. Primary muscle cultures were grown in growth media (DMEM with 10% FBS, 100 U/mL penicillin and 100 µg/mL streptomycin) in 6-cm petri dishes to test for fibroblast contamination. Preliminary studies (data not shown) indicated that approximately 90% of the ovine primary myoblast cells stained positive for the muscle-specific marker desmin. All studies were approved by the Biological Studies Animal Care and Use Committee, Beijing, Peoples Republic of China.

Transduction of primary myoblast cells

For transduction of primary myoblasts with AAV-CFs-FLAG or AAV-GFP control vectors, a ViraDuctin™ AAV Transduction Kit (Cell Biolabs, INC, San Diego, CA, USA) was used according to the manufacturer's instructions as the transduction reagent. Briefly, the OPM cells were seeded at 3×10^3 cells/well in 48-well plates. An AAV virus with the optimal multiplicity of infection (MOI) of 50 was added to the fresh growth medium. The cells were incubated at 37°C for 6 h. Finally, pre-warmed complete medium was added in each well and incubated at 37°C for 48, 72 or 96 h. The green fluorescence excited by the AAV vector was observed using a fluorescence microscope (ECLIPSE TE2000-U, Nikon, Tokyo, Japan) at 48, 72 or 96 h.

RNA isolation and RT-PCR

The total RNA was isolated from primary myoblast cells and ovine fetal skeletal muscle (20 mg) using a Thermo Scientific GeneJET RNA Purification Kit (Fermentas, Life Sciences, Thermo Fisher) according to the manufacturer's instructions. The quality of RNA was verified using a spectrophotometer (Astra Gene, Astra net, Cambridge, USA) at 260/280 nm (OD260/OD280 = 1.8–2.0). The cDNA was prepared from 1 μg of total RNA using a Maxima First Strand cDNA Synthesis Kit (Fermentas, Life Sciences, Thermo Fisher) according to the manufacturer's protocol. The sequences of primers, amplicon size and annealing temperature are shown in Table 2. The PCR reactions and conditions were carried out as described above.

Real-time quantitative RCR analysis

Myoblast cells (3.0×10^3 cells/well) were cultured in GM in 48-well plates. Total RNA was isolated after 96 h from the OPM transduced with AAV-CFS-FLAG and the non-transduced OPMs (control). Moreover, an AAV-GFP control vector was also selected as a negative control for these experiments. The mRNA concentrations for all of the genes, which are listed in Table 2, were quantified via real-time quantitative PCR (RT-qPCR). Primer sequences were designed using the Vector NTI Advance™ 11. The sequences of primers for PCR amplification are shown in Table 2. In this procedure, GAPDH was used as a house-keeping gene for normalizing the gene expression data of the RT-qPCR. RT-qPCR was performed using a BIO-RAD iQ5 Real-time PCR Detection System (Bio-Rad, Hercules, CA, USA) with a Power SYBR-Green PCR Master Mix Kit (Applied Biosystems, Foster City, CA, USA). RT-qPCR was carried out with a final volume of 20 μL containing 1 μL RT product, 10 μL SYBR-Green PCR master mix (2×), 2.5 μM forward and reverse primers and nuclease-free water. The PCR cycling programs were as follows: 3 min of UNG incubation at 55°C, followed by 10 min at 95°C, 40 cycles at 95°C for 10 s, annealing at an optimized annealing temperature for 15 s (temperatures given in Table 2), extension at 72°C for 15 s, plate-reading every 0.2°C from

Table 2 Primers used for RT-PCR and real time quantitative PCR

Gene	Primer sequence (5′–3′)	Anneal. T (°C)	Amplicon size	GenBank accession number
Myostatin	F: 5′-CTGGCCCAGTGGATCTGAATGAG –3′	60	151 bp	NM_001009428
	R: 5′- TTCCAGGCGAAGCTTACTGAGGA-3′			
Follistatin	F: 5′-TGGACCGAGGAGGACGTGAA-3′	60	160 bp	NM_001257093.1
	R: 5′-AGACGCAGCGGGGTTTGTTC-3′			
ActRIIB	F: 5′-TCCCTCACGGATTACCTCAAG-3′	58	149 bp	NM_174495
	R: 5′-TCCCTGTGGGCAATAGATGG-3′			
Myf5	F: 5′-ACCAGCCCCACCTCAAGTTG-3′	60	150 bp	NM_174116.1
	R: 5′-GCAATCCAAGCTGGATAAGGAG-3′			
MyoD	F: 5′-CGACTCGGACGCTTCCAGT-3′	60	181 bp	NM_001040478.2
	R: 5′-GATGCTGGACAGGCAGTCGA-3′			
P21	F: 5′-GCAGACCAGCATGACAGATTT-3′	58	70 bp	NM_001098958
	R: 5′-GGATTAGGGCTTCCTCTTGGA-3′			
P57	F: 5′- CGCCGCACCTTTCCCATG-3′	59	166	NM_001142510
	R: 5′- CCCGCAGTGGCATGTCCA-3′			
CDK2	F: 5′-CATGGATGCCTCTGCACTCACTG-3′	60	102 bp	NM_001142509.1
	R: 5′-AGGACCCGATGAGAGTGGCAGAA-3′			
Akt1	F: 5′-CAGCATCGTGTGGCAGGAC-3′	60	138 bp	NM_001161857.1
	R: 5′-TCTTGGTCAGGTGGCGTAA-3′			
GAPDH	F: 5′-CAAGTTCCACGGCACAGTCA-3′	60	249 bp	NM_001190390
	R: 5′-TGGTTCACGCCCATCACAA-3′			

60°C to 94°C for calculation of the melting curves, and a final extension step of 72°C for 10 min. Finally, fluorescence data were analyzed using the iQ5 Optical System Software (Bio-Rad). Gene expression was normalized to GAPDH expression using the $2^{-\Delta\Delta C}{}_{T}$ method [29].

Western blot analysis

Total protein was extracted from the normal myoblast cells and the transduced myoblast cells to determine MSTN and FST protein expression. Total protein was obtained from hind limb fetal skeletal muscle as a control. Protein concentrations were determined according to the manufacturer's instructions (Bio-Rad, Hercules, CA, USA). A total of 15 μg protein was separated via SDS-PAGE (12%) and transferred to a nitrocellulose membrane via electro-blotting. SDS-PAGE and western blot transfer were performed using standard techniques. The gels run under non-reducing conditions. The following primary antibodies were used in this study: MSTN, goat polyclonal anti-MSTN antibody (sc-6884, 1:500; Santa Cruz); FST-FLAG, mouse monoclonal anti-FLAG M2 antibody (F-3165; 1:500; Sigma); FST, rabbit polyclonal anti-FST antibody (sc-30194, 1:500; Santa Cruz) and GAPDH, mouse monoclonal anti-GAPDH antibody (G8795, 1:5000; Sigma). HRP-conjugated goat anti-mouse IgG antibody (115-035-174, 1:10,000; Jackson Immuno Research, West Grove, PA, USA), rabbit anti-goat IgG-HRP (sc-2922, 1:5000; Santa Cruz) and HRP-conjugated goat anti-rabbit IgG antibody (111-035-003, 1:10,000; Jackson Immuno Research) were used as secondary antibodies.

For myostatin protein detection, protein (20 μg) was loaded for electrophoresis on a 10% gel and transferred to nitrocellulose membrane (Millipore, Billerica, MA, USA) by wetting transferring. Membranes were stained with Ponceau stain to verify loading. Membranes were blocked in 3% BSA-TBST and incubated overnight at 4°C with primary antibodies. Then, Membranes washed five times for 5 min each time with TBST, and incubated for 60 min at room temperature with secondary antibodies.

For GAPDH-loading controls, the membrane was blocked in 3% BSA-TBST and incubated with anti-GAPDH antibody in 3% vBSA-TBST for 3 h at room temperature, followed by incubation with secondary antibody in TBST +5% milk for 60 min at room temperature. Horseradish peroxidase activity was detected using ECL with western blotting reagents (RPN2132, Amersham Biosciences). Finally, the protein expression quantified by densitometry by the Gel Quant NET software, and normalized to GAPDH.

Proliferation assay

The OPM cells were grown prior to the assays in growth media. Cell proliferation assays were conducted in 48-well plates. After a 16-h attachment period, the cells were transduced with the AAV-CFs-FLAG vector. The plates were then incubated at 37°C and 5% CO2 for48, 72 and 96 h. After the incubation period, the cell proliferation assay was performed by adding 20 μl of the Cell-Counting Kit-8 (CCK-8) reagents (Dojindo Molecular Technologies, Maryland, USA) to each well of the plate for 2 h. Finally, the absorbance at 450 nm was measured using a Spectra Max M5 microplate reader (Molecular Devices, CA, USA).

Cell cycle analysis

The cell cycle analysis was performed with cells seeded in 48-well plates (3×10^3 cells/well) in growth media. The cells were allowed to attach overnight. The cultures were transduced with the AAV-CFs-FLAG vector and incubated for 72 and 96 h. The collected cells were then washed in PBS and fixed in 75% ethanol. The cells were stained with 50 μg/mL propidium iodide containing 10 μg/mL RNase A and then analyzed in a FACS Calibur flow cytometer (BD Biosciences, CA, USA). The percentages of cells in the G1, S, and G2-M phases were determined.

Statistical analysis

The data are expressed as the mean ± SEM. The results were analyzed via one-way analysis of variance (ANOVA) (general linear model) using SPSS software (version 13.0). The positions of the samples on the plate were randomly assigned, and all samples were run in replicates of eight. Significant differences between treatments were tested using a Student's t- test. Differences are reported at two significance levels, 0.05 or 0.01.

Availability of supporting data

The mRNA-seq data were deposited at NCBI with the Gen Bank accession number KF833357 (http://www.ncbi.nlm.nih.gov/nuccore/KF833357).

Abbreviations

OPM: Ovine primary myoblast; TGF-β: Transforming growth factor β; DMEM: Dulbecco's modified Eagle's medium; FBS: Fetal bovine serum; CDK: Cyclin-dependent kinase; PBS: Phosphate buffered solution; p21: Cyclin-dependent kinase inhibitor 1; Akt I: Alpha serine/threonine-protein kinase; Rb: Retinoblastoma.

Competing interests

The authors declare that they have no competing interests.

Authors' contributions

MN (the first author) performed the majority of the experiments and analyzed the data. FS helped with RT-qPCR. LZ, FZ and CW performed some experiments and helped with the writing of the paper. LD and LZ designed the experiments, analyzed the data, wrote the paper, and coordinated the study. All authors read, participated in the writing of and approved the final manuscript.

Acknowledgements

This project was supported by Chinese Academy of Agricultural Sciences Foundation (2014ywf-yb-7), Transgenic Major Projects (2009ZX8008-003B),

National Modern Agricultural Industry Technology Fund in Sheep Industry System (CARS-39-04B) and National High Technology Research and Development Program of China (No.2011AA100307-02). It is necessary to acknowledge Prof. Dr. Mohamad Tagi Beigi Nassiri from Ramin Agriculture and Natural Resources University for helpful discussions and assistance with manuscript preparation.

Author details

[1]National Center for Molecular Genetics and Breeding of Animal, Institute of Animal Sciences, Chinese Academy of Agricultural Sciences, Beijing 100193, People's Republic of China. [2]Transgenic and Stem Cell Core, Institute of Animal Science, Chinese Academy of Agricultural Sciences, Beijing 100193, People's Republic of China.

References

1. Honaramooz A, Megee S, Zeng W, Destrempes M, Overton S: **Adeno-associated virus (AAV)-mediated transduction of male germ line stem cells results in transgene transmission after germ cell transplantation.** *FASEB J* 2008, **22:**374–382. http://www.ncbi.nlm.nih.gov/pubmed/17873102.
2. Naldini L, Blomer P, Gallay D, Ory R, Mulligan FH, Gage I, Verma M, Trono D: **In vivo gene delivery and stable transduction of nondividing cells by a lentiviral vector.** *Science* 1996, **272:**263–267. http://www.ncbi.nlm.nih.gov/pubmed/8602510.
3. Ragot T, Vincent N, Chafey P, Vigne E, Gilgenkrantz H, Couton D: **Efficient adenovirus-mediated transfer of a human minidystrophin gene to skeletal muscle of mdx mice.** *Nature* 1993, **361:**647–650. http://www.nature.com/nature/journal/v361/n6413/abs/361647a0.html.
4. Russell DW, Kay MA: **Adeno-associated virus vectors and hematology.** *Blood* 1999, **94:**864–874. http://www.ncbi.nlm.nih.gov/pmc/articles/PMC3739711/.
5. Srivastava A: **Obstacles to human hematopoietic stem cell transduction by recombinant adeno-associated virus 2 vectors.** *J Cell Biochem* 2002, **38**(Suppl):39–45. http://www.ncbi.nlm.nih.gov/pubmed/12046848.
6. Flotte TR, Barraza X, Solow R, Afione SA, Carter BJ, Guggino WB: **An improved system for packaging recombinant adenoassociated virus vectors capable of in vivo transduction.** *Gene Ther* 1995, **2:**29–37. http://www.ncbi.nlm.nih.gov/pubmed/7712331.
7. Flotte TR, Carter BJ: **Adeno-associated virus vectors for gene therapy.** *Gene Ther* 1995, **2:**357–362. http://www.ncbi.nlm.nih.gov/pubmed/7584109.
8. Xiao X, Li J, Samulski RJ: **Efficient long-term gene transfer into muscle tissue of immunocompetent mice by adeno-associated virus vector.** *J Virol* 1996, **70:**8098–8108. http://www.ncbi.nlm.nih.gov/pmc/articles/PMC190884/pdf/708098.pdf.
9. Haidet AM, Rizo L, Handy C, Umapathi P, Eagle A, Shilling C: **Long-term enhancement of skeletal muscle mass and strength by single gene administration of myostatin inhibitors.** *Proc Natl Acad Sci U S A* 2008, **105:**4318–4322. http://www.ncbi.nlm.nih.gov/pubmed/18334646.
10. Kota J, Handy CR, Haidet AM, Montgomery CL, Eagle A, Mendell JR, Kaspar BK: **Follistatin gene delivery enhances muscle growth and strength in nonhuman primates.** *Sci Transl Med* 2010, **1**(Suppl 6):1–17. http://www.ncbi.nlm.nih.gov/pmc/articles/PMC2852878/.
11. Amthor H, Nicholas G, McKinnell I, Kemp CF, Sharma M, Kambadur R, Patel K: **Follistatin complexes myostatin and antagonizes myostatin-mediated inhibition of myogenesis.** *Dev Biol* 2004, **270:**19–30. http://www.ncbi.nlm.nih.gov/pubmed/15136138.
12. Lee SJ: **Sprinting without myostatin: a genetic determinant of athletic prowess.** *Trends Genet* 2007, **23:**475–477. http://www.sciencedirect.com/science/article/pii/S0168952507002697.
13. Han DS, Huang HP, Wang TG, Hung MY: **Transcription activation of myostatin by trichostatin A in differentiated C2C12 myocytes via ASK1-MKK3/4/6-JNK and p38 mitogen-activated protein kinase pathways.** *J Cell Biochem* 2010, **111:**564–573. http://onlinelibrary.wiley.com/doi/10.1002/jcb.22740/pdf.
14. Philip B, Lu Z, Gao Y: **Regulation of GDF-8 signaling by the p38 MAPK.** *Cell Signal* 2005, **17:**365–375. http://www.sciencedirect.com/science/article/pii/S0898656804001603.
15. Trendelenburg AU, Meyer A, Rohner D, Boyle J, Hatakeyama S, Glass DJ: **Myostatin reduces Akt/TORC1/p70S6K signaling, inhibiting myoblast differentiation and myotube size.** *Am J Cell Physiol* 2009, **296:**C1258–C1270. http://ajpcell.physiology.org/content/296/6/C1258.
16. McPherron AC, Lawler AM, Lee SJ: **Regulation of skeletal muscle mass in mice by a new TGF-beta superfamily member.** *Nature* 1997, **387:**83–90. http://www.ncbi.nlm.nih.gov/pubmed/9139826.
17. Grobet L, Martin LJ, Poncelet D, Pirottin D: **A deletion in the bovine myostatin gene causes the double muscled phenotype in cattle.** *Nat Genet* 1997, **17:**71–74. http://www.nature.com/ng/journal/v17/n1/pdf/ng0997-71.pdf.
18. Wagner KR, McPherron AC, Winik N, Lee SJ: **Loss of myostatin attenuates severity of muscular dystrophy in mdx mice.** *Ann Neurol* 2002, **52:**832–836. http://www.ncbi.nlm.nih.gov/pubmed/12447939.
19. Zhang N, Zhang X, Liu M, Tan L: **Ovine Follistatin gene expression and functional analysis of Follistatin domains.** *Chin J Biotech* 2010, **26:**1050–1056. http://www.ncbi.nlm.nih.gov/pubmed/21090108.
20. Benabdallah BF, Bouchentouf M, Rousseau J, Tremblay JP: **Overexpression of Follistatin in human myoblasts increases their proliferation and differentiation, and improves the graft success in SCID mice.** *Cell Transplant* 2009, **18:**709–718. http://www.ncbi.nlm.nih.gov/pubmed/19520047.
21. Keutmann HT, Schneyer AL, Sidis Y: **The role of Follistatin domains in Follistatin biological action.** *Mol Endocrinol* 2004, **18:**228–240. http://www.ncbi.nlm.nih.gov/pubmed/14563935.
22. Sharma M, Kambadur R, Matthews KG, Somers WG, Devlin GP: **Myostatin, a transforming growth factor-beta superfamily member, is expressed in heart muscle and is upregulated in cardiomyocytes after infarct.** *J Cell Physio* 1999, **180:**1–9. http://www.ncbi.nlm.nih.gov/pubmed/10362012.
23. Thomas M, Langley B, Berry C, Sharma M, Kirk S, Bass J, Kambadur R: **Myostatin, a negative regulator of muscle growth, functions by inhibiting myoblast proliferation.** *J Biol Chem* 2000, **275:**40235–40243. http://www.jbc.org/content/275/51/40235.long.
24. Li ZB, Kollias HD, Wagner KR: **Myostatin directly regulates skeletal muscle fibrosis.** *J Biol Chem* 2008, **283:**19371–19378. http://www.ncbi.nlm.nih.gov/pubmed/18453534.
25. Zhu X, Hadhazy M, Wehling M, Tidball JG, McNally EM: **Dominant negative myostatin produces hypertrophy without hyperplasia in muscle.** *FEBS Lett* 2000, **474:**71–75. http://www.ncbi.nlm.nih.gov/pubmed/10828454.
26. Taylor WE, Bhasin S, Artaza J, Byhower F, Azam M: **Myostatin inhibits cell proliferation and protein synthesis in C2C12 muscle cells.** *Am J Physiol Endocrinol Metab* 2001, **280:**E221–E228. http://www.ncbi.nlm.nih.gov/pubmed/11158924.
27. Sambrook J, Michael RG: *Molecular Cloning: A Laboratory Manual.* Cold Spring Harbor Laboratory Press; 2012. http://www.cshlpress.com/pdf/sample/2013/MC4/MC4FM.pdf.
28. Hembree JR, Hathaway MR, Dayton WR: **Isolation and culture of fetal porcine myogenic cells and the effect of insulin, IGF-1, and sera on protein turnover in porcine myotube cultures.** *J Anim Sci* 1991, **69:**3241–3250. http://www.journalofanimalscience.org/content/69/8/3241.full.pdf.
29. Livak KJ, Schmittgen TD: **Analysis of relative gene expression data using real-time quantitative PCR and the 2(−Delta Delta C (T)) Method.** *Methods* 2001, **25:**402–408. http://www.sciencedirect.com/science/article/pii/S1046202301912629.

18

Improved amplification efficiency on stool samples by addition of spermidine and its use for non-invasive detection of colorectal cancer

Jean-Pierre Roperch[1,2*], Karim Benzekri[3], Hicham Mansour[4] and Roberto Incitti[5]

Abstract

Background: Using quantitative methylation-specific PCR (QM-MSP) is a promising method for colorectal cancer (CRC) diagnosis from stool samples. Difficulty in eliminating PCR inhibitors of this body fluid has been extensively reported. Here, spermidine is presented as PCR facilitator for the detection of stool DNA methylation biomarkers using QM-MSP. We examined its effectiveness with *NPY*, *PENK* and *WIF1*, three biomarkers which we have previously shown to be of relevance to CRC.

Results: We determined an optimal window for the amplification of the *albumin* (*Alb*) gene (100 ng of bisulfite-treated stool DNA added of 1 mM spermidine) at which we report that spermidine acts as a PCR facilitator (AE = 1680%) for SG RT-PCR. We show that the amplification of methylated *PENK*, *NPY* and *WIF1* is considerably facilitated by QM-MSP as measured by an increase of CMI (Cumulative Methylation Index, i.e. the sum of the three methylation values) by a factor of 1.5 to 23 fold in individual samples, and of 10 fold in a pool of five samples.

Conclusions: We contend that spermidine greatly reduces the problems of PCR inhibition in stool samples. This observed feature, after validation on a larger sampling, could be used in the development of stool-based CRC diagnosis tests.

Keywords: Spermidine, Colorectal cancer, QM-MSP, Methylated biomarkers

Background

Colorectal cancer (CRC) is one of the most common forms of cancer in the world [1]. CRC can be cured if diagnosed at early stage using endoscopic examination [2,3], making early non-invasive screening a crucial aim. The development of CRC results from the progressive accumulation of genetic and epigenetic alterations leading to the transformation of normal colonic epithelium to colon adenocarcinoma [4-6]. Fecal-occult blood test (FOBT) is the most widely used method of screening for CRC [7,8]. However, FOBT lacks sensitivity as well as specificity for screening an average risk population. Epigenetic alterations have been found frequently in neoplastic diseases [9,10]. It has been reported that the analysis of DNA methylation carried out in body fluids represents a valuable source for the discovery of cancer biomarkers [11]. Prior studies showed that the hypermethylation can be detected in tumor-derived DNA found in the serum [12-14] and stool [14-17] of patients with CRC. More recently, we proposed a panel of three hypermethylated genes (*NPY, PENK* and *WIF1*) as potential biomarkers for the early diagnosis of CRC in tissue and blood samples, based on QM-MSP assay [18]. However, these studies show that the sensitivity of detection must be improved for the application in diagnostic routine. For this reason, the analysis of aberrant methylation in stool DNA might provide a novel strategy for early detection of CRC. The Ahquist's team demonstrated that blood invasion is more common in advanced stages of CRC where an earlier exfoliation of adenoma and/or tumor cells into the colonic lumen [19]. Moreover, Davies RJ and colleagues showed that the number of colonocytes in the stool following exfoliation from malignant lesions is about 4–5 fold greater than from normal tissue [20] with a mean concentration of 100 ng/g stool, corresponding to 0.01% of the total DNA [21]. However, the composition of feces is highly complex including PCR inhibitors (i.e., bile salts and

* Correspondence: roperch@oncodiag.fr
[1]Profilome, Paris Biotech 24 rue du Faubourg St Jacques, Paris 75014, France
[2]OncoDiag, Agoranov 96 Bis, Boulevard Raspail, Paris 75006, France
Full list of author information is available at the end of the article

polysaccharides). It has been reported that the presence of inhibitors can dramatically reduce the sensitivity and amplification efficiency of PCR [22]. As a consequence, consistent extraction of high-quality DNA from fecal samples can be quite challenging, because of the presence of PCR inhibitors that are co-extracted with DNA. Spermidine is a polyamine that has previously been reported to facilitate stool DNA amplification by inhibiting PCR inhibitors [23,24].

Here, we investigated the stimulating effect of spermidine as PCR facilitator for detecting tumor-specific methylated markers in stool DNA.

Results

Using the *Alb* gene to test PCR specificity to bisulfite sequencing; checking about absence of interference with PCR ampilication of the *NPY, PENK* and *WIF* genes

We performed bisulfite sequencing of the entire amplification products in presence and absence of spermidine using nucleic sequences obtained from the *albumin* (*Alb*) gene (data not shown). In Figure 1A, we have represented the sequencing electrophoregram of the *Alb* promoter, as assessed by using SG RT-PCR from C_1 and S_1 with 1 mM spermidine. We observed that thymidine are detected instead of cytosine, as expected after DNA bisulfite modification of unmethylated amplicon products, since they correspond to a region of *Alb* which does not contain CpG sites. Those findings indicate that all cytosine are converted to thymine as a result of the DNA modification step being performed successfully and that the spermidine do not interferes in the specificity of PCR. We also verified that spermidine does not interferes the PCR amplification of *NPY, PENK* and *WIF1* genes into of CpG rich regions (data not shown).

Comparing *Alb* PCR products with and without utilization of spermidine

On agarose gel electrophoresis, we observed, in Control (C) and Sample (S) DNA, the 76 bp band confirming the existence of *Alb* gene with or without the presence of spermidine (data not shown). Figure 1B, shows the DNA migration from C (C_0-C_1) and S (S_0-S_1) with and without 1 mM spermidine addition to the reaction mixture. As expected, we noted a correspondence between band intensities and Ct values with S_0 (Ct = 28.16) and S_1 (Ct = 25.11) and not with C_0 (Ct = 20.05) and C_1 (Ct = 20.11) (Table 1, 1st serial). The negative template control (NTC) was negative, indicating that it was not nonspecific primer binding or contamination using 1 mM spermidine and also in presence of various concentrations of spermidine, ranging from 1 mM to 10 mM (data not shown).

Impact of spermidine on the melting curves of PCR products of *Alb* gene

In Figure 2 are represented the melting curves of amplicons of the *Alb* gene using spermidine in SG RT-PCR from C (Figure 2A) and S (Figure 2B). We highlight in Figure 2C the melting curves of C and S in presence (C_1, S_1) and absence (C_0, S_0) of 1 mM spermidine. Both, C_0 and S_0 showed a similar temperature of melting (Tm) of 77.7°C and 77.4°C as expected, while for C_1 and S_1, we obtained a Tm near 79.2°C (ΔTm = +1.5°C) and 78.9°C (ΔTm = +1.4°C), respectively. (The full results are

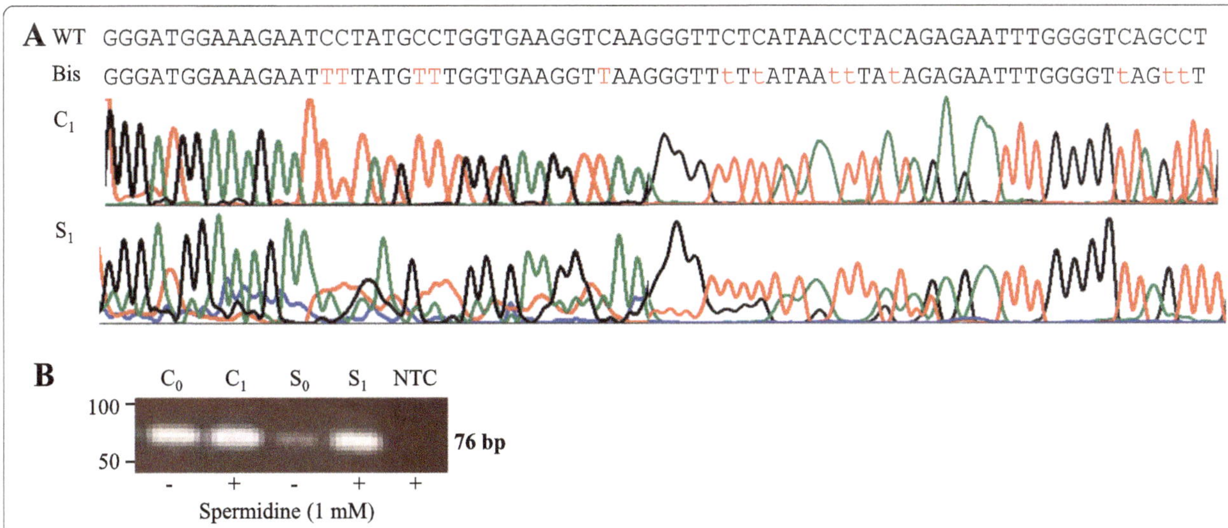

Figure 1 Verification and comparison of PCR amplification products of the *albumin* gene in presence and absence of spermidine. **A**: representative bisulfite sequencing electrophoregram of the *Alb* promoter using SG RT-PCR in presence of 1 mM spermidine from universal methylated human DNA (C_1) and stool DNA sample (S_1). All cytosine are converted to thymine noted in red resulting entirely from DNA modification. This follows after sodium bisulfite treatment (Bis) when referring to wild-type (WT) *Alb* gene sequence and **B**: the same PCR products of the *Alb* were analysed by agarose gel electrophoresis and revealed single amplification fragment of the predicted size (76 pb) when spermidine is present (C_1, S_1) and absence (C_0, S_0); NTC as negative control.

Table 1 PCR efficiencies in presence and absence of spermidine

1st experiment	Sperm.	Ct	ΔCt	AE (%)	Effects ?
1st serial					
Control	0	20.05 ± 0.17		100	
	1	20.11 ± 0.02	0.06	96	=
	2	20.77 ± 0.17	0.72	61	-
	3	21.12 ± 0.03	1.07	48	-
	4	21.70 ± 0.32	1.65	32	-
	5	22.66 ± 0.16	2.61	16	-
	10	NA	ND	0	Total inhibition
Sample	0	28.16 ± 0.85		100	
	1	25.11 ± 0.11	−3.06	831	+
	2	25.35 ± 0.03	−2.81	701	+
	3	25.73 ± 0.11	−2.44	541	+
	4	26.53 ± 0.16	−1.63	310	+
	5	28.08 ± 0.13	−0.08	106	+
	10	NA	ND	0	Total inhibition
2nd serial					
Control	0	19.86 ± 0.33		100	
	0.05	19.61 ± 0.10	−0.25	119	+
	0.10	19.78 ± 0.13	−0.09	106	+
	0.50	19.87 ± 0.15	0	100	=
	1	19.91 ± 0.14	0.05	97	=
Sample	0	28.17 ± 0.26		100	
	0.05	26.63 ± 0.49	−1.54	290	+
	0.10	25.98 ± 0.11	−2.19	456	+
	0.50	25.09 ± 0.01	−3.08	843	+
	1	25.22 ± 0.11	−2.95	773	+

Abbreviations: Sperm., spermidine (mM); Ct, mean of cycle threshold value ± standard deviation value; AE, amplification efficiency; NA, non amplification; ND, not determined; =, equal; −, inhibitor; +, facilitator.
The amplification efficiency of the albumin gene at each spermidine concentration points was calculated using $2^{-\Delta Ct}$ where $\Delta Ct = (Ct_{\text{with spermidine}}) - (Ct_{\text{without spermidine}})$. For example in the 1st study, Control DNA with 1 mM of spermidine showing a ΔCt = 0.06, we recovered 96% of the true yield (100%, ΔCt = 0).

presented in the supplementary data, Additional file 1: Table S1).

Checking about PCR inhibition of *Alb* gene when adding spermidine at different concentration

We tested different concentrations of spermidine, ranging from 1 mM to 10 mM (1st serial) and 0.05 mM to 1 mM (2nd serial). In Figure 3 are represented the amplification curves of *Alb* gene (1st serial) illustrating the observed efficiency-Ct shift relationships using spermidine in SG RT-PCR from C (Figure 3A) and S (Figure 3B). In Table 1, the observed result is that low concentrations of spermidine have opposite effects on PCR efficiency of C and S, with a negative effect on C and a positive effect or PCR facilitator on S, (we hypothetise that this effects depend on the purity of DNA samples, assuming C more pure than S) while an excess of spermidine (10 mM) inhibits amplification and this regardless of the nature of DNA. Taken together, those findings suggest that the addition of 1 mM spermidine during PCR cycling might be an optimum for obtaining the highest amplification efficiency on C with for both studies AE = 800 ± 39% (ΔCt = −3.00) of the true yield (AE = 100%, ΔCt = 0 without spermidine) on the detection of *Alb* gene. No significant difference was found on control with AE = 96 ± 0.7% (ΔCt = +0.06). Exceeding 10 mM of spermidine leads to PCR inhibition.

Varying stool concentration and spermidine concentration

The results are shown in Table 2. They suggest that the 1 mM spermidine together with 100 ng of S DNA provides an optimal amplification efficiency condition, with AE amounting to 1680%.

Measuring promoter methylation of *NPY, PENK,* and *WIF1*

The results are shown in Table 3. We evaluated in triplicate assays the performance of QM-MSP to quantify the methylation levels of *NPY, PENK*, and *WIF1*. For co-amplifying two methylation-specific DNA targets in real-time, we used the associations of Fam-*Alb*/Vic-*WIF1* and Ned-*NPY*/Vic-*PENK*. QM-MSP were performed from ID1 to ID5 plus S (100 ng) and C (50 ng) with (+) and without (−) the addition of 1 mM spermidine and the methylation percentage was determined by the $2^{-\Delta\Delta Ct}$ method. We summarize the results by using the cumulative methylation index (CMI), which is the sum of the three values. As indicated in Table 3, the results suggest that the addition of spermidine to the PCR mixture allows a sensitive analysis of DNA methylation biomarkers, relative to that not containing spermidine (e.g., on S, the CMI is 4.65 ± 0.67% with (+) and 0.51 ± 0.12 with (−), respectively and reflects a near 10-fold factor of CMI ($P < 0.01$, Student-test)). Furthermore, when we do not observe amplification both in the presence or absence of spermidine, the most likely hypothesis is that the promoter region of this gene is unmethylated (e.g., *NPY* gene from ID 3).

Discussion

Detection of precancerous and early-stage CRC is central to improving patient prognosis. Noninvasive colon cancer screening by testing feces for the presence of occult blood still shows poor sensitivity. Consequently, a number of assays for detection of cancer-specific DNA alterations from fecal DNA have been proposed as a new approach for screening and detecting early-stage colon cancer [25-27].

PCR is a powerful technique for the detection of target DNA, but its application to stool specimens is always

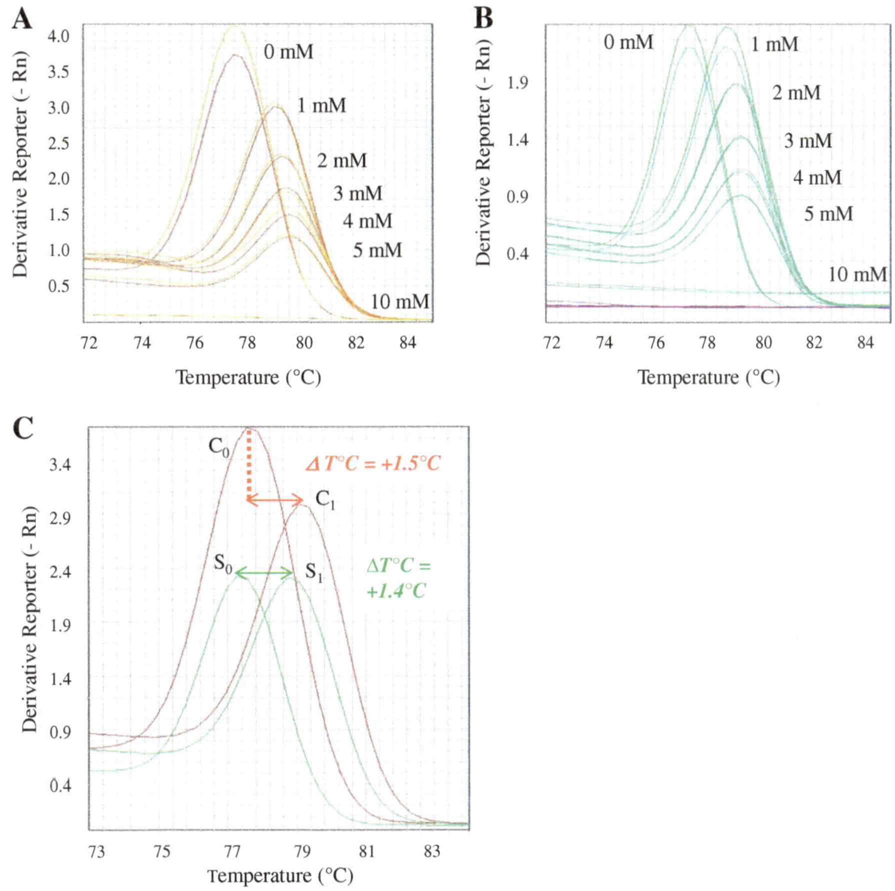

Figure 2 Comparison of melting curves of amplicons of the *Alb* gene using spermidine. We used a mixture of primers to amplify *Alb* gene with an amount of 50 ng of DNA templates at various spermidine concentrations ranging from 0 up to 10 mM. Melting curves of products are shown from C **(A)** and S **(B)**, respectively. In **(C)**, we highlight the melting curves of C and S in presence (C_1, S_1) and absence (C_0, S_0) of 1 mM spermidine.

limited, due to the presence of several components (e.g. bile salts, hemoglobin degradation product, and complex polysaccharides), that are known to often inhibit PCR [28-30]. Problems in eliminating PCR inhibitors from stool specimens have been extensively reported and, for many situations, dilution of inhibited samples proves necessary to provide a rapid and straightforward way of permitting amplification [31]. However, dilution is only possible if the amount of DNA is sufficiently high. Hence, for applications involving low-copy targets and in presence of high background (i.e., bacterial DNA), the dilution solution is often undesirable, and indeed sometimes impossible, due to the further reduction of already reduced target amount [32,33]. Others methods have been used to alleviate the effects of inhibitory substances of PCR, such as the use of separation columns. They present a number of problems too, ranging from reduced DNA yields, leading to decrease DNA targets, to decreased amplification capacity [34-36], and for all these reasons, we have not chosen these methods.

Spermidine, a polyamine compound, has been reported to have a high affinity for plant and stool DNA and could be used as PCR facilitators by its addition to the reaction mixture during the PCR steps [23,24].

In the present study, spermidine concentrations were tested to alleviate PCR inhibition associated with DNA isolated from stool samples. For the first time, we showed that spermidine can act both as an activator on stool DNA or PCR inhibitors on a high purity DNA. Indeed, we showed that, at increasing concentrations, we have 1) on DNA samples extracted from stool: increasing efficiency up to an optimum reached at about 1 mM and then decreasing up to total inhibition and 2) on a set of highly pure DNA: same as 1), but with optimum reached at a much lower concentration, namely 0.05 mM. (We have no explanation for the former.) The hypothesis is that spermidine can block the action of PCR inhibitors (possibly by binding them and or making them more thermolabile, or, alternatively, that by binding DNA at low concentrations it can drastically decrease the action of the PCR inhibitors). The optimum that we observe might be explained

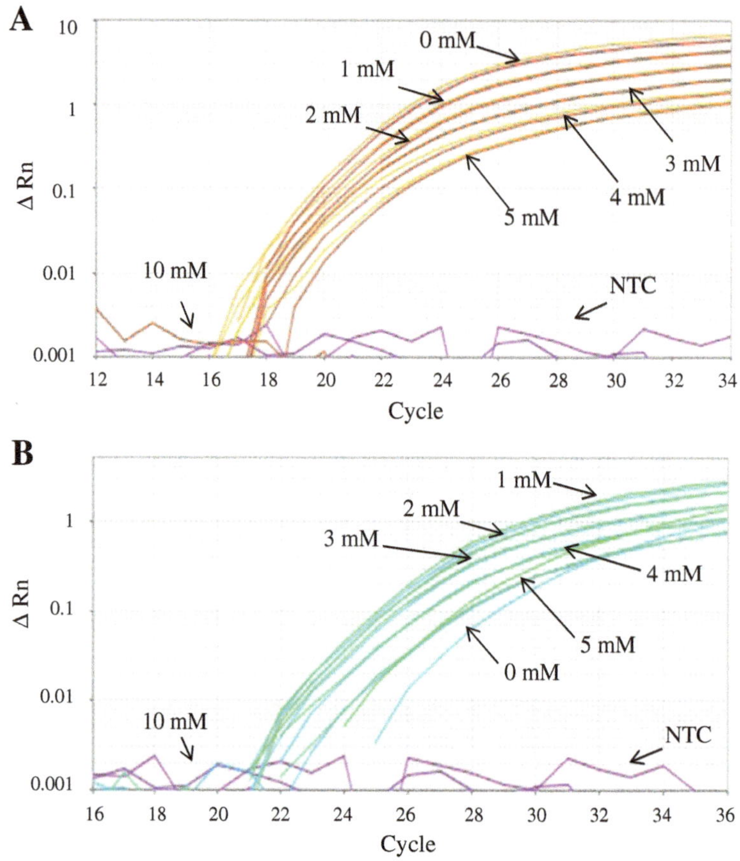

Figure 3 Interference of PCR amplification by spermidine. From 50 ng of DNA, the amplification curves of PCR products of the *Alb* gene using spermidine in SG RT-PCR are shown from C **(A)** and S **(B)**. The spermidine has been used at various concentrations, ranging from 1 mM to 10 mM.

by the existence, for each specific sample type, of a concentration threshold above which spermidine no longer blocks PCR inhibitors, may be due to steric effects or saturation, and starts massively binding to DNA, so inhibiting PCR.

Interesting, we observed that the addition of spermidine causes a positive shift of the melting temperature of the stool and control DNA. This observation may be due to the interaction of the spermidine with DNA as it has been described previously by Ahokas and Erkkila [23]. We have determined a range of spermidine concentrations that counteract the PCR inhibitors co-extracted with DNA, so facilitating the amplification efficiency of methylation markers. We applied that finding to assessing the methylation of *NPY, PENK*, and *WIF1*, (whose detection is of interest in CRC, as shown in [17]) while maintaining sufficient DNA yield. We showed the advantage of our method in the quantification of methylation values of CRC markers *NPY, PENK* and *WIF1*, where, on undiluted stool DNA (100 ng) and by using the QM-MSP in presence of 1 mM spermidine, we globally enhanced detection by a near 10-fold factor, as assessed by summing up the three values. In the future, it would be interesting to evaluate our method with other biomarkers such as Septin 9, which is used as a marker of blood-based methylation requiring improved accuracy for a clinical practise [12].

In summary, we performed a comparative study on the effect of spermidine onto PCR efficiency reporting that spermidine addition is easier and more useful than dilution or purification methods and that it can dramatically improve the quantification of methylation values. We also highlighted a possible mechanism for its action. Our QM-MSP using the presence of 1 mM spermidine and 100 ng of stool DNA could be used as a potential PCR facilitator for stool-based detection of CRC. Our methodology is also a serious candidate for being developed into a robust technology, as it has been optimized with several primer pair and reaction buffer.

Conclusions

In this study, we present a proof of principle for using spermidine to allow alleviation of the PCR inhibitors frequently encountered in DNA amplification from stool samples. We also demonstrated that spermidine, an inexpensive chemical, is useful for sensitive stool-based

Table 2 Relationship between spermidine concentration and amount of stool DNA

	Sperm.	Ct	ΔCt	AE (%)
2nd experiment				
50 ng	0	28.32 ± 0.15		100
	0.5	25.04 ± 0.11	−3.28	968
	1	25.23 ± 0.17	−3.09	849
	5	28.39 ± 0.21	0.07	95
100 ng	0	27.04 ± 0.16		100
	0.5	23.74 ± 0.12	−3.31	988
	1	22.97 ± 0.24	−4.07	1680
	5	25.49 ± 0.17	−1.55	293
250 ng	0	NA	ND	ND
	0.5	28.67 ± 0.27	ND	ND
	1	23.70 ± 0.37	ND	ND
	5	23.65 ± 0.10	ND	ND
500 ng	0	NA	ND	ND
	0.5	NA	ND	ND
	1	28.00 ± 0.56	ND	ND
	5	26.92 ± 0.30	ND	ND

Abbreviations: Sample, stool DNA pooled; Sperm, spermidine (mM); Ct, mean of cycle threshold value ± standard deviation value; AE, amplification efficiency; NA, non amplification; ND, not determined.

detection of methylation-specific markers for CRC tumors using QM-MSP. These results, after corroboration in a large cohort, can lead to the elaboration of a method to be used in clinical practice as a aid in preselecting the patients for colonoscopy.

Table 3 Quantitative DNA methylation analysis

		Methylation (%)			CMI (%)
3rd experiment		**NPY**	**PENK**	**WIF1**	
ID1	-	0.01 ± 0.01	0.13 ± 0.03	NA	0.14 ± 0.04
	+	0.86 ± 0.26	1.10 ± 0.08	1.22 ± 0.41	3.18 ± 0.75
ID2	-	1.22 ± 0.21	1.31 ± 0.53	0.07 ± 0.18	2.60 ± 0.92
	+	1.86 ± 0.11	1.35 ± 0.09	0.76 ± 0.15	3.97 ± 0.35
ID3	-	NA	NA	0.17 ± 0.07	0.17 ± 0.07
	+	NA	0.10 ± 0.04	1.59 ± 0.22	1.69 ± 0.26
ID4	-	0.17 ± 0.08	0.32 ± 0.10	NA	0.49 ± 0.18
	+	2.33 ± 0.32	1.88 ± 0.44	1.71 ± 0.27	5.65 ± 0.67
ID5	-	0.09 ± 0.08	0.17 ± 0.05	0.07 ± 0.01	0.33 ± 0.14
	+	1.33 ± 0.22	1.44 ± 0.16	1.01 ± 0.07	3.78 ± 0.45
Sample (pooling	-	0.21 ± 0.06	0.27 ± 0.06	0.03 ± 0.03	0.51 ± 0.12
of five patients)	+	1.74 ± 0.30	2.10 ± 0.18	0.81 ± 0.1	4.65 ± 0.67

Methods

Human stool samples

We analyzed human stool samples from 5 colonoscopy-negative subjects from the Valihybritest's collection, registered under the number NCT01270360 (*Clinical Trials.gov*). All of the patients provided informed consent for the research use of their samples.

Use of these samples for this study was approved by the ethical committee of the Val de Marne Paris-Est medical district, registered under code CCP-IDF-IX-11-010.

Experimental design

This study included three related experiments: the first experiment was designed to study the interference of PCR amplification by spermidine addition at various concentrations ranging from 0.05 mM to 10 mM both 50 ng of universal methylated DNA (Control, C) and 50 ng of stool DNA pooled (Sample, S), each DNA previously modified by sodium bisulfite. The second experiment determined the optimum condition giving the best performance of PCR amplification by modulating both the amount of stool DNA (from 50 ng to 500 ng of modified DNA) and the concentration of spermidine (from 0.5 mM to 5 mM). In these studies, we used spermidine-containing reaction solutions to assess the amplification of the *albumin (Alb)* gene using SYBER Green real-time PCR (SG RT-PCR). In the third experiment, the quantitative methylation-methylation specific PCR (QM-MSP) was performed, with the optimum condition previously defined in 1st and 2nd experiments, to measured the degree of methylation of a three-gene panel consisting of *NPY*, *PENK*, and *WIF1* and calculated the cumulative methylation index (CMI) value of each extract of stool DNA, i.e., ID 1 to ID 5 and Sample.

Stool DNA isolation and quantification

About 5 g stool were collected from each individual. DNA was isolated from stool samples (200 mg) using the QiAamp DNA stool mini kit (Qiagen) according to the manufacturer's protocol. DNA concentrations were determined by measurement at 260 nm using BioPhotometer (Eppendorf). Isolated DNA was stored at −20°C.

Bisulfite modification

1 µg of each DNA (C and S) were modified by sodium bisulfite overnight at 50°C using the EZ DNA Methylation kit (Zymo Research) and eluted in 100 µl of TE buffer (10 mM Tris–HCl (pH 8.0), 1 mM EDTA). Bisulfite treatment converts all unmethylated cytosine residues to uracil (later replicated as thymidine during PCR cycling), while leaving methylcytosines unchanged.

Primers and probes

For SG RT-PCR, the primers targeting the albumin (*Alb*) gene were reported in Table 1. PCR reactions specific for the *Alb* gene promoter region, which not containing CpG sites. For QM-MSP analysis we used the same primers as those described using SG RT-PCR with in addition the probes targeting *Alb* as Control gene, *NPY*, *PENK* and *WIF1* as CRC-specific genes were reported previously [17] (Additional file 2: Table S2). Primers and probes are designed by Life Technologies company.

Quantitative real-time PCR analysis

All PCR reactions were carried out in a 96-well reaction plate in a StepOne Plus Real-Time PCR system (Life Technologies) in a final volume of 20 μl. We used the universal methylated human DNA standard (Zymo Research) and positive control (Control) and stool DNA as sample (Sample).

1) SG RT-PCR analysis

Modified DNA was analyzed in duplicate by SYBR-Green qPCR master mix (Life Technologies) to determine the optimum concentration of spermidine (Sigma Aldrich), based on albumin amplification product as it accumulates during real-time PCR. The range of spermidine tested is from 50 μM to 10 mM. Five hundred nM of primers (forward and reverse) were also present. The thermal cycling conditions included an initial denaturation at 95°C for 10 min followed by 48 cycles (95°C for 15 s and 60°C for 1 min). The melting curve was determined by heating the PCR product from 60°C to 95°C and monitoring the fluorescence at a transition rate of 0.5°C. The melting temperature or Tm was calculated using the StepOne plus software (Life Technologies), based on the initial fluorescence curve by plotting the negative derivative of fluorescence-reporter (−Rn) over temperature versus temperature (−Rn/T).

2) QM-MSP analysis

QM-MSP was performed in triplicate to detect and quantify simultaneously three methylated markers in control and DNA sample using the TaqMan MBG probes technology (Life Technologies). We chose this technique because it allows accurate quantitative assessment of DNA methylation. For each PCR run, a KAPA PROBE master mix (Kapa Biosystems) was prepared, spermidine (1 mM, used as optimum concentration), primers (500 nM) and probes (250 nM) for *Alb*, *WIF1*, *NPY* and *PENK* have been designed (Life Technologies). For co-amplifying two methylation-specific DNA targets in QM-MSP, we used the combinations of Fam/Vic and Ned/Vic fluorophores probes as each probe presents a strong individual spectral intensity with limited overlapping absorption spectra. The PCR cycling parameters were initial denaturation at 95°C for 10 min followed by 95°C for 15 s, 60°C for 1 min, repeated 48 times.

Bisulfite genomic sequencing

The PCR products of albumin gene were purified before submission to the sequencing process of both strands by using BigDye Terminator Cycle Sequencing kit (Life Technologies) according to the manufacturer's instructions. The sequence reactions were run and analyzed on an ABI 3100 Genetic Analyzer (Life Technologies). Sequence analyses were performed using ChromasPro software (Technelysium).

DNA electrophoresis

Five microliters of each reaction were run on conventional 2.5% agarose gel electrophoresis with ethidium bromide (0.5 μg/ml) in TAE buffer (40 mM Tris-Acetate (pH 8.3), 2 mM EDTA). Electrophoresis was done for 2 h in electric field strength of 40 V/cm gel and the DNA was visualized under UV light-transilluminator light (Bio-Rad). The GeneRuler 100 bp DNA ladder (Fermentas) was run on each gel to estimate the size of the PCR products.

Amplification efficiency measurement

For simplicity, the value (AE) is referred to in the text as amplification efficiency, with any deviations from 100% due to the effects of PCR inhibitors or facilitators in the template DNA (Control and Sample). The Ct of each reaction was yielded by StepOne plus System (Life Technologies) using the amplification-based threshold-determination algorithm; shifts Ct (ΔCt) were measured as the difference between the average Ct template (with spermidine) and the average Ct template (without spermidine). To quantify the inhibition and facilitation effects, we calculated the (AE) value, where $AE = 2^{-\Delta Ct} \times 100\%$.

Calculation of the methylation percentage

The level of methylation (percentage of methylated reference (PMR)) is quantified according to the calculation of delta-delta Ct (ΔΔCt). We calculate the PMR of each gene by taking $2^{-\Delta\Delta Ct}$ as described see below.

ΔΔCt = [(Ct target, Sample)-(Ct ref, Sample)]-[(Ct target, Control)-(Ct ref, Control)] where:

Ct target, Control = Ct value of gene of interest in control DNA

Ct ref, Control = Ct value of reference gene in control DNA

Ct target, Sample = Ct value of gene of interest in tested sample

Ct ref, Sample = Ct value of reference gene in tested sample

Additional files

Additional file 1: Figure S1. Effects of spermidine on the temperature melting of the albumin products.

Additional file 2: Figure S2. Oligonucleotides.

Competing interests
The authors declare that they have no competing interests.

Authors' contributions
Participated in research design: All authors. Conducted experiments: JPR. Performed data analysis: JPR and RI. Wrote or Contributed to the writing of the manuscript: all authors. All authors read and approved the final manuscript to be published.

Acknowledgements
We wish to thank Pr. Iradj Sobhani and Assistance Publique-Hôpitaux de Paris for providing the samples; Hanane El Asri and Amine Amoura for valuable advice (Department of Gastroenterology, Laboratoire d'Investigation Clinique, Henri Mondor Hospital, Créteil, France).

Author details
[1]Profilome, Paris Biotech 24 rue du Faubourg St Jacques, Paris 75014, France. [2]OncoDiag, Agoranov 96 Bis, Boulevard Raspail, Paris 75006, France. [3]Centre d'Investigation Clinique (CIC), Henri Mondor Hospital, Créteil, France. [4]King Abdullah University of Science and Technology (KAUST), Bioscience Core Laboratory Research Department, Thuwal 23955-6900, Saudi Arabia. [5]King Abdullah University of Science and Technology (KAUST), Computational Biology Research Center, Thuwal 23955-6900, Saudi Arabia.

References
1. Jemal A, Siegel R, Xu J, Ward E. Cancer Statistics, 2010. Cancer J Clin. 2010;60(5):277–300.
2. Desch CE, Benson AB, Somerfield MR, Flynn PJ, Krause C, Loprinzi CL, et al. Colorectal cancer surveillance: 2005 update of an American Society of Clinical Oncology practice guideline. J Clin Oncol. 2005;23(33):8512–9.
3. Winawer S, Fletcher R, Bond RD, Desch CE, Benson AB, Somerfield MR, et al. Colorectal cancer surveillance: 2005 update of an American Society of Clinical Oncology practice guideline. J Clin Oncol. 2005;23(33):8512–9.
4. Esteller M. DNA methylation and cancer therapy: new developments and expectations. Curr Opin Oncol. 2005;17(1):55–60.
5. Nosho K, Yamamoto H, Takahashi T, Mikami M, Taniguchi H, et al. Genetic and epigenetic profiling in early colorectal potential in pT1 (early invasive) colorectal cancers. Carcinogenesis. 2007;28(6):1364–70.
6. Wong JJ, Hawkins NJ, Ward RL. Colorectal cancer: a model for epigenetic tumorigenesis. Gut. 2007;56(1):140–8.
7. Osborn NK, Ahlquist DA. Stool screening for colorectal cancer: molecular approaches. Gastroenterology. 2005;128(1):192–206.
8. Sobhani I, Alzahouri K, Ghout I, Charles DJ, Durand-Zaleski I. Cost-effectiveness of mass screening for colorectal cancer: choice of fecal occult blood test and screening strategy. Dis Colon Rectum. 2011;54(7):876–86.
9. Wong JJ, Hawkins NJ, Ward RL. Colorectal cancer: a model for epigenetic tumorigenesis. Gut. 2007;56(1):140–8.
10. Jones PA. DNA methylation and cancer. Oncogene. 2002;21(35):5358–60.
11. Muller HM, Widschwendter M. Methylated DNA as a possible screening marker for neoplastic disease in several body fluids. Expert Rev Mol Diagn. 2003;3(4):443–58.
12. Grützmann R, Molnar T, Devos C, Pilarsky JK, Habermann PM, Schlag HD, et al. Sensitive detection of colorectal cancer in peripheral blood by septin 9 DNA methylation assay. Plos One. 2008;3:3759.
13. Lofton-Day C, Model T, Devos R, Tetzner J, Distler M, Schuster X, et al. DNA methylation biomarkers for blood-based colorectal cancer screening. Clin Chem. 2008;54:414–23.
14. Mansour H. Cell-free nucleic acids as noninvasive biomarkers for colorectal cancer. Front Genet. 2014;5:182.
15. Model F, Osborn N, Ahlquist D, Gruetzmann R, Molnar B, Sipos F, et al. Identification and validation of colorectal neoplasia-specific methylation markers for accurate classification of disease. Mol Cancer Res. 2007;5:153–63.
16. Lenhard K, Bommer GT, Asutay S, Schauer R, Brabletz T, Goke B, et al. Analysis of promoter methylation in stool: a novel method for the detection of colorectal cancer. Clin Gastroenterol Hepatol. 2005;23:142–9.
17. Huang ZH, Li LH, Yang F, Wang JF. Detection of aberrant methylation in fecal DNA as a molecular screening tool for colorectal cancer and precancerous lesions. World J Gastroenterol. 2007;13:954–4.
18. Roperch JP, Incitti R, Forbin S, Bard F, Mansour H, Maesli F, et al. Aberrant methylation of NPY, PENK, and WIF1 as a promising marker for blood-based diagnosis of colorectal cancer. BMC Cancer. 2013;13:566.
19. Ahlquist DA. Molecular detection of colorectal neoplasia. Gastroenterology. 2010;138(6):2127–39.
20. Davies RJ, Miller R, Coleman N. Colorectal cancer screening: prospects for molecular stool analysis. Nat Rev Cancer. 2005;5(3):199–2009.
21. Zou H, Harrington J, Rego RL, Ahlquist DA. A novel method to capture methylated human DNA from stool: implications for colorectal cancer screening. Clin Chem. 2007;53(9):1646–51.
22. Ramakers CJ, Ruijter JM, Deprez RH, Moorman AF. Assumption-free analysis of quantitative real-time polymerase chain reaction (PCR) data. Neurosci Lett. 2003;339:62–6.
23. Ahokas H, Erkkila MJ. Interference of PCR amplification by the polyamines, spermine and spermidine. PCR Methods Appl. 1993;3:65–8.
24. Kikuchi A, Sawamura T, Kawase N, Kitajima Y, Yoshida T, et al. Utility of Spermidine in PCR Amplification of Stool Samples. Biochem Genet. 2010;48:428–32.
25. Sidransky D, Tokino T, Hamilton S, Kinzler K, Levin B, Frost P, et al. Identification of ras oncogene mutations in the stool of patients with curable colorectal tumors. Science. 1992;256:102–5.
26. Ahlquist DA, Shuber AP. Stool screening for colorectal cancer: evolution from occult blood to molecular markers. Clin Chim Acta. 2002;315:157–68.
27. Ahlquist DA. Molecular detection of colorectal neoplasia. Gastroenterology. 2010;138(6):2127–39.
28. Lantz PG, Matsson M, Wadström T, Rådstrom P. Removal of PCR inhibitors from human faecal samples through the use of an aqueous two-phase system for sample preparation prior to PCR. J Microbiol Methods. 1997;28:159–67.
29. Nagasaka T, Tanaka N, Cullings HM, et al. Analysis of fecal DNA methylation to detect gastrointestinal neoplasia. J Natl Cancer Inst. 2009;101(18):1244–58.
30. Monteiro L, Bonnemaison D, Vekris A, Petry KG, Bonnet J, Vidal R, et al. Complex polysaccharides as PCR inhibitors in feces: Helicobacter pylori model. J Clin Microbiol. 1997;35:995–8.
31. Makristathis A, Pasching E, Schütze K, Wimmer M, Rotter ML, Hirschl AM. Detection of Helicobacter pylori in stool specimens by PCR and antigen enzyme immunoassay. J Clin Microbiol. 1998;36:2772–4.
32. King CE, Debruyne R, Kuch M, Schwarz C, Poinar HN. A quatitative approach to detect and overcome PCR inhibition in ancient DNA extracts. Biotechniques. 2009;47:941–9.
33. Wilson IG. Inhibition and facilitation of nucleic acid amplification. Appl Environ Microbiol. 1997;63:3741–51.
34. Olive DM. Detection of enterotoxigenic Escherichia coli after polymerase chain reaction amplification with a thermostable DNA polymerase. J Clin Microbiol. 1989;27:261–5.
35. Dowd SE, Gerba CP, Enriquez FJ, Pepper IL. PCR amplification and species determination of microsporidia in formalin-fixed feces after immunomagnetic separation. Appl Environ Microbiol. 1998;64:333–6.
36. Ahlquist DA, Skoletsky JE, Boynton KA, Harrington JJ, Mahoney DW. al. Colorectal cancer screening by detection of altered human DNA stool: feasibility of a multitarget assay panel. Gastroenterology. 2000;119:1219–27.

Cytotoxicity of purified listeriolysin O on mouse and human leukocytes and leukaemia cells

Radosław Stachowiak[1], Marcin Łyżniak[2], Maja Grabowska[1], Katarzyna Roeske[1], Tomasz Jagielski[1], Jacek Bielecki[1], Bożena K Budziszewska[3], Grażyna Hoser[2] and Jerzy Kawiak[2*]

Abstract

Background: Listeriolysin O (LLO) is the main virulence factor of *Listeria monocytogenes* and facilitates the intracellular survival of the pathogen. Some of its characteristics endorse the growing popularity of LLO for use in biotechnology, particularly in the development of novel vaccines. Here, we evaluate the use of LLO to eradicate leukaemia cells.

Results: A purified LLO preparation was obtained by affinity chromatography. The LLO preparation procedure was optimized and purified LLO was tested for optimal conditions of storage including temperature, application of proteinase inhibitors and serum components. We demonstrated the possibility of regulating LLO activity by adjusting cell membrane cholesterol content. The LLO preparation had haemolytic activity and had a cytotoxic effect on the human T-leukaemia Jurkat cell line as well as mouse and human peripheral blood mononuclear cells.

Conclusions: LLO has a very potent cytotoxic activity towards human leukocytes. Importantly, the cytotoxic activity was easily regulated *in vitro* and could be restricted to areas containing malignant cells, raising the possibility of future clinical application of LLO for leukaemia treatment.

Keywords: Listeriolysin O, Protein purification, Cytotoxicity, Leukocytes, Leukaemia

Background

Most bacterial pathogens produce toxins, which are important virulence factors. Some toxins act on cytoplasmic membranes, whereas others are receptor-targeted toxins or membrane damaging toxins. The latter toxins are referred to as cytolysins, and are produced by numerous Gram-positive and Gram-negative bacteria. Some of these toxins display enzymatic activity, while others have cytolytic capacity without enzymatic activity. Occasionally those two mechanisms may act together. For example some phospholipases facilitate the action of pore-forming cytolysins by hydrolysing membrane lipids [1].

Pore-forming cytolysins are a class of membrane-damaging toxins without enzymatic activity. They act by the insertion of their hydrophobic regions into the cell membrane phospholipid bilayer, effectively disrupting the target cells. The most homogenous and numerous group of membrane pore-forming cytolysins are Cholesterol Dependent Cytolysins (CDC). CDCs are produced mostly by Gram-positive bacteria and share similar amino acid sequences and biochemical properties [2]. *Listeria monocytogenes* is an intracellular bacterial pathogen whose important virulence determinants are secreted toxins, one of which is listeriolysin O (LLO), a cytolysin encoded by the *hly* gene belonging to the CDC family. This protein is crucial for pathogen survival within the cytoplasm of the infected cell [3,4].

There have been several attempts to develop effective LLO purification methods using recombinant *Escherichia coli* strains [5-9]. Previously, we attempted to use modified bacteria species enclosed within capillary membranes using *in vitro* experiments with human T leukaemia Jurkat cells [10]. Here we present an improved LLO purification protocol and the results of *in vitro* experiments to determine the haemolytic and cytotoxic activity of purified LLO on peripheral blood leukocytes. The concentration-dependent activity of purified LLO was tested on a human T cell leukaemia cell line (Jurkat) and on normal peripheral blood mononuclear cells (PBMC). This study might provide useful information for future *in vivo* testing of LLO.

* Correspondence: jkawiak@cmkp.edu.pl
[2]Medical Centre of Postgraduate Education, Marymoncka 99/103, 01-813 Warsaw, Poland
Full list of author information is available at the end of the article

Results

Purification of Listeriolysin O

The synthesis and affinity purification of His-tagged LLO from *E. coli* harbouring a pET29b-hly plasmid was optimized with a set of buffers with the following gradient: buffer pH ranging from 5 to 8 and NaCl concentration from 0 to 0.5 M. The concentration of imidazole ranged from 0 to 100 mM (for the column buffer) and from 0.25 to 1 M (for the elution buffer). Final protocol and optimal buffer composition are described in the Methods. Notably, lowering the pH of the elution buffer from 8 to 6 allowed a four-fold reduction of the imidazole concentration without a detectable loss of efficiency as compared to the original elution buffer (1 M imidazole, pH 8). The analysis of electrophoretically separated *E. coli* lysates and purified LLO preparation was performed. The results of SDS-PAGE for LLO purified fractions and western blotting results with anti-LLO antibodies are presented in Figure 1. Total protein electrophoresis (Figure 1A) of *E. coli* sonicate (lane 1), purified LLO preparation (lane 2), and western blotting of purified LLO preparation with anti-LLO antibodies (right B 1 and 2 lanes) suggested the presence of a highly uniform protein preparation.

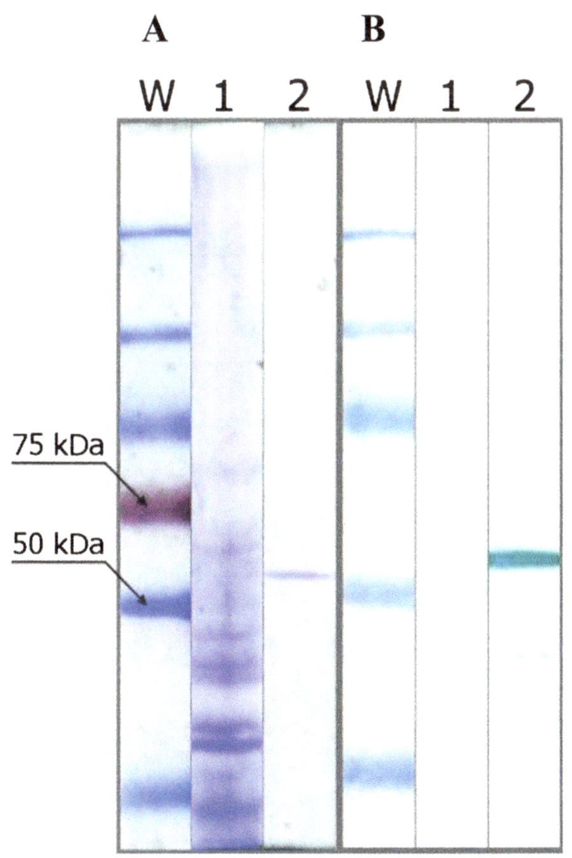

Figure 1 Electrophoresis analysis of LLO samples. Total protein (1), electrophoresis **(A)**, and western blot **(B)**. The purified LLO preparation (2) shows one band similar to immunochemical reaction with anti-LLO antibodies.

The presence of a single protein of approximately 58 kDa, equivalent to the LLO molecular mass was observed. The cytotoxic potential of the purified LLO preparation, prior to application on a Jurkat cell line, was tested on SRBC (sheep red blood cells) for haemolytic activity assay. For most experiments preparations were standardized as follows: the LLO concentration was set at 1.5 μg/ml and samples (3000 HU/ml) were stored at −70°C. To facilitate its application, conditions for longer storage were optimized. The following compounds at various concentrations were tested: glycerol (from 0 to 20%), buffer pH (from 5 to 8), cysteine (from 0 to 10 mM), EDTA (0 to 20 mM) and AEBSF - fluoro 4-(2-aminoethylo)-benzenesulphonyl.HCl (0 to 2 mM). The results for the effects of glycerol and pH on LLO stability are presented in Figures 2 and 3, respectively.

The most potent concentration of glycerol (15%) and pH value (6) were used for all further experiments. For the remaining reagents tested, the differences in their ability to preserve LLO activity were less distinguishable (data not shown). However, the presence of protease inhibitors was necessary to preserve optimum LLO activity. We concluded that the following conditions for LLO preparation storage were optimal for the purified LLO preparation: 1 mM AEBSF, 10 mM EDTA, 15% glycerol, 5 mM cysteine-HCl, pH 6 and a temperature of −70°C. We observed that even after several months of storage, LLO samples retained significant activity.

Cytotoxic activity

The cytotoxicity of the purified LLO preparation was tested on a Jurkat cell line at 22°C and 37°C (Figure 4). The cytotoxic activity of LLO was concentration-dependent and was similar at both temperatures under the test conditions employed. Storage of the LLO preparation even in the presence of protease inhibitors and 10% glycerol at −70°C significantly decreased the preparation activity ($p < 0.0001$) (Figure 5). However, even the addition of 3% glycerol stored purified LLO preparation killed 50% of Jurkat cells and the presence of 30% glycerol purified LLO preparation killed 95% of the cells.

LLO activity in culture medium containing cholesterol

The cholesterol dependency of LLO was tested. The cytotoxicity of the LLO preparation was tested on Jurkat cell line in NCS (neonatal calf serum) medium containing cholesterol (390 μM) (Figure 6). The results showed significant inhibition ($p < 0.0001$) of the killing properties of purified LLO in medium composed of serum containing cholesterol.

The dependence of LLO activity on cholesterol is well known. Cholesterol in cell membranes is assumed to act

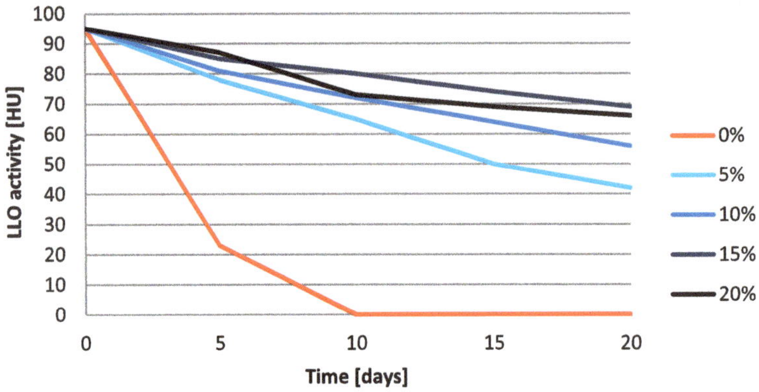

Figure 2 Effect of glycerol concentrations on LLO stability. LLO samples were supplemented with varying GOL concentrations (0–20%) and frozen. The haemolytic activity was checked regularly at 5 days intervals.

as a receptor for LLO, and an insufficient amount of cholesterol content within cell membranes may be inhibitory. Similarly, the presence of cholesterol in incubation medium may bind to LLO, which in turn may prevent its binding to cell membranes. The cell membrane cholesterol content may be diminished by the addition of methyl-β-cyclodextrin (MβCD) to culture medium. This was tested by preincubation of Jurkat cells for 1 h at 37°C in RPMI1640 containing 26 μM cholesterol or 5 mM MβCD or both of these compounds. The effect of these reagents on membrane cholesterol content was verified by cholesterol assay. As anticipated, the addition of cholesterol increased its membrane content while incubation with MβCD completely removed cholesterol from cellular membranes (Figure 7).

Jurkat cells preincubated with modified cholesterol content were subsequently incubated with purified LLO for 30 min the survival rate of the cells was assessed (Figure 8).

The cholesterol concentration dependence on LLO was compared with RPMI1640, and RPMI1640 containing 26 μM cholesterol or 5 mM of MβCD. Statistically significant ($p < 0.004$) differences were noted between cholesterol or MβCD samples and the controls. Interestingly, in medium containing 5% LLO and cholesterol, up to 54% of Jurkat cells survived. A similar percentage of cell survival (51.5%) was observed in the presence of MβCD in the medium as compared with control medium where only 2% of Jurkat cells survived. When both factors were present almost all cells survived, and the cytotoxic activity of LLO was completely blocked. Here the MβCD molar concentration was about 200 times higher than the cholesterol concentration.

Sensitivity of human and mouse PBMC to the purified LLO preparation

Human or mouse PBMC were isolated, diluted to a concentration of 1×10^6 cells/ml, incubated in RPMI 1640-10% NCS with protease inhibitors and their survival rate tested after incubation with the purified LLO preparation. Concentration dependence survival was observed (Figure 9) in a cytometric test with propidium iodide (PI). Statistically significant differences for human and mouse PBMC sensitivity to LLO was observed ($p < 0.005$).

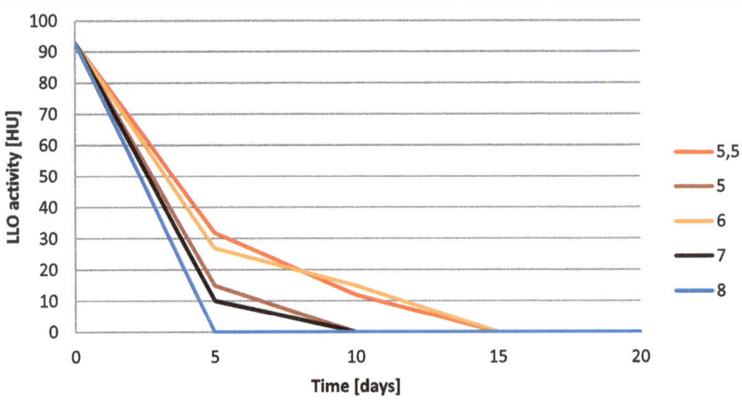

Figure 3 Effect of pH on LLO stability. LLO samples were set to different pH values (from 5 to 8) and frozen. The haemolytic activity was checked regularly at 5 days intervals.

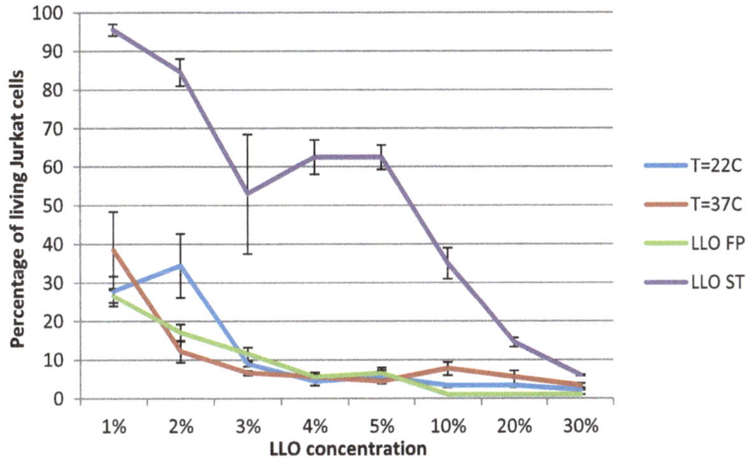

Figure 4 Activity of purified LLO on Jurkat cell line. LLO concentration dependence, temperature dependence and conditions of storage were compared. Presented values show the percentage of living cells after exposure to LLO. Cytotoxicity of LLO at 22°C and 37°C were compared on preparations without protease inhibitors, and the influence of storage for 25 days (at −70°C) of LLO preparations on cytotoxicity in the presence of protease inhibitors was assessed. FP –freshly prepared LLO preparation; ST- stored LLO preparation. Median values; P_{25} and P_{75}, n = 3, p < 0.0001 between FP and ST. Differences in cytotoxic activity at 22°C and 37°C were not significant.

Human cells were more sensitive than mouse cells: at 5% purified LLO concentration 54% of mouse PBMC survived, in contrast to only 14% of human cells.

Discussion

In this study, an *E. coli* strain expressing his-tagged LLO was used, similar to previous reports. However, in contrast to those studies, which used standard elution buffer (pH 8) with one exception [7], here we proposed a modified procedure for LLO purification. To the best of our knowledge, this is the first application of an acidic buffer for the purification of LLO with nickel - nitrilotriacetic acid (Ni-NTA) resin. The use of low pH buffer offers two advantages over commonly used pH 8 elution buffer: (i) it effectively reduces the ability of Ni-NTA resin to bind proteins, facilitating the elution of a higher protein yield with lower imidazole concentrations, and (ii) it preserves the activity of the LLO protein. Given that all CDCs share similar biochemical features and have a similar overall structure [2], the purification method described here should be applicable for other CDCs. Storage conditions are, however, most likely restricted to Listerial cytolysin due to its unique characteristics [3,4]. The storage conditions that promoted the stability of LLO haemolytic activity and cytotoxic activity were confirmed using sheep erythrocytes and Jurkat

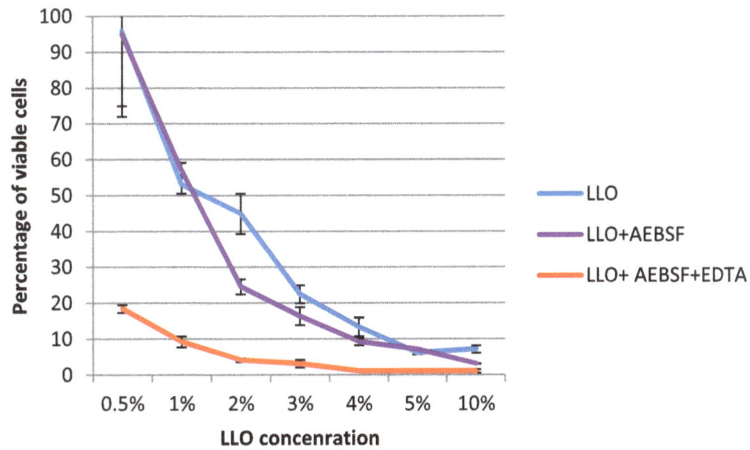

Figure 5 LLO stability in the presence of protease inhibitors tested on Jurkat cells incubated in RPMI 1640 medium containing 4% inactivated NCS. Presented values show percentage of living cells after exposure to LLO. Median value, P_{25} and P_{75}, n = 3, p < 0.0001 compared with controls without inhibitors.

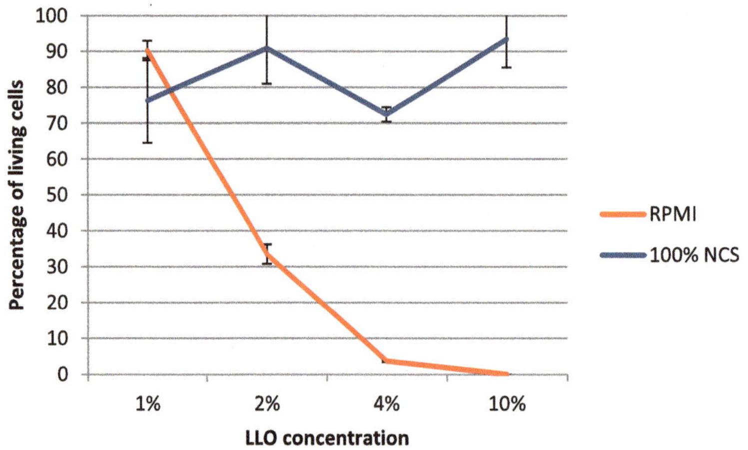

Figure 6 Effect of serum containing cholesterol on LLO cytotoxicity. Cytotoxicity assay on a Jurkat cell line in RPMI-1640 without NCS and in 100% inactivated NCS with PI. NCS contained 390 μM (150 mg/dl) of cholesterol. Results show percentage of living cells. LLO concentration dependence was compared. Median values; P_{25} and P_{75}, $n = 4$, $p < 0.0001$.

cells, respectively. However, the positive effect of protease inhibitors on the stability of the purified LLO preparation suggested the presence of active proteases in the purified preparation of LLO.

It was shown that purified LLO demonstrates the properties of an authentic *Listeria monocytogenes* protein. Listeriolysin O is crucial for pathogen survival within the cytoplasm of infected cells [3,4]. Less well known is the activity dependence of purified LLO when applied outside the cell, on the cell membrane. The cell membrane might not be as uniform as predicted by the classical fluid mosaic model, and could contain differently arranged regions. The lipid raft hypothesis describes membrane areas stabilized by cholesterol within a liquid-ordered phase. These areas are engaged in the regulation of cell adhesion, transport, sorting of lipids and proteins and signal transduction. Membrane lipid rafts are regions where cholesterol and marker proteins such as CD59 are concentrated [11,12]. While cholesterol acts as a receptor for all CDCs, CD59 is considered an additional receptor for two CDC toxins [13]. Possibly, other CDCs also require additional factors, which might explain some long-observed differences in different cells susceptibility to CDCs activity. Human leukocytes were significantly more sensitive than mouse cells to LLO activity. This might be simply a result of the different overall sensitivity of human and mouse cells or a specific difference between cellular receptors, possibly lipid rafts. Very subtle differences at the molecular level might be of great importance, i.e. the difference of a single amino acid for Internalin (Inl) A–E-cadherin interactions. Human cadherin is preferentially targeted by the InlA surface protein [14]. *L. monocytogenes* is better adopted to infect humans than mice and might explain its high

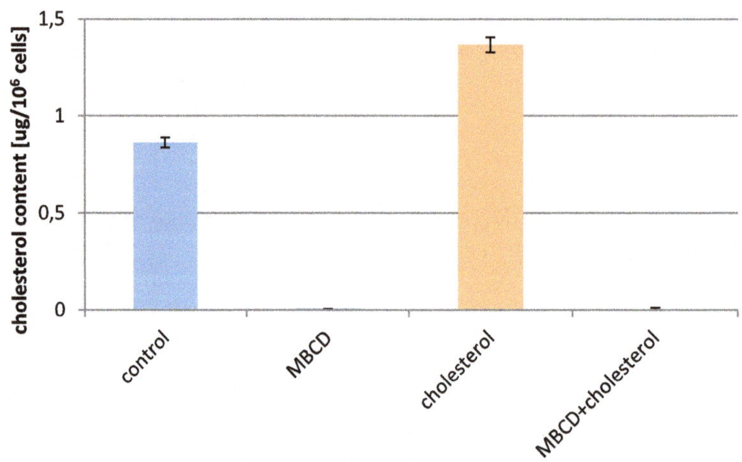

Figure 7 Cholesterol content in Jurkat cells after cholesterol and MβCD treatment. Jurkat cells were incubated for 1 h at 37°C in RPMI1640 containing 26 μM cholesterol or 5 mM MβCD or both these compounds. Results show cholesterol content in μg per 10^6 cells.

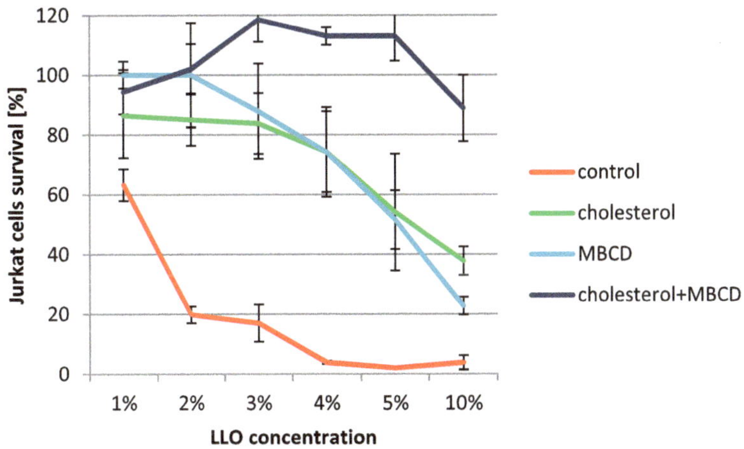

Figure 8 Effect of cholesterol and MβCD pre-treatment on Jurkat cells survival. Jurkat cells were preincubated in RPMI 1640 for 1 h at 37°C with cholesterol (26 μM; 10 μg/ml) or MβCD (5 mM) or a mixture of both. Then LLO was added and the sample was incubated for 30 min. LLO cytotoxicity was tested on a Jurkat cell line with PI. Values represent percentage of living cells. LLO concentration dependence was compared. Median values; P_{25} and P_{75}, $n = 4$ $p < 0.004$ between control and cholesterol or MβCD groups.

activity towards leukocytes that normally threaten *L. monocytogenes* survival in human cells.

The cytotoxic activity of purified LLO was observed when applied to human leukocytes in the presence of a known concentration of cholesterol and/or MβCD in the medium, which removed cholesterol from cell membranes. We concluded that the presence of cholesterol in cell membranes and in the culture medium modified the cytotoxic activity of purified LLO. During *in vivo* conditions, the presence of cholesterol is expected in peripheral blood plasma; however, it may also be present within the tissue matrix [15]. The LLO preparation may have application when locally administered directly into tissues, e.g. subcutaneously, intraperitoneally, intraorganally or intratumourally. We demonstrated the possibility of simple regulation of LLO cytotoxic activity *in vitro* by the additional application of cholesterol or MβCD. This might also have practical significance *in vivo* although the safety of such a treatment might be an issue. However, it might be possible since cyclodextrins are already in clinical use [16]. The LLO preparation has haemolytic activity and cannot be applied intravenously; however its intraperitoneal or subcutaneous application is possible. The relatively low activity and stability of LLO in physiological conditions should be advantageous and allow the restriction of cytotoxic activity to selected regions, thus limiting possible side-effects. However, the specificity of cytotoxin-based anti-cancer therapy remains a great challenge. This could be resolved by combining toxic components with antibodies, producing so called immunotoxins. The most

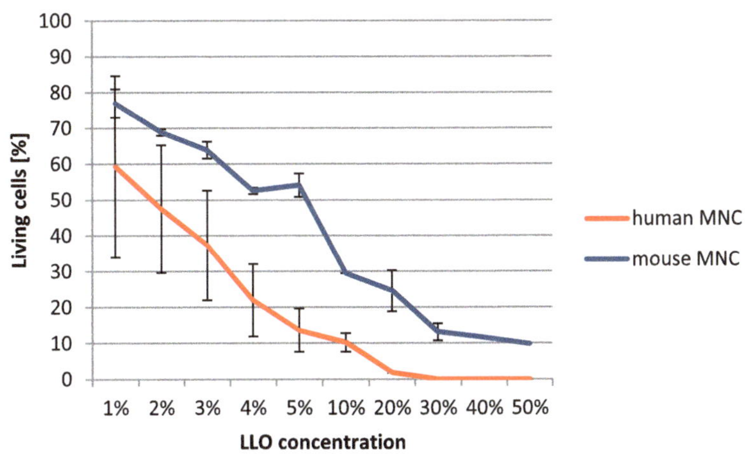

Figure 9 Comparison of human and mouse PBMC sensitivity to LLO cytotoxicity. Results of purified LLO cytotoxicity tested on human and mouse PBMC (1×10^6/ml) in RPMI 1640 containing 4% NCS with PI. The percentage of living cells is shown. LLO concentration dependence was compared. Median values; P_{25} and P_{75}, $n = 3$, $p < 0.005$.

popular immunotoxin components of bacterial origin are exotoxin A derived from *Pseudomonas aeruginosa* and diphtheria toxin from *Corynebacterium diphtheriae* [17,18]. Although its mode of action might appear to preclude the combination of CDCs with antibodies, it was recently shown that directing cytotoxicity using antibodies is applicable for LLO [19]. The cytotoxic features of LLO and the possibility of controlling its activity make it a good immunotoxin candidate, although the direct use of LLO for clinical treatment must be preceded by a detailed description of its activity under the form of purified preparation.

Conclusions

In this study, we purified LLO by affinity chromatography demonstrating the usefulness of acidic buffer to remove LLO from Ni-NTA resin. The satisfactory storage conditions of the purified LLO preparation were also demonstrated to be an important step before planning use of the toxin. This study underscores the role of cryopreservative glycerol and a thiol-reducing agent, cysteine, for LLO stability and activity. The storage additives in the doses used were not toxic to eukaryotic cells, while LLO displayed potent *in vitro* activity towards leukaemia cells. LLO activity was easily regulated *in vitro* and possibly *in vivo*, and this information might facilitate its future clinical application.

Methods

Escherichia coli strains and growth conditions

The source of purified LLO was *E. coli* strain BL21(DE3) (Novagen, Madison, Wis) with plasmid pET29B (Novagen) carrying cloned *hly* gene, kindly provided by Dr. Higgins [6]. The bacteria were grown in LB broth (Sterbios, Warsaw, Poland), supplemented with 30 μg/ml kanamycin (Merck, Darmstadt, Germany) at 37°C with 120 rpm shaking. Bacteria were cultured until the optical density reached 1.5. To induce LLO-His synthesis, IPTG (Sigma, Taufkirchen, Germany) was added to a final concentration of 1 mM.

Eukaryotic cell isolation and growth conditions

The target *Eukaryotic* cells were: SRBC - sheep red blood cells (Biomed, Warsaw, Poland), human acute T cell leukaemia cell line (Jurkat, ATCC TIB 152), and PBMC isolated from humans or mice.

Jurkat cells were grown in RPMI 1640 culture medium (Gibco, Gaithersburg, MD, USA) with 10% of inactivated NCS containing 100 U of penicillin and 0.1 mg of streptomycin per 1 ml or in DMEM medium (Sigma) with 10% of inactivated NCS. Serum contained 150 mg/dl cholesterol, and the final cholesterol concentration in the culture medium was 15 μg/ml (39 μM).

PBMC were cultured, after isolation, *in vitro* for 2–7 days before a cytotoxicity experiment. For human peripheral blood experiments *in vitro*, patients gave their personal consent and the procedure was accepted by the Bioethical Commission at the Medical Center of Postgraduate Education, Warsaw (Feb 27, 2008).

Isolation of PBMC was performed by an overlayer of peripheral blood diluted with PBS (1:1), on 5 ml of Histopaque 1077 (Sigma) preparation in a test tube and centrifugation for 30 min at 540 × *g*. The interlayer of PBMCs consisting of lymphocytes and monocytes was separated and diluted with 10 ml of culture medium, centrifuged for 5 min at 200 × *g* and the washed cells suspended in fresh culture medium were counted in a Burker camera. The concentration of the cell suspension was corrected to a final value of 2×10^6/ml before culturing. The cultures were performed in Falcon vessels (dishes) or in 6-well flat bottom plates in RPMI 1640 with 10% NCS and antibiotics, and half of the medium volume was changed twice weekly.

For some experiments cell cultures were supplemented with the following compounds to modify the LLO activity: cholesterol (Sigma), NCS containing 150 μg/ml cholesterol (390 μM), and MβCD (Sigma).

The affinity chromatography of LLO

Ten ml of *E. coli* suspension (approximately 10^8/ml) grown in the presence of 1 mM IPTG was centrifuged at 4400 × *g* for 10 min at 4°C, and the supernatant was removed. The pelleted cells were resuspended in 1 ml of sterile column buffer (50 mM Tris, pH 8, 300 mM NaCl, 50 mM imidazole) and were ruptured by sonication in 3 pulses of 30 s each. Then, the bacterial fragments were removed by centrifugation (7000 × *g*, 15 min). The supernatant was collected, diluted with column buffer and applied onto the affinity column (His-Bind Ni Column, Novagen) previously equilibrated with the column buffer. The column was washed with the same buffer, and then LLO was desorbed with the elution buffer (50 mM MES, pH 6, 300 mM NaCl, 250 mM imidazole). The collected fraction was subsequently dialysed to remove imidazole (50 mM MES, pH 6, 300 mM NaCl, 5 mM cysteine HCl). Protease inhibitors: fluoro 4-(2-aminoethylo)-benzenesulphonyl.HCl (AEBSF, Sigma), EDTA (Sigma), and glycerol (Merck) were added to a final concentration of 1 mM, 10 mM, and 15% (v/v), respectively. Protein concentrations were assayed with NanoOrange® Protein Quantitation Kit (Molecular Probes) using an infinite M200 PRO reader (TECAN, Goring on Thames, UK).

Western blot analysis of LLO preparations

Electrophoretically resolved proteins (SDS-PAGE) were transferred onto a nitrocellulose membrane using the BIO-RAD Trans-Blot system, according to a protocol recommended by the manufacturer. The membrane was blocked using 5% skimmed milk in Tris-buffered saline buffer (TBS), pH 7.6. Primary rabbit polyclonal anti-LLO antibody (Abcam, Cambridge, MA, USA) and secondary goat polyclonal antibody (Abcam) conjugated with alkaline phosphatase were used at 1:1500 and 1:10,000 dilutions, respectively. The membrane was developed using the NBT/BCIP reagent (Merck).

Assay of haemolytic activity

Haemolytic activity was assayed using SRBC. The erythrocytes were washed three times with PBS and suspended to a final concentration of 20% (v/v) in PBS (pH 7.4). The LLO preparation was diluted by adding 10 µl of its preparation to a final volume of 1 ml of 2% SRBC suspension in PBS. The prepared solution was incubated for 30 min at 37°C and then centrifuged for 3 min at $150 \times g$. The released haemoglobin was measured spectrophotometrically at $\lambda = 410$ nm. Haemolytic Units (HU) were calculated after setting 0 HU as the activity of a negative control, and 100 HU for total haemolysis, observed in samples of erythrocytes lysed with 0.01% SDS.

Cholesterol assay

Jurkat cells were resuspended in fresh RPMI medium without any additives (control), supplemented with 26 µM cholesterol, with 5 mM MβCD or with both these compounds. The cells were incubated for 1 h at 37°C, washed with PBS three times, counted and disrupted by sonication. Membrane cholesterol was extracted with isopropanol and quantified by a cholesterol assay kit (R-Biopharm, Darmstadt, Germany).

Cytometric assays

The activities of the LLO preparations were tested on Jurkat cells or PBMC. The Jurkat cell suspension in a culture medium (1×10^6/ml) of 20 µl, was mixed with 30 µl of LLO preparation diluted with physiological salt solution at different proportions, and immediately incubated for 30 min at room temperature (20–22°C). The time of the incubation was assumed to be the same as for the haemolytic assays. The final cell concentration was 0.4×10^6/ml. The effect of the LLO preparation was titrated in an assay volume of 50 µl containing cell suspension (20 µl) and 30 µl of physiological salt solution containing the LLO preparation. The LLO preparation content investigated was: 0.5, 1, 2, 3, 4, 5, 10, 20, 30, 40, and 50% of the original LLO concentration (1.5 ng/µl) with activity set at 3 HU/µl. After cell sample incubation, physiological saline containing PI (1 µg/ml) was added before acquisition by a cytometer. PI penetrates the cell membrane of damaged cells and binds to the cell nucleus. In the fluorescence quadrant readings of forward scatter (FSC)/PI fluorescence, the PI unstained events indicate the percentage of living cells. In the FACSCalibur BD cytometer, 5000 events were collected from each sample and analysed by Cell Quest software (Becton Dickinson, Franklin Lakes, NJ, US). Experiments were repeated at least three times.

Statistical analysis

The flow cytometry results were presented as median values and percentile values P_{25} and P_{75}. Statistical analysis of flow cytometric experiments was in the global linear model (GLM) after logit transformation of values. The *post hoc* test of Tukey was used. Results of haemolytic assays were presented as mean values ± standard deviation (SD). Analysis of variance was performed by ANOVA. Statistically significant values of $p < 0.05$ were assumed. Statistical software SAS 9.2 was used.

Competing interests

The authors declare no competing interests.

Authors' contributions

RS contributed to experimental strategy, analysed data, conducted experiments, wrote and submitted the manuscript. MŁ performed cytometric assays of eukaryotic cells and drafted the manuscript. MG managed all aspects of the protein purification and optimised the method. KR conducted SDS-PAGE and western-blot analysis. TJ contributed to development of the LLO purification method and edited the manuscript. JB supervised the experimental design and execution and guided manuscript preparation. BB isolated PBMC from patients and cultured the cells. GH isolated and cultured mouse PBMC and managed tissue cultures. JK conceptualized the experimental strategy, supervised cytometric assays and significantly contributed to the writing of the manuscript. All authors read and approved the final manuscript.

Acknowledgements

This work was supported by the European Union structural funds and the Innovative Economy Operational Program POIG.01.01.02-00-109/09-00. The study was partly supported by the Polish Ministry of Science and Higher Education grant N N401 015936 and CMKP grant 2008/2009. We thank Dr. Krystyna Stepien from Diagnostic Laboratory CSK MSWiA and MSc Marek Gryzik for cholesterol assays and Dr. Darren Higgins for the pET29b-*hlyA* plasmid.

Author details

[1]Department of Applied Microbiology, Institute of Microbiology, Faculty of Biology, University of Warsaw, I. Miecznikowa 1, 02-096 Warsaw, Poland. [2]Medical Centre of Postgraduate Education, Marymoncka 99/103, 01-813 Warsaw, Poland. [3]Department of Hematology, Institute of Hematology and Transfusion Medicine, I. Gandhi 14, 02-776 Warsaw, Poland.

References

1. Stachowiak R, Bielecki J: **Contribution of hemolysin and phospholipase activity to cytolytic properties and viability of *Listeria monocytogenes*.** *Acta Microbiol Pol* 2001, **50**(3–4):243–250.
2. Alouf JE, Billington SJ, Jost BH: **Repertoire and general features of the family of cholesterol-dependent cytolysins.** In *The comprehensive sourcebook of bacterial protein toxins.* 3rd edition. Edited by Alouf JE, Popoff MR. London – San Diego: Academic Press; 2005:643–658.
3. Vazquez-Boland JA, Stachowiak R, Lacharme L, Scortti M: **Listeriolysin.** In *The comprehensive sourcebook of bacterial protein toxins.* 3rd edition. Edited by Alouf JE, Popoff MR. London – San Diego: Academic Press; 2005:700–716.
4. Portnoy DA, Jacks PS, Hinrichs DJ: **Role of hemolysin for the intracellular growth of *Listeria monocytogenes*.** *J Exp Med* 1988, **167**(4):1459–1471.
5. Giammarini C, Andreoni F, Amagliani G, Casiere A, Barocci S, Magnani M: **Purification and characterization of a recombinant listeriolysin O expressed in *Escherichia coli* and possible diagnostic applications.** *J Biotechnol* 2004, **109**(1–2):13–20.

6. Gedde MM, Higgins DE, Tilney LG, Portnoy DA: **Role of listeriolysin O in cell-to-cell spread of *Listeria monocytogenes*.** *Infect Immun* 2000, **68**(2):999–1003.
7. Giammarini C, Andreoni F, Amagliani G, Casiere A, Barocci S, Magnani M: **High-level expression of the *Listeria monocytogenes* listeriolysin O in *Escherichia coli* and preliminary characterization of the purified protein.** *Protein Expr Purif* 2003, **28**(1):78–85.
8. Kohda C, Kawamura I, Baba H, Nomura T, Ito Y, Kimoto T, Watanabe I, Mitsuyama M: **Dissociated linkage of cytokine-inducing activity and cytotoxicity to different domains of listeriolysin O from *Listeria monocytogenes*.** *Infect Immun* 2002, **70**(3):1334–1341.
9. Churchill RL, Lee H, Hall JC: **Rapid purification of recombinant listeriolysin O (LLO) from Escherichia coli.** *J Ind Microbiol Biotechnol* 2005, **32**(8):355–363.
10. Stachowiak R, Granicka LH, Wisniewski J, Lyzniak M, Kawiak J, Bielecki J: **Cytotoxicity of listeriolysin O produced by membrane-encapsulated *Bacillus subtilis* on leukemia cells.** *J Microbiol Biotechnol* 2011, **21**(11):1193–1198.
11. Motoyama K, Toyodome H, Onodera R, Irie T, Hirayama F, Uekama K, Arima H: **Involvement of lipid rafts of rabbit red blood cells in morphological changes induced by methylated beta-cyclodextrins.** *Biol Pharm Bull* 2009, **32**(4):700–705.
12. Nicolau DV Jr, Burrage K, Parton RG, Hancock JF: **Identifying optimal lipid raft characteristics required to promote nanoscale protein-protein interactions on the plasma membrane.** *Mol Cell Biol* 2006, **26**(1):313–323.
13. Hamon MA, Ribet D, Stavru F, Cossart P: **Listeriolysin O: the Swiss army knife of Listeria.** *Trends Microbiol* 2012, **20**(8):360–368.
14. Lecuit M, Dramsi S, Gottardi C, Fedor-Chaiken M, Gumbiner B, Cossart P: **A single amino acid in E-cadherin responsible for host specificity towards the human pathogen Listeria monocytogenes.** *EMBO J* 1999, **18**(14):3956–3963.
15. Iwai I, Han H, den Hollander L, Svensson S, Ofverstedt LG, Anwar J, Brewer J, Bloksgaard M, Laloeuf A, Nosek D, *et al*: **The human skin barrier is organized as stacked bilayers of fully extended ceramides with cholesterol molecules associated with the ceramide sphingoid moiety.** *J Invest Dermatol* 2012, **132**(9):2215–2225.
16. Challa R, Ahuja A, Ali J, Khar RK: **Cyclodextrins in drug delivery: an updated review.** *AAPS PharmSciTech* 2005, **6**(2):E329–E357.
17. Pastan II, Kreitman RJ: **Immunotoxins for targeted cancer therapy.** *Adv Drug Deliv Rev* 1998, **31**(1–2):53–88.
18. Kreitman RJ: **Recombinant immunotoxins for the treatment of chemoresistant hematologic malignancies.** *Curr Pharm Des* 2009, **15**(23):2652–2664.
19. Bergelt S, Frost S, Lilie H: **Listeriolysin O as cytotoxic component of an immunotoxin.** *Protein Sci* 2009, **18**(6):1210–1220.

Antimicrobial activity and safety evaluation of *Enterococcus faecium* KQ 2.6 isolated from peacock feces

Wei Zheng[1], Yu Zhang[1], Hui-Min Lu[1], Dan-Ting Li[1], Zhi-Liang Zhang[1], Zhen-Xing Tang[2] and Lu-E Shi[1*]

Abstract

Background: The objective of this paper was to study antimicrobial activity and safety of *Enterococcus faecium* KQ 2.6 (*E. faecium* KQ 2.6) isolated from peacock feces.

Methods: Agar well diffusion method was adopted in antimicrobial activity assay. Disk diffusion test was used to determine the antibiotic resistance. The identification and virulence potential of *E. faecium* KQ 2.6 were investigated using PCR amplification.

Results: The results indicated that cell free supernatant (CFS) of the strain had the good antimicrobial activity against selected gram-positive and gram-negative bacteria. The biochemical characteristics of antimicrobial substances were investigated. The results indicated that the antimicrobial substances were still active after treatment with catalase and proteinase, respectively. Moreover, the stability of antimicrobial substances did not change after heat treatment at 40, 50, 60, 70 and 80°C for 30 min, respectively. The activity of antimicrobial substances remained stable at 4 and −20°C after long time storage. The antimicrobial activity of CFS was compared with that of the buffer with similar strength and pH. The inhibitory zone of the buffer was apparently smaller than that of CFS, which meant that the acid in CFS was not the only factor that was contributed to antibacterial activity of CFS. The antibiotic resistance and virulence potential were evaluated using disk diffusion test and PCR amplification. The results showed that *E. faecium* KQ 2.6 did not harbor any tested virulence genes such as *gelE*, *esp*, *asa1*, *cylA*, *efaA* and *hyl*. It was susceptible to most of tested antibiotics except for vancomycin and polymyxin B.

Conclusion: *E. faecium* KQ 2.6 may be used as bio-preservative cultures for the production of fermented foods.

Keywords: *E. faecium* KQ 2.6, Antimicrobial activity, Safety evaluation, Antibiotics resistance, Virulence genes

Background

Enterococci belong to lactic acid bacteria (LAB), which are widespread in foods and environment. In aspect of food fermentation, it is considered that enterococci play an important role in the development of the sensory characteristics of fermentation foods such as sausages and cheeses [1]. Certain cheese-makers have suggested that enterococci can be utilized as starter cultures in the production of Mediterranean cheese [2,3]. Furthermore, some enterococcal strains have been successfully used as preservatives to inhibit the growth of food spoilage microorganisms. One of reasons that these enterococcal strains with antimicrobial activity, produce lactic acid [4]. Lactic acid reduces the pH that can cause the disruption of cellular substrate transport systems through altering the cell membrane permeability or collapsing the electrochemical proton gradient [5]. In addition, enterococci also can produce other antimicrobial substances such as hydrogen peroxide, bacteriocin and bacteriocin like inhibitory substances (BLIS). In past few years, bacteriocin has been increasingly concerned due to its diversity and novelty. Bacteriocins are ribosomally synthesized, extracellularly released low-molecular-mass peptides or proteins [6,7]. Generally, most known bacteriocins produced by *E. faecium*, are small (<10 kDa), membrane-active and unmodified peptides. One of the

* Correspondence: shilue@126.com
[1]College of Life and Environmental Sciences, Hangzhou Normal University, 310016 Hangzhou, Zhejiang, China
Full list of author information is available at the end of the article

most obvious traits of these bacteriocins is sensitive to proteolytic enzymes. For example, enterocin A and enterocin B from *E. faecium* MMT21 are both sensitive to trypsin, proteinase K and pronase E [8-11].

Enterococci have been used safely in foods for a long history. However, in past few years, the concerns on the safety of enterococci in food or feed industries have been raised. Many studies have reported that enterococci are associated with nosocomial infections like bacteraemia, endocarditis, urinary tract infections and diarrhea [12,13]. The main reasons that cause nosocomial infections, are the resistance of the strains to a board range of antibiotics and the presence of virulence factors in the strains [14]. The multiple antibiotic resistant strains often cause serious infections which can't be cured well. In particular, vancomycin-resistant enterococci (VRE) have produced serious problems in public health [15]. Virulence factors have been well studied in recent years, and some virulence factors have been reported in detail. The main described factors are those are involved in adhesion, damaging tissues and evasion of immune responses (capsular polysaccharides) [16]. Additionally, it should be mentioned that enterococci may acquire antibiotic resistance and virulence factors from other enterococci, since mobile genetic elements like plasmids and transposons, can contribute to the distribution of antibiotic resistance and virulence factors between enterococcal strains [17,18]. Therefore, the safety evaluation of the enterococci should be carried out before the application.

In present study, one enterococcal strain isolated from peacock feces was identified as *E. faecium* KQ 2.6 by PCR and 16S rRNA gene sequencing. Antimicrobial activity and safety of this strain was mainly studied. The production and biochemical properties of antimicrobial substances were also investigated.

Methods

All chemicals were purchased from Sangon (Shanghai, China). Indicator strains and antibiotic-containing disks were obtained from Binhe Microorganism Reagent Co. Ltd (Hangzhou, China). Participants in the study agreed to carry out the following studies. No human subjects including human material or human data, were contained in present study.

Bacterial isolation and identification

Peacock feces were collected in an animal centre located in Hangzhou Normal University. Ten-fold dilutions of feces in sterile water were plated onto de Man, Rogosa and Sharpe (MRS). The plates were incubated at 37°C for 24 h. Twelve of colonies were randomly picked and used for the study of physiological and biochemical characteristics. Meanwhile, the antimicrobial activity of the strains against *Escherichia Coli* was studied using the agar spot method [19]. The strains displaying an inhibition zone were selected, and maintained as stock cultures in MRS broth supplemented with 30 % (v/v) glycerol at −20°C.

Primers, 27 F (5′-AGAGTTGATCCTGGCTCAG-3′) and 1492R (5′- GGTTACCTTGTTACGACTT-3′) based on conserved regions of 16SrRNA gene were used to direct the amplification. The program consisted of: denaturation at 94°C for 5 min, then 35 cycles of 94°C for 1 min, 55°C for 1 min and 72°C for 1 min followed by a final extension at 72°C for 5 min. Amplified PCR products were separated by 1.0 % (w/v) agarose gel electrophoresis, and then purified with the StarPrep Gel Extraction Kit (GenStar, Beijing, China) according to manufacturer's instruction. 16S rRNA gene sequencing was carried out by Sunny Biotechnology Co., Ltd (Shanghai, China).

Antimicrobial activity assay of *E. faecium* KQ 2.6

The antimicrobial activity of *E. faecium* KQ 2.6 against pathogenic bacteria was investigated. Pathogenic bacteria included *Bacillus subtilis, Bacillus cereus, Streptococcus pyogenes, Staphylococcus aureus, Staphylococcus epidermidis, E. faecalis, Escherichia coli, Pseudomonas aeruginosa, Klebsiella pneumoniae, Salmonella paratyphi, Candida albicans* and *Aspergillus niger*. The antimicrobial assay was performed using agar well diffusion method [20]. Firstly, *E. faecium* KQ 2.6 was grown overnight in MRS broth at 37°C. Cells in the culture were discarded by centrifugation at 10, 000 g at 4°Cfor 20 min. 60 μL of indicator bacteria (final concentration of 10^8 CFU/mL) cultured in 20 mL soft agar containing 0.80 % (w/v) agar was poured onto a solid agar plate containing 1.5 % (w/v) agar. Afterwards, wells (8 mm in diameter) were made on agar plate, and filled with 100 μL of cell free supernatant (CFS) of *E. faecium* KQ 2.6. Plates were incubated at 37°C for 24 h after being kept for 3–4 h at 4°C. Finally, the antimicrobial activity was analyzed by observing the clear zones around the wells containing CFS. The clear zones were regarded as inhibitory zones, and recorded in mm.

Growth kinetics and antimicrobial activity of *E. faecium* KQ 2.6

100 mL of MRS broth was inoculated with 1.0 % (v/v) of the culture of *E. faecium* KQ 2.6 and incubated at 37°C. Optical density at 600 nm (OD_{600}) and pH values were monitored at 2 h intervals during 24 h. The antimicrobial activity assay was also performed every two hours. To quantify the antimicrobial activity, CFS was serially diluted 2-folds and 10 μL of each dilution was added into the wells. The titer was defined as 2^n, which is the reciprocal of the highest dilution showing inhibition of

indicator strain. Thus, the arbitrary unit (AU) of antimicrobial activity per milliliter was defined as $2^n \times$ (1,000 μL/10 μL) [21].

Effect of the biochemical factors on antimicrobial activity

E. faecium KQ 2.6 was cultivated in MRS broth at 37°C for 16 h. CFS was obtained by centrifugation at 10,000 g at 4°C for 20 min, and used to carry out the following studies.

Antimicrobial activity of CFS at different temperatures was investigated. CFS was treated at 40, 50, 60, 70 and 80°C for 30 min and 3 h, respectively, and at 121°C for 20 min. Storage stability of CFS at 4 and −20°C for 24, 48 h, 7 days and 15 days, was also performed.

The sensitivity of antimicrobial substances towards catalase and proteinase was studied. 1.0 mL of CFS was added to 1.0 mL of 1.0 mg/mL catalase, trypsin and pepsin, respectively. Afterwards, samples were incubated at 37°C for 30 min, and heated at 95°C for 5 min.

All treated samples were tested against *Bacillus cereus* using agar well diffusion method. Each experiment was performed at least two times. In addition, the antimicrobial activity was done using hydrogen phosphate/citric acid buffer which had a similar pH and strength to CFS of *E. faecium* KQ 2.6.

Antibiotic resistance

Disk diffusion test was used to determine the susceptibility of *E. faecium* KQ 2.6 to antibiotics [22]. Antibiotic-containing disks were those of penicillin, vancomycin, chloramphenicol, tetracycline, erythromycin, rifampicin, ofloxacin, polymyxin B and ciprofloxacin. 20 mL of MRS broth containing 1.5 % agar was seeded with 200 μL of a culture of *E. faecium* KQ 2.6 (10^6-10^7 CFU/mL), and poured into a plate. Then antibiotic-containing disks were added onto the plates according to the manufacturer's instructions. Inhibition zone diameters with/without vancomycin-containing disks were measured (mm) at 37°C after 24 and 18 h incubation, respectively. According to the recommendation of Clinical and Laboratory Standards Institute (CLSI), the strain was considered to be resistant to antibiotics if the inhibition zone was equal or smaller than 16 mm for rifampicin, 15 mm for ciprofloxacin, 14 mm for penicillin, vancomycin and tetracycline, 13 mm for erythromycin and ofloxacin, and 12 mm for chloramphenicol.

PCR for the detection of virulence genes

PCR amplification was used to detect virulence genes *gelE* (gelatinase), *esp* (enterococcal surface protein), *asa1* (aggregation substance), *cylA* (cytolysin), *efaA* (cell-wall adhesion) and *hyl* (hyaluronidase). Primers are listed in Table 1. The following PCR conditions were used: 94°C for 5 min; followed by 35 cycles of 94°C for 1 min, 52°C (for *gelE*, *efaA*), 56°C (for *cylA*, *asa1*, *esp*) and 58°C (for *hyl*) for 30 s, 72°C for 1 min; a final extension at 72°C for 5 min. The DNA from *E faecalis* ATCC 29212 (*asa1*$^+$, *cylA*$^+$, *gelE*$^+$, *efaA*$^+$ and *hyl*$^+$) was used as a positive control. The amplified products were analyzed by electrophoresis on 1.0 % (w/v) agarose gels in 1× TAE buffer.

Table 1 Primer pairs used for detection of virulence genes

Gene	Primers (5'-3')	Size (bp)	References
gelE	F: TATGACAATGCTTTTTGGGAT	213	[37]
	R: AGATGCACCCGAAATAATATA		
esp	F: AGATTTCATCTTTGATTCTTGG	510	[37]
	R: AATTGATTCTTTAGCATCTGG		
asa1	F: GCACGCTATTACGAACTATGA	375	[37]
	R: AAGAAAGAACATCACCACGA		
cylA	F: ACTCGGGGATTGATAGGC	688	[37]
	R: GCTGCTAAAGCTGCGCTT		
efaA	F: GACAGACCCTCACGAATA	705	[18]
	R: AGTTCATCATGCTGTAGTA		
hyl	F: ACAGAAGAGCTGCAGGAAATG	276	[37]
	R: GACTGACGTCCAAGTTTCCAA		

Results

Isolation and identification of LAB strains with antimicrobial activity

Antimicrobial activity of twelve strains isolated from peacock feces, were studied using the agar spot method. The results indicated that only two isolates had obvious antimicrobial activity against *Escherichia coli* (data not shown). According to the studies of physiological and biochemical characteristics, one of two isolates could produce gas through glucose fermentation. It was not convenient to control the fermentation process easily. Therefore, the strain with good antimicrobial activity and gas-negative property, was chosen for this study. The sequencing of the partial 16S rRNA of the strain showed 99 % homology to that of *E. faecium* 3-2-31, so it was identified as *E. faecium* KQ 2.6.

Spectrum of antimicrobial activity

As shown in Figure 1, CFS of *E. faecium* KQ 2.6 could exert inhibiting activity to the growth of *Bacillus subtilis*, *Bacillus cereus* and *Escherichia coli*. The growth of a panel of pathogenic gram-positive and gram-negative bacteria including *Bacillus subtilis*, *Bacillus cereus*, *Streptococcus pyogenes*, *Staphylococcus epidermidis*, *Pseudomonas aeruginosa*, *Salmonella paratyphi* and *E. faecalis*, was also inhibited by CFS of *E. faecium* KQ 2.6. However, it was not active against fungi like *Candida albicans* and *Aspergillsu niger* (Table 2).

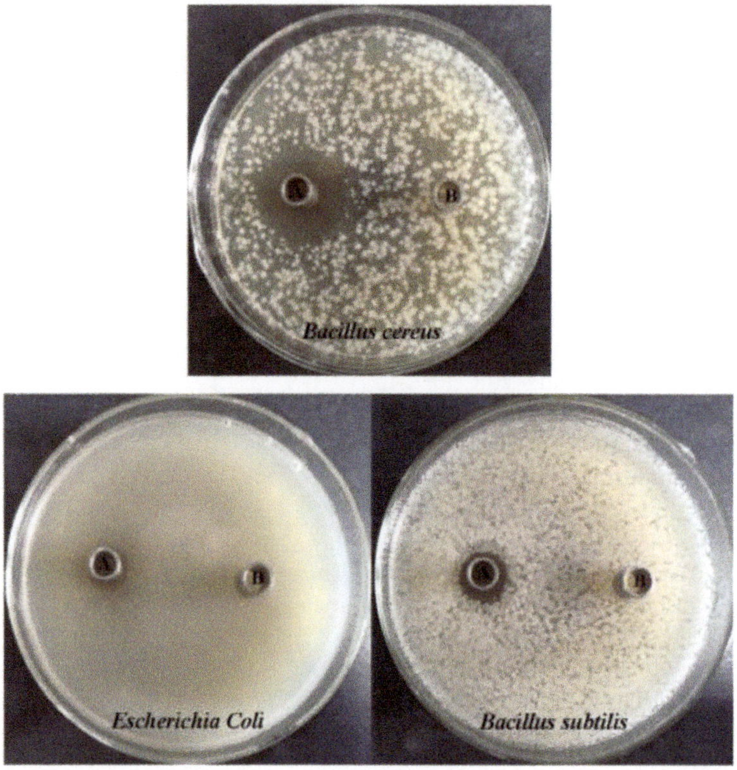

Figure 1 Antimicrobial activity of CFS against *Bacillus cereus, Escherichia coli* and *Bacillus subtilis.* A: CFS of *E. faecium* KQ 2.6, B: Luria-Bertani broth.

Table 2 Antimicrobial activity of CFS produced by *E. faecium* KQ 2.6

Indicator strains	Medium	Incubation temperature(°C)	Antimicrobial activity [a]
Gram-positive			
Bacillus subtilis	LB	37	+
Bacillus cereus	LB	37	+++
Streptococcus pyogenes	LB	37	++
Staphylococcus aureus	LB	37	-
Staphylococcus epidermidis	LB	37	++
E. faecalis	MRS	37	+
Gram-negative			
Escherichia Coli	LB	37	+
Pseudomonas aeruginosa	LB	37	++
Klebsiella pneumoniae	LB	37	-
Salmonella paratyphi	LB	37	+++
Fungi			
Candida albicans	PDA	28	-
Aspergillus niger	PDA	28	-

[a]Results of antimicrobial activity were recorded in the diameter of inhibition zones around the wells (8 mm in diameter): –, no inhibition zone; +, zone < 5 mm; ++, zone < 5–10 mm; +++, zone > 15 mm.

Production of antimicrobial substances and growth kinetics

The results of the cell density, pH of the media and production of antimicrobial substances were obtained during 24 h of growth at 37°C (Figure 2). During this period, the cell density of *E. faecium* KQ 2.6 increased from 0.03 to 1.37 (OD_{600}). pH of the media dropped down to 4.5. *E. faecium* KQ 2.6 began to produce antimicrobial substances (200 AU/mL) after 4 h of growth. Maximum values (1600 AU/mL) of antimicrobial activity was reached at the early stationary phase (16 h), and remained un-change in the following 8 h of growth.

Characterization of antimicrobial substances

Except for heat treatment at 121°C for 20 min, the substances remained stable after heating at 40, 50, 60, 70 and 80°C for 30 min, respectively. Meanwhile, antimicrobial activity did not change when CFS was stored at low temperatures(4 and –20°C) for 24, 48 h, 7 and 15 days (Table 3). It showed that storage conditions did not led to the decrease of antimicrobial activity significantly. Additionally, the addition of catalase, trypsin and pepsin to CFS had no effect on antimicrobial activity of CFS (Table 3). The inhibitory zone of hydrogen phosphate/

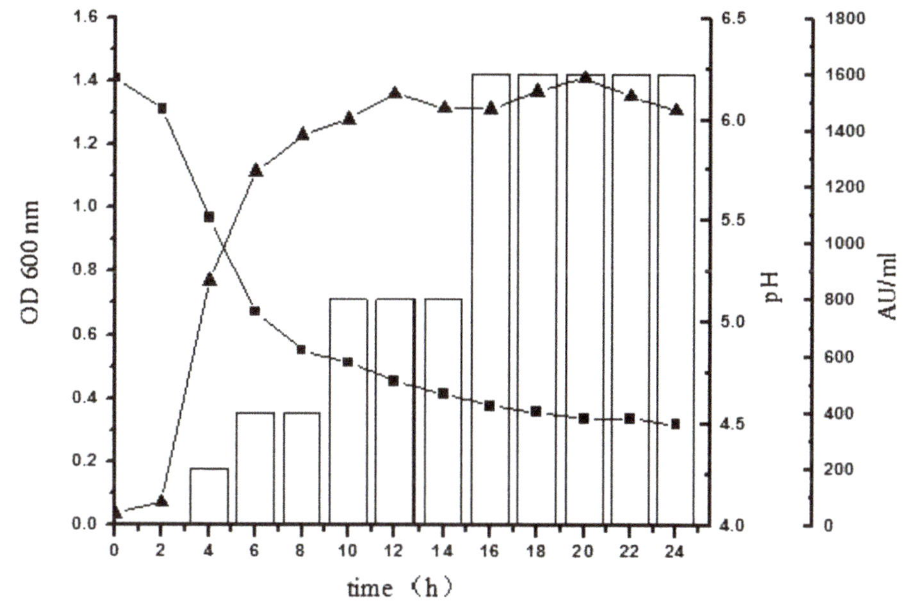

Figure 2 Kinetics growth curves and production of antimicrobial substances by *E. faecium* KQ 2.6. ▲: OD_{600}; ■: pH of the culture medium; black histograms: antimicrobial activity against *Bacillus cereus*.

citric acid buffer was apparently smaller than that of CFS (Figure 3).

Detection of antibiotic resistance and potential virulence factors

Phenotypic results from disk diffusion test demonstrated that *E. faecium* KQ 2.6 was highly susceptible to most of tested antibiotics such as penicillin, chloramphenicol, tetracycline, erythromycin, rifampicin, ofloxacin and ciprofloxacin. However, it was also found that the strain was resistant to vancomycin and polymyxin B (Table 4).

Table 3 Effect of temperature and enzymes on the activity of CFS of *E. faecium* KQ 2.6

Treatments	Antimicrobial activity[a]
Temperature	
40°C for 30 min	+
50°C for 30 min	+
60°C for 30 min	+
70°C for 30 min	+
80°C for 30 min	+
121°C for 20 min	-
Enzymes	
Catalase	+
Trypsin	+
Pepsin	+

[a]+, presence of antimicrobial activity; –, absence of antimicrobial activity; the indicator strain, *Bacillus cereus*

Whether the presence of virulence genes encoding *gelE*, *esp*, *asa1*, *cylA*, *efaA* and *hyl* in the strain was investigated. The results from agarose gel electrophoresis showed that *E. faecium* KQ 2.6 did not harbor virulence genes including *gelE* (213 bp), *esp* (511 bp), *asa1* (328 bp), *cylA* (688 bp), *efaA* (704 bp) and *hyl* (276 bp) (Figure 4).

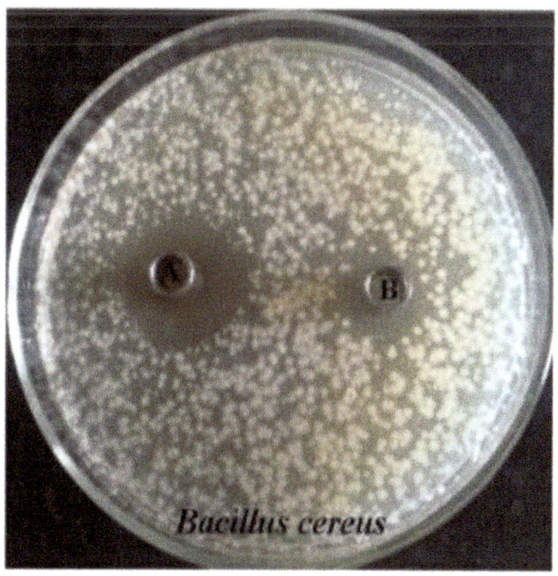

Figure 3 Antimicrobial activity of CFS and buffer against *Bacillus cereus*. A: CFS of *E.faecium* KQ 2.6, B: hydrogen phosphate/citric acid buffer.

Table 4 Antibiotic resistant profile of *E. faecium* KQ 2.6

Antibiotics	Drug concentration per disk (μg)	Susceptibility[a]
Penicillin	10	S
Vancomycin	30	R
Chloramphenicol	30	S
Tetracycline	30	S
Ofloxacin	5	S
Erythromycin	15	S
Rifampicin	5	S
Polymyxin B	30	R
Ciprofloxacin	5	S

[a]The antibiotic resistance was determined by disk diffusion test. The sensitive was analyzed by the recommendation of CLSI (2008). S: sensitive; R: resistant.

Discussion

Enterococci occur in many different environments such as in air, soil, water and the gastrointestinal tract of animals and humans. Due to the association of enterococci with the gastrointestinal tract, it is an ordinary and efficient method to screen enterococci from animal feces. In the last decades, the benefic role of enterococci from animal and human feces in food and animal industries has been well studied [1,23,24]. In this study, twelve isolates were screened from peacock feces, and two of them displayed good antimicrobial properties. The highest antimicrobial activity and gas-negative strain was named as *E. faecium* KQ 2.6.

Antimicrobial activity of *E. faecium* KQ 2.6 was evaluated. The results showed that this strain was able to inhibit gram-positive and gram-negative bacteria. It should be pointed out that many enterococci can produce bacteriocins, which exhibit activity towards gram-positive and gram-negative bacteria [25]. Therefore, the hypothesis that antimicrobial activity of *E. faecium* KQ 2.6 is due to the produced bacteriocin, may be established. However, the activity did not lost after CFS of *E. faecium* KQ 2.6 was treated by proteinase. It demonstrated that the antimicrobial factors were not protein components such as bacteriocin or BLIS. The resistance of CFS to catalase indicated that antimicrobial substance was not hydrogen peroxide. Regarding this phenomenon, some reports have been indicated that the antimicrobial activity may be due to the produced acid [26,27]. Anyogu et al. [28] also indicated that the acid substances produced by *E. faecium* was an important factor to deter the growth and survival of pathogens in the process of submerged cassava fermentation. Therefore, the antimicrobial activity of enterococci in this study may be due to the production of organic acids. Our results showed that the produced acid was not the only factor that contributed to antimicrobial activity of CFS of *E. faecium* KQ 2.6, since the inhibitory zone of CFS was significantly bigger than that of the buffer with similar pH and strength. Thus, we believed that another type of antimicrobial substance should be in CFS of *E. faecium* KQ 2.6.

To study antimicrobial substances of *E. faecium* KQ 2.6 more specially, the heat stability and storability were investigated. The activity could be kept stably after a long time storage or high temperature treatment. It indicated that storage conditions did not lead to the decrease of antimicrobial activity significantly. The high stability of antimicrobial activity can be a good criterion

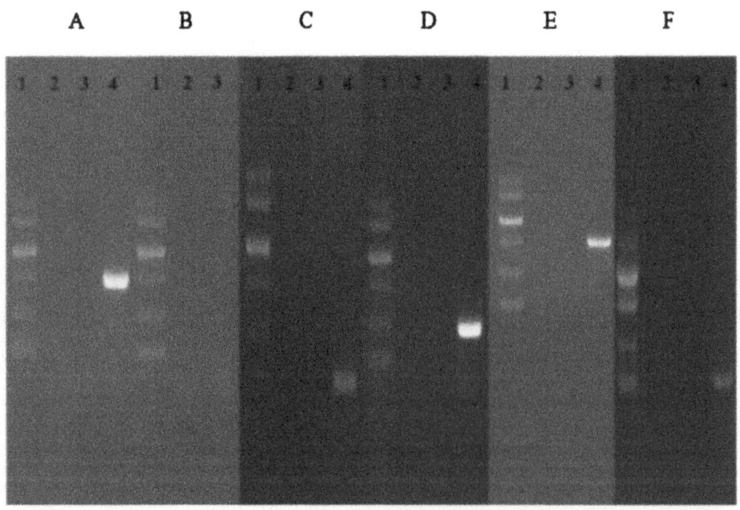

Figure 4 Results of *E. faecium* KQ 2.6 using primers directed against **(A)** 688 bp fragment of the *cylA* gene, **(B)** 510 bp fragment of the *esp* gene, **(C)** 213 bp fragment of the *gelE* gene, **(D)** 375 bp fragment of the *asa1* gene, **(E)** 705 bp fragment of the *efaA* gene and **(F)** 276 bp fragment of the *hyl* gene. Lane 1: standard molecular weight (2000 kb); lane 2: negative control; lane 3: *E. faecium* KQ 2.6; lane 4: positive control (*E. faecalis* ATCC 29212).

for its use as a bio-preservative under complicated conditions of food processing.

The incidence of antibiotic resistance has been received high attention as it is of vital point for the safe use of the strains in foods. It is clear that in hospital environment, multiple antibiotic resistant strains may lead to infections or super-infections. Enterococci are the fourth prevalent strains causing blood infections in European hospital, and the proportion of enterococcal infections continues to increase, mainly because of an increasing number of antibiotic resistant *E. faecium* [29]. In our study, *E. faecium* KQ 2.6 had resistance to vancomycin and polymyxin B. The results indirectly agreed with the study of Messi et al. [30]. Vancomycin-resistance enterococci (VRE) are not restricted to clinical strains, but can be obtained from animal organs and environment. In last few years, the numbers of VRE have been increasing [31]. VRE have brought treatment difficulty, as vancomycin is the last few therapeutic options for enterococcal infections [32,33]. The mechanism of the high resistance to vancomycin is the replacement of the terminal D-Ala of peptidoglycan precursors with D-lactate, which can prevent or destroy the combination between vancomycin and peptidoglycan precursors [34]. Fortunately, *E. faecium* KQ 2.6 was sensitive to the most common antibiotics such as penicillin, tetracycline, chloramphenicol and ciprofloxacin. Therefore, the strain was not multiple antibiotic resistant enterococci.

The investigation of antibiotic resistance alone can't evaluate the safety of enterococci completely. Virulence factors are greatly contributed to enhance infection risks, so potential virulence genes of *E. faecium* KQ 2.6 need to be evaluated. It was reported that the genes encoding adhesion-associated protein were rarely detected in *E. faecium* strain from foods [18]. The absence of full *Cyl* operon in *E. faecium* has also been reported [31]. Our results indicated that this strain did not harbor tested virulence genes *gelE*, *esp*, *asa1*, *cylA*, *efaA* and *hyl*, which was in agreement with the above conclusions. In general, the clinical enterococci harbor more virulence factors than *E. faecium* KQ 2.6.

However, it should be noted that mobile genetic elements like plasmids and transposons, may contribute to the distribution of virulence factors between enterococcci isolated from different sources [17,18]. The virulence genes acquisitions in *E. faecium* have been reported. Clonal complex 17 lineage, a kind of *E. faecium* genetic lineage, can obtain an *esp* gene from other clinical enterococci. And this lineage not only occurs in hospital but also is found in foods [35,36]. Another study indicated that less than 40 % of *E. faecalis* proteins have been found in *E. faecium* draft genome. So, *E. faecium* may harbor additional virulence factors from *E. faecalis* [16]. Furthermore, Sex pheromones or gene transfer pheromones may promote acquisition of virulence genes from other enterococci. Even it is not a common trait that enterococci produce sex pheromones or gene transfer pheromones [18], the work on detecting the presence of sex pheromones or gene transfer pheromones will contribute to assess the safety of the strain.

Conclusion

To our knowledge, this is the first report on the study of *E. faecium* isolated from peacock feces. *E. faecium* KQ 2.6 not only inhibited the growth of gram-positive bacteria, but also had antimicrobial activity towards gram-negative bacteria. The antimicrobial substance was not hydrogen peroxide or protein components. Part inhibitory effect of *E. faecium* KQ 2.6 might be due to the produced acid. Another antimicrobial substance should be in CFS of *E. faecium* KQ 2.6. *E. faecium* KQ 2.6 may be considered safely for its susceptibility to most common antibiotics and absence of the most studied virulence genes. Therefore, this strain has potential to be used as a food preservative in our daily life. However, it should be further evaluated for its ability of virulence genes acquisitions before this strain is applied in the food and/or feed industries.

Abbreviation

E. faecium KQ 2.6: *Enterococcus faecium* KQ 2.6; *E. faecalis*: *Enterococci faecalis*; CFS: Cell free supernatant; BLIS: Bacteriocin like inhibitory substances; LAB: Lactic acid bacteria; VRE: Vancomycin-resistant enterococci; LB: Luria-Bertani broth; MRS: de Man Rogosa Sharpe agar; PDA: Potato Dextrose Agar; CLSI: Clinical and Laboratory Standards Institute.

Competing interests

The authors declare that they have no competing interests.

Authors' contributions

LES participated in the design of the study, carried out the experiments, analyzed the results. WZ participated in the experiments and wrote the manuscript. The rest authors participated in analyzing the results and corrected the manuscript. All authors read and approved the final manuscript.

Acknowledgements

This study was financially supported by the Xinmiao Talent Program of Zhejiang Province (2012R421003, 2013R421006).

Author details

[1]College of Life and Environmental Sciences, Hangzhou Normal University, 310016 Hangzhou, Zhejiang, China. [2]College of Light Industry Science and Engineering, Nanjing Forestry University, 210037 Nanjing, Jiangsu, China.

References

1. Sánchez J, Basanta A, Gómez-Sala B, Herranz C, Cintas LM, Hernández PE. Antimicrobial and safety aspects, and biotechnological potential of bacteriocinogenic enterococci isolated from mallard ducks (*Anas platyrhynchos*). Int J Food Microbiol. 2007;117:295–305.
2. Centeno JA, Menéndez S, Rodríguez-Otero JL. Main microbial flora present as natural starters in Cebreiro raw cow's-milk cheese (Northwest Spain). Int J Food Microbiol. 1996;33:307–13.
3. Parente E, Villani F, Coppola R, Coppola S. A multiple strain starter for water-buffalo Mozzarella cheese manufacture. Lait. 1989;69:271–9.

4. Daeschel MA. Antimicrobial substances from lactic acid bacteria for use as food preservatives. Food Technol. 1989;43:164–7.
5. Ammor S, Tauveron G, Dufour E, Chevallier I. Antibacterial activity of lactic acid bacteria against spoilage and pathogenic bacteria isolated from the same meat small-scale facility: 1—Screening and characterization of the antibacterial compounds. Food Control. 2006;17:454–61.
6. Foulquié Moreno M, Callewaert R, Devreese B, Van Beeumen J, De Vuyst L. Isolation and biochemical characterisation of enterocins produced by enterococci from different sources. J Appl Microbiol. 2003;94:214–29.
7. Klaenhammer TR. Bacteriocins of lactic acid bacteria. Biochimie. 1988;70:337–49.
8. Abriouel H, Lucas R, Ben Omar N, Valdivia E, Maqueda M, Martínez-Cañamero M, et al. Enterocin AS-48RJ: a variant of enterocin AS-48 chromosomally encoded by *Enterococcus faecium* RJ16 isolated from food. Syst Appl Microbiol. 2005;28:383–97.
9. Ghrairi T, Frere J, Berjeaud J, Manai M. Purification and characterisation of bacteriocins produced by *Enterococcus faecium* from Tunisian rigouta cheese. Food Control. 2008;19:162–9.
10. Gutiérrez J, Criado R, Citti R, Martín M, Herranz C, Nes IF, et al. Cloning, production and functional expression of enterocin P, a sec-dependent bacteriocin produced by *Enterococcus faecium* P13, in *Escherichia coli*. Int J Food Microbiol. 2005;103:239–50.
11. Snyder AB, Worobo RW. Chemical and genetic characterization of bacteriocins: antimicrobial peptides for food safety. J Sci Food Agr. 2014;94:28–44.
12. Morrison D, Woodford N, Cookson B. Enterococci as emerging pathogens of humans. J Appl Microbiol. 1997;83:89–99.
13. Omar NB, Castro A, Lucas R, Abriouel H, Yousif NM, Franz CM, et al. Functional and safety aspects of enterococci isolated from different Spanish foods. Syst Appl Microbiol. 2004;27:118–30.
14. Franz CM, Huch M, Abriouel H, Holzapfel W, Gálvez A. Enterococci as probiotics and their implications in food safety. Int J Food Microbiol. 2011;151:125–40.
15. Werner G, Coque T, Hammerum A, Hope R, Hryniewicz W, Johnson A, et al. Emergence and spread of vancomycin resistance among enterococci in Europe. Eurosurveillance. 2008;13:1–11.
16. Ogier JC, Serror P. Safety assessment of dairy microorganisms: *The Enterococcus genus*. Int J Food Microbiol. 2008;126:291–301.
17. Cocconcelli PS, Cattivelli D, Gazzola S. Gene transfer of vancomycin and tetracycline resistances among *Enterococcus faecalis* during cheese and sausage fermentations. Int J Food Microbiol. 2003;88:315–23.
18. Eaton TJ, Gasson MJ. Molecular screening of enterococcus virulence determinants and potential for genetic exchange between food and medical isolates. Appl Environ Microbiol. 2001;67:1628–35.
19. Touré R, Kheadr E, Lacroix C, Moroni O, Fliss I. Production of antibacterial substances by bifidobacterial isolates from infant stool active against *Listeria monocytogenes*. J Appl Microbiol. 2003;95:1058–69.
20. Cheikhyoussef A, Pogori N, Chen H, Tian F, Chen W, Tang J, et al. Antimicrobial activity and partial characterization of bacteriocin-like inhibitory substances (BLIS) produced by *Bifidobacterium infantis* BCRC 14602. Food Control. 2009;20:553–9.
21. Yamamoto Y, Togawa Y, Shimosaka M, Okazaki M. Purification and characterization of a novel bacteriocin produced by *Enterococcus faecalis* strain RJ-11. Appl Environ Microbiol. 2003;69:5746–53.
22. Favaro L, Basaglia M, Casella S, Hue I, Dousset X, Gombossy de Melo Franco BD, et al. Bacteriocinogenic potential and safety evaluation of non-starter *Enterococcus faecium* strains isolated from home made white brine cheese. Food Microbiol. 2014;38:228–39.
23. Jennes W, Dicks L, Verwoerd D. Enterocin 012, a bacteriocin produced by *Enterococcus gallinarum* isolated from the intestinal tract of ostrich. J Appl Microbiol. 2000;88:349–57.
24. Toit MD, Franz C, Dicks L, Holzapfel W. Preliminary characterization of bacteriocins produced by *Enterococcus faecium* and *Enterococcus faecalis* isolated from pig faeces. J Appl Microbiol. 2000;88:482–94.
25. Belguesmia Y, Choiset Y, Prévost H, Dalgalarrondo M, Chobert JM, Drider D. Partial purification and characterization of the mode of action of enterocin S37: a bacteriocin produced by *Enterococcus faecalis* S37 isolated from poultry feces. J Environ Public Health. 2010;2010:1–8.
26. Amoa-Awua WK, Owusu M, Feglo P. Utilization of unfermented cassava flour for the production of an indigenous African fermented food, agbelima. World J Microbiol Biotechnol. 2005;21:1201–7.
27. Mante ES, Sakyi-Dawson E, Amoa-Awua WK. Antimicrobial interactions of microbial species involved in the fermentation of cassava dough into agbelima with particular reference to the inhibitory effect of lactic acid bacteria on enteric pathogens. Int J Food Microbiol. 2003;89:41–50.
28. Anyogu A, Awamaria B, Sutherland J, Ouoba L. Molecular characterisation and antimicrobial activity of bacteria associated with submerged lactic acid cassava fermentation. Food Control. 2014;39:119–27.
29. Bhavnani SM, Drake JA, Forrest A, Deinhart JA, Jones RN, Biedenbach DJ, et al. A nationwide, multicenter, case–control study comparing risk factors, treatment, and outcome for vancomycin-resistant and-susceptible enterococcal bacteremia. Diagn Microbiol Infec Dis. 2000;36:145–58.
30. Messi P, Guerrieri E, De Niederhaeusern S, Sabia C, Bondi M. Vancomycin-Resistant Enterococci (VRE) in meat and environmental samples. Int J Food Microbiol. 2006;107:218–22.
31. Hadji-Sfaxi I, El-Ghaish S, Ahmadova A, Batdorj B, Le Blay-Laliberté G, Barbier G, et al. Antimicrobial activity and safety of use of *Enterococcus faecium* PC4. 1 isolated from Mongol yogurt. Food Control. 2011;22:2020–7.
32. Cetinkaya Y, Falk P, Mayhall CG. Vancomycin-resistant enterococci. Clin Microbiol Rev. 2000;13:686–707.
33. Huycke MM, Sahm DF, Gilmore MS. Multiple-drug resistant enterococci: the nature of the problem and an agenda for the future. Emerg Infect Dis. 1998;4:239–49.
34. Arias CA, Murray BE. The rise of the enterococcus: beyond vancomycin resistance. Nat Rev Microbiol. 2012;10:266–78.
35. López M, Sáenz Y, Rojo-Bezares B, Martínez S, del Campo R, Ruiz-Larrea F, et al. Detection of vanA and vanB2-containing enterococci from food samples in Spain, including *Enterococcus faecium* strains of CC17 and the new singleton ST425. Int J Food Microbiol. 2009;133:172–8.
36. Nallapareddy SR, Singh KV, Okhuysen PC, Murray BE. A functional collagen adhesin gene, *acm*, in clinical isolates of *Enterococcus faecium* correlates with the recent success of this emerging nosocomial pathogen. Infect Immun. 2008;76:4110–9.
37. Vankerckhoven V, Van Autgaerden T, Vael C, Lammens C, Chapelle S, Rossi R, et al. Development of a multiplex PCR for the detection of *asa1*, *gelE*, *cylA*, *esp*, and *hyl* genes in enterococci and survey for virulence determinants among European hospital isolates of *Enterococcus faecium*. J Clin Microbiol. 2004;42:4473–9.

21

Preparation and characterization of nano liposomes of *Orthosiphon stamineus* ethanolic extract in soybean phospholipids

Abdalrahim FA Aisha[1,3], Amin Malik Shah Abdul Majid[2] and Zhari Ismail[1*]

Abstract

Background: *O. stamineus* is a medicinal herb with remarkable pharmacological properties. However, poor solubility of the active principles limits its medicinal value. This study sought to prepare nano liposomes of OS ethanolic extract in unpurified soybean phospholipids in order to improve its solubility and permeability. OS liposomes were prepared by the conventional film method, and were characterized for solubility, entrapment efficiency, Fourier transform infrared spectroscopy (FTIR), transmission electron microscopy (TEM), particle size and zeta potential, release, absorption in everted rat intestinal sacs, and DPPH scavenging effect.

Results: OS liposomes showed substantial enhancement of extract's solubility from 956 ± 34 to 3979 ± 139 µg/ml, with entrapment efficiency of 66.2 ± 0.9%. FTIR study indicates interaction between soybean phospholipids and OS extract. TEM and dynamic light scattering showed presence of round anionic nano liposomes with particle size and zeta potential of 152.5 ± 1.1 nm and −49.8 ± 1.0 mV, respectively. A study using the fluorescent probe pyrene showed the critical micellar concentration is 9.2 ± 2.9 µg/ml. Release studies showed 94 ± 0.1% release in non-formulated extract and 62.4 ± 0.1% in OS liposomes. Released extract from OS liposomes showed improvement in DPPH scavenging effect, IC50 = 23.5 ± 1.1 µg/ml compared to 32.4 ± 0.5 µg/ml in non-formulated extract. OS liposomes were stable at pH 5.5 and 7.4, but showed reversible agglomeration at pH 1.6. Absorption in everted rat intestinal sacs showed substantial improvement in permeability of 3′-hydroxy-5, 6, 7, 4″-tetramethoxyflavone, sinensetin, eupatorin, and 3 other unknown compounds.

Conclusions: Enhanced solubility, absorption and antioxidant effect may improve the overall pharmacological effects and medicinal value of OS ethanolic extract.

Keywords: *Orthosiphon stamineus*, Soybean lecithin, Soybean phospholipids, Liposomal drug delivery system

Background

Orthosiphon stamineus Benth. (Lamiaceae) is a medicinal herb widely distributed in Southeast Asia. Leaves of this plant are commonly used in Southeast Asia and Europe as herbal tea. OS is used as a traditional medicine for treatment of some angiogenesis- related diseases such as rheumatism, tumorous edema, obesity, diabetic retinopathy and psoriasis [1]. Research on OS has increased recently due to its various pharmacological properties such as diuretic and hypouricemic [2], hepatoprotective [3], anti-hyperglycemic [4], antioxidant and antimicrobial [5-7], antiapoptotic [8], antiangiogenic and anticancer effects [1]. Pharmacological effects of OS are attributed to presence of polyphenolics, glycosides, lipophilic flavones [9], rosmarinic acid (RA) and caffeic acid derivatives, triterpenes [10], and diterpens [11-13]. The lipophilic flavones of OS including sinensetin (SIN), eupatorin (EUP) and 3′-hydroxy-5, 6, 7, 4′-tetramethoxyflavone (TMF) (Figure 1) have been given considerable interest as markers of pharmacological activity by several researchers [1,9,14]. However; these lipophilic compounds suffer poor aqueous solubility which limits the OS oral bioavailability and therapeutic applications. The OS ethanolic extract is rich in the lipophilic flavones such as EUP, SIN and TMF while containing very low concentration of proteins and

* Correspondence: zhari@usm.my
[1]Department of Pharmaceutical Chemistry, School of Pharmaceutical Sciences, Universiti Sains Malaysia (USM), Minden 11800, Pulau Pinang, Malaysia
Full list of author information is available at the end of the article

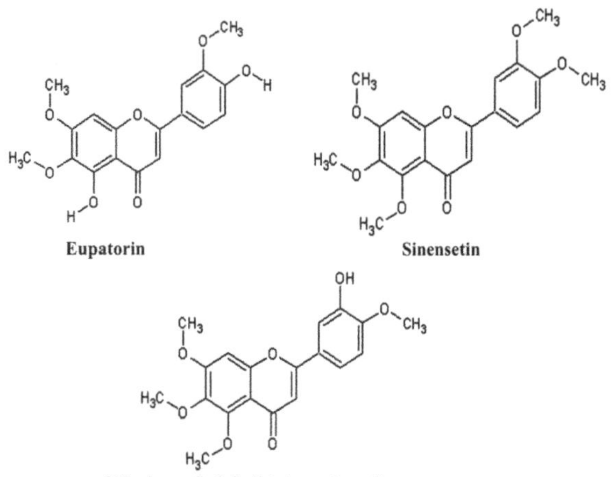

Figure 1 Chemical structure of 3′-hydroxy-5, 6, 7, 4′-tetramethoxyflavone (TMF), sinensetin (SIN) and eupatorin (EUP).

polysaccharides. Therefore, improving solubility of the lipophilic flavones may improve the bioavailability and hence the overall pharmacological activity of OS extracts.

Liposomes are artificial vesicles formed by one or more concentric lipid bilayers separated by water compartments [15]. They have a unique ability of encapsulating hydrophobic, hydrophilic and amphiphilic compounds [16,17]. Drugs with varying hydrophobicities can be encapsulated in the phospholipid bilayer, in the entrapped aqueous core, or at the bilayer interface [18]. Liposomal drug delivery systems have several advantages such as improving solubility, bioavailability and efficacy, reduced toxicity, and increased product stability and patient compliance [15,16,18]. Phospholipids from soybean lecithin are widely used in liposomal drug delivery systems due to their safety, and wide availability at relatively low cost for upscale production. Crude soybean lecithin contains 65–75% phospholipids, together with triglycerides and smaller amounts of other substances such as carbohydrates, pigments, sterols and sterol glycosides [19]. The major phospholipids in soybean lecithin are phosphatidylcholine (PC), phosphatidylethanolamine (PE), and phosphatidylinositol (PI) [20]. Liposomes of herbal extracts or isolated compounds are usually prepared in pure phospholipids [21,22], or in crude phospholipid extracts [17,23]. However, the cost of pure phospholipids is a limiting factor in upscale production of liposomal drug delivery systems particularly those containing herbal extracts or dietary supplements [24]. It was estimated that one kilogram of pure natural phospholipids costs around 980 Euro [17], whereas the cost of crude soybean phospholipids is only 5% of the pure one, which makes unpurified soybean phospholipids a good alternative and attractive choice.

This study sought to produce a cost effective nano liposomes of OS ethanolic extract in crude soybean phospholipids obtained from food grade soybean lecithin, in order to improve the extract's solubility and permeability as major factors for improving oral bioavailability. The liposomes were prepared by the film method and were evaluated for solubility, entrapment efficiency, stability under various pH conditions, critical micellar concentration, particle size and zeta potential, FTIR spectroscopy, TEM, drug release, absorption through everted rat's small intestinal sacs, and free radical scavenging effect.

Results and discussion

Analysis of soybean phospholipids and OS extract

The crude soybean lecithin was found to contain 62 ± 0.2% acetone insoluble phosphatides, and HPLC analysis indicates phosphatidylcholine (PC) is the main phospholipid ingredient in the phospholipids prepared (Figure 2). The percentage of PC (based on peak area) in crude soybean phospholipids, soybean phospholipid extract prepared in ethanol (PH-Et), soybean phospholipid extract prepared in acetone (PH-Ac), and phospholipid fraction prepared by column chromatography (PH-Fr) was 13.4%, 21.6%, 10.6% and 42%, respectively.

Phytochemical analysis of OS extract revealed presence of high content of phenolics (26%) and glycosaponins (17%). UV–vis spectrum showed maximum absorption at 326 and 286 nm (Figure 3), and HPLC analysis (Figure 4) revealed presence of RA (2.2%), EUP (1.7%), SIN (0.23%) and TMF (0.1%).

Optimization of preparation method

Selection of phospholipids to extract ratio

The optimum phospholipids to extract ratio was selected based on the improvement in aqueous solubility of OS extract. At first, the formulations were prepared using PH-Et phospholipid extract at various w/w ratios with OS extract. Solubility of nonformulated OS extract (OS-E) was determined by UV–vis spectrophotometry and was found to be 956 ± 34 μg/ml. The formulations were prepared by mixing ethanolic solution of the extract with chloroform solution of phospholipids followed by evaporation of solvents and hydration of the resulting film. Aqueous solubility results showed significant improvement in formulations prepared at phospholipid: extract ratio of 50:50, 150:50 and 100:50 (Table 1). The formulation prepared at 100:50 w/w ratio showed the highest improvement in solubility; therefore this formulation was selected for further optimization.

Selection of solvent to dissolve the OS extract

At the beginning ethanol was selected to dissolve the extract since OS ethanolic extract is used in this study. However, solubility results of formulated extract were not very encouraging. Therefore, methanol was used to

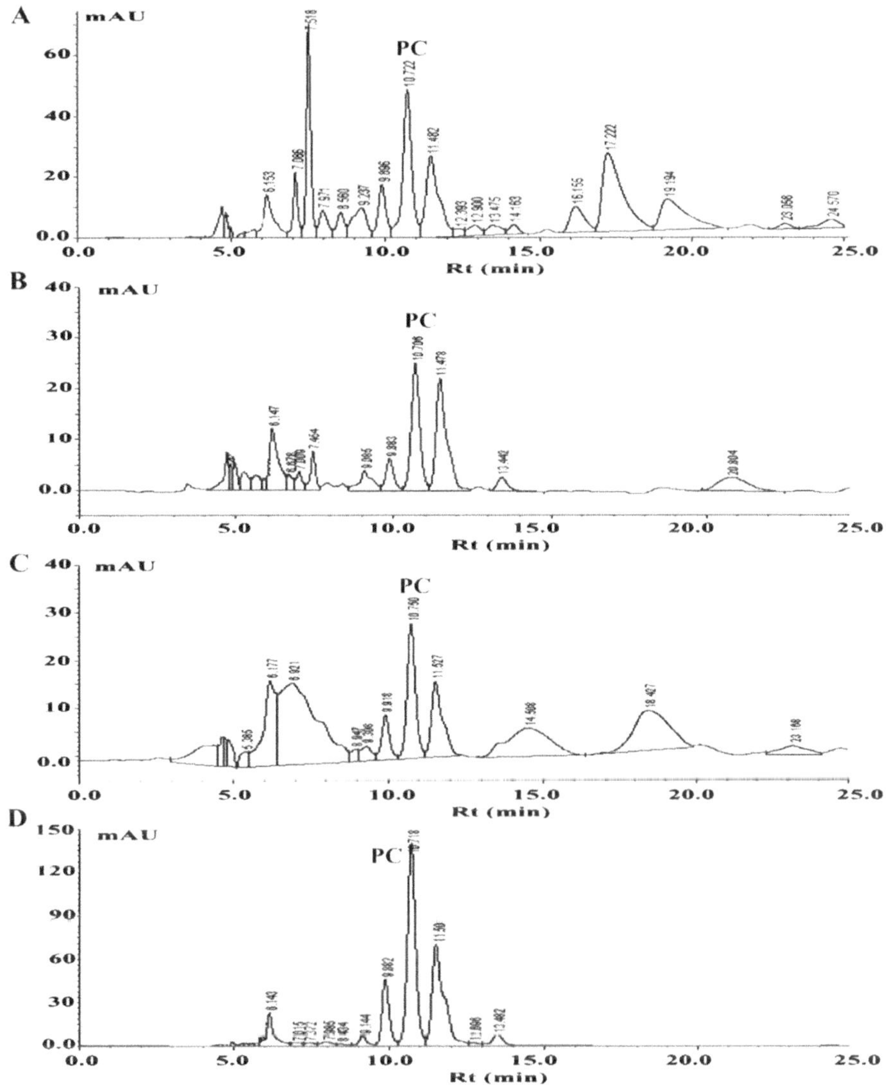

Figure 2 HPLC chromatograms of different types of soybean phospholipids. Crude soybean phospholipids **(A)**, soybean phospholipid extract prepared in ethanol (PH-Et) **(B)**, soybean phospholipid extract prepared in acetone (PH-Ac) **(C)**, and phospholipid fraction prepared by column chromatography (PH-Fr) **(D)**.

dissolve the extract. Aqueous solubility results showed substantial improvement of OS extract from 1402 ± 66 μg/ml (ethanol) to 3979 ± 139 μg/ml (methanol) (P = 0.000). A possible explanation of these results is that chloroform evaporates (boiling point 61.2°C) at higher rate than ethanol (bp, 78.4°C), leaving behind an ethanolic solution of the extract and causing precipitation of phospholipids, leading to phase separation and the formation of a non-homogeneous mixture of extract and phospholipids. On the other hand, chloroform and methanol evaporate at almost the same rate due to the narrow difference between their bp (bp of methanol = 65°C) leaving behind a homogeneous film of the extract and phospholipids. Consequently, this made the water dispersion of the mixture much easier and lead to substantial improvement in solubility.

Selection of phospholipids extract

Three formulation types were prepared in 3 phospholipid extracts including PH-Et, PH-Ac, and PH-Fr. The formulations were prepared at phospholipid: extract ratio of 100:50, and using methanol to dissolve the extract. Aqueous solubility was determined by UV–vis spectrophotometry, and the highest solubility (3862 ± 155 μg/ml) was obtained in the formulation prepared in PH-Et phospholipid extract. Formulation prepared in PH-Ac phospholipid extract showed a solubility of 2997 ± 367 μg/ml, and the lowest solubility was obtained when PH-Fr phospholipid fraction (highest in PC content) was used (2082 ± 62 μg/ml) (P = 0.000). Therefore, the final formulation was prepared in PH-Et phospholipid extract that was used in all subsequent analysis.

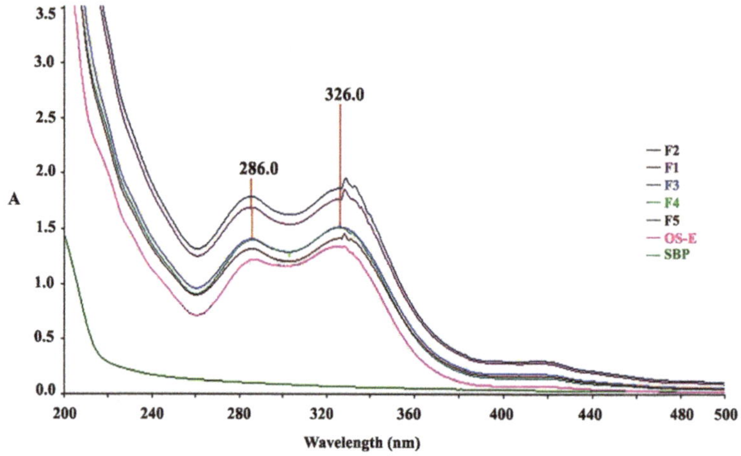

Figure 3 UV-vis spectrophotometry of various formulations of OS ethanolic extract prepared in soybean phospholipids. The OS-E extract showed maximum absorption at 286 nm and 326 nm in both formulated and non-formulated extract. OS-E refers to ethanolic extract, SBP refers to soybean phospholipid, and F1-F5 refers to different formulations prepared at the following phospholipids to extract (mg/mg) ratios: 150/50, 100/50, 50/50, 25/50 and 25/75, respectively. **(A)** refers to absorbance.

HPLC analysis of OS liposomes

HPLC analysis of 4 marker compounds and 4 unknown compounds revealed significant improvement in their aqueous solubility compared to that obtained in non-formulated extract (P values = 0.000, Student's t-test) (Table 2).

Fourier transform infrared spectroscopy

FTIR spectra of soybean phospholipid, OS extract, and OS liposomes were studied in order to get insights into occurrence of interaction between OS extract and phospholipids (Figure 5). In PH-Et phospholipids extract the broad band centered at 3387 cm^{-1} represents the OH stretching, the principal bands at 2854 cm^{-1} and 2924 cm^{-1} correspond to the symmetric and anti-symmetric stretching in the CH_2 groups of alkyl chains, the strong band centered at 1741 cm^{-1} corresponds to the stretching vibrations of the ester carbonyl groups, the band centered at 1650 cm^{-1} is assigned to C = O stretching, and the scissoring vibrations of the CH_2 groups are represented by the band at 1465 cm^{-1}. The characteristic phosphate group vibrational band assigned to the PO_2− anti-symmetric stretching mode is centered at 1232 cm^{-1} and the PO_2− symmetric stretching mode PO_2− at 1075 cm^{-1} , these results are similar to previously reported results [25]. In OS extract, the broad band centered at 3385 cm^{-1} corresponds to OH stretching, vibrations at 2924 cm^{-1} and 2854 cm^{-1} correspond to C-H stretching, the bands centered at 1695 cm^{-1} and 1607 cm^{-1} correspond to stretching vibrations of carbonyl groups, the bands at 1600–1420 cm^{-1} correspond to phenyl groups, the bands at 1267 cm^{-1} corresponds to ester carbonyl groups, and the band at 1060 cm^{-1} corresponds to primary OH groups. These results are consistent with previous reports [26].

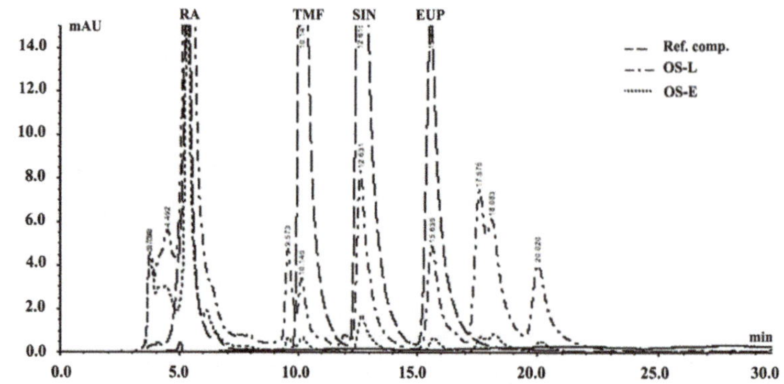

Figure 4 HPLC chromatograms of OS ethanolic extract in formulated and non-formulated forms. The chromatograms show presence of RA, TMF, EUP and SIN; the solubility of these compounds are improved significantly in the formulated OS extract compared to non-formulated extract. OS-E refers to *O. stamineus* extract, and OS-L refers to *O. stamineus* liposomes. RA, TMF, SIN and EUP refer to rosmarinic acid, 3'-hydroxy-5, 6, 7, 4'-tetramethoxyflavone, sinensetin and eupatorin, respectively.

Table 1 Aqueous solubility of various OS liposomes

No.	PH-Et (mg)	OS-E (mg)	Solubility (μg/ml)	*P* value
F1	150.0	50.0	1296.5 ± 18	0.000
F2	100.0	50.0	1401.7 ± 66	0.000
F3	50.0	50.0	1188.8 ± 169	0.001
F4	25.0	50.0	1018.8 ± 15	0.526
F5	25.0	75.0	1072.6 ± 6	0.084

Results are shown as AV ± SD, and the P value was calculated by One way ANOVA with Dunnett's Post Hoc test versus extract's solubility in water (n = 3).

Remarkable changes were observed in the infrared absorption spectra as a result of incorporation of OS extract in phospholipids; the broad band corresponding to OH groups is shifted from 3387 cm^{-1} to 3421 cm^{-1}, the C = O bands at 1695 cm^{-1} and 1607 cm^{-1} disappeared, the band corresponding to C = O stretching in phospholipids is shifted from 1650 cm^{-1} to 1634 cm^{-1}, the band at 1267 cm^{-1} is shifted to lower frequency (1225 cm^{-1}) with much lower oscillation strength, the band at 1232 cm^{-1} disappeared, and band at 1075 cm^{-1} is shifted to 1071 cm^{-1}. The most pronounced spectral changes can be observed in the spectral region corresponding to the polar heads of phospholipids. Such changes may represent hydrogen bonding between the polar heads of phospholipids with the hydroxyl groups in OS extract. Hydrogen bonding can also occur between the keto groups of flavonoids and oxygen groups of phospholipids. Hydrophobic interaction may also occur between the flavone's methoxy groups and the phospholipid tails.

Effect of pH on stability of liposomes

Stability of the OS liposomes at pH 1.6, 5.5 and 7.4 was studied in order to predict their stability in the gastrointestinal tract. Agglomeration and precipitation of OS liposomes occurred immediately after mixing with PBS at pH 1.6, but not at pH 5.5, or 7.4 or water. The percentage of the soluble fraction (stable), relative to that in water, was 66 ± 1.0% (pH 1.6), 94 ± 1.0% (pH 5.5) and 93 ± 3.0% (pH 7.4). These results indicate that OS liposomes are highly stable at pH 5.5 and 7.4, but relatively unstable at pH 1.6.

Table 2 HPLC analysis of 4 marker and 4 unknown compounds in OS extract (OS-E), and OS liposomes (OS-L)

Compounds	RT (min)	OS-E (mAU)	OS-L (mAU)
RA	5.3 ± 0.01	301.5 ± 1.3	1933.9 ± 7.7
TMF	10.2 ± 0.03	10.1 ± 1.7	67.5 ± 0.4
SIN	12.7 ± 0.02	49.5 ± 1.4	254.8 ± 1.0
EUP	15.7 ± 0.03	19.1 ± 1.6	148.8 ± 19.9
5	9.6 ± 0.01	10.4 ± 1.3	85.8 ± 0.8
6	17.6 ± 0.07	13.0 ± 1.3	194.5 ± 0.4
7	18.2 ± 0.08	14.8 ± 1.2	223.4 ± 16.0
8	20.1 ± 0.05	18.1 ± 0.1	145.5 ± 25.9

Results are shown as average peak area (mAU) ± SD (n = 3).

In order to verify whether the agglomeration observed at pH 1.6 is reversible, the precipitate obtained at this pH was re-dispersed in PBS at pH 1.6, 5.5 and 7.4. The precipitate was totally dissolved at pH 5.5 and 7.4 but not at pH 1.6. These findings indicate that the agglomeration observed at pH 1.6 is reversible at pH 5.5 and 7.4. Under GIT pH, the OS liposomes are expected to agglomerate under gastric conditions and re-dissolve under intestinal pH. It is noteworthy that soybean phospholipids and OS extract separately did not show any precipitate at pH 1.6.

Entrapment efficiency

OS liposomes are unstable at pH 1.6, a property which was used to study the entrapment efficiency of OS liposomes. Entrapment efficiency was found to be 66.2 ± 0.9% (n = 3). This value is considered satisfactory when compared to published data on liposomal drug delivery systems of herbal extracts or compounds from herbal origin [15]. Entrapment efficiency of RA, TMF, SIN and EUP was 54.1 ± 0.2%, 28.2 ± 0.2%, 43 ± 0.2% and 38.3 ± 5.1% respectively. The average entrapment efficiency (EE) of the known fraction is almost 41%, as measured by HPLC. The known fraction of the extract (including TMF, RA, EUP and SIN) is only 4% of the total components, and the exact EE of the unknown fraction is unknown and cannot be measured accurately (theoretically, it can be higher than 66% for some compounds and lower for others). These two points may help to explain the difference in EE observed. Basically, the lipophilic compounds have poor solubility in water, and they have higher tendency to be entrapped in the lipid bilayers of liposomes. Whereas the hydrophilic drugs may be entrapped inside the aqueous cores of liposomes, or located in the external water phase [16]. It is noteworthy that the unentrapped extract can be removed by centrifugation after acid treatment of the liposomes, or by dialysis method without pretreatment, or can be kept in the formulation. In this study the free extract was kept in the formulation in order to avoid extract loss, and reduce the production expanses.

The n-octanol: water partition

When a mixture of hydrophilic and hydrophobic compounds is partitioned between n-octanol and water, the hydrophilic compounds are distributed in water and the hydrophobic compounds are concentrated in n-octanol layer. Since OS liposomes have improved the extract's solubility, it is expected that the OS extract will be more concentrated in the aqueous layer. Our results show that 40 ± 1.0% of the non-formulated extract is distributed in the aqueous layer which indicates relatively low hydrophilicity of the OS extract. On the other hand, this ratio

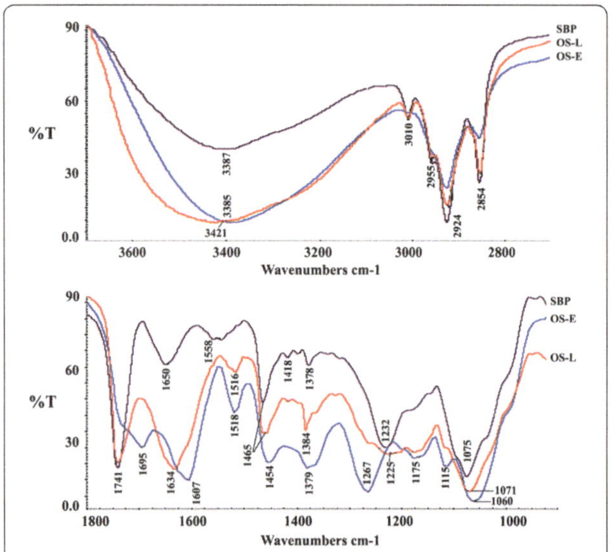

Figure 5 FTIR spectra of *O. stamineus* ethanolic extract (OS-E), soybean phospholipids (SBP), and liposomes (OS-L). Remarkable changes can be seen in the infrared absorption spectra due to incorporation of OS extract in phospholipids; the broad band corresponding to OH groups is shifted from 3387 cm^{-1} to 3421 cm^{-1}, the C = O bands at 1695 cm^{-1} and 1607 cm^{-1} disappeared, the band corresponding to C = O stretching in phospholipids is shifted from 1650 cm^{-1} to 1634 cm^{-1}, the band at 1267 cm^{-1} is shifted to 1225 cm^{-1}, the band at 1232 cm^{-1} disappeared, and band at 1075 cm^{-1} is shifted to 1071 cm^{-1}.

is increased to 64 ± 1.0% in the OS liposomal formulation, which indicates increased extract's hydrophilicity.

Particle size analysis and transmission electron microscopy (TEM)

Particle size measurement by Photon Correlation Spectroscopy (PCS) indicates presence of structures with dynamic diameter of 153–177 nm with narrow size distribution. Zeta potential measurements showed presence of anionic liposomes with a surface charge from −40 to −49 mV (Table 3). The presence and morphology of liposomes were further confirmed by TEM which verified the presence of round liposomes of <200 nm. Furthermore, the lipid bilayer of the OS liposomes can be seen clearly at high magnification (Figure 6).

Critical micellar concentration

The critical micellar concentration (CMC) was determined using the fluorescent probe pyrene. The results showed that addition of pyrene to aqueous solution containing OS liposomes caused substantial quenching of pyrene's fluorescence in a dose dependent manner (Figure 7A). Increasing liposomes concentration also caused a redshift in emission intensity, and the intensity ratio I384/I375 of pyrene was increased accordingly. The CMC was estimated from the graph (Figure 7B) and was found to be 9.2 ± 2.9 μg/ml.

In vitro release and antioxidant effect

Despite of the immediate dissolution of OS liposomes in water, the percentage cumulative release was 62% after 24 h at 37°C and pH 6.8. On the other hand, the percentage cumulative release of non-formulated extract under the same conditions was 94% (Figure 8). This result gives more insights into stability of the liposomal structures under the test conditions, most likely due to the interaction between OS extract and the phospholipid carrier, and consequently leading slow release of the active principles. The released extract was then studied for DPPH scavenging effect. The results showed potent scavenging of DPPH in both OS liposomes and non-formulated extract; however with stronger effect obtained in OS liposomes (IC_{50} = 23.5 ± 1.1 μg/ml) than in non-formulated extract (IC_{50} = 32.4 ± 0.5 μg/ml), P = 0.000.

Absorption through everted intestinal sacs

Absorption studies were carried out using everted rat small intestinal sacs under physiologic conditions (Tyrodes solution, pH 7.4 and 37°C and aeration). Intestinal sacs were recovered after 1 h incubation, and the contents were analyzed by HPLC. Significant improvement in absorption of 7 compounds was achieved, with a minor reduction in absorption of rosmarinic acid (RA). The largest improvement was obtained in eupatorin (EUP) which showed 9.8 ± 0.60 folds increment in the compound's absorption (Table 4). These results indicate higher intestinal permeability of liposomal extract than non-formulated extract, which can be explained due to encapsulation in phospholipids, increased solubility, presence of nanoscale liposomes, increased negativity and colloidal stability [27]. Previous research on anionic liposomes has shown enhancement of colloidal stability and absorption [23], and even with higher cellular uptake rate than neutral and cationic liposomes [28,29].

Conclusions

Collectively, the data presented in this study provide evidence about feasibility of using unpurified soybean phospholipids for preparation of nano liposomes of *O.*

Table 3 Analysis of particle size and zeta potential by PCS

Items	Particle diameter (nm)	Intensity (%)	Count rate (Kcps)	Polydispersity index (PdI)	Zeta potential (mV)
SBP	177.2 ± 1.2	100.0 ± 0.1	543.8 ± 2.8	0.236 ± 0.004	−40.0 ± 2.2
OS-L	152.5 ± 1.1	99.0 ± 1.7	307.0 ± 17.3	0.233 ± 0.011	−49.8 ± 1.0

Results are displayed as AV ± SD (n = 3).

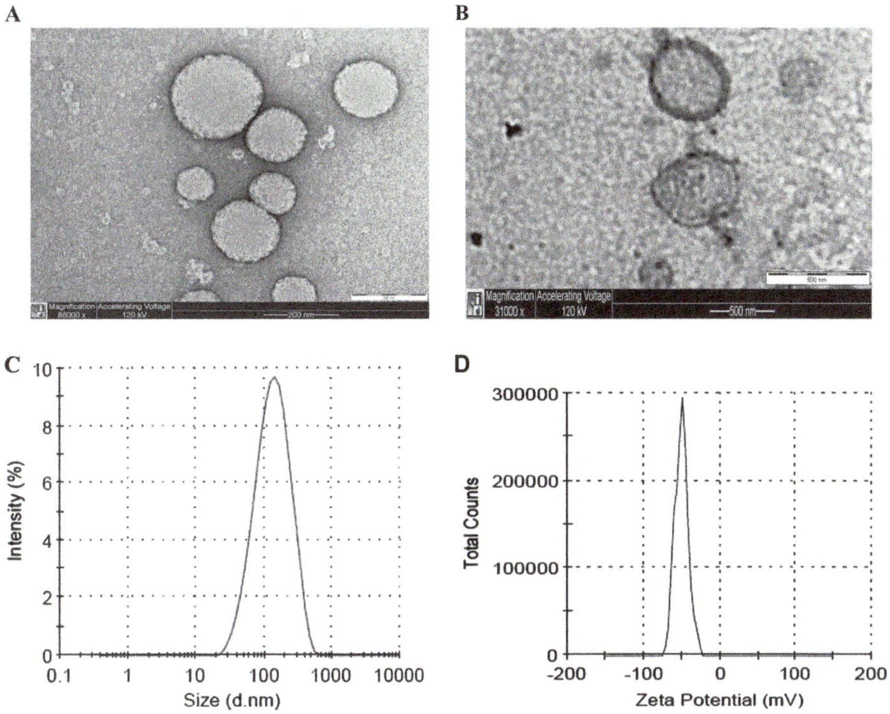

Figure 6 Morphology and particle size analysis of the OS liposomes. Transmission Electron Microscopy (TEM) photograph of OS liposomes **(A)** and their lipid bilayer **(B)**, particle size distribution **(C)**, and zeta potential distribution **(D)**. This figure confirms the presence of round liposomes with narrow size distribution (<200 nm) and a negative surface charge.

stamineus ethanolic extract. The OS liposomes demonstrated significant enhancement of intestinal absorption of the active principles, which can be attributed to the improvement in aqueous solubility and permeability, presence of nanoscale anionic liposomes, and colloidal stability of OS liposomes. The sustained release profile along with the improvement in free radical scavenging effect

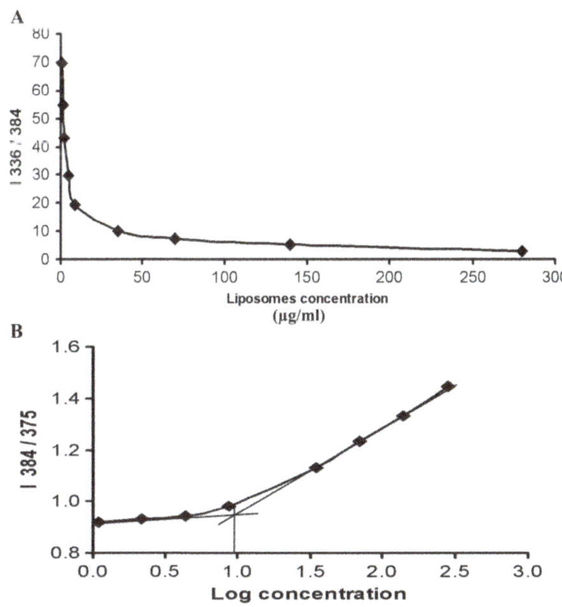

Figure 7 Dose dependent quenching of pyrene fluorescence in OS liposomes. Addition of pyrene to aqueous solution containing OS liposomes caused substantial quenching of pyrene's fluorescence in a dose dependent manner **(A)**; increasing liposomes concentration caused a redshift in emission intensity and increased the I384/I375 intensity ratio **(B)**. The critical micellar concentration (CMC) is estimated from graph B and was found to be 9.2 ± 2.9 µg/ml.

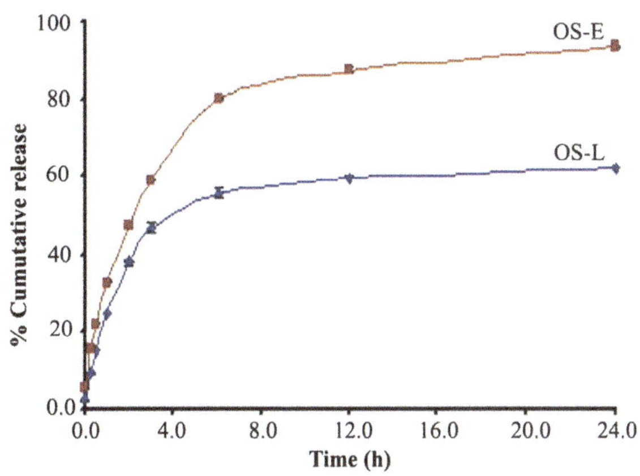

Figure 8 Percentage cumulative release of OS liposomes (OS-L) and OS non-formulated extract (OS-E) at pH 6.8 and 37°C after 24 h. Concentration of released extract was determined by UV–vis spectrophotometry. The percentage cumulative release in non-formulated extract was 94% and reduced to 62% in the formulated extract.

Table 4 Absorption through everted intestinal sacs after 1 h incubation

Compounds	RT (min)	OS-L/OS-E ratio
Unknown 1	4.9 ± 0.04	3.0 ± 0.40
RA	5.3 ± 0.06	0.9 ± 0.01
TMF	10.0 ± 0.16	4.0 ± 0.70
SIN	12.3 ± 0.22	2.3 ± 0.20
EUP	15.2 ± 0.32	9.8 ± 0.60
Unknown 2	17.6 ± 0.37	3.6 ± 0.10
Unknown 3	19.4 ± 0.44	3.4 ± 0.20

Results are presented as the AV ± SD of absorption ratio of OS liposomes (OS-L) to non-formulated extract (OS-E) (n = 3–4).

may provide a prolonged protection effect against oxidative stress related diseases. Therefore, this study may provide a basic formulation of *Orthosiphon stamineus* ethanolic extract for preparation of oral drug delivery system, and possibly topical delivery systems to the skin.

Methods

Preparation and standardization of *O. stamineus* ethanolic extract

OS dried leaves were obtained from specialized supplier of herbal products in Malaysia (Herbagus Trading, Kepala Batas, Pulau Pinang). OS ethanolic extract was prepared by the maceration method as the following; 100 g of OS powder was added to 1 L of 96% ethanol, mixed continuously on a magnetic stirrer for 48 h, filtered, concentrated at 60°C by rotavapor and further freeze-dried to 10.6 g of solid material. Total glycosaponins content of the extract was analyzed as previously described [30], and total phenolics content was estimated as previously described by Aisha et al., [31]. Concentration of 4 marker compounds including rosmarinic acid (RA), sinensetin (SIN), eupatorin (EUP), and 3′-hydroxy-5,6,7,4′-tetramethoxyflavone (TMF) was determined by reverse phase HPLC as previously described [14]; analysis was carried out using Agilent 1100 HPLC system, using Nucleosil C18 column (250 mm × 4.6 mm, 5 μm), column temperature was 25°C, injection volume was 20 μl, the mobile phase was isocratic which consisted of methanol (55%): tetrahydrofuran (5%): 0.1% H_3PO_4 (40%). Flow rate was 0.7 ml/min, separation time 25 min, and wavelength was 330 nm.

UV–vis spectrophotometry was carried out in the wavelength range 500–200 nm using Perkin Elmer UV spectrophotometer.

Preparation of soybean phospholipids

The unpurified soybean phospholipids were prepared from food grade soybean lecithin as previously described with minor modifications [32]. In brief; crude lecithin (500 g) was refluxed in 96% ethanol (2.5 L) for 30 min and cooled to RT. Subsequently, the supernatant was collected by decantation and concentrated at 60°C using rotavapor. The residue was then washed 5× with acetone (1 L) to give 50 g of semi solid material; this phospholipid was called PH-Et. In another experiment, 500 g of the food grade lecithin was refluxed for 30 min in acetone (2.5 L), cooled down to RT, and the precipitate was saved and further washed with acetone (5×) to give 220 g and named PH-Ac. In addition, a fraction containing higher concentration of phosphatidylcholine was prepared as mentioned previously [32]; this phospholipid fraction was named PH-Fr.

The crude lecithin was characterized by measuring concentration of the acetone insoluble phosphatides and by RP-HPLC [20,32]. The soybean phospholipid fractions were analyzed by RP-HPLC as previously described with some modifications [32]; Agilent 1100 HPLC system was used, and separation was achieved on Agilent Eclipse C18 column (250 mm × 4.6 mm, 5 μm) at 35°C, the mobile phase was isocratic which consisted of isopropyl alcohol, methanol and water (70:22:8 v/v) at 0.5 ml/min. The samples were prepared in methanol at 1 mg/ml and filtered through 0.45-μm syringe filter, injection volume was 10 μl, and detection was done at 205 nm.

Preparation of OS liposomes

Liposomes of OS extract (OS-L) were prepared by the film method as the following; soybean phospholipids was dissolved in chloroform and OS extract was dissolved in ethanol or methanol, the solutions were mixed, and the solvent was evaporated under vacuum using rotary evaporator at 45°C for 30 min, followed by drying in oven at 60°C for 1 h.

Determination of aqueous solubility

Solubility was evaluated by UV–vis spectrophotometry; OS extract, and OS liposomes were dissolved at theoretical concentration of 4 mg/ml in deionized water, vortexed for 2 min, and sonicated for 10 min. Subsequently, the solutions were centrifuged at 8000 rpm and 25°C for 10 min. Concentration of OS extract in the supernatant was determined at 286 nm. A calibration curve was prepared at the same time in order to calculate concentration of soluble OS extract ($y = 0.0135x - 0.0423$, $R2 = 0.9999$). All experiments were carried out in triplicates. Concentration of 4 marker and 4 unknown compounds was also determined by HPLC as described previously.

Effect of pH on liposomes stability

Effect of pH on liposome's stability was studied as previously described with some modifications [33]. OS liposomes were dissolved in water at 4 mg/ml and further diluted at 1:4 ratio in phosphate buffered saline (PBS) at pH 1.6, 5.5, 7.4 and water, and incubated for overnight (16 h) at 37°C. Subsequently, the solutions were centrifuged at 8000 rpm and 25°C for 10 min, and concentration of OS extract in the supernatant was determined by UV spectrophotometry at 286 nm. The precipitate obtained at pH 1.6 was resuspended in PBS (pH 1.6, 5.5 and 7.4), vortexed for 5 min, centrifuged, and the supernatant was analyzed by UV spectrophotometry. The results are presented as percentage of soluble fraction relative to liposomes diluted in water.

Determination of entrapment efficiency

Entrapment efficiency was determined by precipitation of OS liposomes at pH 1.6. It is noteworthy that this pH caused precipitation of OS liposomes, but not the free extract. The OS liposomes was dissolved in water at 4 mg/ml, sonicated for 10 min, and centrifuged at 8000 rpm for 10 min. The supernatant (1 ml) was added to the same volume of PBS at pH 1.6, mixed thoroughly by vortex, centrifuged at 8000 rpm for 10 min, the supernatant containing the free extract was decanted, the precipitate was washed with PBS (pH 1.6) and resuspended in 1 ml PBS (pH 7.4). After sonication for 10 min, the extract content was determined by UV spectrophotometry at 286 nm. Concentration of RA, TMF, SIN and EUP was measured by HPLC as described previously [14]. The entrapment efficiency was calculated as the following:

$$\text{Entrapment efficiency} = (\text{concentration in precipitate} / \text{theoretical concentration}) \times 100 \ (n = 3).$$

Partition between n-octanol and water

OS extract and OS liposomes were prepared in deionized water at 1 mg/ml and mixed with the same volume of n-octanol. The mixtures were then mixed by several manual inversions (30 times), allowed to settle for 4 h, and centrifuged at 5000 rpm for 5 min to allow separation of the 2 layers. Concentration of OS extract in both layers was determined by UV–vis spectrophotometry at 286 nm. The results are presented as the ratio of extract concentration in aqueous phase relative to that in n-octanol.

Fourier transform infrared spectroscopy

FTIR analysis of OS extract, soybean phospholipids and OS liposomes was carried out using Spectrum 400 spectrometer (Perkin Elmer, USA). The IR spectra were recorded in the range of 4000 – 400 cm^{-1} (n = 3).

Measurement of particle size and zeta potential

Particle size, polydispersity index (PDI) and zeta potential (ζ) were determined by Photon Correlation Spectroscopy (PCS) using a Zetasizer nano zs (Malvern Instruments Ltd, UK). The samples were dissolved in ultra pure water (18 MΩ) at 1 mg/ml and filtered through 0.45-μm syringe filters to remove any insoluble matter. Measurements were carried out in triplicates.

Determination of critical micellar concentration

Critical micellar concentration (CMC) of OS liposomes was estimated using the fluorescent probe pyrene [33]. Pyrene was dissolved in dimethylsulfoxide (DMSO) at 10 mM and diluted in PBS (pH 6.8) to 10 μM. Concentration of liposomes was maintained in the range 1–280 μg/ml in PBS. Pyrene was added at a final concentration of 0.1 μM and incubated for 30 min in the dark at RT. Fluorescence intensity was measured at excitation of 336 nm and emission of 375 and 384 nm by LS 45 fluorescence spectrometer (Perkin Elmer, USA). The intensity ratio (I384/I375) was calculated and plotted versus log concentration. The resulting curve was used to calculate the CMC (n = 3).

Transmission electron microscopy

Studies in transmission electron microscopy (TEM) were carried out in order to confirm presence of the liposomal structures. One drop of OS liposomes (1 mg/ml in water) was deposited on a 400 mesh copper grid coated with 5 nm layer of carbon, air-dried at RT for 3 min, and stained with 2% uranyl acetate for 1 min. The samples were dried and studied using CM12 TEM (Philips, Netherlands).

In vitro release and DPPH scavenging effect

In vitro release study was performed by dialysis method using dialysis bags with molecular weight cut off value 8200 Dalton. The experiment was carried out using magnetic stirrer in PBS at pH 6.8, and 37°C with continuous stirring at 100 rpm. Briefly, 10 ml of OS liposomes or non-formulated OS extract (in water) containing 10 mg of extract was filled in dialysis bag. The bags were hermetically sealed and kept in the receiver compartment containing 200 ml of same medium. Samples (3 ml) were collected at 0, 0.25, 0.5, 1.0, 2.0, 3.0, 6.0, 12, and 24 h, and immediately replaced with 3 ml fresh medium. Concentration of extract was determined spectrophotometrically at 286 nm. The results are presented as average percentage of cumulative release ± SD (n = 3). DPPH scavenging effect of released extracts was then investigated as described previously [31]; DPPH (9 mg/100 ml) was added to the same volume of released extract, incubated at RT for 30 min, and absorbance was measured at 516 nm. Median inhibitory concentration (IC_{50}) was then calculated from the dose response curves (n = 3).

Absorption through the everted rat intestine

Absorption through everted rat intestinal sacs was carried out as described previously with some modifications [27]. Overnight fasted male Sprague–Dawley rats were anesthetized by diethyl ether and euthanized by cervical dislocation. The first two thirds of the small intestine were collected immediately after euthanasia, everted gently using a glass rod, washed thoroughly, and were kept in Tyrodes solution (NaCl, 8 g; KCl, 0.2 g; $NaHCO_3$, 1 g; $CaCl_2$, 0.2 g; $MgCl_2.6H_2O$, 0.1 g; NaH_2PO_4, 0.05 g; and glucose, 1 g dissolved in 1 L deionized water) at 37°C with aeration. The tissues were cut into 6-cm length, sealed at one end using surgical thread and filled with 1 ml aerated Tyrodes solution. The intestinal sacs were kept for 1 h in a medium containing OS liposomes or OS non-formulated extract (120 μg/ml) in Tyrodes solution at 37°C with continuous agitation at 100 rpm using magnetic stirrer and with continuous aeration with air. Subsequently, sacs were removed, washed thoroughly from exterior with water, the content was collected and centrifuged at 10000 rpm for 5 min, and the supernatant was analyzed by HPLC [14]. Peak area of the marker compounds was recorded, and the fold change in absorption was calculated by dividing peak area of compounds in OS liposomes by that of non-formulated extract (n = 3–4).

Animals

Male Sprague–Dawley rats were obtained from USM animal breeding facility, and were allowed to acclimatize for one week before the experiment. Experiment was performed according to the guidelines of USM Animal Ethics Committee (Ref. No.: USM/Animal Ethics Approval/ 2012/ (78) (399)).

Statistical analysis

Results are presented as average ± SD of triplicate experiments unless otherwise mentioned. Differences between groups were considered significant at $P < 0.05$ using Student's t-test or One way ANOVA.

Abbreviations

OS: *Orthosiphon stamineus*; FTIR: Fourier transform infrared spectroscopy; TEM: Transmission electron microscopy; DPPH: 2,2-Diphenyl-1-picrylhydrazyl; RA: Rosmarinic acid; SIN: Sinensetin; EUP: Eupatorin; TMF: 3′-hydroxy-5,6,7, 4′-tetramethoxyflavone; PC: Phosphatidylcholine; PI: Phosphatidylinositol; PE: Phosphatidylethanolamine; PH-Et: Soybean phospholipid extract prepared in ethanol; PH-Ac: Soybean phospholipid extract prepared in acetone; PH-Fr: Phospholipid fraction prepared by column chromatography; UV–vis: Ultraviolet–visible; RP-HPLC: Reverse phase high performance liquid chromatography; PBS: Phosphate buffered saline; OS-E: *Orthosiphon stamineus* ethanolic extract; OS-L: *Orthosiphon stamineus* liposomes; PCS: Photon correlation spectroscopy; CMC: Critical micellar concentration; PDI: Polydispersity index; RT: Room temperature; SD: Standard deviation; IC_{50}: Median inhibitory concentration.

Competing interests

The authors declare that no competing interests exist.

Authors' contributions

AFAA developed the concept, designed and carried out experimental work, analyzed the results, and prepared the manuscript. ZI and AMSAM developed the concept and revised the manuscript in its final version. All authors read and approved the final manuscript.

Acknowledgements

This project was funded by the Malaysian Ministry of Agriculture and Agro-Based Industry, NKEA Research Grant Scheme (NRGS), Ref. No.: 304/ PFRMASI/650582/K123.

Author details

[1]Department of Pharmaceutical Chemistry, School of Pharmaceutical Sciences, Universiti Sains Malaysia (USM), Minden 11800, Pulau Pinang, Malaysia. [2]Department of Pharmacology, School of Pharmaceutical Sciences, Universiti Sains Malaysia (USM), Minden 11800, Pulau Pinang, Malaysia. [3]Department of Pharmacy, School of Medicine and Health Sciences, An Najah National University, Nablus, West Bank, Palestine.

References

1. Ahamed MBK, Aisha AFA, Nassar ZD, Siddiqui JM, Ismail Z, Omari S, Parish C, Majid AMSA: **Cat's whiskers tea (Orthosiphon stamineus) extract inhibits growth of colon tumor in nude mice and angiogenesis in endothelial cells via suppressing VEGFR phosphorylation.** *Nutr Cancer* 2012, **64:**89–99.
2. Arafat OM, Tham SY, Sadikun A, Zhari I, Haughton PJ, Asmawi MZ: **Studies on diuretic and hypouricemic effects of orthosiphon stamineus methanol extracts in rats.** *J Ethnopharmacol* 2008, **118:**354–360.
3. Alshawsh MA, Abdulla MA, Ismail S, Amin ZA: **Hepatoprotective effects of orthosiphon stamineus extract on thioacetamide-induced liver cirrhosis in rats.** *Evid Based Complement Altern Med* 2011, **2011:** Article ID 103039, 6 pages http://dx.doi.org/10.1155/2011/103039.
4. Mohamed EAH, Mohamed AJ, Asmawi MZ, Sadikun A, Ebrika OS, Yam MF: **Antihyperglycemic effect of orthosiphon stamineus benth leaves extract and its bioassay-guided fractions.** *Molecules* 2011, **16:**3787–3801.
5. Ho C-H, Noryati I, Sulaiman S-F, Rosma A: **In vitro antibacterial and antioxidant activities of orthosiphon stamineus benth. Extracts against food-borne bacteria.** *Food Chem* 2010, **122:**1168–1172.
6. Alshawsh MA, Abdulla MA, Ismail S, Amin ZA, Qader SW, Hadi HA, Harmal NS: **Free radical scavenging, antimicrobial and immunomodulatory activities of orthosiphon stamineus.** *Molecules* 2012, **17:**5385–5395.
7. Hossain MA, Ismail Z, Rahman A, Kang SC: **Chemical composition and anti-fungal properties of the essential oils and crude extracts of orthosiphon stamineus benth.** *Ind Crop Prod* 2008, **27:**328–334.
8. Abdelwahab SI, Mohan S, Mohamed Elhassan M, Al-Mekhlafi N, Mariod AA, Abdul AB, Abdulla MA, Alkharfy KM: **Antiapoptotic and antioxidant properties of orthosiphon stamineus benth (cat's whiskers): intervention in the Bcl-2-mediated apoptotic pathway.** *Evid-Based Complement Altern Med* 2011, 156765. doi:10.1155/2011/156765.
9. Akowuah GA, Zhari I, Norhayati I, Sadikun A, Khamsah SM: **Sinensetin, eupatorin, 3′-hydroxy-5, 6, 7, 4′-tetramethoxyflavone and rosmarinic acid contents and antioxidative effect of orthosiphon stamineus from Malaysia.** *Food Chem* 2004, **87:**559–566.
10. Hossain MA, Mizanur Rahman SM: **Isolation and characterisation of flavonoids from the leaves of medicinal plant orthosiphon stamineus.** *Arabian J Chem* 2011. Article in press.
11. Takeda Y, Matsumoto T, Terao H, Shingu T, Futatsuishi Y, Nohara T, Kajimoto T: **Orthosiphol D and E, minor diterpenes from orthosiphon stamineus.** *Phytochemistry* 1993, **33:**411–415.
12. Awale S, Tezuka Y, Banskota AH, Kouda K, Tun KM, Kadota S: **Five novel highly oxygenated diterpenes of orthosiphon s tamineus from Myanmar.** *J Nat Prod* 2001, **64:**592–596.
13. Awale S, Tezuka Y, Banskota AH, Shimoji S, Taira K, Kadota S: **Norstaminane- and isopimarane-type diterpenes of orthosiphon stamineus from Okinawa.** *Tetrahedron* 2002, **58:**5503–5512.

14. Siddiqui MJA, Ismail Z: **Simultaneous analysis of bioactive markers from orthosiphon stamineus benth leaves extracts by reverse phase high performance liquid chromatography.** *Tropical Journal of Pharmaceutical Research* 2011, **10**:97–103.
15. Munin A, Edwards-Lévy F: **Encapsulation of natural polyphenolic compounds; a review.** *Pharm* 2011, **3**:793–829.
16. Fang Z, Bhandari B: **Encapsulation of polyphenols–a review.** *Trends Food Sci Technol* 2010, **21**:510–523.
17. Yokota D, Moraes M, Pinho S: **Characterization of lyophilized liposomes produced with non-purified soy lecithin: a case study of casein hydrolysate microencapsulation.** *Braz J Chem Eng* 2012, **29**:325–335.
18. Goyal P, Goyal K, Kumar SGV, Singh A, O.M.P K, Mishra DN: **Liposomal drug delivery systems-clinical applications.** *Acta Pharm* 2005, **55**:1–25.
19. Scholfield C: **Composition of soybean lecithin.** *J Am Oil Chem Soc* 1981, **58**:889–892.
20. van Nieuwenhuyzen W, Tomás MC: **Update on vegetable lecithin and phospholipid technologies.** *Eur J Lipid Sci Technol* 2008, **110**:472–486.
21. Gortzi O, Lala S, Chinou I, Tsaknis J: **Evaluation of the antimicrobial and antioxidant activities of Origanum dictamnus extracts before and after encapsulation in liposomes.** *Molecules* 2007, **12**:932–945.
22. Gortzi O, Lalas S, Chinou I, Tsaknis J: **Reevaluation of bioactivity and antioxidant activity of myrtus communis extract before and after encapsulation in liposomes.** *Eur Food Res Technol* 2008, **226**:583–590.
23. Gülseren I, Guri A, Corredig M: **Encapsulation of Tea polyphenols in nanoliposomes prepared with milk phospholipids and their effect on the viability of HT-29 human carcinoma cells.** *Food Digestion* 2012, **3**:36–45.
24. Taylor TM, Weiss J, Davidson PM, Bruce BD: **Liposomal nanocapsules in food science and agriculture.** *Crit Rev Food Sci Nutr* 2005, **45**:587–605.
25. Pawlikowska-Pawlęga B, Misiak LE, Zarzyka B, Paduch R, Gawron A, Gruszecki WI: **Localization and interaction of genistein with model membranes formed with dipalmitoylphosphatidylcholine (DPPC).** *Biochimica et Biophysica Acta (BBA)-Biomembranes* 2012, **1818**:1785–1793.
26. Sim CO, Hamdan MR, Ismail Z, Ahmad MN: **Assessment of Herbal Medicines by Chemometrics–Assisted Interpretation of FTIR Spectra.** *J Analytica Chimica Acta* 2004,1–14.
27. Guan S, Qin X, Zhou Z, Zhang Q, Huang Y: **Investigation of the mechanisms of improved oral bioavailability of bergenin using bergenin-phospholipid complex.** *Drug Development and Industrial Pharmacy* 2013, **40**:163–171.
28. Bajoria R, Sooranna SR, Contractor SF: **Endocytotic uptake of small unilamellar liposomes by human trophoblast cells in culture.** *Hum Reprod* 1997, **12**:1343–1348.
29. Ogiso T, Yamaguchi T, Iwaki M, Tanino T, Miyake Y: **Effect of positively and negatively charged liposomes on skin permeation of drugs.** *J Drug Target* 2001, **9**:49–59.
30. Siddiqui M, Hafizoh S, Ismail Z, Sahib H, Helal M: **Analysis of total proteins, polysaccharides and glycosaponins contents of orthosiphon stamineus benth. in spray and freeze dried methanol: water (1: 1) extract and its contribution to cytotoxic and antiangiogenic activities.** *Pharmacognosy Research* 2009, **1**:320.
31. Aisha AF, Abu-Salah KM, Alrokayan SA, Ismail Z, Abdulmajid AM: **Evaluation of antiangiogenic and antoxidant properties of parkia speciosa hassk extracts.** *Pak J Pharm Sci* 2012, **25**:7–14.
32. Patil VV, Galge RV, Thorat BN: **Extraction and purification of phosphatidylcholine from soyabean lecithin.** *Sep Purif Technol* 2010, **75**:138–144.
33. Aisha AF, Ismail Z, Abu-Salah KM, Majid AM: **Solid dispersions of alpha-mangostin improve its aqueous solubility through self-assembly of nanomicelles.** *J Pharm Sci* 2012, **101**:815–825.

Simultaneous silencing of multiple RB and p53 pathway members induces cell cycle reentry in intact human pancreatic islets

Stanley Tamaki[1], Christopher Nye[2], Euan Slorach[2], David Scharp[3], Helen M Blau[4], Phyllis E Whiteley[5] and Jason H Pomerantz[6*]

Abstract

Background: Human pancreatic islet structure poses challenges to investigations that require specific modulation of gene expression. Yet dissociation of islets into individual cells destroys cellular interactions important to islet physiology. Approaches that improve transient targeting of gene expression in intact human islets are needed in order to effectively perturb intracellular pathways to achieve biological effects in the most relevant tissue contexts.

Results: Electroporation of intact human cadaveric islets resulted in robust and specific suppression of gene expression. Two genes were simultaneously suppressed by 80% from baseline levels. When multiple (up to 5) genes were simultaneously targeted, effective suppression of 3 of 5 genes occurred. Enzymatic pretreatment of islets was not required. Simultaneous targeting of RB and p53 pathway members resulted in cell cycle reentry as measured by EDU incorporation in 10% of islet nuclei.

Conclusions: At least three genes can be effectively suppressed simultaneously in cultured intact human pancreatic islets without disruption of islet architecture or overt alterations in function. This enabled the effective modulation of two central growth control pathways resulting in the phenotypic outcome of cell cycle reentry in postmitotic islet cells. Transient exposure to multiple siRNAs is an effective approach to modify islets for study with the potential to aid clinical applications.

Keywords: Intact human islets, siRNA, Electroporation, RB, p53, Cell cycle reentry

Background

We report the simultaneous suppression of multiple genes in intact adult human pancreatic islets. The study of human islets is hindered by obstacles including difficulty of maintaining islet cells *in vitro* but within their native complex tissue environment, the intact islet, and difficulty modulating gene expression in a majority of islet cells without significantly damaging islet viability, function and architecture. The complexity of islet biology results in part from the physical interactions and communication between the various islet cell types [1-4]. This intercellular network is disrupted when islets are dissociated and cultured, which is a commonly used approach to study islet cells that facilitates transfection or infection with virus (reviewed in [5]). By contrast, modification of islet cell gene expression in the context of intact intercellular relationships should increase the relevance of the information obtained. Thus, the establishment of techniques that expand our ability to study intact islets is merited.

Translational efforts towards clinical applications using intact islets could also benefit from the advances described here. As transplantation of intact islets advances in clinical trials, clinical applications requiring direct modification of gene expression to improve islet survival and engraftment would likely rely on transient modifications that do not result in permanent integration of genetic constructs. In the case of islet transplantation the period of ex vivo culture prior to transplantation affords

* Correspondence: jason.pomerantz@ucsfmedctr.org
[6]Departments of Surgery and Orofacial Sciences, Division of Plastic and Reconstructive Surgery, Craniofacial and Mesenchymal Biology Program, Eli and Edythe Broad Center of Regeneration Medicine, University of California, San Francisco, CA 94143, USA
Full list of author information is available at the end of the article

an opportunity for transient delivery of siRNA to modulate gene expression in efforts to improve transplantation protocols [6,7]. Moreover, the importance of developing protocols that facilitate mechanistic studies using human islets is underscored by the observations that insights gained from studies of rodent islets often do not translate directly when applied to human tissue [8,9].

Islet study also has inherent technical challenges shared by other cell types. Although it has been demonstrated that single molecules can significantly influence human beta cell mitogenic responses [10], the ability to commandeer positive and negative proliferation pathways requires modulation of multiple genes at a time. This is in part due to inherent genetic redundancy that often makes single gene modulation insufficient, and also to the need to evaluate upstream and downstream events relating to an initial perturbation. For example, the retinoblastoma (RB) pathway contains three partially redundant pocket proteins, pRB, p107 and p130 in addition to multiple upstream and downstream modulators and effectors [11]. A generally assumed limitation of RNA interference experiments has been the number of targeted genes that can be directly suppressed at one time. In islets, to date single genes have been efficiently suppressed by electroporation [12]. However, whether it is possible in intact islets to efficiently silence multiple genes simultaneously, which may be obligatory to promote robust biological endpoints remains unknown. A number of potentially complicating factors include delivery of adequate amounts of siRNA molecules into individual cells, saturation of the RISC complex, and toxicity to individual cells throughout the islet.

Here we use electroporation to suppress gene expression using siRNA in intact human islets. We show that suppression of multiple transcripts can occur simultaneously using an approach that preserves morphology, viability and glucose responsiveness. Suppression of genes within the RB and p53 pathways illustrates the capacity to use siRNA to dissect functional interactions among multiple pathway members. These findings should prove useful, as they demonstrate a potent and safe approach to transiently modulate gene expression within the native islet environment.

Results

We sought to determine an optimal method for modulation of gene expression in intact adult human islets that would be transient and would be minimally disruptive in terms of islet architecture, function and viability. In order to be a sufficiently potent approach to achieve biological endpoints, another goal was to successfully modulate more than a single gene at a time. Prior studies by others have shown that adenoviral infection can target the surface cells of the islet, but not the core [13-17], and it has recently been shown that enzymatic partial disruption of islets with accutase followed by electroporation was effective for suppressing expression of a single gene [12].

We initially evaluated suppression of a single gene *RB*, using standard lipid based transfection (Figure 1A, and see Table 1 for PCR primers and probes used throughout). Evaluation of gene expression compared to control using quantitative RT-PCR showed modest suppression of total RNA *RB* transcript levels by about 30% relative to controls. This could reflect poor transfection efficiency or inability of the lipid reagent to penetrate the core of the islet. In light of prior reported success with electroporation we explored this approach. We used a capillary tip based system (Neon) to electroporate islets in suspension. Multiple parameters of voltage and pulse times were evaluated first for their effects on islet morphology and viability, and then for gene suppression (data not shown). Ultimately we found that with this system voltages of 1000-1100 volts and a single 40 msec pulse preserved islet morphology and achieved robust suppression of *RB* gene expression by 78% relative to controls (Figure 1A). This robust silencing was attained without digestion of the islet membrane with proteases. Therefore in subsequent experiments we avoided accutase treatment in order to maximally preserve native architecture and islet cell interactions. This level of suppression of *RB* transcript levels indicates both good efficiency of siRNA penetration into islet cells as well as targeting of a majority of cells within the islets.

Although a range of voltages and pulse times were effective for suppression, voltages in the 1000-1100 range using a single pulse of 40 msec resulted in robust gene knockdown and spared islets from marked morphological and functional changes. To determine whether electroporation significantly altered islet function we evaluated insulin secretion in response to glucose stimulation. Similar to untreated islets, islets electroporated at various voltages including 1100 volts responded to glucose stimulation by secreting insulin, and the insulin secretion returned to baseline upon exposure to 3 mM glucose (Figure 1B). We did observe lower total insulin levels in the electroporated islets compared to non-electroporated controls. It is possible that this could reflect a decrease in insulin production by islets as a result of electroporation. We found that within this relatively narrow range of electroporation parameters, normal islet morphology was essentially preserved (Figure 1C top panels) and the typical cell types remained, including c-peptide-expressing beta cells (Figure 1C middle panels). Finally, we assessed islet viability after electroporation by ethidium bromide incorporation, and found that although electroporation caused a modest increase in the number of non-viable islets, the majority of islets (75-80%) remained impermeable, indicating good viability (Figure 1C bottom panels). Thus, glucose stimulation of insulin secretion, morphology and cell type composition are preserved after

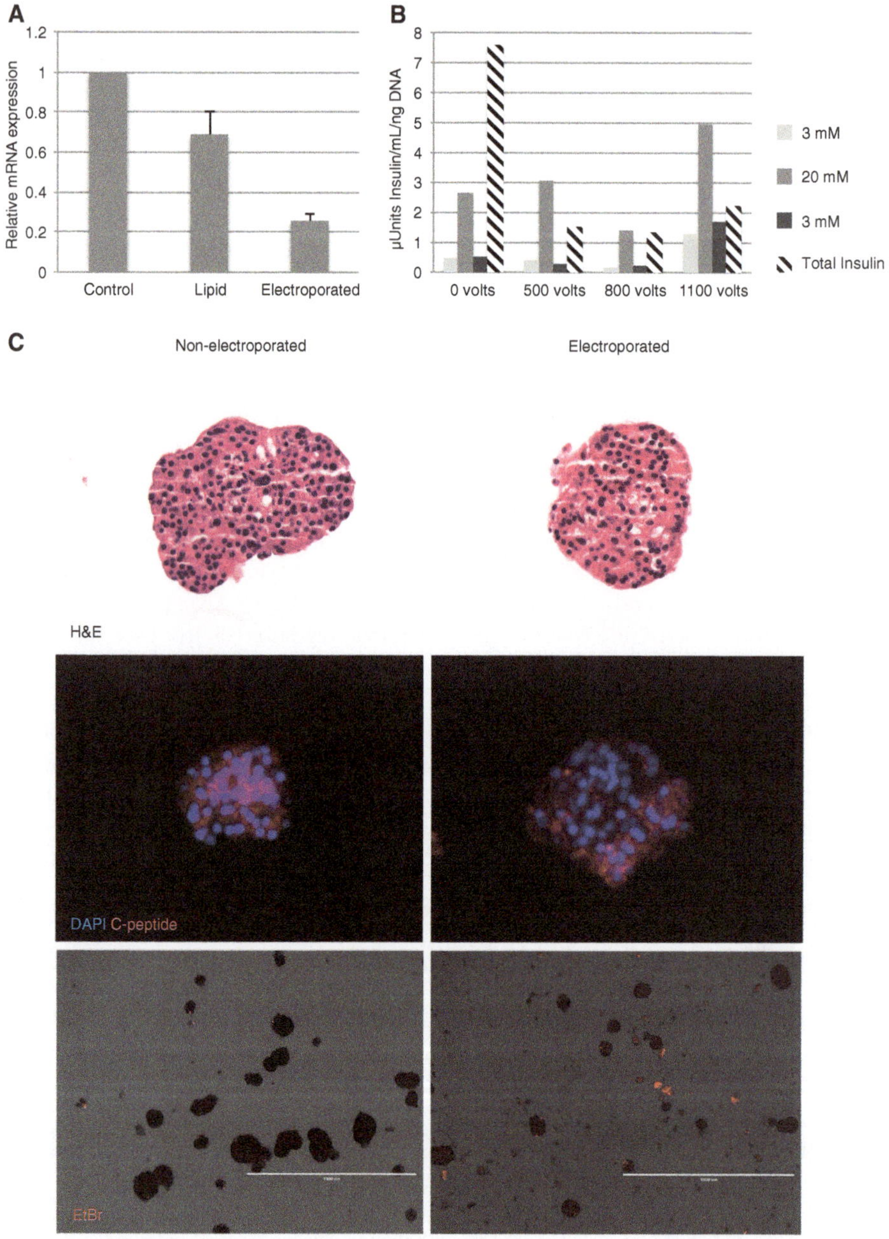

Figure 1 Electroporation and siRNA delivery in intact human islets. (A) Graph showing RT-qPCR results of *RB* knockdown following lipid or electroporation transfection. Intact human islets were transfected with non-targeting control siRNA (D-001810 ON-TARGETplus from Thermo Fisher/Dharmacon) or siRNA targeting *RB* (L-003296) using lipid (Dharmafect 1 from Thermo Fisher/Dharmacon) or electroporation (Neon Transfection System from Life Technologies), cultured for two days, and RT-qPCR was performed to assess knockdown. Error bars indicate SEM (n = 3 separate experiments). **(B)** Graph showing representative results of glucose stimulation of insulin secretion assay. Intact islets were electroporated with control, non-targeting siRNA at a range of voltages (x-axis) and a glucose stimulation of insulin secretion assay was performed 72 hours following electroporation. Media was collected after exposure to 3 mM glucose (light gray bars), then after switch to 20 mM glucose (gray bars), followed by return to 3 mM glucose (black bars), and total insulin (hatched bars). **(C)** Images showing morphological comparison between non-electroporated (left) and electroporated (right) islets. Images of islets stained with hematoxylin and eosin (top), immunostained for for c-peptide (red) and nuclei DAPI (blue)(middle), and after exposure to ethidium bromide to test viability 72 hours after electroporation (bottom).

electroporation, supporting the utilization of this approach as a practical method for gene suppression experiments that is minimally disruptive to islet structure and function.

The potent suppression of a single gene transcript using electroporation prompted us to examine whether this approach would be useful for targeting multiple

Table 1 TaqMan PCR primers and probes

Primer/Probe	Sequence
RB F1	5'- GGAAGCAACCCTCCTAAACC - 3'
RB R1	5' - TTTCTGCTTTTGCATTCGTG - 3'
RB probe	5' - [6-FAM] CATCTCCCAGGAGAGTCCAA [BHQ1a-6FAM] - 3'
p107 F1	5' – AGAATGCCTCCTGGACCTTT - 3'
p107 R1	5' – GGGGTGTCACGAGTGAACTT - 3'
p107 probe	5' - [6-FAM] ACGCAGAAGAGGAAATTGGA [BHQ1a-6FAM] - 3'
p130 F1	5' – ATTTGGCATGGAAACCAGAG - 3'
p130 R1	5' – GTCACCCTTCTGGGAGTCAA - 3'
p130 probe	5' - [6-FAM] AGAACCTGGAAAGGGCAGAT [BHQ1a-6FAM] - 3'
p21 F1	5' – AGAGGAGGCGCCATGTCAG - 3'
p21 R1	5' – CATTAGCGCATCACAGTCGC - 3'
p21 probe	5' - 6-FAM] CAGAACCCATGCGGCAGCAA [BHQ1a-Q] - 3'
p53 F1	5' – GTGGAAGGAAATTTGCGTGT - 3'
p53 R1	5' – CCAGTGTGATGATGGTGAGG - 3'
p53 probe	5' - [6-FAM] ACATAGTGTGGTGGTGCCCT [BHQ1a-Q] - 3'
ARF F1	5' – AGGGTTTTCGTGGTTCACAT - 3'
ARF R1	5' – CTGCCCATCATCATGACC - 3'
ARF probe	5' - [6-FAM] CGATCCAGGTCCATGATGATG [BHQ1a-6FAM] - 3'
GAPDH F1	5' – GAGTCAACGGATTTGGTCGT - 3'
GAPDH R1	5' TTGATTTTGGAGGGATCTCG 3'
GAPDH probe	5' - [6-FAM] CTGAGAACGGGAAGCTTGTC [BHQ1a-6FAM] - 3'

transcripts simultaneously and thereby enable the perturbation of multiple interacting gene products in the RB and p53 pathways. We found 25 nM to be the optimal concentration of each siRNA for individual gene products. When two distinct gene products were targeted simultaneously, each with a 25 nM concentration of specific siRNA, suppression of mRNA transcript levels was similarly robust as when a single gene was targeted (Figure 2A). Specific siRNAs targeting the pocket protein gene products *RB* and *p130* when combined each at a concentration of 25 nM, and electroporated into intact human islets (1000 volts 40 msec single pulse) resulted in suppression of *RB* and *p130* mRNA levels by approximately 80% of baseline levels when compared to islets electroporated with control siRNA at the same concentration.

Major growth control pathways such as RB and p53 contain multiple upstream and downstream members as well as redundant family members that can compensate for loss of one another [18]. Such complexity significantly complicates experiments designed to study pathway function and presents a challenge for developing practical applications that require modulation of one or more pathways. We targeted various combinations of the RB and p53 pathway members, including *RB, p130, p107, p53,* and *p21* each with validated siRNAs (Dharmacon) at 25 nM or 50 nM concentrations, and also measured *p14ARF* transcript levels for changes in response to modulation of other genes. Indeed, we detected a modest induction of *ARF* gene expression after suppression of RB and p53 pathway members, an expected effect based on the known function of *ARF* to detect aberrant RB/p53 pathway signaling. Quantitative RT-PCR showed that at least 3 out of 5 targeted gene products were suppressed to levels considered standard in siRNA experiments. For example *RB, p130* and *p21* levels were suppressed by approximately 80% compared to control levels in samples treated for *RB, p130, p107, p53* and *p21* (Figure 2B). In multiple experiments (Figure 2B and data not shown) suppression of 4-5 genes resulted in variability in terms of reduction of each targeted mRNA. Likely contributing factors include saturation of the intracellular RNA interference machinery or induction of gene expression by activation of compensating pathway members in response to suppression of others (known to occur among the pocket proteins as well as among the p53 pathway members). We did not observe any significant increased toxicity when using higher total siRNA concentrations (up to 300 nM) required for these experiments. To confirm that electroporation of siRNA and gene suppression measured by mRNA was an indicator of decreased protein levels, Western blotting for RB protein was performed after treatment with control siRNAs and after combinations of targeting siRNAs. We found expected suppression of RB protein as suggested by the mRNA data, confirming the validity of the PCR data (Figure 2C).

The ability to suppress multiple genes simultaneously using electroporated siRNA indicated that this approach might be useful for functional modulation of the RB and p53 pathways. We tested this by attempting to target multiple interacting genes simultaneously in the RB and p53 pathways to achieve the phenotypic endpoint of cell cycle reentry in mature post mitotic islet cells. This would indicate reversal of relatively stringent maintenance of the islet postmitotic state. After exposure to various siRNA combinations (Figure 3), islets were incubated in 5-ethynyl-2′-deoxyuridine (EdU) for 5 days. Whereas islets treated with control siRNA did not incorporate EDU and those treated for suppression of only *RB* and *p53* siRNA rarely contained cells that incorporated EDU, those that had targeted multiple pathway members often contained clusters of EDU positive nuclei (Figure 3A). Up to 10% of nuclei incorporated EDU when 5 or 6 genes were targeted (Figure 3B). Together, the experiments targeting multiple members of the RB and p53 pathways clearly demonstrate that by simultaneously targeting several genes,

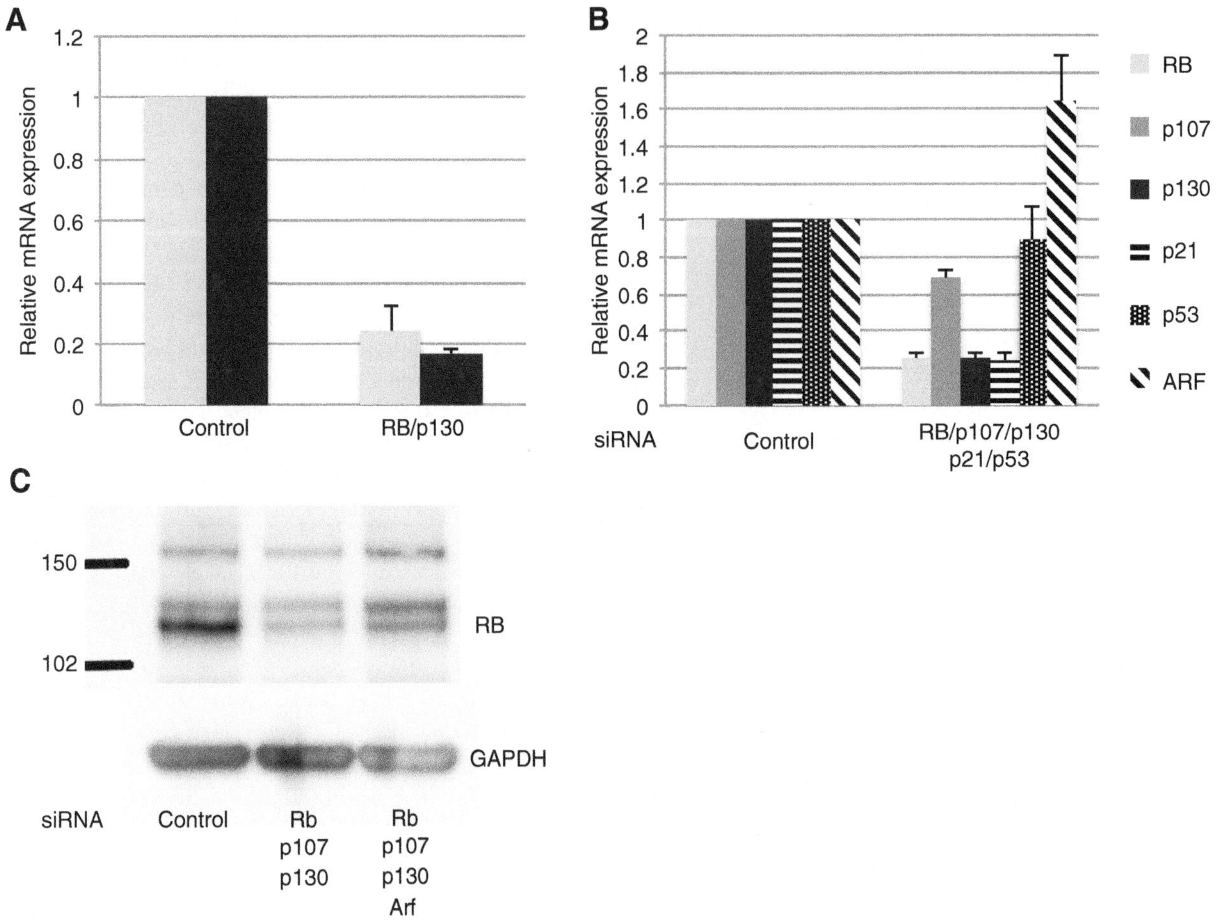

Figure 2 Simultaneous knockdown of multiple genes in primary human islets following electroporation. (A) Graph shows RT-qPCR results of islets electroporated with non-targeting control siRNA or siRNA targeting *RB* and *p130* (L-003299). **(B)** Graph shows RT-qPCR results of islets electroporated with non-targeted control siRNA or siRNAs targeting the indicated gene products. The housekeeping gene *GAPDH* was used as the control. Representative experiment shown from 3 separate experiments. Error bars indicate standard deviation. **(C)** Western blot for RB protein in human islet lysates after electroporation of control or targeting siRNA. GAPDH was used as a loading control. RB appears as a characteristic doublet representing the hypo and hyperphosphorylated protein. Representative experiment shown from 3 different experiments.

the postmitotic state of human islet cells can be reversed to induce cell cycle reentry as evidenced by DNA synthesis and EDU incorporation. This biological effect was not achieved by suppression of single genes in these pathways, a demonstration of the potential utility of this siRNA combination approach.

Discussion

One challenge facing the development of therapeutics to treat disorders of pancreatic islets is the difficulty of studying intact human islets. Although many important insights have and will continue to come from studies of rodent islets, the significant differences between islets from rodents and humans and the failure of many rodent findings to be conserved in human islet biology underscores the need for studies using human tissue. For example, adult rodent islets can readily undergo islet expansion in different circumstances including increased beta cell stress from Type 2 Diabetes [19]. But adult human islets are more refractory, and efforts to demonstrate adult beta cell expansion have met with great difficulty [20]. Moreover, rodents are genetically tractable and germline targeting of genes of interest is routine, but the limited lifespan of human islets in culture makes such studies using primary human tissue not possible at present. Therefore there is an unmet need for approaches to modify gene expression in intact human islets for translational application. The present study furthers that effort by demonstrating culture and transfection parameters that permit suppression of multiple genes at once, resulting in a phenotypic outcome not achievable with single gene manipulations.

This study builds on previously established approaches to transfect primary cells including islets using electroporation. In contrast to previous work [12], we used intact human islets transfected without accutase digestion

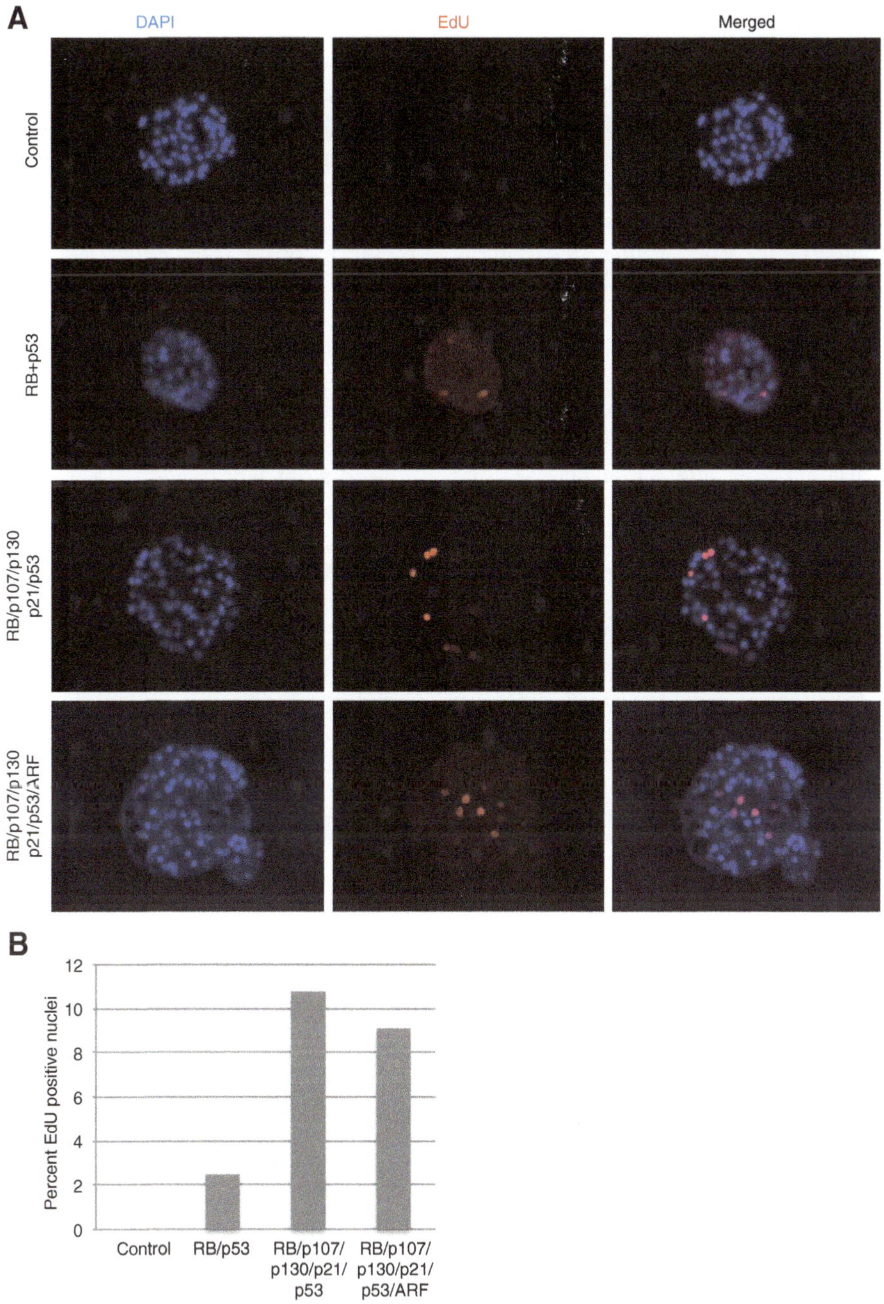

Figure 3 Cell cycle reentry in primary human islets following knockdown of RB and p53 pathway members. Primary human islets were electroporated with siRNAs against non-targeting control, or combinations of RB, p107 (L-003298), p130 (L-003299), p21 (L-003471), p53 (L-003329) and ARF/CDKN2A (L-011007). Islets were cultured in the presence of 5-ethynyl-2′-deoxyuridine (EdU) for 7 days following electroporation then fixed and stained with DAPI and for EdU. **(A)** Images showing EdU incorporation in Islets with nuclei stained with DAPI (blue, left), EdU (red, middle) and a composite image (right). **(B)** Graph showing percentage of EdU incorporating cells in islets following transfection. Total cells counted in the representative experiment presented are: siControl (245); Rb + p53 (535); Rbfamily + p53 + p21 (456); Rbfamily + p53 + p21 + Arf (834). The experiment was repeated 3 separate times.

and by modifying the electroporation parameters of voltage and frequency, pervasive suppression of targeted genes could be achieved without enzymatic disruption of the islet architecture. This advance is potentially beneficial because of the comparatively lesser manipulation of islets. Survival of electroporated islets and the maintenance of function after electroporation may have been aided in our study by the use of developed culture conditions and media designed to optimize islet survival and resistance to stress [21].

This study extends previous work [12] by showing that electroporation is a superior means of delivering siRNA into islets than liposome-based transfection. We compared the two approaches directly and electroporation caused a significantly greater reduction in targeted mRNA compared to lipofection. We show here that after optimization of siRNA concentrations and electroporation parameters two simultaneous knockdowns were possible to levels comparable to standards for easily transfected immortalized cell lines. Remarkably, we found that significant robust suppression of at least three genes simultaneously is possible using this method. Despite the preservation of islet structure and glucose responsiveness achieved with the present approach, it is likely that this system could be further optimized, for example by decreasing voltage in combination with use of newer siRNA modifications that endow greater transfection efficiency and potency [22,23]. It is possible that subtle alterations in islet viability or function not detected in the assays used in this study would be better preserved at lower voltages.

Multiple gene suppression in the absence of damage to islet structure or glucose responsiveness allowed us to effectively target two central growth control pathways with inherent redundancies in each. Simultaneous compromise of the RB and p53 pathways led to entry of postmitotic islet cells into the cell cycle. This phenotypic outcome was a result of specific inhibition of the targeted genes and required suppression of multiple pathway members. These findings demonstrate the potential of transient suppression of multiple genes to have significant functional impact in situations where single gene targeting is inadequate. Future experiments will address global changes in islet differentiated and metabolic state after RB/p53 suppression by electroporation. Electroporation of human islets should prove a useful approach to aid the study of other important pathways containing multiple members.

Conclusions

The application of electroporation in optimized culture conditions is an effective approach to deliver siRNA molecules targeting multiple genes into a majority of cells within intact human islets. Benefits are that it is transient, avoids genetic integration and is well tolerated by islets. The capacity to significantly impact molecular pathways by targeting multiple genes within them should prove to be a valuable tool for both basic study and for practical applications.

Methods

Primary Islets cells and culture conditions

Primary human cadaveric islets were obtained from Prodo Labs. Typically islets were shipped overnight on ice 2-4 days after isolation. Upon receipt islets were diluted 50% in prewarmed 37°C PIM(S) media supplemented with PIM(ABS) and PIM(G) (complete) (all purchased from Prodo Laboratories) and centrifuged for 2 min 1200 rpm. Islets were washed and resupended at a concentration of 300-400 islet equivalents (IEQ) per ml in complete PIM(S) media in non-tissue culture treated flasks. Islets were incubated at 37°C in 5% CO_2 with 50% media change every other day. For the samples used in our experiments the purity of the islets ranged between 85 and 90% in all samples. The average islet size was: 50-100 microns (74%); 100-200 microns (20%); 200-300 microns (5%); 300 microns and higher (1%).

siRNAs

All siRNAs were ON-TARGETplus SMARTpools purchased from Dharmacon, Thermo Scientific. The following siRNAs were used:

siControl (D-001810)
RB (L-003296)
p107 (L-003298)
p130 (L-003299)
p53 (L-003329)
p21 (L-003471)
ARF/CDKN2A (L-011007)

Lipid Transfection of islets

Islet transfection with cationic lipid was conducted by diluting 1 μl of Dharmafect 1 (T-2001 Life Techonogies) in 50 μl of PIM(S) followed by addition of siRNAs in 50 μl of PIM(S), vortexing and then allowing the mixture to sit at room temperature for 20 min. The lipid-siRNA mix was added to 400 IEQ in 400 μl complete media in 24-well low adhesion plates and incubated at 37°C in 5% CO_2. The next day 1.5 ml of complete media was added to each well. The plates were then returned to the incubator to allow the islets to settle to the bottom and 1.25 ml of media was removed with a pipet. This process was repeated 3 times to remove the cationic lipid-siRNA following which the islets were returned to the incubator.

Electroporation of islets

Electroporation of islets was conducted using the Neon Transfection System (Life Technologies). Islets were centrifuged for 2 min 1200 rpm and then transferred to a 1.5 ml Eppendorf tube and washed with PBS and re-centrifuged. Islets were resuspended at a density of 400 IEQ per 10 μl in R buffer (Life Technologies) and typically electroporated once at approximately 1000-1100 Volts for 40 msec. For knockdown and EdU incorporation experiments the following concentrations of siRNA were used 25 nM RB, 25 nM p130, 25 nM p21, 50 nM ARF/CDKN2A, 50 nM p53, 50 nM p107 and siControl ranged from 25 nM-300 nM. Each batch of electroporated Islets was cultured individually in a single well of a 24-well low adhesion plates (per manufacturer recommendations) in 0.5 ml complete PIM (S) media. Viability was evaluated using ethidium bromide (Molecular Probes) diluted to 1:1000.

Evaluation of gene expression

Typically, cellular RNA was isolated from 400 IEQ 72 hours after electroporation using an RNeasy Mini Kit (Qiagen) and used to generate cDNA with SuperScript VILO Master Mix (Invitrogen). 2 µl cDNA was used as template for TaqMan PCR in conjunction with TaqMan Fast Advanced Master Mix (Invitrogen). TaqMan PCR was performed using a StepOnePlus™ Real Time PCR system (Life Technologies) with the following cycling parameters;

$$\text{Cycle } 1 = 95^{\circ}\text{C–2 mins} \times 1$$

$$\text{Cycle } 2 = 95^{\circ}\text{C–1 sec}, \; 60^{\circ}\text{C–20 sec} \times 40$$

All PCR data between samples were normalized to GAPDH expression levels and the level of gene knockdown was expressed relative to siCONTROL sample.

Western blots

Islet samples for protein analysis were pelleted by centrifugation lysed in RIPA buffer and sonicated. After protein concentration was determined, 30 micrograms of protein were loaded per lane. For detection of Retinoblastoma protein blots were probed with 0.1 µg/mL of Human RB1 (MAB6495, R&D Systems) followed by HRP-conjugated Anti-Mouse IgG Secondary Antibody (R &D Systems). GAPDH was similarly done with 0.05 µg/mL of Mouse Anti-Human GAPDH Monoclonal Antibody (MAB5718, R&D Systems).

EdU or Insulin staining

Islet assay for DNA synthesis was initiated by switching the culture media 48 hr after electroporation into PIM (R) media supplemented with PIM(ABS) and PIM(G) and 10% FBS (all purchased from Prodo Laboratories) containing 5 µM EdU (Life Technologies). Half the media was replaced each day containing 5 µM EdU. 1000 IEQ equivalents for each sample were harvested 7 days after electroporation by centrifuging for 2 min at 1200 rpm followed by aspiration of the media. The islet pellet was resuspended in the residual media and 25 µl of histogel (60°C) was added, mixed with the islets, and quickly placed on ice. The plug was fixed in 4% paraformaldhye-PBS overnight and then dehydrated in alcohol prior to paraffin embedding (Histotech Lab). Paraffin blocks were sectioned at 5 microns onto glass slides. Sections were deparaffinzed using orange oil and passaged through an alcohol series. EdU incorporation was assayed by staining for EdU using a Click-iT EdU Alexa Fluor 568 Imaging Kit (Invitrogen) according to manufacturer's instructions. In the case of insulin staining deparaffinzed sections were blocked with 5% goat serum in 0.1% Triton PBS for 30 min and anti-pro-insulin C-peptide (Millipore, CBL94) 1:500 dilution and then incubated overnight. Sections were washed and stained with secondary antibody, goat-anti-mouse Alexa 556 (1:500) was used. For both EdU and antibody staining, cover-slips were mounted on sections with Vectashield Mounting Medium containing Dapi (Vector Lab). Digital photographs were taken using a 20X objective on an EVOS FL Cell Imaging System (Life Technologies).

Glucose Stimulation Insulin Secretion assay (GSIS)

Approximately 100 IEQ were placed into inserts. A glucose stimulation insulin secretion assay was conducted on islets 48 hr after electroporation. Approximately 100 IEQ were placed into millicell (Millipore Inc.) cell culture inserts in a 24 well plate. The minicell inserts were transferred into new wells containing 3 mM Glucose Media (50% Hams F10 (110 mg/dl glucose) 50% DMEM (no glucose), 7 mM Nabicarbonate, 0.75 mM $CaCl_2$ $2H_2O$ and 10 µg/ml of Ciprofloxan. The inserts with islets were transferred to wells containing 3 mM Glucose Media for 1-2 min to wash out the original media. To remove insulin from the cultures, the islets were washed three separate times with 3 mM Glucose Media and then incubated overnight in the same media at 37 degrees with 5%CO2. The following morning islets were again washed by two 1 hr incubations in 3 mM Glucose Media before initiation of media collection to assay secretion of insulin into the media. For the first collection of the assay, the inserts plus islets were placed into 1 ml fresh 3 mM Glucose Media and incubated for one hour, and then the media was collected. Further samples were collected after 1 hr incubations in 20 mM Glucose Media. Finally the inserts plus islets were washed 2 times with 3 mM Glucose media and then incubated for a final 1 hr in 3 mM Glucose media and collected. The insert plus islets were then transfer to 10 mM Tris-1 mM EDTA pH 8.0 and islets collected from inserts and centrifuged at 180 g for 5 min. The islets were lysed by resuspension in 10 mM Tris-1 mM EDTA pH 8.0 plus 1% Triton-X containing protease inhibitors and vortexed. All samples collected were stored at -80°C until analysis.

ELISA to detect insulin was conducted using Millipore Human Insulin ELISA kit. The two PIM(S) samples from each matrix condition were pooled and assayed as a single sample. The protocol was followed as suggested by the manufacturer with the following modifications. Samples from 3 mM Glucose media were diluted 1:2, 12 mM Glucose media were diluted 1:5, 20 mM Glucose media diluted 1:10 and lysed islets diluted 1:20 prior to testing in ELISA assay. Plates were read at 450 nm absorbance on an Epoch Biotek plate reader.

DNA from lysed islets was quantitated using Invitrogen Quant-iT PicoGreen dsDNA assay kit. Samples were measured on a PerSeptive Biosytems Cyto-fluor and plotted against a standard curve of lambda DNA.

Competing interests

ST, CN and ES were employees, PW was CEO and founder, and HB and JP were scientific founders and advisors of Didimi Inc., a biotechnology startup focused on regenerative medicine, where the experiments were performed. DS is CEO and founder of Prodo Laboratories.

Authors' contributions

ST, CN and ES designed and performed the experiments, and analyzed the data. ST and CN contributed to manuscript writing. ST designed and performed the majority of the experiments. DS contributed to experimental design, interpretation of results and to manuscript revision. HB contributed to research conception, interpretation of results and manuscript revision. PW contributed to research conception, experimental design, data interpretation, research oversight and manuscript revision. JP contributed to research conception, experimental design, data interpretation, and wrote the manuscript. All authors read and approved the final manuscript.

Acknowledgements

We gratefully acknowledge the constructive suggestions of Peter Lomedico, Richard Insel, Julia Greenstein and Laura Papuga of the Juvenile Diabetes Research Foundation. This work was funded in part by Juvenile Diabetes Research Institute grant 17-2012-515.

Author details

[1]Present Address: Department of Surgery, Division of Plastic and Reconstructive Surgery, University of California, San Francisco, CA 94143, USA. [2]QB3 East Bay Innovation Center, 2929 7th St, Berkeley, CA 94710, USA. [3]Prodo Laboratories, 32 Mauchly, Irvine, CA 92618, USA. [4]Baxter Laboratory for Stem Cell Biology, Department of Microbiology and Immunology, Institute for Stem Cell Biology and Regenerative Medicine, Stanford University School of Medicine, Stanford, CA 94305, USA. [5]Present Address: Mohr Davidow Ventures, Menlo Park, CA 94025, USA. [6]Departments of Surgery and Orofacial Sciences, Division of Plastic and Reconstructive Surgery, Craniofacial and Mesenchymal Biology Program, Eli and Edythe Broad Center of Regeneration Medicine, University of California, San Francisco, CA 94143, USA.

References

1. Efrat S, Russ HA: **Making beta cells from adult tissues.** *Trends Endocrinol Metab* 2012, **23**(6):278–285.
2. Parnaud G, Hammar E, Ribaux P, Donath MY, Berney T, Halban PA: **Signaling pathways implicated in the stimulation of beta-cell proliferation by extracellular matrix.** *Mol Endocrinol* 2009, **23**(8):1264–1271.
3. Halban PA, German MS, Kahn SE, Weir GC: **Current status of islet cell replacement and regeneration therapy.** *J Clin Endocrinol Metab* 2010, **95**(3):1034–1043.
4. Yang SN, Shi Y, Yang G, Li Y, Yu J, Berggren PO: **Ionic mechanisms in pancreatic beta cell signaling.** *Cell Mol Life Sci* 2014.
5. Hughes A, Jessup C, Drogemuller C, Mohanasundaram D, Milner C, Rojas D, Russ GR, Coates PT: **Gene therapy to improve pancreatic islet transplantation for Type 1 diabetes mellitus.** *Curr Diabetes Rev* 2010, **6**(5):274–284.
6. Shapiro AM: **Strategies toward single-donor islets of Langerhans transplantation.** *Curr Opin Organ Transplant* 2011, **16**(6):627–631.
7. Li F, Mahato RI: **RNA interference for improving the outcome of islet transplantation.** *Adv Drug Deliv Rev* 2011, **63**(1–2):47–68.
8. Jiao Y, Le Lay J, Yu M, Naji A, Kaestner KH: **Elevated mouse hepatic betatrophin expression does not increase human beta-cell replication in the transplant setting.** *Diabetes* 2014, **63**(4):1283–1288.
9. Bernal-Mizrachi E, Kulkarni RN, Scott DK, Mauvais-Jarvis F, Stewart AF, Garcia-Ocana A: **Human beta-cell proliferation and intracellular signaling part 2: still driving in the dark without a road map.** *Diabetes* 2014, **63**(3):819–831.
10. Fiaschi-Taesch NM, Salim F, Kleinberger J, Troxell R, Cozar-Castellano I, Selk K, Cherok E, Takane KK, Scott DK, Stewart AF: **Induction of human beta-cell proliferation and engraftment using a single G1/S regulatory molecule, cdk6.** *Diabetes* 2010, **59**(8):1926–1936.
11. Mulligan G, Jacks T: **The retinoblastoma gene family: cousins with overlapping interests.** *Trends Genet* 1998, **14**(6):223–229.
12. Lefebvre B, Vandewalle B, Longue J, Moerman E, Lukowiak B, Gmyr V, Maedler K, Kerr-conte J, Pattou F: **Efficient gene delivery and silencing of mouse and human pancreatic islets.** *BMC Biotechnol* 2010, **10**:28.
13. Barbu AR, Bodin B, Welsh M, Jansson L, Welsh N: **A perfusion protocol for highly efficient transduction of intact pancreatic islets of Langerhans.** *Diabetologia* 2006, **49**(10):2388–2391.
14. Csete ME, Afra R, Mullen Y, Drazan KE, Benhamou PY, Shaked A: **Adenoviral-mediated gene transfer to pancreatic islets does not alter islet function.** *Transplant Proc* 1994, **26**(2):756–757.
15. Giannoukakis N, Mi Z, Gambotto A, Eramo A, Ricordi C, Trucco M, Robbins P: **Infection of intact human islets by a lentiviral vector.** *Gene Ther* 1999, **6**(9):1545–1551.
16. Rehman KK, Wang Z, Bottino R, Balamurugan AN, Trucco M, Li J, Xiao X, Robbins PD: **Efficient gene delivery to human and rodent islets with double-stranded (ds) AAV-based vectors.** *Gene Ther* 2005, **12**(17):1313–1323.
17. Takahashi R, Ishihara H, Takahashi K, Tamura A, Yamaguchi S, Yamada T, Katagiri H, Oka Y: **Efficient and controlled gene expression in mouse pancreatic islets by arterial delivery of tetracycline-inducible adenoviral vectors.** *J Mol Endocrinol* 2007, **38**(1–2):127–136.
18. Harris SL, Levine AJ: **The p53 pathway: positive and negative feedback loops.** *Oncogene* 2005, **24**(17):2899–2908.
19. Dalboge LS, Almholt DL, Neerup TS, Vassiliadis E, Vrang N, Pedersen L, Fosgerau K, Jelsing J: **Characterisation of age-dependent beta cell dynamics in the male db/db mice.** *PLoS One* 2013, **8**(12):e82813.
20. Efrat S: **Recent progress in generation of human surrogate beta cells.** *Curr Opin Endocrinol Diabetes Obes* 2013, **20**(4):259–264.
21. Scharp D, Arulmoli G, Taeidi P, Albert E, Avakian A, Sunshine H, Kucera C, McRaney D, Bondurant D, Haight B, Magaldi M: **Survival and Function of Human Islets Cultured for 4 Weeks Post-Processing.** In *International Pancreas and Islet Transplantation Association: 9/27/2013; Monterey California.* 2013.
22. Corey DR: **Chemical modification: the key to clinical application of RNA interference?** *J Clin Invest* 2007, **117**(12):3615–3622.
23. Kenski DM, Butora G, Willingham AT, Cooper AJ, Fu W, Qi N, Soriano F, Davies IW, Flanagan WM: **siRNA-optimized modifications for enhanced In Vivo activity.** *Mol Ther Nucleic Acids* 2012, **1**:e5.

Microbial synthesized biodegradable PHBHHxPEG hybrid copolymer as an efficient intracellular delivery nanocarrier for kinase inhibitor

Xiao-Yun Lu[1*], Ming-Chuan Li[2], Xin-Liang Zhu[1], Fan Fan[1], Lei-Lei Wang[1] and Jian-Gang Ma[1]

Abstract

Background: Protein Kinases are key regulators of cell function and play essential roles in the occurrence and development of many human diseases. Many kinase inhibitors have been used for molecular targeted treatment of those diseases such as cancer and inflammation. However, those highly hydrophobic kinase inhibitors shared the common features of poor bioavailability and limited in vivo half-life, which strongly impeded their practical applications. Our previous study demonstrated that microbial synthesized biodegradable polyester poly(3-hydroxybutyrate-*co*-3-hydroxyhexanoate) (PHBHHx), a member of polyhydroxyalkanoates (PHAs) family, could serve as a promising delivery nanocarrier for those hydrophobic kinase inhibitors. Recently, a novel natural synthesized hybrid copolymer, PEG200 end-capped PHBHHx (PHBHHxPEG) was produced by *Aeromonas hydrophila* fermentation. In this study, the novel PHBHHxPEG NPs were prepared and investigated to serve as intracellular delivery nanocarriers for sustained release of hydrophobic kinase inhibitors.

Results: PHBHHxPEG nanoparticles (NPs) prepared by an emulsification–solvent evaporation method were spherical with a diameter around 200 nm. The entrapment efficiency on rapamycin in PHBHHxPEG NPs was 91.9% and the sustained release of rapamycin from PHBHHxPEG NPs could be achieved for almost 10 days. The cellular uptake of PHBHHxPEG NPs was significant higher than that of PHBHHx NPs. The anti-proliferation effect and mTOR inhibition ability of rapamycin-loaded PHBHHxPEG NPs was stronger than that of drug-loaded PHBHHx NPs and free rapamycin.

Conclusions: PHBHHxPEG NPs could achieve the efficient entrapment and sustained release of rapamycin. The novel biodegradable PHBHHxPEG appeared a promising nanocarrier for sustained delivery of hydrophobic kinase inhibitors with improved cellular uptake and kinase inhibition efficiency.

Keywords: Polyhydroxyalkanoate, PEG, Rapamycin, Nanoparticle, Drug delivery

Background

Protein kinases play essential roles in signal transduction and regulation by phosphorylating other proteins to alter their enzymatic activity, cellular location, or association with other proteins. Phosphorylation of the target protein is a necessary and key step in the occurrence and development of some cancers and inflammatory diseases. Therefore, protein kinase inhibitors which specifically block the action of one or more protein kinases have been used as drugs for molecular targeted treatment of cancer and inflammatory diseases. However, poor bioavailability and limited in vivo half-life are rather common features of many kinase inhibitors that usually are highly hydrophobic compounds. For example, rapamycin (RAP, also known as Sirolimus) isolated from *Streptomyces hygroscopicus* is such kind of hydrophobic kinase inhibitor which has already been approved by US Food and Drug Administration (FDA) as an immunosuppressive drug to prevent the rejection of solid organ transplants [1]. RAP is a natural inhibitor of the mammalian target of rapamycin (mTOR) which is a critical regulator of cell growth, proliferation and survival. Dysregulation of mTOR activities often occurs in a variety of human malignant diseases, making it a crucial

* Correspondence: luxy05@mail.xjtu.edu.cn
[1]Department of Biological Science and Bioengineering, Key Laboratory of Biomedical Information Engineering of Ministry of Education, School of Life Science and Technology, Xi'an Jiaotong University, Xi'an 710049, Shaanxi, P. R China
Full list of author information is available at the end of the article

and validated target for the cancer intervention [2]. The anti-proliferation activity of RAP was evaluated and it was found that RAP could also inhibit the proliferation of a number of tumor cell lines [3]. However, since it is a strongly hydrophobic compound and only slightly soluble in several acceptable excipients such as ethanol, propylene glycol and polyethylene glycol 400 [4], RAP is only available in oral but with a relative low bioavailability of around 17% [5]. Therefore, conventional dosage formulations of RAP could not fulfill the demand for anti-proliferation therapy.

Biodegradable polymeric nanoparticles are more and more frequently used in drug delivery systems and represent one of the most rapidly developing areas, which have now attracted growing interest of chemists, pharmacists and biologists. Nanosized polymeric nanoparticles provide a comprehensive platform for achieving the enhanced drug solubility/stability, improving the effectiveness of drug therapy and reducing the side effects of the loaded drug to healthy tissues. Many different natural and synthetic polymers have been developed as drug delivery carriers. Polyhydroxyalkanoates (PHAs) are a family of biodegradable aliphatic polyesters which are synthesized by a wide range of bacteria. PHAs have been extensively studied as implantable tissue repair/regeneration devices and other biomedical devices such as sutures and suture fasteners, because of their good biocompatibility and biodegradability [6,7]. Recently, PHAs have also been explored for controlled drug-release applications [8,9]. Based on their polyester aliphatic properties, PHAs are more efficient in encapsulating the hydrophobic compounds and thus would be especially beneficial for hydrophobic drugs. Several commercialized PHAs, for example, polyhydroxybutyrate (PHB) and poly(3-hydroxybutyrate-*co*-3-hydroxy-hexanoate) (PHBHHx), have been successfully developed into nanoparticles and investigated for the entrapment and release profiles of hydrophobic compounds. The drug-loaded PHA nanoparticles demonstrated sustained release of hydrophobic drugs and enhanced bioavailability in cell-based assays [8].

In our previous studies, a novel naturally synthesized hybrid polyester, polyethylene glycol 200 (PEG200) end-capped poly(3-hydroxybutyrate-*co*-3- hydroxyhexanoate) (PHBHHxPEG) was directly produced by *Aeromonas hydrophila* fermentation [10]. In this study, the novel PHBHHxPEG NPs were prepared and investigated to serve as the nanocarrier for sustained release of a hydrophobic kinase inhibitor like RAP. RAP cellular uptake of PHBHHxPEG NPs, *in vitro* release of RAP from the NPs, cellular proliferation inhibition and kinase inhibition of RAP-loaded NPs were investigated. The results obtained in this study indicated that the microbial synthesized PHBHHxPEG hybrid copolymer represents a promising material to serve as an intracellular drug delivery carrier for sustained release of hydrophobic kinase inhibitors.

Results

Preparation and characterization of PHBHHx and PHBHHxPEG NPs

Two kinds of free nanoparticles (PHBHHx and PHBHHxPEG), two kinds of rhodamine-loaded nanoparticles (PHBHHx and PHBHHxPEG) and three kinds of rapamycin-loaded nanoparticles (PLA, PHBHHx and PHBHHxPEG) were prepared by an emulsification–solvent evaporation technique. The sizes and polydispersities of the nanoparticles were listed in Table 1. Figure 1 showed the TEM photographs and the size distribution of PHBHHx and PHBHHxPEG NPs, respectively. The sizes of nanoparticles ranged from 100 to 300 nm after 6 h stirring with relative low polydispersity index (PDI). At the drug/material feeding ratio of 1:10, entrapment efficiency (EE) of rapamycin in both PHBHHx and PHBHHxPEG NPs were higher than 90%. The drug loading content (DLC) of RAP-loaded PHBHHx and PHBHHxPEG NPs were 8.52% and 8.47%, respectively. PLA NPs demonstrated lower entrapment ability compared with PHBHHx and PHBHHxPEG NPs, which was consisted with our previous report.

In vitro release of rapamycin

Three kinds of RAP-loaded NPs were prepared based on PLA, PHBHHx and PHBHHxPEG, respectively. Figure 2 shows the *in vitro* release profiles of RAP-loaded NPs in water at 37°C. PLA and PHBHHxPEG NPs showed

Table 1 The diameters and polydispersities of the free and rapamycin/rhodamine-loaded nanoparticles

NPs	D_{mean}±SD (nm)	PDI	EE±SD (%)	DLC±SD (%)
PHBHHx	196.2±39.6	0.023	—	—
PHBHHxPEG	178.3±31.5	0.045	—	—
PLA-rapamycin	203.0±37.2	0.025	64.1±2.2	6.01±0.3
PHBHHx-rapamycin	211.4±36.5	0.048	92.7±3.4	8.52±0.7
PHBHHxPEG-rapamycin	206.7±33.1	0.044	91.9±3.8	8.47±0.5
PHBHHx-rhodamine	202.7±42.9	0.091	—	—
PHBHHxPEG-rhodamine	202.2±34.2	0.018	—	—

Abbreviations: NPs nanoparticles, D_{mean}, mean diameter, *SD* standard deviation, *PDI* polydispersity index, *EE* Entrapment efficiency.

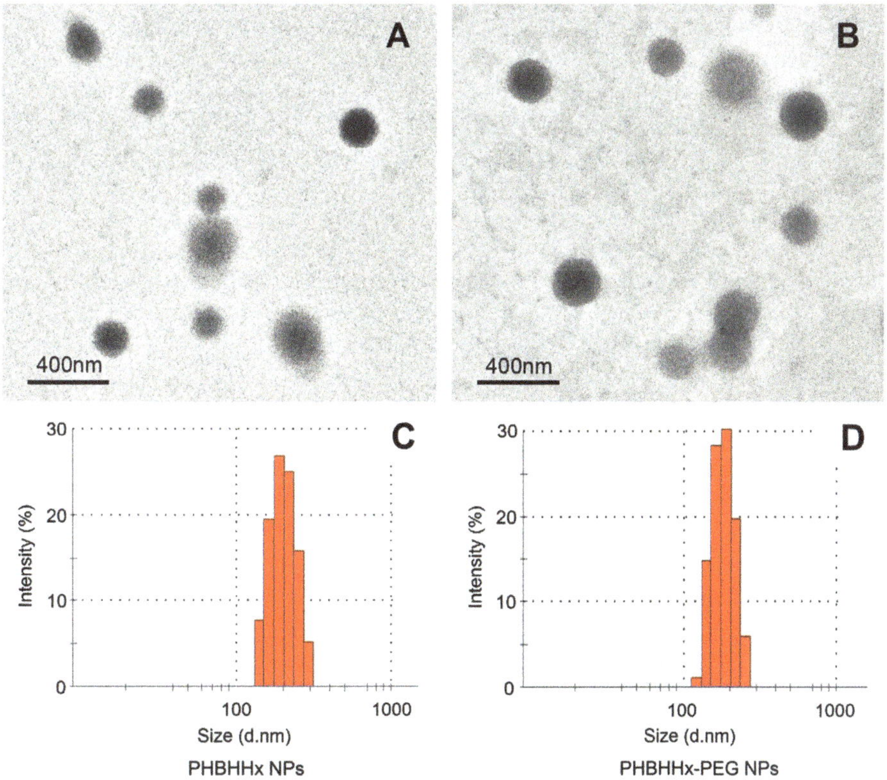

Figure 1 TEM photographs (A, B) and the NPs size distribution (C, D) of PHBHHx NPs (A, C) and PHBHHx-PEG NPs (B, D).

similar profiles of an initial burst release followed by a sustained discharge stage, which was a little bit faster than that from PHBHHx NPs. About 40% of the drug could be released within the first 24 hrs from PLA NPs and PHBHHx-PEG NPs and only about 30% of RAP was released from PHBHHx NPs during the same period. At 75 hours, the percentage of retained drug in PHBHHx and PHBHHxPEG NPs were 42% and 25%, respectively, which demonstrated that the release of RAP from PHBHHxPEG NPs was faster than that from PHBHHx NPs.

Interestingly, the release of RAP from PHA-based NPs was more efficient than that from PLA NPs. Indeed, more than 95% of the drug could be released from the PHBHHx and PHBHHxPEG NPs after 10 days. On the

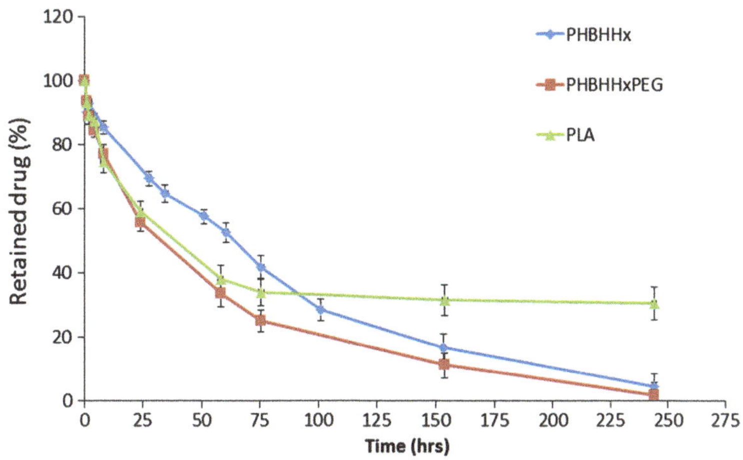

Figure 2 In vitro release profiles of RAP-loaded PHBHHx, PHBHHxPEG and PLA nanoparticles.

contrary, more than 30% RAP still remained in the PLA NPs even after 10 days, which indicated that this 30% of drug was strongly encapsulated in the PLA NPs and could not be released. Such behavior might be explained by the relativly poor crystalline nature of PHBHHx and PHBHHxPEG materials, thus allowing RAP to diffuse from the loose particle core. These results thus indicated that the PHBHHx and PHBHHxPEG nanoparticles could efficiently and effectively extend the release profile of RAP and could be eventually used as controlled delivery carriers for hydrophobic drugs.

Cellular uptake of PHBHHx and PHBHHxPEG NPs

Human prostate cancer cell line PC3 and a murine macrophage cell line RAW264.7 were used to investigate the *in vitro* endocytosis of PHA-based NPs. Cells were treated with the rhodamine B solution, rhodamine B-loaded PHBHHx NPs and PHBHHxPEG NPs in the same fluorescence intensity, respectively. After 3 hrs treatment, the intracellular fluorescence of the rhodamine B solution treated group was weak in both PC3 and RAW264.7 cells, which suggested that rhodamine B could not be easily absorbed by both these two cell lines (Figure 3A). However, significant intracellular fluorescence signal could be detected either by the fluorescence microscopy observation (Figure 3A) or by fluorescence intensity analysis (Figure 3B) in both PC3 and RAW264.7 cells 3 hrs after the addition of rhodamine B–loaded NPs. The internalized NPs were mainly accumulated in the cytoplasm and didn't reach the cell nucleus area, indicated by the weak fluorescence intensity area inside the cells. Moreover, the rhodamine B–loaded PHBHHxPEG NPs treating group showed much stronger intracellular fluorescence intensity compared to the rhodamine B–loaded PHBHHx NPs treating group, indicating that PHBHHxPEG NPs showed higher cell affinity and easier cellular uptake.

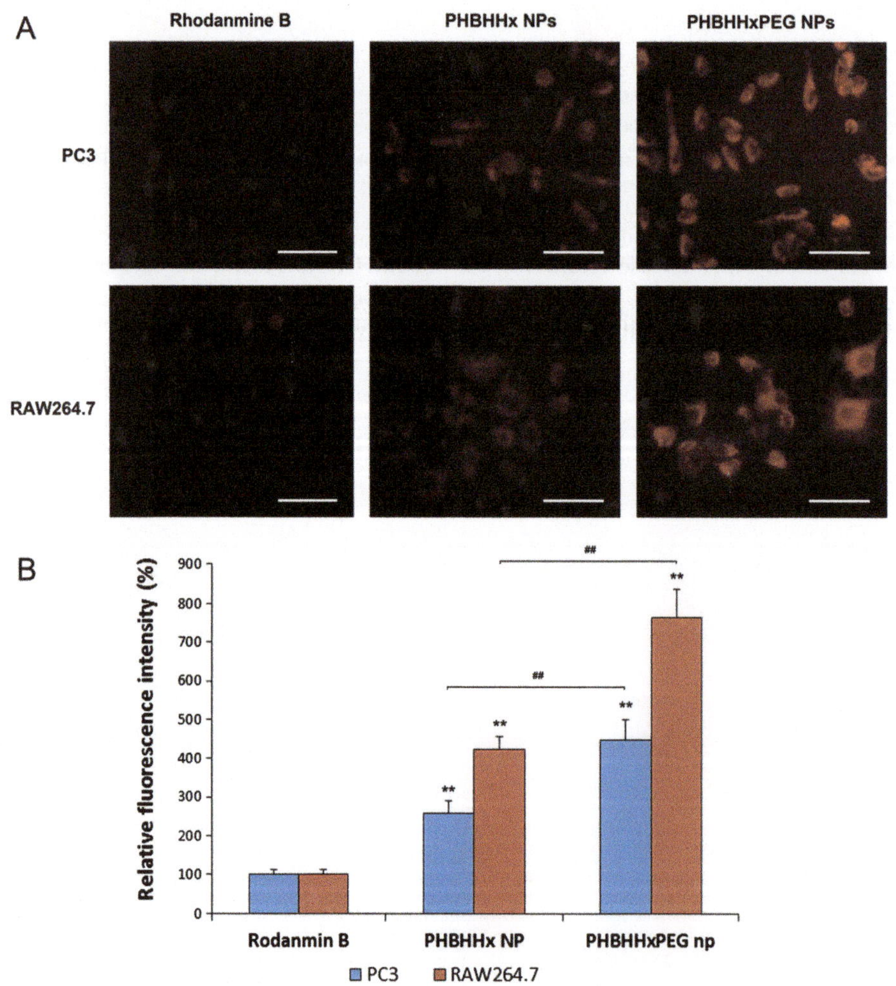

Figure 3 Uptake of PHBHHx and PHBHHxPEG nanoparticles by PC3 and RAW264.7 cells. A: Fluorescence microscopy pictures; B: Intracellular fluorescence intensities quantification by fluorescence spectrophotometer.

In vitro cytotoxicity evaluation of PHBHHx and PHBHHxPEG NPs

The cytotoxic effect of PHBHHx and PHBHHxPEG NPs on PC3 was evaluated *in vitro* by MTT assay. Figure 4 shows relative proliferation percentage of PC3 incubated with different NPs at doses of 100, 500 and 1000 μg ml^{-1} for 24 h, respectively. PHBHHx and PHBHHxPEG NPs could be well tolerated at the concentration of 100 μg ml^{-1}. However, decreased cell viability was observed in the groups treated with 1000 μg ml^{-1} and 500 μg ml^{-1} NPs, which resulted in 40% and 20% decrease of cell viability, respectively. PHBHHx NPs and PHBHHxPEG NPs-treated groups did not show significant difference in cell viability.

Effects of rapamycin-loaded NPs on proliferation of PC3 cells

The anti-proliferation effect of RAP-loaded NPs on PC3 cells was evaluated by the MTT assay. Cells were treated with 10 μM free RAP, RAP-loaded PHBHHx (100 μg ml^{-1}) and PHBHHxPEG NPs (100 μg ml^{-1}) which contained equal amounts of drug, respectively, for 48 h. PC3 cells showed a dramatic inhibition of cellular proliferation when treated with RAP-loaded NPs compared to cells treated with free RAP at the same concentration (Figure 5). The anti-proliferative effects persisted after 48 hours without further addition of NPs, potentially because of the sustained intracellular release of RAP from the NPs. In line with the efficient cellular uptake and the fast voiding rate of PHBHHxPEG NPs, RAP-loaded PHBHHxPEG NPs inhibited PC3 cell proliferation more efficiently than RAP-loaded PHBHHx NPs.

Effects of RAP-loaded NPs on mTOR activity

p70S6K (p70 S6 ribosomal protein kinase) is a main downstream mediator of mTOR and phosphorylation level of p70S6K at Thr389 could be used as a probe for mTOR activity. Different from the anti-proliferation effect on PC3, phospho-p70S6K signaling could be significant inhibited by RAP at a very low concentration (Figure 6), as previosly reported [11,12]. The inhibition on mTOR activity in PC3 could also be achieved by adjusting drug-loaded NPs dosage (10–50 ng ml^{-1}) to reach a final RAP concentration of either 1 or 5 nM (Figure 7). Treatment with free RAP, RAP-loaded PHBHHxPEG NPs and RAP-loaded PHBHHxPEG NPs equally resulted in decreased p70S6K phosphorylation (Thr389), without affecting total p70S6K and total mTOR levels. This indicated that the decreased phosphorylation level of p70S6K was only due to the inhibition of mTOR activity but not because of the downregulation of mTOR or p70S6K. In keeping with their anti-proliferative effects, drug-loaded PHBHHx and PHBHHxPEG NPs demonstrated to better inhibit mTOR than free RAP. RAP-loaded PHBHHxPEG NPs showed even stronger inhibition effects than RAP-loaded PHBHHx NPs.

Discussion

Because poor bioavailability and limited *in vivo* half-life are rather common features of many kinase inhibitors that usually are highly hydrophobic compounds, the development of polymer-based nanocarrier for kinase inhibitors provided a promising strategy to improve their therapeutic activity by releasing drugs in a sustained manner, and consequently enhancing their bioavailability, improving their solubility, and reducing the toxic side effects [13]. In addition, drug-carrying nanoparticles could be passively accumulated in tumor or inflammatory tissues through the enhanced permeability and retention (EPR) effect and thus could further increase the effectual concentration of hydrophobic drugs [14]. Polymeric nanoparticles have been extensively studied as drug carriers [15]

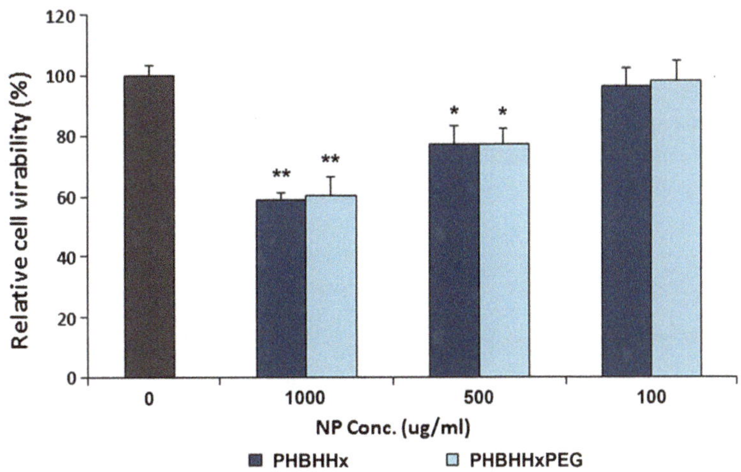

Figure 4 In vitro cytotoxicity of PHBHHx and PHBHHxPEG NPs on PC3.

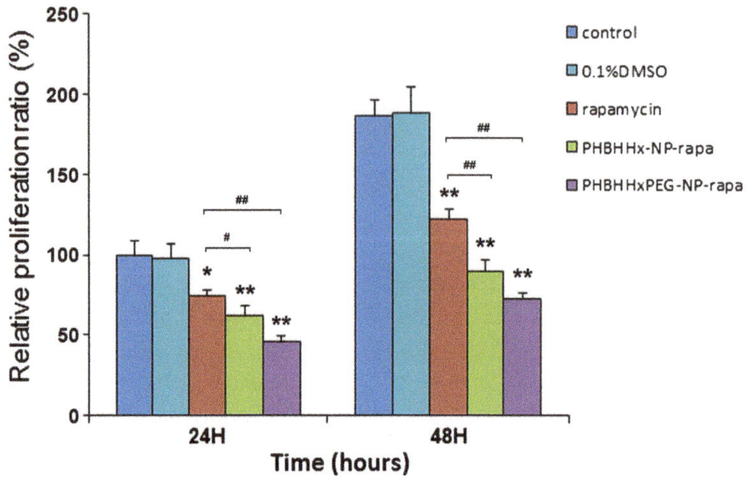

Figure 5 **The effect of RAP-loaded NPs on the proliferation of PC3.**

and several polymer-based, nanoparticulated formulations have been approved by the FDA for clinical applications [16]. PHAs are linear biopolyesters produced as energy- and carbon-storage materials by many bacteria. The medical applications of PHA have been extensively explored in recent years for implant biomedical applications [6]. Several members of PHAs, such as PHB and PHBHHx, have also been developed into nanoparticles and investigated for controlled drug-release applications [8,9,17]. Our previous data indicated that they were efficient in encapsulating hydrophobic compounds such as the phosphoinositide-3-kinases (PI3Ks) inhibitor TGX221 (Mw = 364.4). In this study, rapamycin, a mTOR inhibitor with relative high molecular mass of 914.17, was used to investigate the entrapment and release abilities of PHA NPs for high molecular weight kinase inhibitors. Both PHBHHx and PHBHHxPEG NPs showed higher rapamycin entrapment efficiency (more than 90%) and more complete drug release than that of PLA NPs, which was consisted with the study of TGX221-loaded PHA NPs. This indicated the advantage of PHA-based NPs in the entrapment and sustained release of hydrophobic kinase inhibitors. The results of this study also showed that release of rapamycin from PHBHHxPEG NPs was faster than that from PHBHHx NPs, which might be due to the incorporation of PEG fragments and the relativly low molecular weight of PHBHHxPEG.

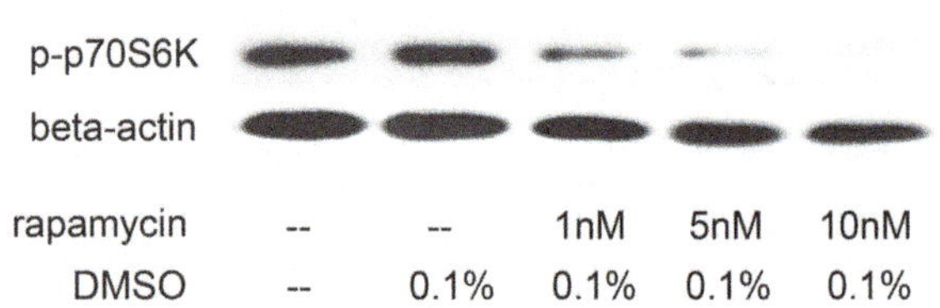

Figure 6 **Analysis of p70 S6K phosphorylation after treated with different concentrations of RAP.**

The incorporation of PEG could also confer more hydrophilicity to PHBHHxPEG NPs. This might explain the enhanced cellular uptake of PHBHHxPEG NPs, since a hydrophobic surface is known to disturb cell interaction. Yang et al. reported that the surface of PHA film demonstrated higher hydrophilicity and better cell compatibility after NaOH treatment [18]. This can be explained by the ability of NaOH treatment to break down the ester bond of PHA backbone, leaving more hydroxyl group and carboxyl group on the surface of PHA film. Our previous study also indicated that the NaOH treatment increased the surface hydrophilicity of PHBHHx film and the neural stem cell-PHBHHx film interaction (data not published). The hydrophilic PEG represents a biocompatible material commonly used to enhance the biocompatibility of an object with good cell compatibility. Therefore, the emulsification/solvent evaporation method was used to expose PEG at the surface of PHBHHxPEG NPs and hence to increase the hydrophilicity of NPs surface as well as the cell-NPs interaction. In agreement, our results showed that cell contact was maximal in PHBHHxPEG NPs.

PEGylation of nanocomposites was reported to be able to resist nonspecific protein adsorption and have prolonged circulating half-life *in vivo* [19]. However, in this study, the PHBHHxPEG NPs could not prevent the endocytosis by RAW264.7 murine macrophage-like cells. This might be explained by an insufficient PEG density on the colloidal surface of PHBHHxPEG NPs. PEG-modification of PLGA NPs is known to improve circulation time, depending on the molecular weight (chain length) and molar ratio (grafting efficiency) of PEG incorporation [20]. After 5 min from administration, the remaining PLGA particles in circulation are about 5%,

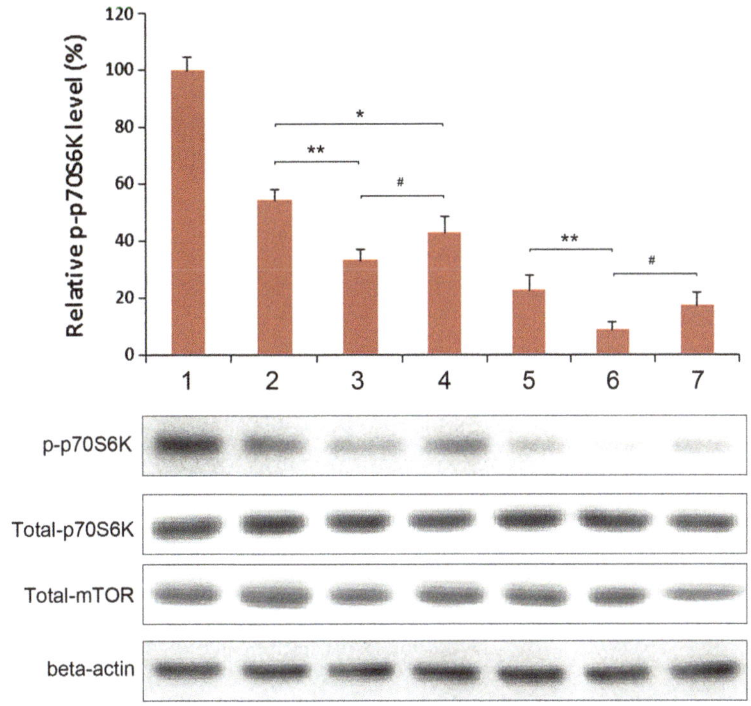

Figure 7 Analysis of p70 S6K phosphorylation after treated with different concentrations of RAP-loaded NPs.

25% and 50% for unmodified, PEG5000 modified and PEG20000 modified PLGA particles, respectively. This shows that the long PEG chain is necessary for increased anti-clearance properties. In our study, the PHBHHxPEG copolymer was synthesized by *A. hydrophila* by adding PEG200 in the culture medium. PEG200 attacks the carbonyl carbon of the PHBHHx chain directly and forms the end-capped PHBHHxPEG copolymer [10]. Thus, in our experiments the incorporated PEG fragment was no longer than 3–4 units in each chain. While this was insufficient to promote the formation of a dense, hydrophilic cloud on the surface of NPs, potentially preventing macrophage-dependent opsonization, the use of PEG200 was sufficient to optimize NPs hydrophilicity and bioavailability.

Conclusions

Overall, this study represented the potential of PHBHHxPEG NPs as a promising kinase inhibitor delivery carrier in anti-cancer study. Our results indicated that the entrapment of rapamycin into PHBHHxPEG NPs could sufficiently enhance its bioavailability and anti-proliferation effect. Nonetheless, *in vivo* evaluation of circulating time and anti-proliferation ability of the rapamycin-loaded PHA NPs in animal models is still needed. Conversely, the proven effectiveness in cell based assays of this novel intracellular delivery nanocarrier opens the way to similar formulation of other kinase inhibitors. Since the PHBHHxPEG NPs based drug delivery carrier could also be internalized by other cell types such as macrophage, rapamycin-loaded PHBHHxPEG NPs might also have alternative applications, for example, in the field of immunosuppressant therapeutic tregimens.

Methods

Materials

Poly(3-hydroxybutyrate-*co*-3-hydroxyhexanoate) (PHBHHx) (Mw =4 × 10^5) was donated by Lab of Microbiology, Department of Biological Science and Biotechnology, Tsinghua University (Beijing, P. R. China). PHBHHxPEG (Mw =1.8 × 10^5) was produced by our group [10]. Poly (vinyl alcohol) (PVA) (P1763), poly(DL-lactide) (PLA) (P1691) and rhodamine B (83689) were purchased from Sigma-Aldrich (USA). Rapamycin (S1039) was purchased from selleckchem (USA).

Preparation of drug-Loaded PHBHHx-PEG nanoparticles

The RAP-loaded NPs and rhodamine B-loaded NPs were fabricated by a modified emulsification/solvent evaporation method [8,9,21]. Briefly, 20 mg of polymers and RAP (2 mg) or rhodamine B (1 mg) were added into 1 ml chloroform and the mixture was stirred to ensure that all materials were dissolved. Then the polymer-drug organic solution was slowly dropped into 20 ml of 1% PVA (w/v) under sonication. The double emulsion was then deal with sonication using a probe sonicator

(Sonics & Materials, Newtown, CT, USA) for 5 min, followed by moderately stirred with a magnetic mixer for 10 h to solidify the nano-droplets. Chloroform was removed by volatilization at room temperature. The NPs were collected by centrifugation at 12000 rpm for 30 min and then washed twice with phosphate buffered saline (PBS) solution. Three types of hydrophobic polymer NPs (PHBHHx, PHBHHxPEG and PLA) were prepared in this study.

Characterization of nanoparticles

The particle size distribution and polydispersity index (PDI) were measured by a laser light scattering machine (Zetasizer Nano ZS, Malvern, UK). The samples were diluted into proper concentration with deionized water and examined to determine the average particle diameters and PDI. The shape of resulting NPs was observed by JEM-2100 transmission electron microscope (JEOL Ltd. Japan).

Entrapment efficiency and drug-loading content measurement

The drug entrapment efficiency refers to the amount of RAP loaded into the NPs as compared with the total amount fed at the beginning. The RAP entrapment efficiency and drug loading content (DLC) was calculated according to the equations (1) and (2).

$$\text{Entrapment efficiency }(\%) = (M_t/M_i) \times 100\% \qquad (1)$$

$$\text{DLC}(\%) = (M_t/M_p) \times 100\% \qquad (2)$$

Where M_i was the mass of RAP fed initially, M_t represented for the total amount of RAP in NPs and M_p was the mass of the resulting NPs, respectively [22]. The amount of RAP was analyzed by high performance liquid chromatography (HPLC) with a SinoChrom ODS-BP column (5 μm, 4.6 mm × 15 mm; Elite Analytical Instruments, Dalian, P.R. China). To measure the amount of entrapped or retained RAP in NPs, 50 microliters of resuspended RAP-loaded NPs were dissolved in 450 μl acetonitrile at room temperature and then stored at 4°C for 4 hours to precipitate polymer. The supernatant was used to perform the HPLC analysis after filtrated with 0.2 μm filter.

In vitro drug release study

In vitro release studies of RAP from PHBHHx, PHBHHxPEG and PLA NPs were performed by the dialysis bag method [22]. Briefly, 10 ml NPs suspension was placed in a dialysis membrane bag with a molecular weight cutoff of 8000 ~ 14000 g/mol. The bag was tied and immersed into a 500 ml beaker with sustained running water to ensure the sufficient dissolution of RAP in the water outside the bag. The entire system was kept at 37°C with continuous stirring. Then 50 microliters of the NPs solution was taken out from the dialysis bag at predetermined time intervals. The amount of retained RAP (Cr) and the initial amount of RAP (Ci) in NPs were determined by HPLC as mentioned in the previous section. The released percentage was calculated according to the equation (3).

$$\text{Released percentage }(\%) = (\text{Ci-Cr})/\text{Ci} \times 100 \qquad (3)$$

Cell culture and proliferation assay

Human prostate cancer cell line PC3 was used to evaluate the *in vitro* proliferation inhibition effect of RAP-loaded NPs. PC3 cells were maintained in RPMI 1640 (Invitrogen, USA) supplemented with 10% fetal bovine serum (Gibico) and 1% penicillin-streptomycin at 37°C in humidified environment of 5% CO_2. The medium was replenished every other day and the cells were subcultured after reached confluence.

For measurement of proliferation, cells were seeded in triplicate at 2×10^3 cells/well in 96-well culture plates and incubated overnight to allow cell attachment. To evaluate the cytotoxicity of NPs, cells were treated with empty PHBHHx NPs and PHBHHxPEG NPs with the concentration of 100, 500 and 1000 μg/mL for 24 hours. Cells were incubated with the RAP-loaded PHBHHx NPs, PHBHHxPEG NPs or free RAP for 24 and 48 hours to analyze the anti-proliferation effects of RAP-loaded NPs on PC3 cell lines. Cell viability was quantified by a 3-(4,5-dimethylthiazol-2-yl)-2,5-diphenyltetrazolium bromide (MTT) assay. The absorbance of viable cells was measured at 570 nm using a microplate reader (Anthos 2020, Anthos Labtech Instruments, Wals., Austria). Relative cell viability was calculated with regard to that of the cell control at 24 hour, which was set to 100% viability.

Cellular uptake of nanoparticles

Human prostate cancer cell line PC3 and a murine macrophage cell line RAW264.7 cells were used to investigate the *in vitro* endocytosis of NPs. Cells were seeded in 96-well plates at 1×10^4 cells/well and incubated at 37°C with 5% CO_2 in a humidified incubator overnight to allow cell attachment, followed by treating with rhodamine-loaded NPs (100 μg/mL) and free rhodamine B (with equal fluorescence intensity) of for 3 h. Medium was then discarded and cells were washed with PBS twice to remove the extracellular rhodamine B and rhodamine B-loaded NPs. New medium was added to each cell cultures again and PC3 and RAW264.7 cells were observed by fluorescence microscopy to evaluated cellular uptake of various NPs. The intracellular fluorescence intensities were quantified by fluorescence spectrophotometer (Infinite M200 PRO, Tecan Group Ltd. Switzerland).

mTOR inhibition assay and western blot analysis

Cells were seeded in 6-well plates at 2×10^5 cells/well and incubated at 37°C with 5% CO_2 in a humidified incubator overnight to get attached and then treated with free RAP or RAP-loaded NPs for 24 h. After washing with cold PBS for two times, cells were lysated by adding RIPA lysis Buffer (50 mM Tris–HCl (pH 7.4), 150 mM NaCl, 1% NP-40), followed by incubating at 4°C for 30 min. Cell lysates were centrifuged at 10000 rpm for 5 min and supernatant was then loaded into 10% SDS–PAGE gels. After electrophoresis, proteins were transferred onto PVDF membranes, which were then blocked with 5% bovine serum albumin in TBST buffer for 1 h at room temperature. PVDF membranes were incubated overnight with an anti-phospho-p70S6 Kinase (Thr 389) antibody (Cell Signaling Technology Inc., USA, #9205), anti-p70S6 Kinase antibody (Cell Signaling Technology Inc., USA, #9202), anti-mTOR antibody (Cell Signaling Technology Inc., USA, #2972) and anti-β-actin antibody from Beijing Biosynthesis Biotechnology (Beijing, China), respectively, and subsequently hybridized with secondary HRP conjugated anti-rabbit IgGs (Zhongshan Goldenbridge Biotechnology, Beijing, China, #ZB-2301). The hybrid signals were detected by incubating with an enhanced chemiluminescence solution (ECL) purchased from Thermo scientific (#LK151361). The amount of phospho-p70S6K, total p70S6K, total mTOR and β-actin were analyzed by Quantity One software (Bio-Rad laboratory Inc., USA) and the relative phospho-p70S6K level was normalized according to the amount of total p70S6K.

Statistical analysis

All results were expressed as mean ± standard deviation of at least three independent experiments performed in triplicates. Statistical analysis was performed using a one-way analysis of variance (ANOVA) with Bonferroni post-hoctest. Differences were considered to be statistically significant at a level of $p < 0.05$.

Competing interests

The authors declare that they have no competing interests.

Authors' contributions

XYL designed and conducted all experiments and prepared the manuscript. MCL prepared the copolymer materials and fabricated the nanoparticles. XLZ and FF participated to characterize the nanoparticles and analyzed the in vitro release profile of rapamycin. LLW carried out the cell culture and proliferation assay as well as the western blot assay of mTOR inhibition. JJM helped to analyze the data and also contributed to the manuscript preparation. All authors read and approved the final manuscript.

Acknowledgments

Prof. Lu Shemin (College of Medicine, Xi'an Jiaotong University) is acknowledged for providing the murine macrophage cell line RAW264.7 cells. The authors would also like to express their thanks to Prof. Wu Daocheng and Prof. Liu Jiankang (both from School of Life Science and Technology, Xi'an Jiaotong University), for assistance with the characterization of nanoparticles and biological assay of mTOR inhibition, respectively. This work was supported by grants from the National Natural Science Foundation of China (81172170, 81371288), Science and Technology Research and Development Program of Shaanxi Province (2013KW32-04) and the Fundamental Research Funds for the Central Universities from Xi'an Jiaotong University.

Author details

[1]Department of Biological Science and Bioengineering, Key Laboratory of Biomedical Information Engineering of Ministry of Education, School of Life Science and Technology, Xi'an Jiaotong University, Xi'an 710049, Shaanxi, P. R China. [2]Molecular Biotechnology Center, Universita di Torino, 10126 Torino, Italy.

References

1. Kahan BD, Camardo JS: **Rapamycin: clinical results and future opportunities.** *Transplantation* 2001, **72**(7):1181–1193.
2. Meric-Bernstam F, Gonzalez-Angulo AM: **Targeting the mTOR Signaling Network for Cancer Therapy.** *J Clin Oncol* 2009, **27**(13):2278–2287.
3. Douros J, Suffness M: **New Anti-Tumor Substances Of Natural Origin.** *Cancer Treat Rev* 1981, **8**(1):63–87.
4. Simamora P, Alvarez JM, Yalkowsky SH: **Solubilization of rapamycin.** *Int J Pharm* 2001, **213**(1–2):25–29.
5. Dias VC, Yatscoff RW: **An in vitro method for predicting in vivo oral bioavailability of novel immunosuppressive drugs.** *Clin Biochem* 1996, **29**(1):43–49.
6. Chen GQ, Wu Q: **The application of polyhydroxyalkanoates as tissue engineering materials.** *Biomaterials* 2005, **26**(33):6565 6578.
7. Wu Q, Wang Y, Chen GQ: **Medical application of microbial biopolyesters polyhydroxyalkanoates.** *Artif Cells Blood Substit Biotechnol* 2009, **37**(1):1–12.
8. Lu XY, Ciraolo E, Stefenia R, Chen GQ, Zhang YL, Hirsch E: **Sustained release of PI3K inhibitor from PHA nanoparticles and in vitro growth inhibition of cancer cell lines.** *Appl Microbiol Biotechnol* 2011, **89**(5):1423–1433.
9. Xiong YC, Yao YC, Zhan XY, Chen GQ: **Application of polyhydroxyalkanoates nanoparticles as intracellular sustained drug-release vectors.** *J Biomater Sci-Polym Ed* 2010, **21**(1):127–140.
10. Zhang YL, Lu XY, Liu QQ, Li MC, Yang ZQ, Ma JG: **Production and characterization of poly(3-hydroxybutyrate-co-3-hydroxyhexanoate)-poly (ethylene glycol) hybird copolymer with adjustable molecular weight.** *Chin J Polymer Sci* 2012, **30**(1):101–111.
11. Blaser B, Waselle L, Dormond-Meuwly A, Dufour M, Roulin D, Demartines N, Dormond O: **Antitumor activities of ATP-competitive inhibitors of mTOR in colon cancer cells.** *BMC Cancer* 2012, **12**:86.
12. Foster DA, Toschi A: **Targeting mTOR with rapamycin: one dose does not fit all.** *Cell Cycle* 2009, **8**(7):1026–1029.
13. Shi JJ, Votruba AR, Farokhzad OC, Langer R: **Nanotechnology in Drug Delivery and Tissue Engineering: From Discovery to Applications.** *Nano Lett* 2010, **10**(9):3223–3230.
14. Langer R: **Drug delivery and targeting.** *Nature* 1998, **392**(6679):5–10.
15. Mora-Huertas CE, Fessi H, Elaissari A: **Polymer-based nanocapsules for drug delivery.** *Int J Pharm* 2010, **385**(1–2):113–142.
16. Wang M, Thanou M: **Targeting nanoparticles to cancer.** *Pharmacol Res* 2010, **62**(2, Sp. Iss. SI):90–99.
17. Yao YC, Zhan XY, Zhang J, Zou XH, Wang ZH, Xiong YC, Chen J, Chen GQ: **A specific drug targeting system based on polyhydroxyalkanoate granule binding protein PhaP fused with targeted cell ligands.** *Biomaterials* 2008, **29**(36):4823–4830.
18. Yang XS, Zhao K, Chen GQ: **Effect of surface treatment on the biocompatibility of microbial polyhydroxyalkanoates.** *Biomaterials* 2002, **23**(5):1391–1397.
19. van Vlerken LE, Vyas TK, Amiji MM: **Poly(ethylene glycol)-modified nanocarriers for tumor-targeted and intracellular delivery.** *Pharm Res* 2007, **24**(8):1405–1414.
20. Gref R, Minamitake Y, Peracchia MT, Trubetskoy V, Torchilin V, Langer R: **Biodegradable Long-Circulating Polymeric Nanospheres.** *Science* 1994, **263**(5153):1600–1603.

21. Perez C, Sanchez A, Putnam D, Ting D, Langer R, Alonso MJ: **Poly(lactic acid)-poly(ethylene glycol) nanoparticles as new carriers for the delivery of plasmid DNA.** *J Control Release* 2001, **75**(1–2):211–224.
22. Li Q, Wang Y, Feng NP, Fan ZZ, Sun J, Nan YL: **Novel polymeric nanoparticles containing tanshinone IIA for the treatment of hepatoma.** *J Drug Target* 2008, **16**(10):725–732.

Mesenchymal stem cell spheroids exhibit enhanced *in-vitro* and *in-vivo* osteoregenerative potential

Yuichiro Yamaguchi[1], Jun Ohno[2*], Ayako Sato[1], Hirofumi Kido[1] and Tadao Fukushima[3]

Abstract

Background: Mesenchymal stem cells (MSCs) are a favored cell source for regenerative medicine because of their multilinage potential. However, the conventional monolayer technique used to culture MSCs, inadequately overcomes their low differentiation capacity. Culture of MSCs in multicellular spheroids, more accurately mimics the *in-vivo* microenvironment; thus, resolving this problem. In this study, we assessed whether the osteoregenerative potential of MSC spheroids is greater than that of monolayer MSCs.

Results: MSC spheroids were generated from rat MSCs (rMSCs) using low-binding plates. Real-time reverse transcription-polymerase chain reaction and immunocytochemical analysis indicated that osteogenic properties were accelerated in MSC spheroids compared with monolayer rMSCs when treated with an osteoblast-inducer reagent for 7 days. Moreover, increased calcium deposition was visualized in MSC spheroids using Alizarin red staining. In a rat calvarial defect model, micro-computed tomography and histological assays showed that MSC spheroid-engrafted defects experienced enhanced bone regeneration.

Conclusions: Our *in-vitro* and *in-vivo* results reveal that MSCs in the spheroid culture exhibit enhanced osteoregenerative efficiency compared with monolayer MSCs.

Keywords: Spheroid, Mesenchymal stem cell, Osteogenesis, Rat calvarial defect, Bone regeneration

Background

Adult stem cells are widely used as a cell source for regenerative medicine because of their multilineage potential. These cells are easily harvested from bone marrow [1,2], adipose tissue [3,4], and other sites [5-9]. Bone marrow contains hematopoietic stem cells, mesenchymal stem/stromal cells (MSCs), and multipotent adult progenitor cells [10-12]. MSCs are capable of self-renewal and differentiation into several mesenchymal lineages *in-vitro* and *in-vivo*, including bone, fat, cartilage, and skeletal muscles [12,13]. These are considered suitable for use in tissue regeneration. As culture in the presence of osteogenic supplements facilitates MSCs to undergo differentiation into the osteogenic phenotypes, MSCs have been utilized as a cell source for osteogenic tissue regeneration [10,11]. Bone marrow-derived MSCs have great potential for bone regeneration in future clinical applications.

MSCs are commonly cultured as a two-dimensional (2D) monolayer using conventional tissue-culture techniques. These 2D-monolayer techniques inadequately reproduce the *in-vivo* microenvironment of stem cells, established by extrinsic and intrinsic cellular signals and have a profound influence on their biological functions [14]. Culturing multipotent MSCs in a 2D adherent monolayer can alter their normal physiological behavior, resulting in the loss of replicative ability, colony-forming efficiency, and the differentiation capabilities over time [15,16]. Replication of this complex *in-vivo* microenvironment *in-vitro* requires highly sophisticated cell-culture systems.

To mimic the *in-vivo* microenvironment more accurately *in-vitro*, various three-dimensional (3D) culture systems have been developed. Spheroids, spherical clusters of cells formed by self-assembly, comprise one of the best models for the 3D culture [17,18]. Our understanding of

* Correspondence: johno@college.fdcnet.ac.jp
[2]Department of Morphological Biology, Section of Pathology, Fukuoka Dental College, 2-15-1 Tamura, Sawara-ku, Fukuoka 814-0193, Japan
Full list of author information is available at the end of the article

the spheroid cell biology mainly derives from the *in-vitro* culture of cancer cell lines. Many studies have highlighted significant differences between 2D and 3D cultures, with the latter better reflecting the *in-vivo* microenvironment in terms of cellular heterogeneity, nutrient and oxygen gradients, cell-cell interactions, matrix deposition, and gene expression profiles [19-24]. Previous studies have reported several methods of generating 3D MSC spheroids. Many of these methods involve the use of a cell-suspension system or nonadherent surface to induce the spheroid formation [25-27]. In general, these MSC spheroids possess a greater differentiation capacity.

In this study, we examined the osteogenerative potential of the spheroids of rat MSCs (rMSCs) isolated from bone marrow compared with monolayer rMSCs in both *in-vitro* assays and a rat calvarial defect model, to elucidate whether MSC spheroids exhibit enhanced bone regeneration.

Results

Mesenchymal stem cell spheroid formation

First, we examined the generation of MSC spheroids from rMSCs (Figure 1A) at a density of 1×10^4 cells/well using round-bottom, low-binding plates. Culture of rMSCs in the low-binding plates generated a single spheroid per well. At 2 h after seeding 100 μl of rMSC cell suspension into the plates, cells spontaneously aggregated in the medium (Figure 1B). Cell aggregates formed compact multicellular spheroids 24 h after suspension culture (Figure 1C).

Furthermore, we investigated the effects of seeding density (10,000, 1,000, or 100 per well) on the spheroid formation. After the spheroids had been allowed to aggregate for 24 h, the spheroid diameter was quantified via bright field microscopy ($n = 5$ per seeding density). The spheroid diameter increased with increasing cell number, ranging from 83.2 ± 6 μm (6 micrometer) for the 100-cell spheroid to 530 ± 34.3 μm (34.3 micrometer) for the 10,000-cell spheroid (Figure 2A and B). These results indicate that the spheroid size can be controlled by cell seeding density.

Increased expression of osteogenic genes in mesenchymal stem cell spheroids treated with an osteoblast-inducer reagent

After treatment with an osteoblast-inducer reagent (OIR) for 7 days, no cytological changes were observed in both monolayer MSCs and MSC spheroids (Figure 3). The spheroid diameter slightly decreased for 7-day treatment with OIR, ranging from 327.8 ± 17.8 μm on day 0 to 290.3 ± 33.8 μm on day 7. To investigate whether osteogenic genes are induced in MSC spheroids treated with an OIR, the relative mRNA expression levels of *runt-related transcription factor-2* (*RUNX-2*), *osterix* (*OSX*), *osteopontin* (*OPN*), and *bone sialoprotein* (*BSP*) mRNA expression were analyzed using quantitative reverse transcription-polymerase chain reaction (RT-PCR). Compared with monolayer rMSCs treated with an OIR, the relative mRNA expression levels of *OSX* (normalized to that of *glyceraldehyde 3-phosphate dehydrogenase* [G3PDH] mRNA) increased by 8.27-fold (Figure 4A). Similarly, the mRNA expression levels of *RUNX-2*, *OPN*, and *BSP* increased by 1.57-, 1.94-, and 1.33-fold, respectively (Figure 4B,C, and D). In contrast, *alkaline phosphatase* (ALP) mRNA expression remains unchanged in both MSC spheroids and monolayer rMSCs (Figure 4E). The levels of expression of these osteogenic genes were significantly upregulated in MSC spheroids compared with monolayer rMSCs.

Immunocytochemical expression of osterix in differentiated mesenchymal stem cell spheroids

Furthermore, we performed immunocytochemical analysis to examine the protein expression of the osteogenic marker OSX in spheroid and monolayer rMSCs, cultured with medium containing an OIR (Figure 5). MSC spheroids exhibited red-colored nuclear fluorescence (Figure 5A, upper). In contrast, monolayer MSCs showed no staining in the nuclei, and only faint staining was evident in the cytoplasm (Figure 5A, lower). A significant increased percentage of total osterix positive cells was observed in MSC spheroids as compared to monolayer MSCs [total: 73.0 ± 4.0 vs 18.5 ± 1.8 ($p < 0.01$)] (Figure 5B).

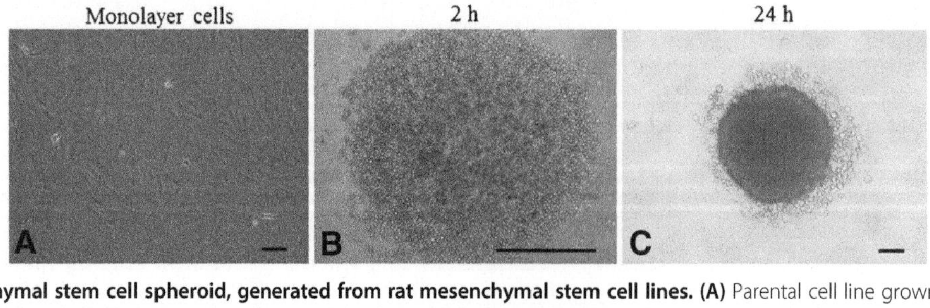

Figure 1 Mesenchymal stem cell spheroid, generated from rat mesenchymal stem cell lines. (A) Parental cell line grown permanently as a monolayer culture. **(B)** and **(C)** The generation of mesenchymal stem cell (MSC) spheroids. Cellular aggregation of MSCs was observed 2 h after a microplate culture **(B)**. A cell aggregation formed a compact spheroid at 24 h **(C)**. Scale bars = 100 μm (a and c) and 1 mm **(B)**.

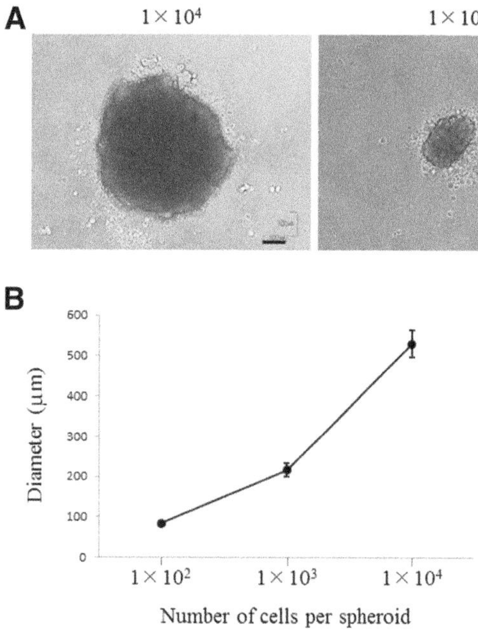

Figure 2 Effect of seeding densities on a spheroid formation. (A) Representative images of mesenchymal stem cell (MSC) spheroids grown on low-binding plates at 24 h after implantation. Scale bar = 100 μm. **(B)** The diameter of MSC spheroids increased linearly with increasing cell number.

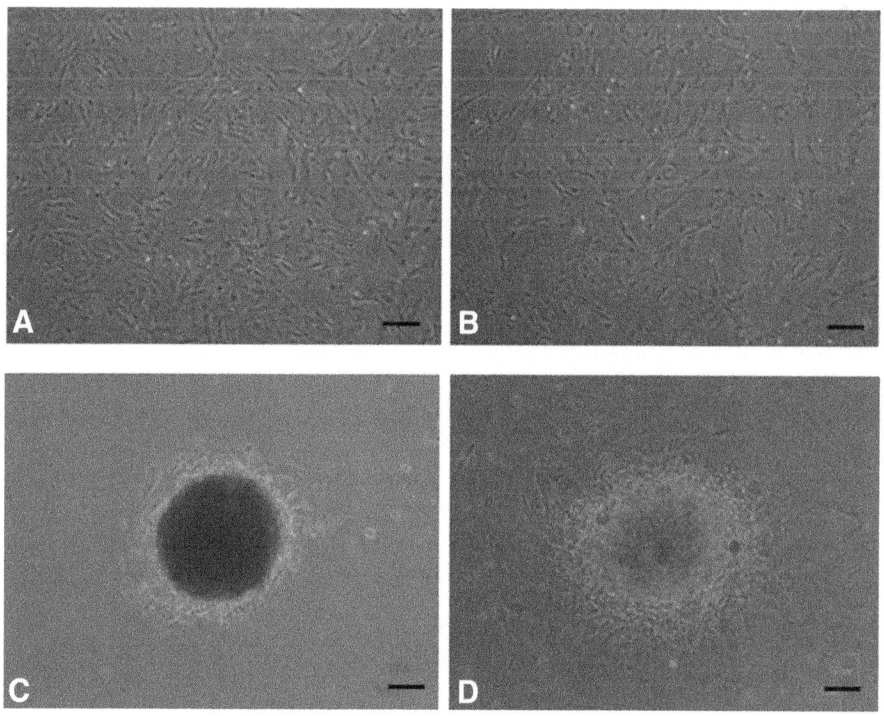

Figure 3 Cytological images of differentiated monolayer mesenchymal stem cells (MSC) and MSC spheroids. No cytological changes were observed inboth monolayer MSCs and MSC spheroids cultured with an osteoblast-induced reagent. Untreated monolater MSCs **(A)**, treated monolayer MSCs **(B)**, untreated MSC spheroids **(C)**, and treated MSC spheroids **(D)**. Bars = 100 μm.

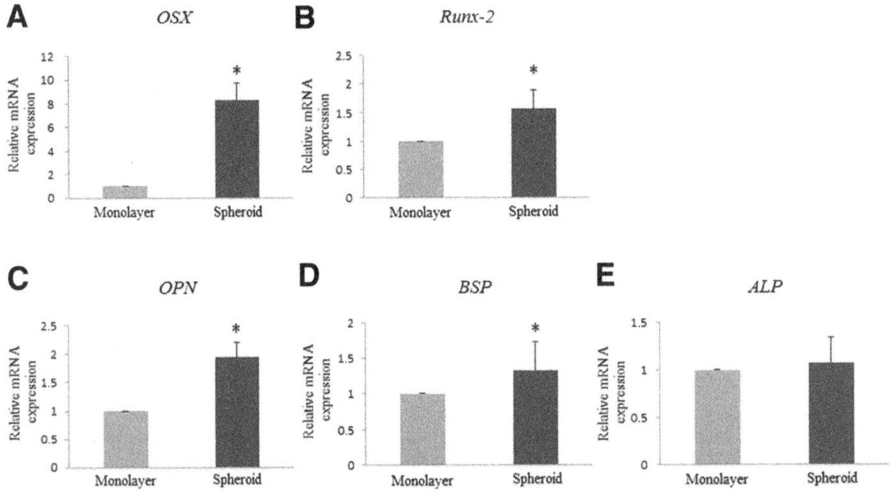

Figure 4 Expression of osteogenic marker genes in both mesenchymal stem cell spheroids and monolayer rat mesenchymal stem cells. Quantitative reverse transcription-polymerase chain reaction results for *runt-related transcription factor-2*, *osterix*, *osteopontin*, and *bone sialoprotein*, and *alkaline phosphatase* mRNA expression in mesenchymal stem cell (MSC) spheroids (black column) and monolayer rat MSCs (rMSCs; gray column) treated with an osteoblast-inducer reagent (OIR). Results are expressed as fold-increases in mRNA expression (normalized to that of *glyceraldehyde 3-phosphate dehydrogenase* mRNA) and compared with results for monolayer rMSCs treated with an OIR on Day 7. mRNA expression levels of osteogenic genes, other than alkaline phosphatase (ALP) **(E)**, in MSC spheroids were increased, compared with those of monolayer MSCs **(A-D)**. Vertical lines represent the mean ± standard deviation of three independent experiments, each performed in triplicate. *Significantly different at $p < 0.05$ compared with monolayer rMSCs. *OSX*, osterix; *RUNX-2*, runt-related transcription factor-2; *OPN*, osteopontin; *BSP*, bone sialoprotein; *ALP*, alkaline phosphatase.

These results reveal that OSX expression is induced in the nuclei of spheroid MSCs treated with an OIR for 7 days.

Increased alkaline phosphatase staining and induction of calcified deposits in mesenchymal stem cell spheroids treated with an osteoblast-inducer reagent

ALP staining intensity and Alizarin red (AR) deposits are well-defined cytochemical markers of osteogenesis and are considered to reflect the differentiation into osteoblasts [28,29]. We compared the staining intensity for ALP and AR in MSC spheroids and monolayer rMSCs, cultured with an OIR for 7 days. ALP staining intensity was evident in both MSC spheroids (Figure 6A) and monolayer rMSCs (Figure 6B) cultured with an OIR for 7 days. These results indicate that both spheroids and monolayer cells can differentiate into progenitors for bone formation after an OIR stimulation.

AR staining of calcified deposits was used to examine the mineralization of the extracellular matrix. We investigated whether MSC spheroids cultured with an OIR for 7 days, exhibit a late period of matrix maturation. The centeral areas of MSC spheroids (Figure 7A) appeared red, whereas no spots of AR staining were evident in monolayer rMSCs (Figure 7B). These results indicate that MSC spheroids treated with an OIR induce more calcification than monolayer rMSCs.

Mesenchymal stem cell spheroids enhance the healing of rat calvarial defects

To determine whether MSC spheroids enhance bone regeneration, we implanted MSC spheroids/Matrigel™ (BD Biosciences, San Jose, CA, USA), monolayer rMSCs/Matrigel™, Matrigel™ alone, or nothing (blank) into rat calvarial bone defects. The area of regenerated bone was investigated by micro-computed tomography (micro-CT) scanning (Figure 8A). At 2 and 4 weeks after implantation, bone defect areas were still evident in the blank, Matrigel™ alone, and monolayer rMSC/Matrigel™ groups. Conversely, the defect areas were small in the MSC spheroid/Matrigel™ group because they were filled with regenerated bones.

We quantified these micro-CT findings to yield the percentage of defect healing by quantifying the pixels in these defects. These results are summarized in Figure 7B. MSC spheroid/ Matrigel™ -implanted rats (49.4 ± 8.7% at 2 weeks; 72.2 ± 7.0% at 4 weeks) showed higher percentages of new bones at 4 weeks compared with rats in the other groups. The area of regenerated bones in rats in the other three groups was less than 55% (blank: 25.6 ± 4.9% and 45.0% ± 8.1%, Matrigel™ alone: 26.3 ± 5.0% and 46.2 ± 5.6%, monolayer rMSC/ Matrigel™: 39.9 ± 6.0% and 52.3 ± 10.2% at 2 and 4 weeks, respectively]. Thus, the MSC spheroids group showed higher bone regeneration in the peripheral area surrounding the defect compared with other groups.

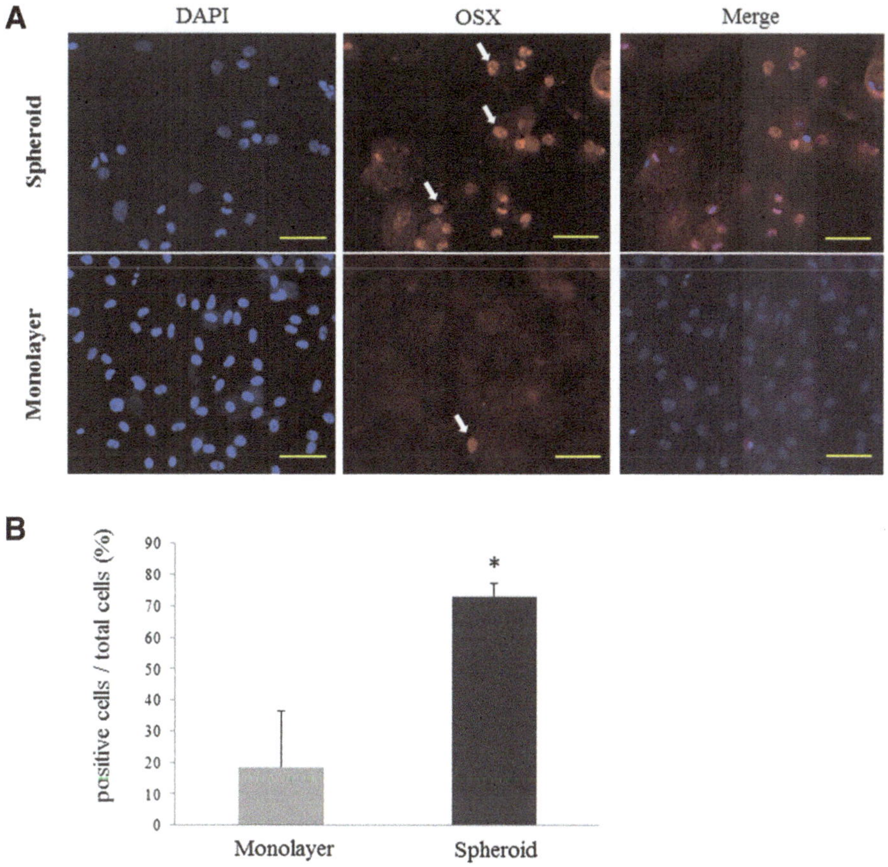

Figure 5 Immunocytochemical analysis of osterix expression in mesenchymal stem cell spheroids treated with an osteoblast-inducer reagent for 7 days. **(A)** Compared with monolayer mesenchymal stem cell (MSC; lower), positive cells (red) with anti- osterix antibody were increased in MSC spheroids (upper). 4',6-diamidino-2-phenylindole (DAPI) was used to label nuclei (blue). Arrows, positive reaction in nuclei. Scale bars = 100 μm. **(B)** Significantly increased percentage of osterix positive cells was observed in MSC spheroids when compared to monolayer MSCs (*, $p < 0.01$).

Histological analysis also revealed that the bones were well generated in the MSC spheroids group compared with the other groups. At 4 weeks after implantation, the bone defects were almost replaced by regenerated bone in the MSC spheroids group. In the other groups, some newly regenerated bones were evident, but considerable areas of the bone defects remained filled with fibrous connective tissue. No inflammatory changes were observed in any group (left and middle columns in Figure 9A). Azan Mallory staining showed that areas of the bone defects were filled with mature collagenous fibers in the blank, Matrigel™ alone, monolayer MSC/ Matrigel™ groups (right column in Figure 9A). A histomorphometric analysis represented that 74.4 ± 1.9% and

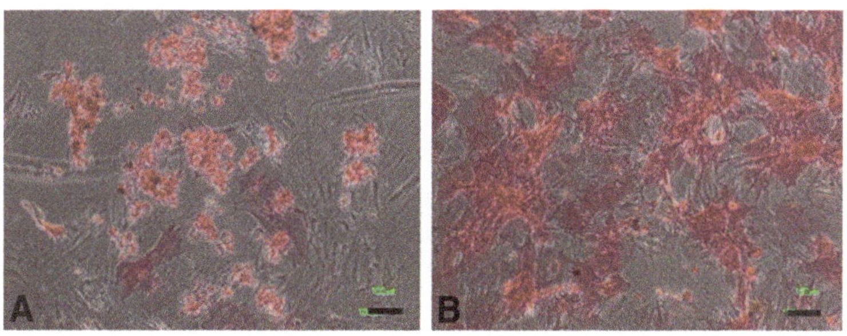

Figure 6 Alkaline phosphatase staining in mesenchymal stem cell spheroids and monolayer rat mesenchymal stem cells. Alkaline phosphatase staining was observed in both mesenchymal stem cell (MSC) spheroids **(A)** and monolayer rat MSCs **(B)**. Scale bar = 100 μm.

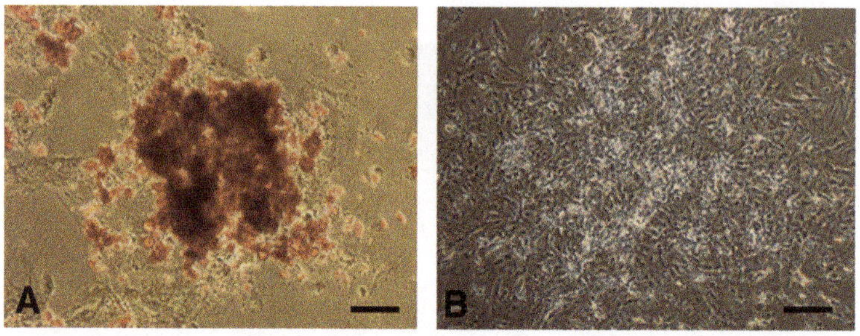

Figure 7 Mineralization of mesenchymal stem cell spheroids cultured with an osteoblast-inducer reagent. Mesenchymal stem cell (MSC) spheroids **(A)** and monolayer rat MSCs **(B)** were incubated for 7 days with an osteoblast-inducer reagent (OIR). Matrix mineralization was visualized in OIR-cultured MSC spheroids using Alizarin red staining. Scale bar = 100 μm.

77.1 ± 4.1% of the bone defects were filled with new bone in the blank and Matrigel™ alone groups, respectively. In the monolayer MSC/Matrigel™ group, 73.9 ± 1.8% of the defects were filled with newly regenerated bone (Figure 9B).

Discussion

The spheroid culture system has some advantages over the standard monolayer culture. The methods of the spheroid culture, allows the cells to adapt to their native morphology, facilitating greater cell-cell contacts and interactions between the cells and the extracellular matrix. In this study, we present two lines of evidence to support the concept that MSC spheroids exhibit enhanced osteogenic potential compared with monolayer rMSCs cultured, using conventional techniques. First, *in-vitro* assays confirmed that MSC spheroids upregulate osteogenic genes and proteins and also exhibit increased calcium deposition. Second, the implantation of MSC spheroids *in vivo* enhanced bone regeneration in rat calvarial defects.

A standardized microplate method used in this study generated a single MSC spheroid in each well. The round-bottom, low-binding plates had the following desirable characteristics: (i) a single spheroid per well, centered for ease of optical imaging; (ii) high reproducibility; (iii) simple harvesting for further analysis [30]. It was possible to form spheroids by altering the initial seeding density of rMSCs. Spheroids size must be important because of limitations in the length of nutrient transport by diffusion. Curcio et al. reported that the spheroid radii greater than 200 μm rendered cells in their core vulnerable to hypoxia

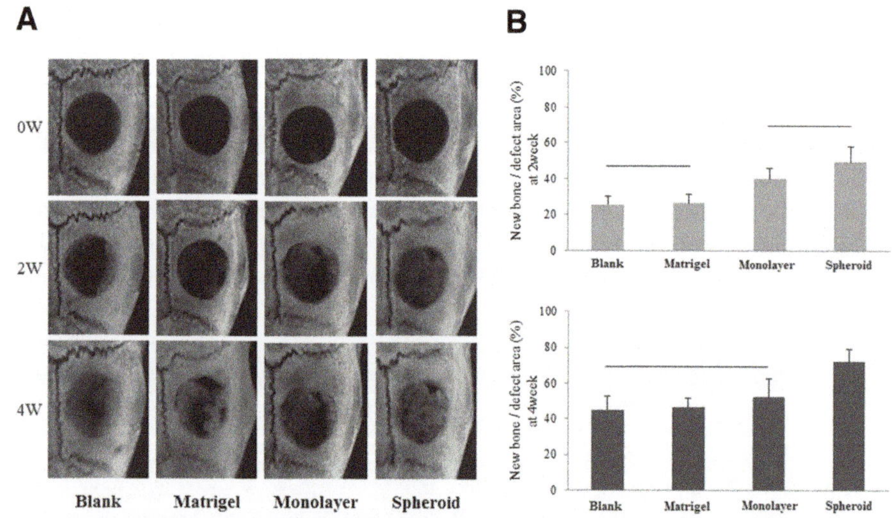

Figure 8 Healing of rat calvarial defects assessed by micro-computed tomography. (A) Micro-computed tomography (micro-CT) images of the calvaria, 0, 2, and 4 weeks after implantation of the indicated materials. Newly regenerated bone in the blank, Matrigel™ alone, monolayer rat mesenchymal stem cells (MSCs)/ Matrigel™, and MSC spheroids/ Matrigel™ groups were evaluated. In the blank group, the defect was left unfilled. **(B)** Quantitative results for micro-CT analysis. Healing of defects was shown as the fraction of new bone with respect to the total defect area as quantified by micro-CT analysis (n = 5 per group). There were statistical differences between the area of new bone formed in the MSC spheroids group and that in the other groups ($p < 0.05$). There are no significant difference in groups of gray columns jointed to horizontal bar.

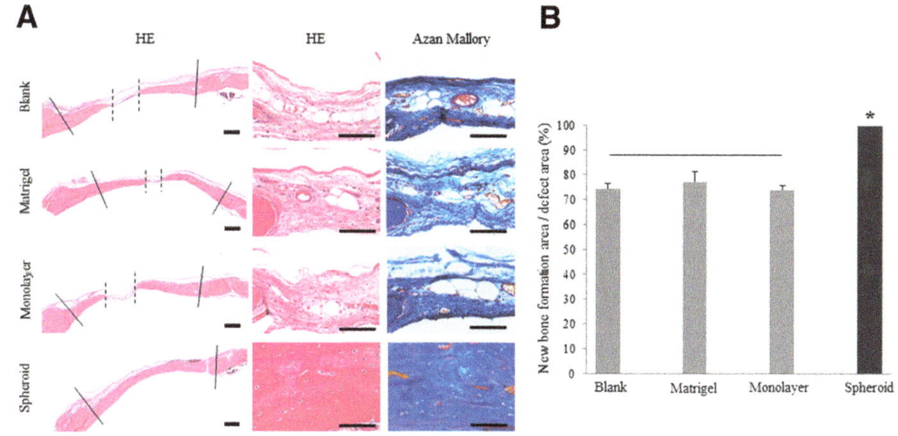

Figure 9 Histological analysis of regenerated bone after implantation. (A) Hematoxylin and eosin (HE) and Azan Mallory staining of calvarial tissue sections was conducted 4 weeks after implantation. Newly regenerated bone in the blank, Matrigel™ alone, monolayer rat mesenchymal stem cells (MSCs)/Matrigel™, and MSC spheroids/Matrigel™ groups were evaluated histologically. The left and middle columns showed HE staining at low and high magnification, respectively. Solid lines indicate the edges of the host bone. Dotted lines indicate the defect filled with the connective tissue. In MSC spheroids/Matrigel™ group, the defect were filled with newly regenerated bone. In the right column showing Azan Mallory staining, the defects of the blank, Matrigel™ alone, monolayer rat mesenchymal stem cells (MSCs)/Matrigel™ groups were filled with mature collagenous fibers. Scale bar in the left column = 1 mm. Scale bars in the middle and right columns = 100 μm. **(B)** Percentages of newly regenerated bone in relation to the size of the original defect evaluated histomorphometricly 4 weeks after implantation. Replacement of newly bone in MSC spheroids/ Matrigel™ group was greater than that in other groups. *, significantly different at $p < 0.05$ compared with other groups. There are no significant difference in groups of gray columns jointed to horizontal bar.

and cell death [31]. In both *in-vitro* and *in-vivo* experiments, we used the spheroids containing approximately 10,000 cells with a mean radius of approximately 200 μm. Conversely, delivery of a substantial number of cells is required for cell-based therapies to drive tissue formation, and larger spheroids facilitate the transplantation of fewer aggregates.

The results of our *in-vitro* assays indicate that osteogenic properties are accelerated in MSC spheroids treated with an OIR. We compared changes the osteogenic gene expression profiles of 3D MSC spheroids vs. 2D monolayer rMSCs by quantitative RT-PCR. Expression of the *RUNX-2, OSX, OPN,* and *BSP* genes were upregulated in MSC spheroids treated with an OIR compared with monolayer rMSCs. The process of osteoblastic differentiation is regulated by multiple factors and signaling pathways. Among the upregulated genes, *RUNX-2,* also called *core-binding factor A1,* and *OSX* are often referred to as the "master switch" of osteogenic differentiation [32,33]. It is established that bone genes are crucial for the generation of a mineralized tissue, and are usually analyzed during the early phases of osteogenic differentiation. MSC spheroids treated with an OIR for 7 days appear to begin toward the osteogenic differentiation. Our immunocytochemical findings showing OSX expression in MSC spheroids support this speculation.

Interestingly, our PCR results also showed an upregulation of the intermediate to late markers of osteogenesis, *OPN* and *BSP. OPN* is a noncollagenous, secreted glycosylation phosphoprotein expressed during both the early and the intermediate stages of osteogenesis [34-36]. *BSP* is also a noncollagenous, acidic phosphoprotein normally expressed in mineralized tissues such as bone and dentin [37]. Upregulation of these genes prompts us to suggest that they may serve as a matrix-associated signal directly promoting osteogenic differentiation and resulting in the increased production of a mineralized matrix in MSC spheroids. Using AR staining, we found increased calcium deposition in MSC spheroids treated with an OIR for 7 days. It is difficult to explain why the mixtures of various differentiation phases are contained within MSC spheroids. We believe that the multicellular spheroids consist of cells at various stages of osteogenic differentiation. Future work will address how osteogenic differentiation is regulated in the cells consisting of MSC spheroids.

In our *in-vitro* experiments, there were no differences in *ALP* gene expression and ALP activity between MSC spheroids and monolayer rMSCs. ALP is another important marker, as an effector protein responsible for osteoprogenitor markers and mineralization of the extracellular matrix during osteogenic differentiation [38]. In this study, we found that the expression levels of *ALP* mRNA in monolayer rMSCs treated with an OIR for 7 days was identical to that in MSC spheroids. ALP staining was of identical intensity between monolayer rMSCs and MSC spheroids treated with an OIR for 7 days, but no AR deposits were observed in monolayer rMSCs. These results suggest that monolayer rMSCs can differentiate into a progenitor phase without the initiation of mineralization. Therefore, MSC spheroids treated with with an OIR exhibit enhanced the differentiation capabililties compared with monolayer rMSCs.

Micro-CT analysis of MSC spheroid-engrafted calvarial defects in rats indicates that these spheroids have a dramatic effect on bone regeneration. Histological findings complemented these micro-CT results, which showed that a bone bridge had almost covered the defect in rats in the MSC spheroids group, although the fibrous connective tissue remained in the defect in rats in the other groups. Based on the histological findings of engrafted defects in the control groups, we suggest that the bone healing process consistes of the extensions of new bone from the edges of the defect. This healing process is supported by our previous report [29]. Our *in-vitro* and *in-vivo* results reveal that MSCs in the spheroid culture exhibit enhanced osteogenic efficiency compared with those in the monolayer culture. However, the detailed mechanisms that enhanced *in-vitro* osteogenesis and *in-vivo* bone regeneration by which MSC spheroids occur remain unclear. Future studies will address the detailed effects of MSC spheroids on microenvironments during the induction of osteogeneis in bone defects.

Conclusion

Our *in-vitro* experiments on osteogenic induction showed that MSC spheroids possessed enhanced osteogenic potential compared with monolayer rMSCs. Moreover, we demonstrated that engrafted MSC spheroids induced efficient bone regeneration in calvarial defects in rats. Based on these results, we conclude that MSCs in the spheroid culture exhibit enhanced the osteogenic capabilities.

Methods

Cell lines

Bone marrow-derived rMSCs, isolated from Fischer 344 rats, were purchased from Lonza Group AG (Basel, Switzerland) and expanded according to the established protocols. Briefly, cryopreserved rMSCs were thawed and plated onto a 10-cm tissue-culture dish in 10 ml Dulbecco's modified Eagle's medium (DMEM; Wako Pure Chemical Industiries,Ltd., Osaka, Japan) supplemented with 10% fetal bovine serum (FBS; Sigma-Aldrich Corporation, St. Louis, MO, USA), and 1% (v/v) penicillin/streptomycin (PS; Invitrogen/GIBCO, Carlsbad, CA, USA) for 7 days at 37°C in a humidified 5% CO_2 atmosphere. The medium was changed every 3 days. After allowing the adherent cells to grow to approximately 80% confluence, they were detached from the tissue culture plate using 0.25% trypsin- 1 mM ethylenediamine tetraacetic acid (EDTA) solution (Sigma-Aldrich Corporation), counted, and either replated for monolayer cultures or used for spheroid formation.

Mesenchymal stem cell spheroid formation

Dissociated rMSC monolayers (Passage 3) were resuspended in medium to obtain a single cell suspension. MSCs (8×10^5 cells/ml, corresponding to approximately 100, 1,000, and 10,000 cells/well) were added to Nunc®Low Cell Binding Surface 96-well plates (Thermo Scientific Nunc A/S, Roskilde, Denmark) and incubated in DMEM medium supplemented with FBS and PS at 37°C for up to 7 days.

Osteogenic induction in mesenchymal stem cell spheroids

To generate osteogenic differentiation, MSC spheroids were cultured with DMEM in the presence of OIR (Takara Bio Inc., Otsu, Japan), which included ascorbic acid, hydrocortisone, and β-glycerophosphate. Culture of MSC spheroid was allowed to grow for 7 days with media exchanged every 2 days. For immunocytochemistry and, ALP and AR staining, MSC spheroids incubated with OIR were removed from the well and plated onto 13-mm cover glass for 2 hours. Parental rMSCs were seeded on to 13-mm cover glasses and cultured with DMEM containing 10% (v/v) FBS and an OIR for 7 days.

Real-time reverse transcription-polymerase chain reaction

The total RNA was isolated from differentiated spheroids or monolayer MSCs cultured with the OIR for 7 days using a High Pure RNA Isolation Kit (Roche Diagnostics, Basel, Switzerland), according to the manufacturer's instructions. Total RNA (1 μg) was transcribed into cDNA using random, oligo (dT) primers, and reverse transcriptase in a total volume of 10 μl (ReverTra Ace® qPCR RT Kit; Toyobo Co., Ltd., Osaka, Japan). Reverse transcription was performed at 37°C for 15 min, at 50°C for 5 min, and then at 98°C for 5 min. The resulting templates were amplified in a LightCycler® Nano RT-PCR system (Roche Diagnostics) according to the manufacturer's protocol. *G3PDH* was used as an internal control. Relative mRNA expression was determined as the ratio of *RUNX-2, OSX, OPN*, or *BSP* mRNA to *G3PDH* mRNA. The sequence of the specific primers and Universal ProbeLibrary (Roche Diagnostics) probe numbers used for the quantitative RT-PCR analysis are listed in Table 1. All reactions were run in triplicate. Results are expressed as fold-increases in mRNA expression (normalized to that of *G3PDH* mRNA) and compared with results for monolayer rMSCs treated with an OIR on Day 7.

Immunocytochemical detection of osterix in mesenchymal stem cell spheroids

After culturing for the indicated time, spheroids and parental cells on 13-mm cover glass were fixed with 4% paraformaldehyde for 10 min and then washed in 0.1% Triton X-100 in phosphate-buffered saline (PBS) for 15 min. They were incubated with a rabbit polyclonal

Table 1 Primers and probes used in this study

Gene	Forward and reverse primers (5′→3′)	Probe no.
G3PDH	AATGTATCCGTTGTGGATCTGA	80
	GCTTCACCACCTTCTTGATGT	
RUNX-2	GGCCCTGGTGTTTAAATGG	83
	AGCACTCACTGACTCGGTTG	
OSX	CCCAACTGTCAGGAGCTAGAG	26
	GATGTGGCGGCTGTGAAT	
OPN	GGCTACAGCATCTGAGTGTTTG	82
	CGGTGAAAGTGGCTGAGTTT	
BSP	TCGGAAGAAAATGGAGATGG	63
	TTCCTCTTCATTTGAAGTCTCCTC	
ALP	ACGAGGTCACGTCCATCCT	89
	CCGAGTGGTGGTCACGAT	

G3PDH, *glyceraldehyde 3-phosphate dehydrogenase*; RUNX-2, *runt-related transcription factor-2*; OSX, *osterix*; OPN, *osteopontin*; BSP, *bone sialoprotein*; ALP, *alkaline phosphatase*.

antibody to Sp7/OSX (1:100; Abcam, Cambridge, UK), at 4°C overnight. After washing with PBS, both spheroids and cells were incubated in a mixture of anti-rabbit Immunoglobulin G antibody conjugated with Alexa Fluor® 568 (1:200; Molecular Probes, Eugene, OR, USA). To visualize the nuclei, immunostained spheroids and cells were counterstained with 4, 6-diamidino-2-phenylindole (DAPI; Vector Laboratories, Inc., Burlingame, CA, USA).

Alkaline phosphatase and Alizarin red staining

ALP and AR staining were assessed in both MSC spheroids and monolayer rMSCs induced after exposure to the OIR for 7 days. ALP staining was performed using an ALP staining kit (Takara Bio Inc., Otsu, Japan), according to the manufacturer's instructions. Calcified deposits were detected by AR staining. Both MSC spheroids and monolayer rMSCs were fixed for 10 min in 4% paraformaldehyde in PBS and was rinsed twice with distilled water. The cells were stained at room temperature for 5 min with AR.

The rat calvarial bone defect model

Ten-week-old male Fischer rats (weight approximately 250 g; $n = 20$) were anesthetized with 2% isoflurane (Abbott Laboratories, Abbott Park, IL, USA) and an air mixture gas flow of 1 l/min using an anesthesia gas machine (Anesthesia Machine SF-B01; MR Technology, Inc., Tsukuba, Japan). After shaving the skin, an incision was made in the skull, and the periosteum was opened, exposing the surface of the calvarial bones. A circular bone defect (full-thickness, 5 mm diameter) was created in the left parietal bone with a trephine drill and was irrigated with saline to remove bone debris. MSC spheroids were then implanted into the defects. Five MSC spheroids or monolayer rMSCs (5×10^4 cells), which had been cultured in medium containing the OIR for 7 days, were loaded into 20 μl Matrigel™ on ice and transplanted into the defects at room temperature. Controls for the transplant experiments included calvarial defects implanted with or without Matrigel™. Our animal experimentation protocols were approved by the Animal Care and Use Committee of Fukuoka Dental College, Fukuoka, Japan (No. 13006).

Evaluation of bone regeneration

Bone regeneration was evaluated using an *in-vivo* micro-CT system (Skyscan-1176 Micro-CT Scanner; Burker MicroCT, Kontich, Belgium) at 50 kVp and 500 μA on rats under anesthesia (as described above), 4 weeks after implantation. Each image data set consisted of a scan size of approximately 35 mm. The percentage of newly formed bone in each calvarial bone defect was calculated as previously described [28,29].

After micro-CT scanning, rats were sacrificed by injecting an overdose of isoflurane. Subsequently, the cranial tissues containing MSC spheroids, monolayer rMSCs, or control substances were immediately excised. Tissue specimens were fixed in 4% paraformaldehyde in PBS, decalcified in 10% EDTA for 4 weeks at 4°C, and then was embedded in paraffin. Paraffin sections were then stained with hematoxylin and eosin and Azan Mallory to visualize any histological changes. In histomorphometric analysis, the percentage of the volume of the newly formed bone per the original volume of the defect was calculated.

Statistical analysis

The results are presented as means ± standard deviation. Group results were compared using one-way analysis of variance and Scheffe's multiple comparison tests. A p-value of <0.05 was considered statistically significant.

Competing interests

The authors declare that they have no competing interests.

Authors' contributions

YY and JO planned the study, performed the experiments and data analysis, and wrote the manuscript. AS performed animal experiments and immunocytochemistry. HK and TF supervised manuscript writing. All authors read and approved the final manuscript.

Acknowledgments

The present study was supported in part by a Grant-in-aid for Scientific Research (B) (23390455) and by a Grant-in-aid for strategic study base formation support business (S1001059) form the Japan Society for the Promotion of Science. The authors would like to thank Enago (www.enago.jp) for the English language review.

Author details

[1]Department of Oral Rehabilitation, Section of Oral Implantology, Fukuoka Dental College, 2-15-1 Tamura, Sawara-ku, Fukuoka 814-0193, Japan. [2]Department of Morphological Biology, Section of Pathology, Fukuoka Dental College, 2-15-1 Tamura, Sawara-ku, Fukuoka 814-0193, Japan. [3]Center for Regenerative Medicine, Fukuoka Dental College, 2-15-1 Tamura, Sawara-ku, Fukuoka 814-0193, Japan.

References

1. Bruder SP, Jaiswal N, Haynesworth SE: **Growth kinetics, self-renewal, and the osteogenic potential of purified human mesenchymal stem cells during extensive subcultivation and following cryopreservation.** *J Cell Biochem* 1997, **64**:278–294.
2. Colter DC, Class R, DiGirolamo CM, Prockop DJ: **Rapid expansion of recycling stem cells in cultures of plastic-adherent cells from human bone marrow.** *Proc Natl Acad Sci U S A* 2000, **97**:3213–3218.
3. Zuk PA, Zhu M, Mizuno H, Huang J, Futrell JW, Katz AJ, Benhaim P, Lorenz HP, Hedrick MH: **Multilineage cells from human adipose tissue: implications for cell-based therapies.** *Tissue Eng* 2001, **7**:211–228.
4. Zuk PA, Zhu M, Ashjian P, De Ugarte DA, Huang JI, Mizuno H, Alfonso ZC, Fraser JK, Benhaim P, Hedrick MH: **Human adipose tissue is a source of multipotent stem cells.** *Mol Biol Cell* 2002, **13**:4279–4295.
5. Rodriguez-Lozano FJ, Bueno C, Insausti CL, Meseguer L, Ramirez MC, Blanquer M, Marin N, Martinez S, Moraleda JM: **Mesenchymal stem cells derived from dental tissues.** *Int Endod J* 2011, **44**:800–806.
6. Pierdomenico L, Bonsi L, Calvitti M, Rondelli D, Arpinati M, Chirumbolo G, Becchetti E, Marchionni C, Alviano F, Fossati V, Staffolani N, Franchina M, Grossi A, Bagnara GP: **Multipotent mesenchymal stem cells with immunosuppressive activity can be easily isolated from dental pulp.** *Transplantation* 2005, **80**:836–842.
7. Kerkis I, Caplan AI: **Stem cells in dental pulp of deciduous teeth.** *Tissue Eng Part B Rev* 2012, **18**:129–138.
8. Erices A, Conget P, Minguell JJ: **Mesenchymal progenitor cells in human umbilical cord blood.** *Br J Haematol* 2000, **109**:235–242.
9. Covas DT, Siufi JL, Silva AR, Orellana MD: **Isolation and culture of umbilical vein mesenchymal stem cells.** *Braz J Med Biol Res* 2003, **36**:1179–1183.
10. Osugi M, Katagiri W, Yoshimi R, Inukai T, Hibi H, Ueda M: **Conditioned media from mesenchymal stem cells enhanced bone regeneration in rat calvarial bone defects.** *Tissue Eng Part A* 2012, **18**:1479–1489.
11. Blum JS, Barry MA, Mikos AG, Jansen JA: **In vivo evaluation of gene therapy vectors in ex vivo-derived marrow stromal cells for bone regeneration in a rat critical-size calvarial defect model.** *Hum Gene Ther* 2003, **14**:1689–1701.
12. Pittenger MF, Mackay AM, Beck SC, Jaiswal RK, Douglas R, Mosca JD, Moorman MA, Simonetti DW, Craig S, Marshak DR: **Multilineage potential of adult human mesenchymal stem cells.** *Science* 1999, **284**:143–147.
13. Pittenger MF: **Mesenchymal stem cells from adult bone marrow.** *Methods Mol Biol* 2008, **449**:27–44.
14. Baraniak PR, McDevitt TC: **Scaffold-free culture of mesenchymal stem cell spheroids in suspension preserves multilineage potential.** *Cell Tissue Res* 2012, **347**:701–711.
15. Park E, Patel AN: **Changes in the expression pattern of mesenchymal and pluripotent markers in human adipose-derived stem cells.** *Cell Biol Int* 2010, **34**:979–984.
16. Baer PC, Griesche N, Luttmann W, Schubert R, Luttmann A, Geiger H: **Human adipose-derived mesenchymal stem cells in vitro: evaluation of an optimal expansion medium preserving stemness.** *Cytotherapy* 2010, **12**:96–106.
17. Yamamoto M, Kawashima N, Takashino N, Koizumi Y, Takimoto K, Suzuki N, Saito M, Suda H: **Three-dimensional spheroid culture promotes odonto/osteoblastic differentiation of dental pulp cells.** *Arch Oral Biol* 2014, **59**:310–317.
18. Salamon A, van Vlierberghe S, van Nieuwenhove I, Baudisch F, Graulus G-J, Benecke V, Alberti K, Neumann H-G, Rychly J, Martins JC, Dubruel P, Peters K: **Gelatin-based hydrogels promote chondrogenic differentiationof human adipose tissue-derived mesenchymal stem cells in vitro.** *Materials* 2014, **7**:1342–1359.
19. Rutherford RB, Moalli M, Franceschi RT, Wang D, Gu K, Krebsbach PH: **Bone morphogenetic protein-transduced human fibroblasts convert to osteoblasts and form bone in vivo.** *Tissue Eng* 2002, **8**:441–452.
20. Yoon HH, Bhang SH, Shin JY, Shin J, Kim BS: **Enhanced cartilage formation via three-dimensional cell engineering of human adipose-derived stem cells.** *Tissue Eng Part A* 2012, **18**:1949–1956.
21. Wiesmann HP, Nazer N, Klatt C, Szuwart T, Meyer U: **Bone tissue engineering by primary osteoblast-like cells in a monolayer system and 3-dimensional collagen gel.** *J Oral Maxillofac Surg* 2003, **61**:1455–1462.
22. Cerwinka WH, Sharp SM, Boyan BD, Zhau HE, Chung LW, Yates C: **Differentiation of human mesenchymal stem cell spheroids under microgravity conditions.** *Cell Regen* 2012, **1**:2.
23. Cheng NC, Wang S, Young TH: **The influence of spheroid formation of human adipose-derived stem cells on chitosan films on stemness and differentiation capabilities.** *Biomaterials* 2012, **33**:1748–1758.
24. Yeh HY, Liu BH, Sieber M, Hsu SH: **Substrate-dependent gene regulation of self-assembled human MSC spheroids on chitosan membranes.** *BMC Genomics* 2014, **15**:10.
25. Bartosh TJ, Ylostalo JH, Mohammadipoor A, Bazhanov N, Coble K, Claypool K, Lee RH, Choi H, Prockop DJ: **Aggregation of human mesenchymal stromal cells (MSCs) into 3D spheroids enhances their antiinflammatory properties.** *Proc Natl Acad Sci U S A* 2010, **107**:13724–13729.
26. Frith JE, Thomson B, Genever PG: **Dynamic three-dimensional culture methods enhance mesenchymal stem cell properties and increase therapeutic potential.** *Tissue Eng Part C Methods* 2010, **16**:735–749.
27. Wang W, Itaka K, Ohba S, Nishiyama N, Chung UI, Yamasaki Y, Kataoka K: **3D spheroid culture system on micropatterned substrates for improved differentiation efficiency of multipotent mesenchymal stem cells.** *Biomaterials* 2009, **30**:2705–2715.
28. Shinozaki M, Yanagi T, Yamaguchi Y, Kido H, Fukushima T: **Osteogenic evaluation of DNA/protamine complex paste in rat cranial defects.** *J Hard Tissue Biol* 2013, **22**:401–408.
29. Toda M, Ohno J, Shinozaki Y, Ozaki M, Fukushima T: **Osteogenic potential for replacing cells in rat cranial defects implanted with a DNA/protamine complex paste.** *Bone* 2014, **67C**:237–245.
30. Vinci M, Gowan S, Boxall F, Patterson L, Zimmermann M, Court W, Lomas C, Mendiola M, Hardisson D, Eccles SA: **Advances in establishment and analysis of three-dimensional tumor spheroid-based functional assays for target validation and drug evaluation.** *BMC Biol* 2012, **10**:29.
31. Curcio E, Salerno S, Barbieri G, De Bartolo L, Drioli E, Bader A: **Mass transfer and metabolic reactions in hepatocyte spheroids cultured in rotating wall gas-permeable membrane system.** *Biomaterials* 2007, **28**:5487–5497.
32. Ducy P, Zhang R, Geoffroy V, Ridall AL, Karsenty G: **Osf2/Cbfa1: a transcriptional activator of osteoblast differentiation.** *Cell* 1997, **89**:747–754.
33. Cao Y, Zhou Z, de Crombrugghe B, Nakashima K, Guan H, Duan X, Jia SF, Kleinerman ES: **Osterix, a transcription factor for osteoblast differentiation, mediates antitumor activity in murine osteosarcoma.** *Cancer Res* 2005, **65**:1124–1128.
34. McKee MD, Nanci A: **Osteopontin at mineralized tissue interfaces in bone, teeth, and osseointegrated implants: ultrastructural distribution and implications for mineralized tissue formation, turnover, and repair.** *Microsc Res Tech* 1996, **33**:141–164.
35. Liu HY, Liu MC, Wang MF, Chen WH, Tsai CY, Wu KH, Lin CT, Shieh YH, Zeng R, Deng WP: **Potential osteoporosis recovery by deep sea water through bone regeneration in SAMP8 mice.** *Evid Based Complement Alternat Med* 2013, **2013**:161976.
36. Chung E, Rylander MN: **Response of preosteoblasts to thermal stress conditioning and osteoinductive growth factors.** *Cell Stress Chaperones* 2012, **17**:203–214.
37. Gordon JA, Tye CE, Sampaio AV, Underhill TM, Hunter GK, Goldberg HA: **Bone sialoprotein expression enhances osteoblast differentiation and matrix mineralization in vitro.** *Bone* 2007, **41**:462–473.
38. Marom R, Shur I, Solomon R, Benayahu D: **Characterization of adhesion and differentiation markers of osteogenic marrow stromal cells.** *J Cell Physiol* 2005, **202**:41–48.

Permissions

All chapters in this book were first published in BMCB, by BioMed Central; hereby published with permission under the Creative Commons Attribution License or equivalent. Every chapter published in this book has been scrutinized by our experts. Their significance has been extensively debated. The topics covered herein carry significant findings which will fuel the growth of the discipline. They may even be implemented as practical applications or may be referred to as a beginning point for another development.

The contributors of this book come from diverse backgrounds, making this book a truly international effort. This book will bring forth new frontiers with its revolutionizing research information and detailed analysis of the nascent developments around the world.

We would like to thank all the contributing authors for lending their expertise to make the book truly unique. They have played a crucial role in the development of this book. Without their invaluable contributions this book wouldn't have been possible. They have made vital efforts to compile up to date information on the varied aspects of this subject to make this book a valuable addition to the collection of many professionals and students.

This book was conceptualized with the vision of imparting up-to-date information and advanced data in this field. To ensure the same, a matchless editorial board was set up. Every individual on the board went through rigorous rounds of assessment to prove their worth. After which they invested a large part of their time researching and compiling the most relevant data for our readers.

The editorial board has been involved in producing this book since its inception. They have spent rigorous hours researching and exploring the diverse topics which have resulted in the successful publishing of this book. They have passed on their knowledge of decades through this book. To expedite this challenging task, the publisher supported the team at every step. A small team of assistant editors was also appointed to further simplify the editing procedure and attain best results for the readers.

Apart from the editorial board, the designing team has also invested a significant amount of their time in understanding the subject and creating the most relevant covers. They scrutinized every image to scout for the most suitable representation of the subject and create an appropriate cover for the book.

The publishing team has been an ardent support to the editorial, designing and production team. Their endless efforts to recruit the best for this project, has resulted in the accomplishment of this book. They are a veteran in the field of academics and their pool of knowledge is as vast as their experience in printing. Their expertise and guidance has proved useful at every step. Their uncompromising quality standards have made this book an exceptional effort. Their encouragement from time to time has been an inspiration for everyone.

The publisher and the editorial board hope that this book will prove to be a valuable piece of knowledge for researchers, students, practitioners and scholars across the globe.

List of Contributors

Amber L Couzens
Department of Biology, York University, 4700 Keele Street, Toronto, ON M3J1P3, Canada
Centre for Systems Biology, Lunenfeld-Tanenbaum Research Institute, Mount Sinai Hospital, Toronto, ON M5G 1X5, Canada

R Montgomery Gill
Department of Biology, York University, 4700 Keele Street, Toronto, ON M3J1P3, Canada

Michael P Scheid
Department of Biology, York University, 4700 Keele Street, Toronto, ON M3J1P3, Canada

Qun Wang
Department of Ophthalmology, General Hospital of People's Liberation Army, No.28 Fuxing Road, Haidian District, Beijing 100853, China

Jing Yang
Institute of Basic Medical Sciences, P.O. Box 130 (3), Taiping Road, Beijing 100850, China

Kun Tang
Department of Ophthalmology, Aerospace center Hospital, No. 15 Yuquan Road, Haidian District, Beijing 100049, China

Longlong Luo
Institute of Basic Medical Sciences, P.O. Box 130 (3), Taiping Road, Beijing 100850, China

Liqiang Wang
Department of Ophthalmology, General Hospital of People's Liberation Army, No.28 Fuxing Road, Haidian District, Beijing 100853, China

Lei Tian
Department of Ophthalmology, General Hospital of People's Liberation Army, No.28 Fuxing Road, Haidian District, Beijing 100853, China

Yanming Jiang
Department of Ophthalmology, The second Artillery General Hospital, No.16 Xinwai Road, Xicheng District, Beijing 100088, China

Jiannan Feng
Institute of Basic Medical Sciences, P.O. Box 130 (3), Taiping Road, Beijing 100850, China

Yan Li
Institute of Basic Medical Sciences, P.O. Box 130 (3), Taiping Road, Beijing 100850, China

Beifen Shen
Institute of Basic Medical Sciences, P.O. Box 130 (3), Taiping Road, Beijing 100850, China

Ming Lv
Institute of Basic Medical Sciences, P.O. Box 130 (3), Taiping Road, Beijing 100850, China

Yifei Huang
Department of Ophthalmology, General Hospital of People's Liberation Army, No.28 Fuxing Road, Haidian District, Beijing 100853, China

Soundarrajan Ilavenil
Grassland and forage division, National Institute of Animal Science, RDA, Seonghwan-Eup, Cheonan-Si, Chungnam 330-801, Korea

Mariadhas Valan Arasu
Grassland and forage division, National Institute of Animal Science, RDA, Seonghwan-Eup, Cheonan-Si, Chungnam 330-801, Korea
Department of Botany and microbiology, College of Science, King Saud University, Riyadh 11451, Saudi Arabia

Jeong-Chae Lee
Research Center of Bioactive Materials, Chonbuk National University, Jeonju 561-756, Republic of Korea

Da Hye Kim
The United Graduate School of Agricultural Sciences, Tottori University, Tottori-Shi 680-8553, Japan

Mayakrishnan Vijayakumar
Grassland and forage division, National Institute of Animal Science, RDA, Seonghwan-Eup, Cheonan-Si, Chungnam 330-801, Korea

Kyung Dong Lee
Department of Oriental Medicine Materials, Dongsin University, Naju 520-714, Korea

Ki Choon Choi
Grassland and forage division, National Institute of Animal Science, RDA, Seonghwan-Eup, Cheonan-Si, Chungnam 330-801, Korea

Quan-Li Li
College & Hospital of Stomatology, Anhui Medical University, 230032 Hefei, China

Tian-Yun Ning
Faculty of Dentistry, the University of Hong Kong, 34 Hospital Road, Hong Kong, China

Ying Cao
College & Hospital of Stomatology, Anhui Medical University, 230032 Hefei, China
Faculty of Dentistry, the University of Hong Kong, 34 Hospital Road, Hong Kong, China

Wei-bo Zhang
College & Hospital of Stomatology, Anhui Medical University, 230032 Hefei, China

May Lei Mei
Faculty of Dentistry, the University of Hong Kong, 34 Hospital Road, Hong Kong, China

Chun Hung Chu
Faculty of Dentistry, the University of Hong Kong, 34 Hospital Road, Hong Kong, China

Eunseok Seo
Division of Integrative Biosciences and Biotechnology, Pohang University of Science and Technology, San 31, Hyoja-dong, Nam-Gu, Pohang, Gyeongbuk 790-784, Korea

Kyung Won Seo
Center for Biofluid and Biomimic Research, Department of Mechanical Engineering, Pohang University of Science and Technology, San 31, Hyoja-dong, Nam-Gu, Pohang, Gyeongbuk 790-784, South Korea

Jung-Eun Gil
Department of Biological Sciences, Center for Cognition and Sociality, Institute for Basic Science, 291 Daehak-ro, Yuseong-gu, Daejeon 305-701, South Korea

Young-Ran Ha
Division of Integrative Biosciences and Biotechnology, Pohang University of Science and Technology, San 31, Hyoja-dong, Nam-Gu, Pohang, Gyeongbuk 790-784, Korea

Eunseop Yeom
Center for Biofluid and Biomimic Research, Department of Mechanical Engineering, Pohang University of Science and Technology, San 31, Hyoja-dong, Nam-Gu, Pohang, Gyeongbuk 790-784, South Korea

Seungchul Lee
Center for Biofluid and Biomimic Research, Department of Mechanical Engineering, Pohang University of Science and Technology, San 31, Hyoja-dong, Nam-Gu, Pohang, Gyeongbuk 790-784, South Korea

Sang Joon Lee
Division of Integrative Biosciences and Biotechnology, Pohang University of Science and Technology, San 31, Hyoja-dong, Nam-Gu, Pohang, Gyeongbuk 790-784, Korea
Center for Biofluid and Biomimic Research, Department of Mechanical Engineering, Pohang University of Science and Technology, San 31, Hyoja-dong, Nam-Gu, Pohang, Gyeongbuk 790-784, South Korea

Raihana Edros
School of Chemical and Bioprocess Engineering and Conway Institute of Biomolecular and Biomedical Research, University College Dublin, Dublin, Ireland

Susan McDonnell
School of Chemical and Bioprocess Engineering and Conway Institute of Biomolecular and Biomedical Research, University College Dublin, Dublin, Ireland

Mohamed Al-Rubeai
School of Chemical and Bioprocess Engineering and Conway Institute of Biomolecular and Biomedical Research, University College Dublin, Dublin, Ireland

Galina Gulis
Institute of Biological Sciences, Department of Cell Biology, University of Brasilia, Campus Universitário Darcy Ribeiro, Brasília, DF 70910-900, Brazil

Kelly Cristina Rodrigues Simi
Institute of Biological Sciences, Department of Cell Biology, University of Brasilia, Campus Universitário Darcy Ribeiro, Brasília, DF 70910-900, Brazil

Renata Rodrigues de Toledo
Institute of Biological Sciences, Department of Cell Biology, University of Brasilia, Campus Universitário Darcy Ribeiro, Brasília, DF 70910-900, Brazil

Andrea Queiroz Maranhao
Institute of Biological Sciences, Department of Cell Biology, University of Brasilia, Campus Universitário Darcy Ribeiro, Brasília, DF 70910-900, Brazil

Marcelo Macedo Brigido
Institute of Biological Sciences, Department of Cell Biology, University of Brasilia, Campus Universitário Darcy Ribeiro, Brasília, DF 70910-900, Brazil

Jianyu Su
College of Light Industry and Food Sciences, South China University of Technology, Guangzhou 510640, China

Xiaochen Liu
College of Light Industry and Food Sciences, South China University of Technology, Guangzhou 510640, China

Hemiao Cui
College of Light Industry and Food Sciences, South China University of Technology, Guangzhou 510640, China

Yanyan Li
State Key Laboratory of Food Science and Technology, Jiangnan University, Wuxi 214122, China

Dingqiang Chen
Department of Laboratory Medicine, First Affiliated Hospital of Guangzhou Medical College, Guangzhou 510120, China

Yanmei Li
Guangzhou Women and Children's Medical Center, Guangzhou 510623, China
Department of Microbial Pathogenesis, University of Maryland, Baltimore 21201, USA

Guangchao Yu
First Affiliated Hospital of Jinan University, Guangzhou 510620, China

Hao Shi
College of Chemical Engineering, Nanjing Forestry University, Nanjing 210037, China
Jiangsu Key Lab of Biomass-Based Green Fuels and Chemicals, Nanjing 210037, China

Huaihai Ding
College of Chemical Engineering, Nanjing Forestry University, Nanjing 210037, China
Jiangsu Key Lab of Biomass-Based Green Fuels and Chemicals, Nanjing 210037, China

Yingjuan Huang
College of Chemical Engineering, Nanjing Forestry University, Nanjing 210037, China
Jiangsu Key Lab of Biomass-Based Green Fuels and Chemicals, Nanjing 210037, China

Liangliang Wang
College of Chemical Engineering, Nanjing Forestry University, Nanjing 210037, China
Jiangsu Key Lab of Biomass-Based Green Fuels and Chemicals, Nanjing 210037, China

Yu Zhang
College of Chemical Engineering, Nanjing Forestry University, Nanjing 210037, China
Jiangsu Key Lab of Biomass-Based Green Fuels and Chemicals, Nanjing 210037, China

Xun Li
College of Chemical Engineering, Nanjing Forestry University, Nanjing 210037, China
Jiangsu Key Lab of Biomass-Based Green Fuels and Chemicals, Nanjing 210037, China

Fei Wang
College of Chemical Engineering, Nanjing Forestry University, Nanjing 210037, China
Jiangsu Key Lab of Biomass-Based Green Fuels and Chemicals, Nanjing 210037, China

Verena Lohr
Max Planck Institute for Dynamics of Complex Technical Systems, Sandtorstr. 1, 39106 Magdeburg, Germany
ProBioGen AG, Goethestr. 54, 13086 Berlin, Germany

Oliver Hädicke
Max Planck Institute for Dynamics of Complex Technical Systems, Sandtorstr. 1, 39106 Magdeburg, Germany

Yvonne Genzel
Max Planck Institute for Dynamics of Complex Technical Systems, Sandtorstr. 1, 39106 Magdeburg, Germany

Ingo Jordan
ProBioGen AG, Goethestr. 54, 13086 Berlin, Germany

Heino Büntemeyer
Cell Culture Technology, Bielefeld University, Universitätsstr. 25, 33615 Bielefeld, Germany

Steffen Klamt
Max Planck Institute for Dynamics of Complex Technical Systems, Sandtorstr. 1, 39106 Magdeburg, Germany

Udo Reichl
Max Planck Institute for Dynamics of Complex Technical Systems, Sandtorstr. 1, 39106 Magdeburg, Germany
Bioprocess Engineering, Otto von Guericke University Magdeburg, Universitätsplatz 2, 39106 Magdeburg, Germany

YaQing Tan
Biotechnology Research Institute, Chinese Academy of Agricultural Sciences, 12 Southern Zhong Guan Cun Road, Beijing 100081, China

Kevin Yueju Wang
Department of Natural Sciences, Northeastern State University, Broken Arrow, Oklahoma 74014, USA

Nan Wang
Biotechnology Research Institute, Chinese Academy of Agricultural Sciences, 12 Southern Zhong Guan Cun Road, Beijing 100081, China

GangQiang Li
Biotechnology Research Institute, Chinese Academy of Agricultural Sciences, 12 Southern Zhong Guan Cun Road, Beijing 100081, China

DeHu Liu
Biotechnology Research Institute, Chinese Academy of Agricultural Sciences, 12 Southern Zhong Guan Cun Road, Beijing 100081, China

Nadezhda A Orlova
Laboratory of Mammalian Cell Bioengineering, Centre "Bioengineering", Russian Academy of Sciences, 60-letija Oktyabrya 7, Moscow 117312, Russia
Laboratory of Biocatalysis, Institute of Bioorganic Chemistry, Russian Academy of Sciences, Miklukho-Maklaya 16/10, Moscow 119971, Russia

Sergey V Kovnir
Laboratory of Mammalian Cell Bioengineering, Centre "Bioengineering", Russian Academy of Sciences, 60-letija Oktyabrya 7, Moscow 117312, Russia
Laboratory of Biocatalysis, Institute of Bioorganic Chemistry, Russian Academy of Sciences, Miklukho-Maklaya 16/10, Moscow 119971, Russia

Julia A Hodak
Laboratory of Mammalian Cell Bioengineering, Centre "Bioengineering", Russian Academy of Sciences, 60-letija Oktyabrya 7, Moscow 117312, Russia
Laboratory of Biocatalysis, Institute of Bioorganic Chemistry, Russian Academy of Sciences, Miklukho-Maklaya 16/10, Moscow 119971, Russia

Ivan I Vorobiev
Laboratory of Mammalian Cell Bioengineering, Centre "Bioengineering", Russian Academy of Sciences, 60-letija Oktyabrya 7, Moscow 117312, Russia
Laboratory of Biocatalysis, Institute of Bioorganic Chemistry, Russian Academy of Sciences, Miklukho-Maklaya 16/10, Moscow 119971, Russia

Alexandre G Gabibov
Laboratory of Biocatalysis, Institute of Bioorganic Chemistry, Russian Academy of Sciences, Miklukho-Maklaya 16/10, Moscow 119971, Russia
Kazan Federal University, Kazan, Republic of Tatarstan 420008, Russia

Konstantin G Skryabin
Laboratory of Mammalian Cell Bioengineering, Centre "Bioengineering", Russian Academy of Sciences, 60-letija Oktyabrya 7, Moscow 117312, Russia

Hao Wang
Department of Pediatric Cardiothoracic Surgery, Shanghai Children's Medical Center, School of Medicine, Shanghai Jiao Tong University, 1678 Dong Fang Road, Shanghai 200127, China

Hao Chen
Department of Pediatric Cardiothoracic Surgery, Shanghai Children's Medical Center, School of Medicine, Shanghai Jiao Tong University, 1678 Dong Fang Road, Shanghai 200127, China

Bei Feng
Department of Pediatric Cardiothoracic Surgery, Shanghai Children's Medical Center, School of Medicine, Shanghai Jiao Tong University, 1678 Dong Fang Road, Shanghai 200127, China Institute of Pediatric Translational Medicine, Shanghai Children's Medical Center, School of Medicine, Shanghai Jiao Tong University, 1678 Dong Fang Road, Shanghai 200127, China

Xiang Wang
Institute of Pediatric Translational Medicine, Shanghai Children's Medical Center, School of Medicine, Shanghai Jiao Tong University, 1678 Dong Fang Road, Shanghai 200127, China Department of Hematology/Oncology, Shanghai Children's Medical Center, School of Medicine, Shanghai Jiao Tong University, 1678 Dong Fang Road, Shanghai 200127, China

Xiaomin He
Department of Pediatric Cardiothoracic Surgery, Shanghai Children's Medical Center, School of Medicine, Shanghai Jiao Tong University, 1678 Dong Fang Road, Shanghai 200127, China

Renjie Hu
Department of Pediatric Cardiothoracic Surgery, Shanghai Children's Medical Center, School of Medicine, Shanghai Jiao Tong University, 1678 Dong Fang Road, Shanghai 200127, China

Meng Yin
Department of Pediatric Cardiothoracic Surgery, Shanghai Children's Medical Center, School of Medicine, Shanghai Jiao Tong University, 1678 Dong Fang Road, Shanghai 200127, China

Wei Wang
Department of Pediatric Cardiothoracic Surgery, Shanghai Children's Medical Center, School of Medicine, Shanghai Jiao Tong University, 1678 Dong Fang Road, Shanghai 200127, China

Wei Fu
Department of Pediatric Cardiothoracic Surgery, Shanghai Children's Medical Center, School of Medicine, Shanghai Jiao Tong University, 1678 Dong Fang Road, Shanghai 200127, China Institute of Pediatric Translational Medicine, Shanghai Children's Medical Center, School of Medicine, Shanghai Jiao Tong University, 1678 Dong Fang Road, Shanghai 200127, China

Zhiwei Xu
Department of Pediatric Cardiothoracic Surgery, Shanghai Children's Medical Center, School of Medicine, Shanghai Jiao Tong University, 1678 Dong Fang Road, Shanghai 200127, China

Harshita Sharma
Department of Functional Materials Science, Graduate School of Science and Engineering, Saitama University, 255 Shimo-okubo, Saitama 338-8570, Japan

Yasunori Kinoshita
Department of Functional Materials Science, Graduate School of Science and Engineering, Saitama University, 255 Shimo-okubo, Saitama 338-8570, Japan

Seiichi Fujiu
Department of Functional Materials Science, Graduate School of Science and Engineering, Saitama University, 255 Shimo-okubo, Saitama 338-8570, Japan
Janusys Corporation, Saitama Industrial Technology Center, 3-12-18 Kamiaoki, Kawaguchi, Saitama 334-0844, Japan

Shota Nomura
Department of Functional Materials Science, Graduate School of Science and Engineering, Saitama University, 255 Shimo-okubo, Saitama 338-8570, Japan

Mizuho Sawada
Janusys Corporation, Saitama Industrial Technology Center, 3-12-18 Kamiaoki, Kawaguchi, Saitama 334-0844, Japan

Shamim Ahmed
Department of Functional Materials Science, Graduate School of Science and Engineering, Saitama University, 255 Shimo-okubo, Saitama 338-8570, Japan
Department of Biochemistry and Molecular Biology, Shahjalal University of Science and Technology, Sylhet, Bangladesh

Masaki Shibuya
Lifetech Co., Ltd., 4074 Miyadera, Iruma City, Saitama 358-0014, Japan

Kosaku Shirai
Lifetech Co., Ltd., 4074 Miyadera, Iruma City, Saitama 358-0014, Japan

Syota Takamatsu
Enplas Corporation, 2-30-1 Namiki, Kawaguchi City, Saitama 332-0034, Japan

Tsuyoshi Watanabe
Enplas Corporation, 2-30-1 Namiki, Kawaguchi City, Saitama 332-0034, Japan

Hitoshi Yamazaki
Finetech Corporation, 1-7-1, Asagaya-minami Suginami-ku, Tokyo, Japan

Ryohei Kamiyama
Department of Functional Materials Science, Graduate School of Science and Engineering, Saitama University, 255 Shimo-okubo, Saitama 338-8570, Japan

Tetsuya Kobayashi
Department of Functional Materials Science, Graduate School of Science and Engineering, Saitama University, 255 Shimo-okubo, Saitama 338-8570, Japan

Hidenao Arai
Department of Functional Materials Science, Graduate School of Science and Engineering, Saitama University, 255 Shimo-okubo, Saitama 338-8570, Japan

Miho Suzuki
Department of Functional Materials Science, Graduate School of Science and Engineering, Saitama University, 255 Shimo-okubo, Saitama 338-8570, Japan

Naoto Nemoto
Department of Functional Materials Science, Graduate School of Science and Engineering, Saitama University, 255 Shimo-okubo, Saitama 338-8570, Japan

Ki Ando
Department of Functional Materials Science, Graduate School of Science and Engineering, Saitama University, 255 Shimo-okubo, Saitama 338-8570, Japan
Department of Electrical and Electronic Engineering, Tokyo Denki University, 5 Senjyu-Asahi-cho, Adachi-ku, Tokyo 120-8551, Japan

Hidekazu Uchida
Department of Functional Materials Science, Graduate School of Science and Engineering, Saitama University, 255 Shimo-okubo, Saitama 338-8570, Japan

Koichiro Kitamura
Janusys Corporation, Saitama Industrial Technology Center, 3-12-18 Kamiaoki, Kawaguchi, Saitama 334-0844, Japan

Osamu Takei
Lifetech Co., Ltd., 4074 Miyadera, Iruma City, Saitama 358-0014, Japan

Koichi Nishigaki
Department of Functional Materials Science, Graduate School of Science and Engineering, Saitama University, 255 Shimo-okubo, Saitama 338-8570, Japan

Manon Fuchs
Technische Universität Braunschweig, Institut für Biochemie und Biotechnologie, Spielmannstr.7, 38106 Braunschweig, Germany

Susanne Kämpfer
Lionex GmbH, Salzdahlumer Str. 196, 38126 Braunschweig, Germany

Saskia Helmsing
Technische Universität Braunschweig, Institut für Biochemie und Biotechnologie, Spielmannstr.7, 38106 Braunschweig, Germany

Ralf Spallek
Lionex GmbH, Salzdahlumer Str. 196, 38126 Braunschweig, Germany

Wulf Oehlmann
Lionex GmbH, Salzdahlumer Str. 196, 38126 Braunschweig, Germany

Wiebke Prilop
Lionex GmbH, Salzdahlumer Str. 196, 38126 Braunschweig, Germany

Ronald Frank
Helmholtz Zentrum für Infektionsforschung, Abteilung Chemische Biologie, Inhoffenstr. 7, 38124 Braunschweig, Germany
Leibniz-Institut für Molekulare Pharmakologie, Robert-Roessle-Str. 10, 13125 Berlin, Germany

Stefan Dübel
Technische Universität Braunschweig, Institut für Biochemie und Biotechnologie, Spielmannstr.7, 38106 Braunschweig, Germany

Mahavir Singh
Lionex GmbH, Salzdahlumer Str. 196, 38126 Braunschweig, Germany

Michael Hust
Technische Universität Braunschweig, Institut für Biochemie und Biotechnologie, Spielmannstr.7, 38106 Braunschweig, Germany

Ryan M Williams
Department of Basic Pharmaceutical Sciences, West Virginia University, 1 Medical Center Drive, PO Box 9530, Morgantown, WV 26506, USA
Memorial Sloan Kettering Cancer Center, Molecular Pharmacology & Chemistry Program, 1275 York Ave., New York, NY 10065, USA

Cyrus J Hajiran
Department of Biology, West Virginia University, 53 Campus Drive, PO Box 6057, Morgantown, WV 26506, USA

Sara Nayeem
Department of Basic Pharmaceutical Sciences, West Virginia University, 1 Medical Center Drive, PO Box 9530, Morgantown, WV 26506, USA

Letha J Sooter
Department of Basic Pharmaceutical Sciences, West Virginia University, 1 Medical Center Drive, PO Box 9530, Morgantown, WV 26506, USA

Mahmood Nazari
National Center for Molecular Genetics and Breeding of Animal, Institute of Animal Sciences, Chinese Academy of Agricultural Sciences, Beijing 100193, People's Republic of China

Fatemeh Salabi
Transgenic and Stem Cell Core, Institute of Animal Science, Chinese Academy of Agricultural Sciences, Beijing 100193, People's Republic of China

Li Zhang
National Center for Molecular Genetics and Breeding of Animal, Institute of Animal Sciences, Chinese Academy of Agricultural Sciences, Beijing 100193, People's Republic of China

Fuping Zhao
National Center for Molecular Genetics and Breeding of Animal, Institute of Animal Sciences, Chinese Academy of Agricultural Sciences, Beijing 100193, People's Republic of China

Caihong Wei
National Center for Molecular Genetics and Breeding of Animal, Institute of Animal Sciences, Chinese Academy of Agricultural Sciences, Beijing 100193, People's Republic of China

Lixin Du
National Center for Molecular Genetics and Breeding of Animal, Institute of Animal Sciences, Chinese Academy of Agricultural Sciences, Beijing 100193, People's Republic of China

Jean-Pierre Roperch
Profilome, Paris Biotech 24 rue du Faubourg St Jacques, Paris 75014, France
OncoDiag, Agoranov 96 Bis, Boulevard Raspail, Paris 75006, France

Karim Benzekri
Centre d'Investigation Clinique (CIC), Henri Mondor Hospital, Créteil, France

Hicham Mansour
King Abdullah University of Science and Technology (KAUST), Bioscience Core Laboratory Research Department, Thuwal 23955-6900, Saudi Arabia

Roberto Incitti
King Abdullah University of Science and Technology (KAUST), Computational Biology Research Center, Thuwal 23955-6900, Saudi Arabia

Radosław Stachowiak
Department of Applied Microbiology, Institute of Microbiology, Faculty of Biology, University of Warsaw, I. Miecznikowa 1, 02-096 Warsaw, Poland

Marcin Łyżniak
Medical Centre of Postgraduate Education, Marymoncka 99/103, 01-813 Warsaw, Poland

Maja Grabowska
Department of Applied Microbiology, Institute of Microbiology, Faculty of Biology, University of Warsaw, I. Miecznikowa 1, 02-096 Warsaw, Poland

Katarzyna Roeske
Department of Applied Microbiology, Institute of Microbiology, Faculty of Biology, University of Warsaw, I. Miecznikowa 1, 02-096 Warsaw, Poland

Tomasz Jagielski
Department of Applied Microbiology, Institute of Microbiology, Faculty of Biology, University of Warsaw, I. Miecznikowa 1, 02-096 Warsaw, Poland

Jacek Bielecki
Department of Applied Microbiology, Institute of Microbiology, Faculty of Biology, University of Warsaw, I. Miecznikowa 1, 02-096 Warsaw, Poland

Bożena K Budziszewska
Department of Hematology, Institute of Hematology and Transfusion Medicine, I. Gandhi 14, 02-776 Warsaw, Poland

Grażyna Hoser
Medical Centre of Postgraduate Education, Marymoncka 99/103, 01-813 Warsaw, Poland

Jerzy Kawiak
Medical Centre of Postgraduate Education, Marymoncka 99/103, 01-813 Warsaw, Poland

Wei Zheng
College of Life and Environmental Sciences, Hangzhou Normal University, 310016 Hangzhou, Zhejiang, China

Yu Zhang
College of Life and Environmental Sciences, Hangzhou Normal University, 310016 Hangzhou, Zhejiang, China

Hui-Min Lu
College of Life and Environmental Sciences, Hangzhou Normal University, 310016 Hangzhou, Zhejiang, China

Dan-Ting Li
College of Life and Environmental Sciences, Hangzhou Normal University, 310016 Hangzhou, Zhejiang, China

Zhi-Liang Zhang
College of Life and Environmental Sciences, Hangzhou Normal University, 310016 Hangzhou, Zhejiang, China

Zhen-Xing Tang
College of Light Industry Science and Engineering, Nanjing Forestry University, 210037 Nanjing, Jiangsu, China

Lu-E Shi
College of Life and Environmental Sciences, Hangzhou Normal University, 310016 Hangzhou, Zhejiang, China

Abdalrahim FA Aisha
Department of Pharmaceutical Chemistry, School of Pharmaceutical Sciences, Universiti Sains Malaysia (USM), Minden 11800, Pulau Pinang, Malaysia
Department of Pharmacy, School of Medicine and Health Sciences, An Najah National University, Nablus, West Bank, Palestine

Amin Malik Shah Abdul Majid
Department of Pharmacology, School of Pharmaceutical Sciences, Universiti Sains Malaysia (USM), Minden 11800, Pulau Pinang, Malaysia

Zhari Ismail
Department of Pharmaceutical Chemistry, School of Pharmaceutical Sciences, Universiti Sains Malaysia (USM), Minden 11800, Pulau Pinang, Malaysia

Stanley Tamaki
Department of Surgery, Division of Plastic and Reconstructive Surgery, University of California, San Francisco, CA 94143, USA

Christopher Nye
QB3 East Bay Innovation Center, 2929 7th St, Berkeley, CA 94710, USA

Euan Slorach
QB3 East Bay Innovation Center, 2929 7th St, Berkeley, CA 94710, USA

David Scharp
Prodo Laboratories, 32 Mauchly, Irvine, CA 92618, USA

Helen M Blau
Baxter Laboratory for Stem Cell Biology, Department of Microbiology and Immunology,
Institute for Stem Cell Biology and Regenerative Medicine, Stanford University School of Medicine, Stanford, CA 94305, USA

Phyllis E Whiteley
Mohr Davidow Ventures, Menlo Park, CA 94025, USA

Jason H Pomerantz
Departments of Surgery and Orofacial Sciences, Division of Plastic and Reconstructive Surgery, Craniofacial and Mesenchymal Biology Program, Eli and Edythe Broad Center of Regeneration Medicine, University of California, San Francisco, CA 94143, USA

Xiao-Yun Lu
Department of Biological Science and Bioengineering, Key Laboratory of Biomedical Information Engineering of Ministry of Education, School of Life Science and Technology, Xi'an Jiaotong University, Xi'an 710049, Shaanxi, P. R China

Ming-Chuan Li
Molecular Biotechnology Center, Universita di Torino, 10126 Torino, Italy

Xin-Liang Zhu
Department of Biological Science and Bioengineering, Key Laboratory of Biomedical Information Engineering of Ministry of Education, School of Life Science and Technology, Xi'an Jiaotong University, Xi'an 710049, Shaanxi, P. R China

Fan Fan
Department of Biological Science and Bioengineering, Key Laboratory of Biomedical Information Engineering of Ministry of Education, School of Life Science and Technology, Xi'an Jiaotong University, Xi'an 710049, Shaanxi, P. R China

Lei-Lei Wang
Department of Biological Science and Bioengineering, Key Laboratory of Biomedical Information Engineering of Ministry of Education, School of Life Science and Technology, Xi'an Jiaotong University, Xi'an 710049, Shaanxi, P. R China

Jian-Gang Ma
Department of Biological Science and Bioengineering, Key Laboratory of Biomedical Information Engineering of Ministry of Education, School of Life Science and Technology, Xi'an Jiaotong University, Xi'an 710049, Shaanxi, P. R China

Yuichiro Yamaguchi
Department of Oral Rehabilitation, Section of Oral Implantology, Fukuoka Dental College, 2-15-1 Tamura, Sawara-ku, Fukuoka 814-0193, Japan

Jun Ohno
Department of Morphological Biology, Section of Pathology, Fukuoka Dental College, 2-15-1 Tamura, Sawara-ku, Fukuoka 814-0193, Japan

Ayako Sato
Department of Oral Rehabilitation, Section of Oral Implantology, Fukuoka Dental College, 2-15-1 Tamura, Sawara-ku, Fukuoka 814-0193, Japan

Hirofumi Kido
Department of Oral Rehabilitation, Section of Oral Implantology, Fukuoka Dental College, 2-15-1 Tamura, Sawara-ku, Fukuoka 814-0193, Japan

Tadao Fukushima
Center for Regenerative Medicine, Fukuoka Dental College, 2-15-1 Tamura, Sawara-ku, Fukuoka 814-0193, Japan